The making of an expert engineer

The making of an expert engineer

How to have a wonderful career creating a better world and spending lots of money belonging to other people

James P. Trevelyan

CRC Press
Taylor & Francis Group
Boca Raton London New York

CRC Press is an imprint of the
Taylor & Francis Group, an **informa** business

A BALKEMA BOOK

Cover photo: Engineers discuss the implications of vibration measurements at SVT Engineering Consultants head office in Perth, Australia, photographed by the author.

Published by:
CRC Press/Balkema
P.O. Box 447, 2300 AK Leiden, The Netherlands
e-mail: Pub.NL@taylorandfrancis.com
www.crcpress.com – www.taylorandfrancis.com

First issued in paperback 2020

© 2014 by Taylor & Francis Group, LLC
CRC Press/Balkema is an imprint of the Taylor & Francis Group, an informa business

No claim to original U.S. Government works

Typeset by MPS Limited, Chennai, India

ISBN 13: 978-0-367-57606-6 (pbk)
ISBN 13: 978-1-138-02692-6 (hbk)

Visit the Taylor & Francis Web site at
http://www.taylorandfrancis.com

and the CRC Press Web site at
http://www.crcpress.com

Library of Congress Cataloging-in-Publication Data

Dedication

Dedicated to Joan Trevelyan, Begum Sarfraz Iqbal and Malik Muhammad Iqbal Khan

Table of contents

Acknowledgements

So many people have contributed to the research for this book that it is not possible to name all of them.

Between 2001 and 2011 Sabbia Tilli supported me as we searched through hundreds of journals, conference proceedings and research books, gradually discovering, with a degree of dismay, just how little had been published describing systematic research on the details of engineering practices in different settings. Sabbia conducted the longitudinal study to follow the early careers of 160 engineering graduates from the University of Western Australia.

My graduate students Leonie Gouws, Adrian Stephan, Vinay Domal, Sule Nair, Ernst Krauss, Zol Bahri Razali, Emily Tan, Edwin Karema, Siong Tang, and Wilson Casado Neto, some of whom are still finishing their research, have helped me with a wealth of questions, research observations and ideas. Between them all they have accumulated about 140 years of first-hand engineering practice experience, a rich source of insight for this research.

Undergraduates, both coursework and research students, are too numerous to mention. However I have to acknowledge distinguished contributions from Nur Azalia Razman, Mohnish Bajaj, Naijiao Bo, Christopher Brown, Michael Comuniello, James Concannon, Michael Crossley, Lana Dzananovic, Adrian Han, Emma Hansma, Kenny Ho, Henry Hui Chean Tan, Ron Jacobs, Xiang Xu Jing, Sawan Jongkaiboonkit, Timothy Maddern, Daniel McErlaine, David Mehravari, Sheena Ong, Sankar Palat, Vincent Peh, Sarah-Jayne Robinson, Quang Vu and Gerarda Wescott.

Sally Male found that her PhD research was closely aligned and joined the team, remaining as a valued research colleague since she graduated. Other University of Western Australia faculty colleagues Caroline Baillie, Mark Bush, John Cordery, Maria Harries, Melinda Hodkiewicz, Tim Mazzarol and Ruza Ostroganac provided help in joint research projects, joint supervision and with research methods. Lesley Jolly and Lydia Kavanagh from the University of Queensland, Keith Willey and Anne Gardiner from the University of Technology, Sydney, and Roger Hadgraft from Central Queensland University (formerly RMIT and Melbourne) have also helped from a distance.

Bill Williams, Juan Lucena, Gary Downey, Diane Bailey, Joyce Fletcher, John Heywood and Russell Korte have all provided help and encouragement from the other side of the world when I needed it most. Others who helped immensely include Mir Zafrullah Jamali, Prof. Chandra Mohan Reddy, Shujaat and Bushra Sheikh, Inder Malholtra, Ambassador (ret) Vijay Kumar and Air Commodore (ret) Jasjit Singh.

Hundreds of engineers in Australia, India, Pakistan, Brunei and several other countries contributed their time to participate in surveys, interviews and to accommodate students conducting field observations. Their employers also contributed by supporting their participation.

Australian taxpayers paid for my salary through the University of Western Australia as well as providing additional research funding support in the form of infrastructure, grants and scholarships.

Many companies provided direct and indirect support: they are not named to protect their commercial interests and the identities of the participants.

Ian Harrington supported by his wife, Alison, critically reviewed the manuscript as an experienced practicing engineer. Marli King edited and proof checked the final manuscript adding so much more clarity and removing many confusing phrases.

My wife, Samina Yasmeen, helped me learn how to write an argument, though she has yet to let me win one. Her love, affection and care have sustained me through the struggle to write this book and carried me through some serious health issues. Her mother, Begum Sarfraz Iqbal, patiently suffered my endless unanswerable questions. Her father, Malik Muhammad Iqbal Khan, managed the Hameed and Ali Research Centre that provided my first window into engineering on the Indian subcontinent.

Finally, I have to thank countless aspiring young engineers, many of who I have taught, others whom I have met and provided encouragement. They represent the future, and they have been at the forefront of my mind all the time I was writing this book.

To all, I owe an immense debt of gratitude.

List of tables

List of figures

Preface: Engineering practice has been invisible

I am fortunate to be an engineer, an inventor, a teacher, and a researcher. I am lucky enough to have received international awards for my research, patents for my inventions, and awards for my teaching. I have seen my designs built and used by satisfied people, and I have had a wonderful career. I have spent enormous amounts of money that belonged to other people to do achieve all that. Determination played a part. So did persistence, good fortune, plenty of encouragement, and being in the right place at the right time.

In the most recent part of my career, I have researched engineering practice: what engineers do and how they do it. For my first 30 years as an engineer, I took the notion of engineering practice for granted. Although I was aware that my daily work was only distantly related to what I had learnt in engineering school, it never occurred to me that this was unknowable to anyone who was not already doing it. Somewhere, someone would have studied what engineers do, surely.

Around 2003, faced with unanswerable questions about engineering practice issues, I started to interview and observe engineers at work. Gradually, I came to realise, with a creeping sense of shock and amazement, that much of engineering practice seemed not to have been understood before and described in ways that made sense to me as an engineer. As I analysed hundreds of pages of interview transcripts and field notes, I surprised myself with the hitherto hidden complexity of what I had been doing myself, and observing others doing. Persistent patterns started to emerge to give this complexity a coherent form that could be understood.

Many of the engineers I met started by telling me how they hardly did any real engineering work: design, calculations, and the solitary technical work that they learnt to do in engineering school. Instead, their lives seemed to be filled with what one described as "random madness", seemingly trivial and routine paperwork, meetings, phone calls, frustrations, confusion, and misunderstandings. In the words of Dilbert, the cartoon engineer created by Scott Adams, "My job involves explaining things to idiots, who make decisions based on misinterpreting what I said. Then, it is my job to fix the massive problems caused by the bad decisions."

Some of the engineers I met and observed were clearly in a class of their own: they were experts in their own fields of practice, acknowledged by their peers to be exceptional performers. In time, I came to understand a little about what distinguished their performances, and I decided to write this book to help engineers and students who would like to become more expert in their practice.

Many work as engineers, but simultaneously wonder why the work is not what they thought it was going to be: why it seems dull, mundane, and does not challenge the technical abilities that they learnt in engineering schools. Many aspiring young engineers graduate into little more than disappointment. Many never even manage to find an engineering job to start off their careers. Many older engineers seem to lie trapped in dead-end careers by assumptions, myths, and misconceptions that are reinforced by endless unquestioned repetition. Much of this disappointment arises from misconceptions about engineering practice, misconceptions that persist in the absence of a comprehensive written account.

Anyone who has graduated from an engineering school has the capacity to become an expert engineer. However, much of what they need to know is neither taught, nor even known, in most engineering schools.[1]

I am confident that the ideas in this book can transform the careers of engineers, both young and old. As far as I know, most of this material has never been presented in a single book like this and some has never been described until now. Most of the ideas and concepts have emerged from research conducted over the last decade to gather material for this book. Through this research, I realised that engineering practice, what we engineers do every day, has been largely unknown even to us. We simply do the work without thinking much about what we are really doing. Only a tiny number of earlier research studies revealed any details before I started, and even then, they only did so in exotic high-technology engineering that only a tiny number of engineers encounter in their careers.

Engineering has been invisible to nearly all of its participants.

I want to change that by explaining the ideas that led to this book, and change the world through the engineering that my young readers will be able to perform with the help of these ideas.

I believe that this book can lead any engineer to a rewarding and exciting career. Engineering is a wonderful profession. With some learning from experts, any engineer can not only enjoy spending lots of other peoples' money, but in doing so, also provide great value, and earn more for themselves in the process.

A minority of people in the world enjoy comfortable lives because engineering enterprises in their regions have been able to provide services and products in an efficient and economical way. In Australia, the driest continent, copious amounts of potable water – clear, clean, and safe to drink – flow from kitchen taps 24 hours a day at a total cost of about $2 per tonne. In much of the world today, potable water costs between $20 and $100 per tonne. Getting the bare minimum needed for survival takes up to a quarter of the economic resources of poor families, more than 10% of the GDP in a country like Pakistan. Good engineering can change that staggering reality for billions of people who live in misery at the moment.

We also know that we are consuming too much of the world's resources, partly because engineering has not yet provided the means for people to use our resources efficiently. While a minority lead comfortable lives, it is only reasonable for the poor majority to aspire to the same living standards. Yet, if everyone on the planet consumed at the same rate as the comfortable minority, the remaining resources of this planet would be exhausted in a short time.

Therefore, the ultimate challenge for today's young engineers is to find a way for all people to live in affordable comfort and safety within the limitations of this planet. To achieve that, we need tens of millions of expert engineers.

Today, there is no single course of study from which you can learn to be an expert engineer. I do not know of a single engineering school or college that enables you to do this.

Part of the frustration encountered by graduates arises because these institutions claim to teach engineering. They do teach engineering science, which is an important part of engineering, but graduates need to learn much more to become experts. That is the part that has largely been invisible. I hope this book will change that by removing Harry Potter's cloak of invisibility that perhaps he carelessly left lying over the secrets of engineering practice. This book provides guidance on how to acquire this critical know-how for engineering.

Why is so much of engineering invisible?

You can see the results of engineering all around the planet. Phones, buildings, roads, vehicles, and aircraft: the list of engineering achievements is almost endless.

However, therein lies the trap: these are all objects, some of them vast systems of man-made structures, while others are almost too small to see.

Why is engineering invisible in these objects?

Engineering is a human performance: it is performed by people. Extraordinary people, in some cases, but most of them are entirely ordinary people. The creation of engineering systems and artefacts relies on human actions; therefore, feasibility is limited by human capabilities, as well as the laws of physics.

It is the evidence of their performances, the objects and the information left behind, that we associate with engineering.

Engineering artefacts, drawings, objects, documents: each represents for the most part what is to be, or what has been built – the finished objects. What they do not represent is the human process that led to their creation and the creation of the objects that they represent.

One of the great controversies of the ancient world concerns the techniques used to construct the great pyramids of Egypt. Even with the prolific hieroglyphic writing that litters the remains of the entire ancient Egyptian empire, no one has been able to find an account that explains how the pyramids were built. Engineers today are no different from their Egyptian forebears. The documents and artefacts we create represent the endpoints of our performances. How these artefacts came into being, the human engineering process, is no more likely to be written down now than it was 4,500 years ago. It remains as it always has been – practiced, yet simultaneously invisible.

Here, I would like to introduce the first important idea about engineering practice. I have labelled each of the main ideas in the book as 'Practice Concepts' in bold headings.

Practice concept 1: The landscape of practice

We can conceive of a map that includes all the engineering possibilities that could provide effective solutions for a particular project and we can also imagine contours of difficulty. The low contours surround possibilities that are easier to achieve, while the high contours include more difficult possibilities, and the boundaries mark the limits of feasibility. What determines this landscape? Which factors shape the contours of practice?

The contours are partly determined by the laws of physical sciences, which are almost always expressed in the language of mathematics. For example, thanks to Shannon's pioneering work on telecommunication theory,[2] we know that carrier frequency bandwidth determines how long it will take you to download a 10-gigabyte high-definition movie. The strength of steel and concrete and theories of fibre-matrix bonding stability determine the height of our tallest buildings.

Even with all that knowledge in place, there are many other factors influencing these contours, factors that only become apparent once you start working as an engineer.

The rights to use certain intellectual property, or the need to protect it, the distribution of human know-how, how much we can remember in our heads, how reliably and how fast we can learn more, how effectively we can explain this to others, technical collaboration capabilities, the time available to complete the project, local, national and international regulations, your reputation as an engineer, how much finance investors are willing to make available, their risk appetite, the capabilities of the people on hand, how they are organised, the state of the local and international economy, political stability, and attitudes of the local community and end users: these are all factors that shape this landscape of practice.

As an engineer, your job is to navigate this complex landscape, steering an engineering project away from the peaks of difficulty towards the plains of practicality. Engineering decisions reflect the shapes of these contours.

Now ask yourself, how well did your engineering degree course prepare you to navigate this landscape?

In this book, along with introducing key concepts that help to explain the landscape of practice, I also point out misconceptions that make it more difficult to appreciate the landscape. These are mental blockages, blinders over our eyes, which get in the way of a clear, unimpeded view of the landscape of practice.

Here's the first.

Misconception 1: These are all non-technical issues

What do we mean by 'non-technical'? Often, this term is used to label any idea beyond the world of objects that we, as engineers, can think about comfortably.

Take communication and collaboration, as an example. As we shall see in later chapters, engineers communicate and collaborate constantly, and most of the time, there are technical issues that frame communication and collaboration. Technical issues influence much of the risk faced by investors, making it easy or difficult to comply with regulations. Technical issues also determine intellectual property constraints and appropriate collaboration arrangements. Political issues, the state of the international economy, and local community attitudes are factors that we could discuss without recourse to any engineering technical understanding, yet they inevitably shape the technical constraints within which we engineers have to work. Furthermore, innovative technical solutions can often create new space and ease these constraints once we can understand the issues.

Most factors that shape the contours of the landscape of practice have technical implications or dependencies. Two of the main attributes that distinguish expert engineers are their abilities to perceive how technical factors and social interactions shape the landscape of practice and their skill at influencing people working with them to preserve the technical intent through multiple reinterpretations by other people.

Practice concept 2: Socio-technical factors shape the landscape of practice

Therefore, to understand engineering, we need to understand human capabilities and social behaviour, as well as the laws of physics. We have to watch what people do, listen to what they say, and understand some of their feelings, both those of frustration and pride. Only by making these actions and feelings visible can we begin to understand. That's the job of researchers, to make visible what was previously hidden and subsequently provide as concise and readable descriptions as possible so that you, the readers, can share their insights.

In the field of science and technology studies, researchers have coined the terms 'socio-technical' and 'heterogeneous' to describe this intrinsic link between people and engineering. In engineering, the social and technical are intertwined, inseparable realities of practice.

Unfortunately, engineering schools rarely help you learn about people and this concept of socio-technical factors. You will be lucky if you can find one that includes any study of people in the core curriculum.

That's precisely why graduates get so frustrated. Engineering schools are only enabling them to learn part of the picture: the other part has remained hidden.

Engineers like to see themselves as agents of change who can shape the future of the world with technology. Engineers produce new technology and technology can change the world …

… but only if people, particularly other engineers, adopt the technology, which requires people to change. Engineers can only change the world if they can influence their colleagues, the technicians who translate ideas into reality, and the financiers who provide the money: it is people who change the world through the technologies they choose to create and use.

Engineering schools don't explain this to graduates … yet. They help students learn some elements of engineering physical science and abstract thinking about objects. While all engineers need these capabilities, they are insufficient by themselves for a person to become an expert engineer.

That's why I'm writing this book.

As an engineering student, a novice engineer, or even an experienced engineer, you still have a lifetime of learning ahead of you. Until recently, however, it has not been clear what you needed to learn. Engineers talk about this simply as "experience" and "practical skills" but have not been able to explain exactly what they have meant by this, only that it is something that has to be learnt "on the job", something that you cannot learn in a formal academic setting. With the help of research, we can now describe much more clearly what you need to learn beyond the engineering school curriculum: engineering practice.

Until the 1950s, engineers learnt engineering by emulating their teachers, either in classes or in the workplace. This practice-based education made it difficult for engineers to keep up with and apply the huge scientific advances of the 20th century. Physical sciences and mathematics came into the curriculum and have now entirely displaced practice. The extraordinary technological advances we have seen in the second half of the 20th century have come from the ability of engineers to exploit fundamental advances in science, mathematics, and computation. This ability was seen as critical when engineering curricula were reformed in the 1950s.[3] Education founded on science and mathematical theory has created this ability: theories enable you to quickly learn what you need to know, exactly when you need it.

Engineers today are still learning practice by trial and error, on the job, and many have never learnt very much that way.

Now, in the 21st century, there are fundamental advances in many other fields, such as the humanities, economics, psychology, learning sciences, social sciences, organisational science, linguistics, even philosophy and history, that can be applied to help engineers learn about most of the human factors that shape the landscape of engineering practice. Complementing these advances, we now have ten years of detailed research on what engineers do every day.[4] This book builds on these advances, and a good portion of it will involve some difficult and challenging reading. You will need to open your mind and understand ideas and theories that may be new and seem baffling at first.

However, just as mathematics and physics theories made it easy for you to learn engineering technical principles quickly when you need them, social and human science theories make it easier for you to learn about socio-technical factors.

These are the ideas in this book that have the potential to help you become an expert engineer much more easily than before.

Many engineering graduates find themselves in a frustrating position: you may be experiencing this for yourself right now. They find that nearly every engineering job advertisement requires applicants to have "experience", but they can't get that experience without being in an engineering job.

Learning with the help of this book cannot completely replace engineering experience. However, by working through the prescribed exercises and practicing the skills, you can prepare yourself and make the process of finding an engineering job much easier. When you start to work as an engineer, you will be able to learn much more from your experiences faster and with less frustration. I am confident that your talents will be greatly appreciated by any enterprise in which you can apply these ideas.

This book can only be a guide for your future learning. Apart from a few basic skills, this book does not contain what you need to learn. Instead, it shows you how to learn from your surroundings. Most of what you have to learn is right there in front of you, even today. You simply need to learn to see it, to pull aside the 'cloak of invisibility'. You will have to do the learning, although other people can certainly help you.

Engineering can be a wonderful career, full of intellectual stimulation and challenges, financial rewards, and fantastic fun.[5] All of the engineers that I've interviewed as part of the research for this book have immensely enjoyed their careers. Many have faced extraordinary frustrations at times, for which the material in this book might have been helpful, had it been available at the time.

This book has been built on extensive research across many disciplines, but it is meant to be a practical guide for engineers. Notes link the text to research publications mainly to help future researchers improve on what is already here. Several of the explanations presented in the book have grown out of valuable philosophical discussions to which I was kindly invited.[6]

Experienced engineers will find useful material in this book: most will find that at least some parts of the book have new ideas that could help them improve their performance. The book can also provide guidance for experienced engineers who are helping younger engineers to develop practice skills. Human resources professionals can benefit from this book: no previous book has provided a comprehensive and research-based description of what engineers really do in their work. Over 200 interviews and extensive field observations by myself and my students provided the data for this research.[7] My own professional work experience over four decades framed the interpretations that were checked and refined by discussing the results with many of the participants. To help the reader, particularly researchers and students, I have also provided notes throughout the book listing references and publications from other researchers that corroborate, amplify, or extend my interpretations and data.

This book is only a start point: it is only a guide to help you learn more from expert engineers. If you read it all at once, you may find some sections repeat earlier material. Each chapter is designed to be read more or less on its own, and so has to summarise relevant ideas from earlier ones. By the end of your lives, I hope, you will have learnt much more than I did. Maybe one or two of you will write even better books than this one. It is my aim that before then, you will have helped to improve the lives of billions of people, allowing most people on this planet to live in reasonable comfort and safety.

NOTES

1. The reasons for tenuous knowledge of practice within engineering schools are explained in Trevelyan (2014b).
2. Shannon (1948).
3. The Grinter Report (1955) advocated that engineering education be founded on mathematics, physics, and chemistry, as well as a set of engineering sciences, such as materials science and thermodynamics. However, at the same time, the report argued that about 20% of the curriculum should be on social sciences, a recommendation that has been almost completely ignored.
4. Trevelyan (2014b), also in the online appendix on research methods.
5. Sam Florman (1976; 1987; 1997) wrote several books on the pleasures of engineering that are still relevant today.
6. I should acknowledge some of the many contributions on philosophy relevant to engineering and technology that helped inspired these explanations (Downey, 2009; Foucault, 1984; Heywood, 2011; Heywood, Carberry, & Grimson, 2011; Koen, 2009; Marjoram, Lamb, Lee, Hauke, & Garcia, 2010; Nussbaum, 2009).
7. Described more fully in the online appendices.

Chapter 1

Why engineer?

What do engineers actually do? The truth is, even many engineers find this question difficult to answer.

Many of the students whom I've taught think that engineers mostly work alone, designing and solving technical problems, while looking forward to a 'practical, hands-on' profession.

Many engineers, perhaps a majority, often think that the amount of 'real engineering' they do is rather small. In fact one engineer suggested that only a small part was engineering and 'the rest would just be random madness'. From their perspective, it often seems that they spend a lot of time focused on what they consider to be non-technical issues and administration.

One of the aims of this book is to help engineers answer this fundamental question by providing new concepts and ideas to explain what exactly it is that they 'do'.

However, first and foremost, I would like you to think about something far more important and fundamental, something that seems to have been forgotten in the majority of engineering schools. It seems so basic that we all tend to take it for granted: **why do we do engineering and why is it so valuable?**

For most of my career, I never thought to question why engineering might be valuable. Until recently it had always seemed self-evident to me. It was only when I started to think a bit deeper about the question that I realised the answers are far from obvious.

Students in law school and medical school have no difficulty explaining the value of their professions. Lawyers and judges help people get out of trouble, provide access to justice, and protect human rights. Doctors heal the sick and help people live longer, healthier lives. On the other hand, when I ask engineering students about the 'value' of engineering, they usually hesitate for a moment before tentatively offering 'new technology' as the underlying value of their future career. 'And how does that help people?' I ask. Most engineering students find that question quite hard to answer.

As engineers, most of us don't get to develop new technology, although finding practical applications for it is something we do all the time. However, if we're not actually *developing* new technology, then what is the value of what we do?

Hence, the title of this chapter: **Why Engineer?**

Try and answer these simple questions (see practice quiz 1 below):

Why should a company employ me as an engineer?
How can I help the enterprise?
What value can I create for an enterprise and my community?

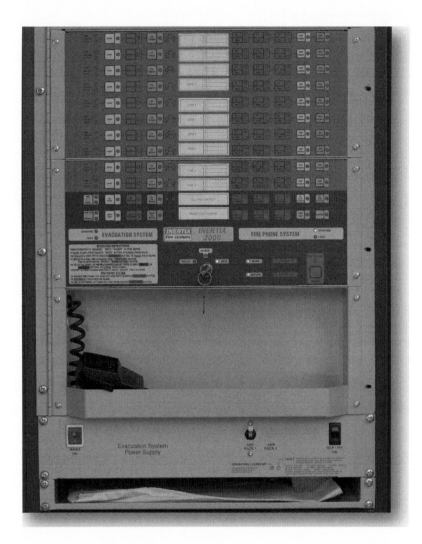

Figure 1.1 Fire alarm panel installed in the entrance of the building where my office is located.

Recently, I was approached by a novice engineer; she was a recent graduate who had obtained her first job with a small family company that employed about 20 other people. The company owner had asked her the same questions. She came to me a few days later explaining that she had been completely unable to answer these simple questions. Somewhat baffled, she asked for my suggestions.

I would like you to try and answer her questions before proceeding any further with this book. First, you need to know a little more about the company in question.

At the time, the company manufactured electronic panels for fire alarm systems. Electronic smoke detectors and other fire sensors fitted to buildings are all wired to a fire alarm control system panel. Electronic circuits in the panel respond to a smoke alarm and cause a fire alarm to sound as a warning to people in the building. The same circuits also automatically raise the alarm with the local fire brigade, providing the address and directions so they are able to reach the building quickly. The panel contains a public address system with a microphone, enabling the fire brigade to make announcements throughout the building when they arrive. The panel provides internal wiring connectors for the smoke detector sensors, a public address amplifier, switching circuits, and communications circuits.

The company manufactured the panels using both locally sourced and imported components. The owner started the company approximately 20 years ago, utilising his practical electrical wiring and circuit board assembly skills. Other members of his family worked for the company in administrative and accounting positions.

Practice quiz 1

If you don't feel comfortable answering the questions for this particular company, think of any other company, enterprise, or organisation that you would like to work for and answer the same questions in that alternate context.

If you have access to the online site for this book,[1] you can find the quiz there: please type your answers in the electronic quiz. Otherwise, write brief notes here for future reference. This first quiz should take 30–90 minutes to complete.

Q1: Why should the company employ an engineer?

1. _____

_____

_____

2. _____

_____

_____

3. _____

_____

_____

Q2: What value would an engineer create for the company that will justify the cost of employment? You don't have to calculate a financial value; a qualitative description is sufficient.

1. _____

_____

_____

_____

2. _____

_____

_____

_____

3. _____

_____

_____

_____

4. _____
   _____
   _____
   _____

5. _____
   _____
   _____
   _____

Use this space for additional comments or questions that might have occurred to you to ask the owners of the enterprise.

_____
_____
_____
_____
_____

Now that you have written your answers, I would like you to self-evaluate your responses.

Evaluating question 1 responses

Perhaps you suggested this for the first question:

> 'I can design a better fire alarm system for you. It would utilise new technology and work more efficiently.'

Another response might have been:

> 'I can solve technical problems affecting your existing systems. For example, if a similar issue often occurs after your panel has been installed in a building, I could work out the underlying cause and rectify the problem by redesigning your system.'

Here's another:

> 'I can improve the efficiency and quality of processes and practices used in the company.'

Or even this:

> 'I work hard and I can offer new insight into old problems.'

Give yourself one point for each response that resembles one of these examples, and then consider this reply from the company:

> 'Solutions on paper and hard work are not much use to us unless they lead to tangible results. We need problems fixed without redesigning our panels. What do

we have to do so problems are no longer an issue for us? How can we retrofit all the systems that we have already installed so similar failures don't occur in those as well?'

For each of your responses that answers one or more of these reasonable requests from the company, award yourself two additional points.

Evaluating question 2 responses

Each of your answers needs to describe the *value and other benefits for the company derived from employing an engineer.*

Carefully compare your responses with the ones below.

Give yourself one point for every response similar to the ones below *and two extra points* if your response described the value to the company, at least in qualitative terms. Notice how each of the responses below includes a description of the value created – in the first few responses, this section is in italics.

Product and process improvement, research and development, and anticipating future developments:

- Work on an application to justify a large proportion of the cost of employing an engineer as research and development, thereby gaining tax concessions. Value: *significantly reduces the cost of making technical improvements to products.*
- Prepare designs for an upgraded product that could be manufactured in the event that a competitor enters the market with a better quality product at a similar price. Also, evaluate the likelihood of this happening. Value: *reduces the risk that a competing product will take a significant portion of the existing market share.*
- Prepare designs for a cheaper product with a similar performance standard and quality that could be manufactured in the event that a competitor enters the market with a similar quality product at a significantly cheaper price. Value: *reduces the risk that a competing product will take a significant portion of the existing market share.* This strategy also helps if the local currency rises in relative value, thereby enabling imported products to cost less in the local currency.
- Improve workforce skills to provide better service to clients in a timelier manner. Value: *improves client satisfaction and customer loyalty.*
- Introduce design modifications to allow for easily installed expansion in panels to minimise the cost when customers need to extend their fire alarm systems. Value: *increases profitability or improves price competitiveness.*
- Investigate self-powered and wireless sensor technology in fire alarm systems in order to lower the cost of installing cabling at new customer sites. Value: *increases the company profitability or improves price competitiveness.*
- Reduce internal power consumption of panel electronics by eliminating ventilation fans. This will reduce dust accumulation inside the panel. Value: *reduces maintenance costs and improves company profitability by reducing warranty and service costs.*

Business development research and understanding customer needs:

- Research competitors' products, product announcements, and sales information in order to anticipate changes in the market. Value: *allows the company to respond accordingly in a timely fashion, thereby reducing the risk of lost market share.* Explore technical similarities that might suggest effective partnerships with other manufacturers.
- Understand client needs so that the company can diversify sales to its existing customers, which will increase customer satisfaction. Value: *strengthens the company's market position.*
- Prepare a feasibility study for the company to manufacture building security systems (or to purchase a competing company). Value: *increases market opportunities by providing a more comprehensive service package to existing customers.*
- Work with fire insurance companies and demonstrate the company's product compliance with international and commercial fire safety standards. Value: *opens new market opportunities by reducing customers' insurance premiums.*
- Work on a design for a low-cost, reliable fire alarm system that can be installed in vulnerable down-market boarding houses to reduce the risk of significant casualties in the event of a fire. Consider working with local authorities so that these affordable alarm systems will eventually become mandatory. Value: *significantly increases the overall size of the market for the company's products and improves the safety for less affluent members of the community.*

Cost monitoring, control, and reduction:

- Reorganise manufacturing processes with a more systematic workflow so that existing products can be tailored more effectively to client needs. Value: *significantly increases profit margins.*
- Investigate and, if necessary, help to rearrange the company's accounting systems to ensure that the different costs of manufacturing electronic panels, the costs involved in servicing and maintaining products installed at customer sites, and the costs of maintaining or replacing production and service facilities and equipment can be accurately monitored. Value: *accurate monitoring of, costs enables the company to identify where technical improvements could provide real value, and demonstrates which previous improvements have provided real benefits in order to improve cost-efficiency.*
- Introduce systematic inventory and configuration management systems. Value: *reduces the number of spare parts held in stock, thereby reducing the need for working capital and storage space.*

Risk management and reducing uncertainties:

- Increase compliance with international standards for all aspects of the product, as well as company operations, in order to increase the likelihood that future company expansion will be financed by the company's bank (rather than family capital) with a lower cost of capital. Banks closely inspect a company's operations when a company requests bank financing. Banks employ trained engineers to perform these investigations. Value: *banks are more likely to provide financing for projects that are well managed and produce reliable outcomes.* Demonstrating compliance with standards often (but not always) indicates sound management practices.

- Develop systematic procedures and an organisational method for maintaining the production and service facilities to ensure that there are no costly interruptions in manufacturing or company service activities caused by equipment breakdowns or the need to repair buildings or equipment. Value: *saves on the cost of lost production due to breakdowns and preventable disruptions.*

- Develop systematic quality assurance procedures for purchasing supplies and components that will ensure that incoming products are thoroughly checked and inspected, even tested, if necessary, in order to eliminate the possibility that a defective component might cause a costly panel failure. This could include working with the company's suppliers to develop customised quality assurance procedures. This will also improve customer perceptions of product reliability. Value: *improves customer loyalty and increases future market opportunities.*

- Automate testing of manufactured panels to eliminate current haphazard manual testing, which is known to be inefficient. Value: *reduces the incidence of panel defects discovered after installation, thereby reducing service and warranty expenses.*

- Increase the reliability of in-house manufacturing processes so that senior staff members can go on holidays without having to worry about what is happening at the factory. Value: *staff member morale improves due to time off, therefore making them more productive upon their return.*

- Increase imported components of the electronic panel systems to balance the effects of changes in international currency exchange rates. Value: *reduces the effect of exchange rate fluctuation on company profitability.*

Here are two completely different, but equally good, responses:

- 'I think that the decision to employ a graduate in this position could be a mistake for both the graduate and your business. It might be much more valuable to consider employing an experienced engineer who would be able to provide guidance on compliance with fire protection standards, quality requirements for components and assembly, design and documentation management, and issues such as product quality assurance systems and factory acceptance testing.' Value: *an experienced engineer would provide significant, measurable value to the company much faster than a graduate.* Also, an experienced engineer would be able to pass on high-quality training and professional development to a subsequently hired graduate engineer. This might be very difficult (if not impossible) for the staff members currently employed by the company (Bonus – 10 points).

- As an alternative, insist on an arrangement in which a young and relatively inexperienced engineer receives 2–3 hours of mentoring and guidance each week from an experienced engineer, including daily telephone access. The company would need to pay for the mentoring and advice, in addition to the engineer's salary. Value: *the engineer becomes more effective and productive much faster* (Bonus – 10 points).

If you managed to devise another way to demonstrate the value of employing an engineer that is different than the responses above, award yourself 10 bonus points and please send us the suggestion.

How did you score?

Less than 10 points out of a possible score of more than 60 points?

Don't panic. Most engineers, even those with several years of experience, might also find these questions hard to answer.[2]

You are an exceptionally smart person: anyone who can pass through a university-level engineering course has to be intelligent. Your family is probably immensely proud that you are an engineer or are preparing to graduate from an engineering school. That's no small achievement, so don't forget to congratulate yourself!

On the other hand, you may be concerned that you found answering these questions harder than you expected.

Perhaps you are beginning to think just like one young engineer that I met recently. He told me that he had been seconded to an engineering firm undertaking a major expansion project for his employer, a large multinational corporation. He described his frustration that he was not doing any 'real engineering work', which is what he had expected the job to entail.

I asked him what he was doing in this position, since he wasn't doing 'engineering work'.

He said, 'They just send me out to check whether contractors have installed electrical junction boxes in the right places, or put the right culverts under the roads where they were supposed to. When it rains, I have to go and watch to see if the water flows through okay. Then, I have to fill in a stack of paperwork back at the office. It's not engineering work at all, definitely not what I expected. I'm also not learning anything that's going to help me in my engineering career.'

I answered with complete honesty, telling him that the work he had described *was* engineering work. In engineering, it is vital to check that the contractors have completed all their assigned work in compliance with the requirements in the contract documents before they are paid.

I said, 'It's just that your engineering lecturers forgot to tell you about the dozens, if not hundreds, of very important elements of engineering work, which are just as important as performing structural design calculations.'

Once the contractors have been paid, it can be very difficult to get them to come back and fix mistakes without additional payment. It can be even more expensive to pay someone else to fix the mistakes, and it usually requires high-level engineering knowledge to spot those mistakes.

This young engineer was contributing value by reducing the chances that mistakes had been missed, as well as avoiding the risk that extra money would be needed to fix them.

You can empathise with another of our graduates who reported feeling 'completely incompetent' in his first job working for a major multinational oil company. He told me that he did not know enough to do anything useful for the firm without having to ask other people a plethora of questions.

Don't forget that you are standing at the start of a lifelong career; it takes time to become an expert engineer. A degree confirms that you an intelligent and talented novice engineer.

If you described a multitude of ways for our young engineer to contribute additional value to the firm, then you may already be well on your way to becoming an expert.

Practice concept 3: Why engineering provides value and what this means for you

Now we can return to the question from the start of the chapter: what do engineers really do?

Looking at the sample responses for the fire alarm panel company's questions, we can see many of the ways in which engineers provide value for their clients, employers, and communities.

Engineers contribute value in two main ways:

- In the time available and to the greatest extent possible, engineers minimise the human effort and the consumption of materials and energy needed to achieve a desired result.
- Engineers provide a reasonably accurate forecast of the technical and economic performance, cost, and time required for a given project, while also being reliably capable of delivering results within these expectations.

A frequently cited quote on this issue reads, 'An engineer can do for one dollar what any fool could do for two'.[3] However, this adage misses the value contributed by the relative certainty of an engineer's predictions and the ability to deliver, compared to other people. In fact, an engineer's solution may actually be more expensive, but it is also more predictable and reliable. Chapter 11 discusses this in more detail.

There are many ways to do this. Engineering might provide a small productivity or reliability improvement in the short term without having to spend much money, or perhaps a much larger improvement in the future that requires more upfront expense. Maybe the improvement is small, but relatively certain, while another improvement might be larger, but less certain. Therefore, engineers can also help people decide how to invest their money in terms of making these improvements.

- Engineers seek to understand, discern, and explain the needs of a client, firm, or community in terms of engineering possibilities.
- Engineers conceive achievable economic solutions and forecast their performance, benefits, and costs to help with investment decisions.
- When investors decide to proceed, engineers arrange, organise, and manage the predictable delivery, installation, and operation of reliable artefacts (man-made objects, materials, and information systems).
- Engineers think ahead to reduce or eliminate risks and uncertainties that could have negative consequences. While many events are intrinsically unpredictable, engineers can help to make sure that negative consequences are minimal or at least

reduced as much as possible. At the same time, engineers are also prepared to take advantage of unpredictable opportunities.

More specifically, especially in the company case study above:

- Engineers can help improve a product or service, which will subsequently improve the way that customers experience the product or service, thereby increasing the apparent and actual value for a customer.
- Engineers can find ways to reduce or eliminate uncertainties, particularly mistakes caused by human error. Mistakes can cause delays or lead to wasted materials and energy; introducing systematic methods to consistently eliminate uncertainty results in a reduction of perceived risk.
- Engineers can reduce the material and energy consumption required to achieve a given outcome without reducing performance.
- Engineers can find ways to save time. Time always has a value, even if it is time for which people are not normally paid.
- Engineers can improve customer confidence by improving the reliability of artefacts and products.
- Engineers can anticipate future costs and benefits of alternative courses of action, providing information with an established level of reliability and uncertainty to better guide decision-makers.

TECHNICAL EXPERTISE

Scanning the contents of this book, it would be easy to get the impression that engineering science and technical knowledge are not considered sufficiently important enough to even have a chapter devoted to them. However, this impression is categorically incorrect. This book explains how expert engineers have integrated technical thinking with all the other important aspects of engineering practice.

The ability to conceive and design a practical achievable solution and predict its technical and economic performance is a critical component of the value contributed by engineers. On the other end of the spectrum, predicting failure is equally valuable. The techniques needed to do this are covered extensively in university courses, engineering science texts, patents, and hundreds of thousands of new journal articles and conference papers published every year.

It is this technical knowledge that distinguishes engineers as an occupational group. Therefore, I would ask you, the reader, to remember throughout this book that technical knowledge remains the primary qualifying attribute that defines an expert engineer.

There is just one other issue concerning technical knowledge. While this knowledge is used for analysis and diagnosis, more than anything else, engineering science knowledge is mostly applied directly when making accurate predictions. These are crucial in establishing an engineer's reputation for foresight.

The advanced engineering science and mathematical techniques learnt at university are rarely applied directly in the workplace. They are most often used indirectly: analysis software packages used by engineers incorporate these techniques and make them available for convenient application. The software packages often originate in universities. Once their practical utility for engineers in commercial practice has been demonstrated, the software is often acquired and reworked by specialist engineering software companies to make it easier to use and more tolerant of the typical mistakes made by users. These companies can also supply high-quality documentation, user manuals, and even training courses.

Another way in which engineering science and mathematics is applied in the workplace is in the form of tacit knowledge and understanding.[4] For example, Julie Gainsburg described structural engineers trying to decide if a simplified two-dimensional mathematical model of a building structure would be sufficient to predict its behaviour when the building is subjected to horizontal loads, for example from an earthquake or strong winds. They were asked to investigate whether the forces in the fasteners that secure the floor of the building to the structure would be within allowable limits. The engineers had a choice of several modelling techniques, including a full three-dimensional model that would have been much more accurate. However, it would have taken much more time, and therefore expense for the client, to prepare all the data so that the building would be accurately represented in this more complex modelling method. By choosing simpler methods that are known to be conservative (in other words, the forces are likely to be overestimated), the engineers were able to perform the calculations in much less time. Choosing which method to use, without having completed a full and detailed analysis, requires judgement based on tacit understanding of the relevant mathematics, conventional practices in structural engineering, and the engineers' personal experience and familiarity with the different techniques.[5]

An expert engineer will be continually building a personal repertoire of engineering science knowledge and techniques, and is likely to be familiar with a growing range of engineering software applications.

In the end, however, technical expertise is not sufficient by itself. In the words of an experienced engineer, 'No amount of reliability calculation achieves anything until maintenance technicians use their tools differently'. As we shall see throughout this book, engineering is a collaborative effort; what really counts is how an engineer uses technical expertise when working with other people.

There is no better illustration of this than the problem of supplying drinking water to a large number of people. Most of the detailed technical problems have been solved long ago. However, as we shall see in the next section, the essential factor is the ultimate economic cost for people to access the drinking water.

AN INDICATOR OF ENGINEERING PRACTICE

In my opinion, there is no better illustration that demonstrates the value of good engineering practice than the economic cost of potable water (safe, clean drinking water). Water supply has been the responsibility of engineers since the earliest urban civilisations. Water is heavy to move or transport, and is therefore not traded on the world market. Engineers design and organise the construction of water supply systems that enable water to run in pipes to any location where it is needed. However, many of these water supply systems don't work as intended. In the megacities of many countries like Haiti, India, Pakistan, and even parts of Europe, water flows from supply pipes for just an hour or so each day and can sometimes be unsafe to drink. When you need safe, clean drinking water in these cities, you usually have to prepare it, carry it to your home, or pay someone else to bring it to you, which can be very expensive.[6]

The economic cost of supplying potable water in many developing countries is very high compared to industrialised countries. A woman wearing colourful clothing and carrying water from a well to her house is often the subject of romantic paintings or photographs from developing countries. A typical earning rate for such a woman might be around USD $0.50 per hour, which sounds like a very low rate of pay. Now consider that she can probably make one round trip to a well in an hour and carry home about 15 litres of water. Assume that she needs another hour to boil and properly sterilise the water.

She does not have to pay cash for the water, but there is an opportunity cost because her water-carrying work prevents her from doing other work that could earn income or save expenses, such as growing food or caring for animals (she does have to pay for the stove and fuel to boil the water, as well as containers to boil, store, and carry it in).

An opportunity cost represents the income that has not been earned because the opportunity to earn the income was lost, either intentionally or unintentionally. Opportunity cost associated with unpaid time can also be referred to as 'shadow-priced cost for unpaid labour' or 'value of time'. It can include things such as waiting in a queue or traveling to and from work, or unpaid time off from work due to injury or illness. In low-income countries, the opportunity costs incurred from domestic work roughly equates to two-thirds of the local female earning rate. In other words, the time-consuming nature of domestic labour prevents women in low-income countries from gaining greater earnings through paid work.

This measure can be determined by making a large number of observations on decisions that people make, for example, when they decide whether to walk, take public transport, or hire a taxi. The difference in cost, when averaged over a large number of such instances, represents one measure of people's 'value of time'.

Once we account for this opportunity cost, the real economic cost of potable water is around USD $40 per tonne. In Australia, the driest continent, copious quantities of potable water come out of a kitchen tap at a cost of about USD $2 per tonne, including the connection charges of basic plumbing. Good-quality water and sanitation costs around 2% of the average family income in Australia, whereas getting barely enough water to survive can take 20–40% of a family's economic resources in a low-income country. Furthermore, sanitation services are often non-existent. If people don't drink clean water, they take an equivalent economic penalty: they get sick and lose several weeks of earnings every year, and also can lose years of life expectancy. When the real economic costs and penalties are taken into account, acquiring the bare minimum of daily drinking water consumes 10% or more of the GDP (Gross Domestic Product) of a country like Pakistan, whereas the cost in an advanced, industrialised country is less than 0.1% of GDP.

Few people realise that the cost of water for families in many low-income countries is this high. Officially, about one out of every six people in the world lacks access to an 'improved' water supply.[7] However, well over half of the world's population lacks continuous, piped, potable water supplies to their homes: these people have to either carry potable water themselves or pay someone else to perform the task.

What we can learn from this simple analysis is that the value of engineering in the industrialised world lies in the enormously reduced cost of essential services like potable water. I would even go so far as to suggest that the real economic cost of obtaining potable water is an indicator of the effectiveness of engineering practices in any given region of the world.

Engineers not only find ways to reduce the cost of products and services, but they also organise and arrange the reliable delivery of these products and services so that people can effectively use them. The value from engineering, therefore, emerges from *reducing* the human effort and resource consumption needed to provide the product or service, which in this case is a supply of reliable, safe, clean drinking water. A reliable water supply in the home liberates women and children from endless backbreaking labour, without which survival is impossible. Women can then provide education for their children and earn supplementary income for themselves.

I will leave further explanations for the failure of public water supply utilities for Chapter 13. The main lesson to learn from this example is the social and economic value of engineering. We see people in low-income countries paying far more for essential services, in equivalent currency terms, than those of us lucky enough to live in wealthy, high-income countries where engineering functions more effectively.

I was shocked when I first realised how expensive essential services and commodities are for the poorest people on our planet. There is no denying that we take them for granted in wealthy countries.

Later, I realised that this issue presents a new generation of expert engineers with wonderful opportunities by creating effective utility services that work well. Chapter 13 explains these opportunities in more detail.

While many engineers see themselves as technical problem solvers, simply finding a solution for a technical problem may not provide any useful value by itself. The value only arises when the technical solution is implemented and applied so that people in the community can experience the benefits. Engineers take part of the responsibility for this; otherwise, their work may be of little intrinsic value.

One of the most important observations arising from our research is that engineers almost always have to rely on other people to implement the results of their work and reliably deliver products and services to the people who are going to use them. Engineers, by themselves, perform very little, if any, hands-on work. Most engineers rarely even meet the people who make use of their products and services. Therefore, the value of engineering work arises indirectly through the actions and behaviour of people other than engineers.

Given this situation, engineers need to be able to influence the behaviour of the people who deliver their products and services if their work is to result in lasting value. In the same manner, the only way that this book will result in any value for humanity is through you, the reader. This book will only achieve any useful value if the techniques that I suggest work out in practice and you choose to adopt and use them.

Therefore, in a sense, an expert engineer has to be an expert educator. Such an individual needs to educate and influence other people to reliably deliver the artefacts and services as the engineer intended in order for others to make effective use of them.

There is one additional, important step in this discussion on the value of engineering. I need to take you through a short argument based on economics to do this, but you will soon appreciate why this is critically important for you, the reader.

DISCOVERING EXPERT ENGINEERS

Labour market economics demonstrates that, in a stable market that responds well to information about supply and demand, employee remuneration reflects the employee's marginal product. In other words, employees are paid in proportion to the additional value that their work creates. At least in theory, your pay as an engineer will reflect the value you create for your employer and clients.

In reality, however, your pay depends on many factors, only a few of which reflect the value created by your contributions. First, as we have seen, the value of engineering work arises indirectly and depends on the contributions of many other people apart from engineers. This makes it difficult to measure the real value created by engineers. Second, and partly because of this, engineers are often paid according to a broadly agreed upon set of conventions. Third, valuable engineering ideas often arise, seemingly by chance, from social interactions with other people, making it even harder to identify any one individual's contributions.

That being said, our research has produced strong supporting evidence for this principle by comparing experts and other engineers in developing countries.

In India and Pakistan, our research revealed that most engineers are paid about one-third of the salary that engineers receive in advanced industrialised countries like the USA, Canada, Germany, and France. The same research demonstrated a critical shortage of engineering skills in India and Pakistan, as evidenced by widespread performance shortfalls in water supply and many other industries. The extraordinarily high economic cost of water reflects this shortage. These two observations seem to contradict one another. If there is a shortage of engineering skill, then why is the price of engineering labour so low when compared to industrialised countries?

The answer lies in what we consider to be engineering skill. Engineering students in India and Pakistan learn much of the same information in their university courses as students in any industrialised country. However, few of them have the real world opportunities to learn the kinds of skills necessary to become an expert engineer, the skills to which this book introduces you. In an industrialised country, novice engineers have more opportunities to learn from engineers with more skills, although this is often due more to chance than intentional design. Many of my former students left engineering in frustration because they ended up in jobs where they were expected to learn by themselves, without expert engineers to imitate or provide guidance.

We found a small number of engineers in India and Pakistan with the kinds of skills that made them experts in their fields. Their skills were recognised and rewarded by their employers, in some cases government-owned organisations. Their employers knew that these individuals could move to highly paid jobs in engineering firms in the Gulf region or even in Europe, Canada, and the USA. They were being paid well above the salary level that they could earn elsewhere, in equivalent currency terms. At the time (2004–2007) they were each earning the full-time equivalent of USD $90,000–150,000, salaries that most engineers in Southern Asia could not even dream of earning.

What were their skills that were recognised as so valuable?

These engineers were highly sought after because they had the capabilities needed to produce high value for their firms. Just as I have explained above, these firms recognised their capability of producing value indirectly, *through the skilled contributions of many other people*. They were no more intelligent than engineers employed for a small fraction of their salary. However, they had learnt how to tailor their engineering performance to deliver consistently valuable results for their firms. By studying their work, I started to understand how they achieved that. Some, but not all, had gained their engineering qualifications in Europe, Canada, or the USA. Some, but not all, had actually worked in Europe or the USA, although not necessarily in the best-managed firms. Some, but again, not all, had studied in business to supplement their engineering qualifications. Some others had none of these opportunities, yet they had still managed to acquire 'expert' engineering capabilities. What they had in common was a combination of determination to succeed, a belief that they could succeed, and an understanding that they could only achieve their own success through the successful contributions of many other people. Helping others to succeed was the critical factor in their own success. Most also explained their contributions in terms of the value created for their employers, which justified their high salaries.

In other words, these select few 'experts' had realised that success in engineering means creating economic and social value by working with and through other people: it cannot come solely from individual technical excellence, no matter how brilliant and insightful it happens to be.

If you live in a developing country like India or Pakistan, this should demonstrate that you can, in fact, succeed and become an expert engineer without having to leave your home country. You can also look forward to earning a salary at least equal to the best in the world. Of course, that all depends on you learning how to become an expert engineer.

If you live in a country like Australia or the USA, this lesson is just as relevant for you. Surveys of engineers have revealed that a large majority are dissatisfied with their earnings, responsibility levels, and opportunities for advancement. American civil engineers reported this in a large survey in 2004.[8] They complained about being passed over for leadership responsibilities and promotions in lieu of people without engineering qualifications. They complained about working with project managers who had little or no technical understanding and less experience, yet they were denied opportunities to lead projects themselves.

Here, then, is the answer to the question in the title of this chapter, 'Why Engineer?' Engineering is a wonderful career, partly because you can have great fun spending lots of money that belongs to other people. However, you need to remember that these people are investing in you, and your ability to consistently provide valuable results. Investors, by and large, don't understand much about engineering, but they spend their money on you (and others) in the expectation that it will result in sufficient value for their shareholders. Your ability to attract their money, confidence, and trust (as well as their patience when things go wrong), depends on your capacity to build relationships and deliver results . . . almost entirely through the actions of other people.

There is a social justice issue here as well: who gets to share the benefits of an engineering investment? While the economic benefits of reliable piped water to the home undoubtedly go to the householder's family, the benefits may not be so evenly distributed with every example of engineering success. Ultimately, if the benefits are not shared fairly and reasonably, people may intervene and bring even the best-planned engineering project to a halt. We will see this in Chapter 12, which discusses sustainability.

You cannot do engineering completely on your own, of course. No single person has all the required knowledge, skills, and capabilities to achieve engineering success without other people. Therefore, what counts is your ability to win the willing and conscientious cooperation of others who have complementary skills to yours. This ability, above all others, is what will underpin your success as an expert engineer.

I hope this explanation will show you that the real challenge as an engineer is to be able not only to devise wonderful technical solutions, but also to make sure that these solutions are delivered predictably, and that they provide real value for other people, both your clients and the people that they serve.

You might be thinking that all of this has to do with management, something that will only become relevant later in your career, at least not until after the first few years. You might think that you can learn all of this by studying for an MBA degree when you feel the need, can afford the fees, and are willing to face the prospect of studying hard once again.

You can see for yourself how much of the material in this book is included in an MBA programme: most is specific engineering material.

Our research demonstrates conclusively[9] that all of this will be relevant in the first year of your first job. Novice engineers find themselves relying on skilled contributions and help from many other people right from the start of their career. Understanding this, and building on this understanding by learning the skills described in this book, can get your engineering career off to a flying start.

PRIOR LEARNING

We're not quite done yet.

In fact, there is a long way to go. Learning has to start with an exploration of what the learner already knows when they begin. If I ask you to learn something that you already know, you will probably put this book down and never read the rest of it. Equally, if I ask you to learn something that relies on some prior knowledge that you don't yet understand, then no amount of explanation is going to be sufficient for you to learn it.

Throughout this book, there are many references to further reading if you need to build up your prior knowledge.

One of the main aims of this book is to provoke you into thinking more about engineering practice. There are many misconceptions about practice that can interfere with the quest to become an expert engineer, so it can be helpful to gauge how much you may have to shift your own perceptions to remove some obstacles to learning.

These are not small issues.

In all, this research has revealed more than 80 significant aspects of engineering practice that are missed by texts and conventional university engineering courses. These omissions allow myths and misconceptions to persist, mainly through unquestioning repetition. The texts were written by engineering educators, so it is quite possible that you unknowingly developed similar knowledge gaps from your experiences at an engineering school.[10]

Practice quiz 2

This quiz consists of a series of propositions that engineers talk about in the context of discussions on engineering practice. Respond to each question by indicating your current beliefs on the proposition. Select one item from the responses provided, ranging from 'strongly disagree' to 'strongly agree'.

Each proposition has been addressed by research that has contributed to this book. When you assess your responses (or receive feedback from the electronic version of the quiz), you will be able to see how well your beliefs align with the results of this research.

The evaluation guide in the online appendix also provides explanations for the scoring: read this carefully to learn more. Your score will provide some indication of the extent to which you need to study this book to help you on your way towards becoming an expert engineer.

Q1: An engineer who achieved higher grades at university tends to perform better in engineering work.

Strongly disagree	Disagree	Neither agree nor disagree	Agree	Strongly agree

Q2: As an engineer, it is critical that you accumulate sufficient technical knowledge by yourself to solve any problem that you are likely to be confronted with.

Strongly disagree	Disagree	Neither agree nor disagree	Agree	Strongly agree

Q3: Engineering is a hands-on practical occupation.

Strongly disagree	Disagree	Neither agree nor disagree	Agree	Strongly agree

Q4: In engineering, many decisions are made on the basis of perceptions that can be inaccurate or incorrect.

Strongly disagree	Disagree	Neither agree nor disagree	Agree	Strongly agree

Q5: Being a successful engineer depends primarily on your technical expertise.

Strongly disagree	Disagree	Neither agree nor disagree	Agree	Strongly agree

Q6:   Facts are more objective and unbiased when stated in terms of numbers rather than words.

Strongly disagree	Disagree	Neither agree nor disagree	Agree	Strongly agree

Q7:   The ability to build collaborative relationships with more experienced engineers, suppliers, and site supervisors has more of an effect on workplace performance in engineering than academic ability.

Strongly disagree	Disagree	Neither agree nor disagree	Agree	Strongly agree

Q8:   Most of what an engineer needs to know is learnt in the workplace.

Strongly disagree	Disagree	Neither agree nor disagree	Agree	Strongly agree

Q9:   You can only learn communication skills by practice; they cannot be taught.

Strongly disagree	Disagree	Neither agree nor disagree	Agree	Strongly agree

Q10:   In engineering, decisions are almost always based on technical facts, computation, analysis, results, and logic.

Strongly disagree	Disagree	Neither agree nor disagree	Agree	Strongly agree

Q11:   Engineers often have to work with vague verbal statements of requirements from their clients.

Strongly disagree	Disagree	Neither agree nor disagree	Agree	Strongly agree

Q12:   Graduate engineers, on average, spend just as much time interacting with other people as senior engineers, who often have management responsibilities.

Strongly disagree	Disagree	Neither agree nor disagree	Agree	Strongly agree

How did it go?

What did you get out of a possible 120 for the second quiz?

Even if you scored less than 60, you're doing okay. However, you still have a lot to learn. If your score was better than 90, consider yourself well on the way to being expert: congratulations!

Practice concept 4: What does it cost to employ you?

The second question asked of my recent graduate at the start of this chapter related to the cost of employing an engineer. Most engineering students, and many working engineers to whom I have explained this, are very surprised when they learn just how much it costs to employ an engineer.

Naturally, the exact costs depend on the particular company and its respective circumstances. The figures below indicate the approximate costs in Australia at the time of writing but will vary from place to place. However, in any given setting, the costs will be more or less similar in proportion.

Obviously, the first cost is the salary that a firm pays you.

Other costs include the following, expressed as an approximate percentage of your salary, although the actual amount will vary from place to place.

The last item (business development) represents the proportion of your time that will be needed to support business development, winning contracts for the firm, helping to write proposals, and tender submissions to gain new work.

With four weeks of annual leave and a nominal 37.5-hour working week, you will be at work for a maximum of 1800 hours each year. Allowing for one week of sickness, time for personal hygiene, work breaks, and moving from one location to another (200 hours), business development activities (200 hours), and training and guidance from supervisors (300 hours), approximately 1100 hours are available for 'billable' work, which are projects for which a client will pay. In practice, it is difficult to achieve this because there are times when there is insufficient work to keep everyone

Table 1.1 Additional employment costs.

Payroll tax, employer's liability insurance	5%
Superannuation/pension contribution	10%
Annual leave	10%
Office accommodation rent (approximately)	15%
Administrative support	15%
Supervision and assistance from senior engineering staff, approximately one hour per day for the first year and about 30 minutes per day thereafter	33%
Health insurance contribution	2%
Electricity, water	3%
Travel, accommodation, visits to engineering sites, attendance at conferences, sales seminars, etc.	15%
Engineering software licence fees	10%
Subscriptions to databases, libraries of standards, and other sources of information	5%
Business development, preparing proposals, and presentations by other company staff	25%
Total	148%

occupied. Let's optimistically assume that a young engineer would be assigned to work on projects for 1000 hours annually. That means that the cost of employment has to be recovered on 1000 hours of work for which the firm is paid by the clients.

With a monthly salary of $5,000 ($32.50 per hour pay, before tax is deducted), the hourly rate needed to cover your salary and all the other costs listed above will be at least $150. After allowing for a profit margin, the cost to a client will be about $170 per hour.

You need to understand this. The important issue here is that, as an engineer, you need to create sufficient value through your work to justify the cost that a client will be asked to pay for your services. You need to think very carefully about how you do your work because the cost can rise very quickly.

Most graduates and many engineers have never seen this simple calculation and therefore find it difficult to understand why their managers might be concerned about the way they spend their time. Recent research has shown that young engineers find it very difficult to describe the precise value created by their work, and most engineers only consider the direct salary cost of employment, not the overhead costs listed above. To be an expert engineer, and also to enjoy your work, you need to be confident that you are creating at least as much value as it costs a client to obtain your services. Adapting designs and plans that have provided value for previous clients is often the best way to create value for new clients. You also need to charge clients a fee that covers your costs and a reasonable profit margin. It is easy to underestimate both of these costs.

IDEAS FROM ECONOMICS

If you are going to spend lots of money, particularly money that belongs to other people, you need to have a basic idea of how money works.

Money has evolved as a means of exchange for goods and human services. The price of goods reflects both their perceived value to the purchaser, their relative scarcity, and the amount of human effort and other resources required to produce them. This much is common knowledge.

Money, therefore, is an approximate measure of human labour.

As an engineer, you will probably be paid an agreed amount for each hour or day that you work, but not everyone works on this basis. A contractor often gets paid an agreed upon amount for completing a task.

Two fundamental difficulties arise in both of these situations.

The first is that what a person can achieve in a given amount of time is highly variable. If the person is focused on the task, free from interruptions, well rested, keen, motivated, and finds the work intrinsically rewarding and satisfying, then that person will often achieve a great deal in a short time.

If, on the other hand, the same person is tired, working in a noisy environment with frequent interruptions, unable to concentrate, thinks the work is seemingly of no consequence, like a report that no one will read or that will be simply filed away forever, then much less will be achieved in the same amount of time.

The same applies even in the case of a person who is working on a 'piece rate' and gets paid an agreed upon amount for completing a task.

If the person is focused on the task, free from interruptions, well rested, keen, motivated, and finds the work intrinsically rewarding and satisfying, then that person will not only finish the task earlier, but will also 'put their heart and soul into the task', working conscientiously to make sure that there are as few faults or defects as possible.

So even though the amount of money paid is predetermined, the quality of the work performed and the amount of work performed by people can be highly variable.

The second difficulty is that the value of money itself changes over time and, more importantly, with human perceptions of value.

Most of the time, the prices of goods and the salaries that people earn slowly increase all the time: this is known as inflation. Most government economic regulatory agencies aim to keep annual inflation rates between 2% and 3%. Because of inflation, the value of money changes with time. Normally, when the inflation rate is low, the effect on the value of money is quite small.

However, much larger changes occur in the value of money because much of it is invested in shares in business enterprises.

The value of shares fluctuates daily, sometimes by a large percentage. It is not unusual for the price of shares in a firm to go up or down by more than 5% in a single day. Yet, on most days, the firm occupies the same buildings, with the same clients, the same customers, and basically the same volume of sales. Why then does the share price rise and fall so much? And why is this important for you, particularly if you do not even own any shares?

The reason why the share price rises and falls so much is because the share price depends not so much on the day-to-day business conducted by the firm, but rather by what prospective buyers of the shares think is going to happen to the share price over the next 6 to 12 months, even over the next few days. What this means is that they are trying to figure out what prospective buyers of the shares are going to think in future, and the future, as we are constantly reminded, is entirely unpredictable. It is like trying to forecast what other people think will be in fashion when they attend next year's fashion parades.

Many people think that a house built from bricks and mortar, constructed on a block of land, and unable to be moved easily, is a rock-solid, reliable investment. Some of the time, house prices change very little from one day to the next. However, this is not always the case. There have been many times when buyers have believed that by offering a price well below what the seller is demanding and then waiting, the seller will eventually compromise and reduce the price. This happens when it is hard to obtain money from the banks to buy houses, particularly in uncertain economic times. The real value of a house, what a buyer will actually pay for it, can be very hard to predict and can fluctuate just as much as the shares in a business. Barring any unforeseen disasters, the house remains exactly the same: it does not change. The only things that change are the expectations of the buyer and the anxiety of the seller to conclude a sale.

As I explained before, engineering depends on spending lots of money that belongs to other people. Those people are investors. They borrow much of the money through investment banks, but if the engineering does not work as expected, many will have to sell their shares, and even their houses and land, to repay the loans from the banks. The banks protect their interests well: they will hire other engineers to analyse the proposed venture in order to ensure that the engineering and market studies have been done thoroughly. Banks pass on risks to other investors. The investors only take on the risk because they are confident that, eventually, the engineering work will provide products or services that are valuable for other people, and that these other people will buy enough products and services at sufficiently high prices so that the investors will get their money back with an additional profit.

Notice, therefore, how much depends on human perceptions of value.

Investors who are providing money for engineering depend on other investors to buy the shares and houses that they sell to raise capital.

The same investors would not be providing money for an engineering enterprise, part of which will pay your salary, unless they (and their bankers) were confident that you and the other engineers involved will be able to deliver the products and services at the expected cost. Nor would they provide money unless they were confident that other people will buy enough of the products and services at high enough prices for them to eventually make a profit.

Investors usually know very little about engineering. While they may hire consulting engineers for advice, in the end, they rely on their own judgement, which is based on their perceptions that the enterprise will succeed.

Therefore, their willingness to spend money, and furthermore your opportunity to perform engineering work for them, critically depends on the perceptions of investors and many other people. In the same way, the cost of engineering work and the quality with which it is performed depends to a large extent on the perceptions and feelings of the people performing the work.

Now, all of this can be very difficult to accept and understand, particularly when one has been taught for several years to work with predictions based on accurate computer models, precise laws of physics, and repeatable experiments. Many engineers yearn for the unerring certainty and feelings of precise control that come from writing their own computer software: once it works, it works every time. Furthermore, if it does not work, the only mistakes are your own and no one else's.

However, within the real world of shifting human perceptions and emotions, engineering becomes much more challenging . . . and interesting. Like a ship at sea, human emotions move and change all the time. Just as the ship's bow may be momentarily submerged by a large oncoming wave, seconds later it will soar above the oncoming trough. In the same way, when a person feels a temporary loss of confidence, when even the smallest obstacles seem overwhelming, you can be confident that this dark depression will soon be forgotten and displaced a few days later by humour, excitement, and optimism.

Among all of this social fluctuation and change, one of the most attractive attributes of an engineer is the ability to clearly distinguish the essential engineering functions needed to satisfy human desires and requirements, especially for end-users, the people who ultimately use the engineered products or services. This means understanding and listening to the client, end-users, and other stakeholders, something that we will explore in greater detail in the coming chapters. However, it also means perpetual dialogue, a continuing conversation to educate the client and other stakeholders about feasible engineering possibilities.

At the same time, an engineer will be searching for the most cost-effective ways to achieve the required function, considering ways that bypass unsolved technical problems and uncertainties, adapting what has been done before, and thinking ahead – how could this be misused, abused, or neglected? Can we line up reliable people to do the work at the time they will be needed? What is the situation facing our preferred contractors that have experienced similar situations before? Are they going to be overcommitted with other work, or at risk of financial default?

At the same time, in the search for feasible choices for the client, an expert engineer will be instinctively thinking ahead and planning for risks and uncertainties to minimise the effect of unpredictable events on the ability to deliver results earlier than required, safely, with better quality, and at a lower cost. Simultaneously, the engineer will need to reassure the client and help build a trusting relationship so that the client feels confident that the engineer can deliver. This takes time and experience in working with people.

You might think that this is something that will await you after several years in your career. Our research proves otherwise: even engineers in the first few months of their employment find themselves dealing with apprehensive and anxious clients, often indirectly through their employees or representatives, which makes it all the more difficult to understand what's actually happening.

Did I know any of this when I started as an engineer?

No, not in the least … none of this. I was shy and apprehensive, and also quite confident that I was one of the least equipped people to deal with human relationships, business, and finance. I focused instead on my technical skills, where I had reasonable confidence in my abilities.

Gradually, I came to appreciate my dependence on other people: no amount of technical expertise could isolate me from the need to work with other human beings. The skills and insights in this book could have helped me understand so much more than I was able to at the time. They could have helped me become an expert engineer much sooner and with less frustration than I alternatively experienced through ignorance.

Even now, I am still learning about people. I admit it: I can be a slow learner. With the help of this book, you will be able to learn far faster than I did, and it will help you become an 'expert' engineer in a few years from now.

NOTES

1. Access to online quizzes and practice exercises is available through the online appendix for this book.
2. Trevelyan (2012b).
3. Wellington (1887).
4. Goold (2014).
5. Gainsburg (2006).
6. Trevelyan (2014a).
7. An 'improved water supply' is a term used in the context of UN projects and discussions on meeting the 'Millennium Goals' for human development, meaning a water supply that does not require people to go to the nearest river, water hole, or other natural water source. It includes water supplies such as wells and piped water supplies, but does not imply safe, clean, potable water. Piped water does not have to be continuous: intermittent service for an hour or so every few days, possibly contaminated by water seeping into the pipes

through broken connections when the pressure is low, still counts as an 'improved water supply'.
8. Reported in American Society of Civil Engineers (2008).
9. We conducted an extensive study of the work and careers of a cohort of 160 engineering graduates from the University of Western Australia (Trevelyan & Tilli, 2008).
10. See online appendix for details.

Chapter 2

What type of engineer?

When I tell people that I am an engineer, the first question they always ask is, 'What type of engineering do you do?' I tell them that I'm actually an engineering academic and that I teach young engineers. The conversation moves on predictably from there.

'Yes, but what type of engineer are you? Are you mechanical, civil, or electrical?'

'That's a difficult question,' I often reply. 'Most of my career has been in mechatronics and robotics. I graduated as a mechanical engineer and moved into aerospace and human factors engineering at the start of my career. Now, I teach design and sustainability and commercialise new air conditioning technology.'

The word 'type' in this context means 'discipline', a specialisation within the field of engineering. This is a relatively new concept; until the middle of the 19th century, there were only two types of engineers: civilian engineers and military engineers. However, by the end of the 19th century, the three main disciplines had emerged. Civil engineers, the original civilian engineers, were responsible for most engineering projects at the time. They designed and organised the construction of railways, roads, water supply and sanitation systems, canals, ports, and buildings. The Industrial Revolution in the 18th and 19th centuries led to the proliferation of manufacturing and machines; mechanical engineers specialised in the design and manufacture of machines, as well as the rapidly growing number of consumer products. The invention of electric machines in the 19th century naturally resulted in the development of electrical engineering as an engineering discipline in its own right. In the early years of the 20th century, the industrialisation of chemical science necessitated the development of chemical engineering as a further major engineering discipline. In the late 20th century, as computer programming increasingly became a significant focus of development, software engineering emerged as another new discipline. Growing concern about the need to maintain the quality of our environment and reduce the destructive impact of human activities, particularly in terms of engineering activities, has recently given rise to environmental engineering as a distinct discipline as well. A vast range of other disciplines have emerged in the last half-century, making it important to take stock of the situation and pose an important question:

What is, and what is not, an engineering discipline?

My research notes have provided me with a list of about 260 engineering disciplines, labels that engineers use to mark out the particular specialised knowledge on which they base their practice. What is the difference between each of these disciplines? What distinguishes an electrical engineer, for example, from an electronics engineer? What characterises a ship engineer, an aerospace engineer, a fire protection engineer, and a machine designer? What defines an electrical power network engineer? Along with what separates these disciplines, what do they have in common?

The research for this book demonstrated that the work performed by engineers in many different disciplines is remarkably similar. Even from a technical point of view, practically all engineers rely on the same basic concepts in engineering science:

- **Equilibrium:** Things stay the same and only change in response to an external influence.
- **Mathematics:** Provides the underlying mathematical logic in which the laws of engineering science are framed.
- **Systems Thinking:** Engineers define a boundary around a cluster of interacting objects and artefacts and then think about what must cross the boundary and what remains inside.

Disciplines other than software engineering also build on two other basic concepts:

- **Conservation Laws:** Energy, momentum, charge, and mass must be conserved. Except in the context of certain nuclear reactions, mass remains constant, which means that Einstein's famous law relating mass and energy stays out of the picture.
- **Continuity:** The notion that what goes in must come out, somewhere. It can't disappear.

Most engineering students learn these fundamentals in their first two years of study, while much of the mathematics and science required for engineering disciplines has already been learnt in high school.

What makes up the rest of an engineering discipline is a language of ideas, confidence in learning about technical issues quickly, and an appreciation of supporting resources, practices, and ideas in a given area of technical specialisation, along with a wealth of detailed technical knowledge; we will learn more about this in Chapter 5.

This book is mainly about ideas that are common to all engineering disciplines, particularly the socio-technical concepts that underpin collaboration, which is an essential feature of all engineering pursuits. It is paradoxical that these ideas seem to have been largely forgotten in engineering schools. Perhaps it is simple to understand what distinguishes us from other academic disciplines like the humanities and social sciences, but it is not so easy to see what is common for all engineers.

As we shall see in the next chapter, collaborating with other people and coordinating technical work takes the majority of the time and effort in real world engineering practice. Remarkably, these practices are common to all disciplines of engineering that we have studied so far. There are only very minor differences in certain disciplines. It is this, the major part of engineering practice, which the rest of this book is all about.

Your university education provides you with the technical language, ideas, and methods used for one of the main engineering disciplines: civil, mechanical, electrical, environmental, chemical, or software/IT. Yet, graduates from these disciplines then go on to enter two to three hundred different areas of engineering practice and disciplines! At the end of this chapter, Table 2.1 lists the disciplines that we have encountered in our research. This table shows which of the main disciplines that engineers in each more specialised discipline may have migrated from. Migration is an appropriate word in this context. While you might start out as a civil engineer, over the course of your career, you may gradually migrate from one discipline to another, often by chance as opportunities arise.

Many, if not most, engineers end up practising in a completely different area from the one they expected to enter at the time of graduation. That is nothing to be concerned about; while the technical knowledge in a particular discipline is always important, it can be learnt remarkably quickly because of the practice that you have had at university. What is common to all engineering disciplines is much harder to learn because it involves concepts and issues that you have not learnt much about at university.

This book is all about these common issues, those challenging ones that take time and real world experience to learn. We will understand more about why they are so difficult to master in the next chapter.

While most of the work that engineers perform is remarkably similar, even in technically very different disciplines, there are different places that engineers need to go to and ways that engineers need to think. Table 2.2 lists some of these fundamental differences. It compares some aspects of the main engineering disciplines, but it is a simplification of reality. Columns in the table describe the following aspects:

Where, apart from the office
Most engineers spend much of their time in offices, while many others spend a lot of time travelling. Working on-site takes you to the other locations listed in this column.
What you can see, feel, smell, and touch
This column briefly lists aspects of the materials and artefacts that you work with that you can perceive with your senses.
What you cannot perceive, where abstract thinking is required
Lists some of the aspects for which abstract thinking is needed, often with the help of mathematical analysis and computer modelling.
3-D spatial thinking
In some disciplines, it is an advantage to be able to think in three dimensions, to be able to visualise three-dimensional objects portrayed with two-dimensional diagrams and sketches. However, this specialised skill is not essential in any discipline.
Predicting hazards and anticipating failure
Being able to forecast possible material and artefact failures is an essential critical thinking ability for engineers. Being able to perceive the early signs that indicate the likelihood of failure is a valuable attribute that will save time and enhance safety.
Public health and social risks
Engineering work in most disciplines, but not all, can have a large influence on public health, as well as social stability and cohesion. For example, the failure of a water supply system can have catastrophic public health and social stability consequences.

Is this something I can show my children?

For many engineers, this is a powerful motivator. 'I helped to build that big ugly ship you can see in the picture!' However, in many aspects of engineering, the contributions of an individual engineer are more difficult to see.

CHOOSING A DISCIPLINE

When students have asked me about different disciplines, usually when they are trying to make up their mind about which discipline to specialise in, they usually ask the following (or similar) questions:

Which discipline has the most jobs for engineers?

> There is no simple answer to this question. Which of more than 200 diverse, dynamic disciplines do you think has the most jobs for engineers? Employment opportunities for engineers are nearly always good, but there are large fluctuations in some areas of engineering. Engineering depends on people spending money long before the benefits are received: engineers spend money that investors are willing to provide. In times of economic uncertainty, investors are usually much more cautious, which means that there is less money to spend on engineering, thereby limiting job opportunities. However, engineers are always in demand, to a certain degree, in order to keep things running. People always need water, food, energy, shelter, transport, communications, and sanitation and there are plenty of opportunities for engineering to provide these necessities. As long as you are prepared to be flexible and have the skills to work on different aspects of engineering at different times, there will always be jobs available for you.

Which discipline is going to be in demand in the future?

> I once went to a lecture by an electronic engineering professor who claimed that his was the only discipline worth studying. All the other disciplines had reached the end of their intellectual development, so all that was left to do was incremental refinement. All the excitement, he claimed, had vanished from the other disciplines. He was clearly mistaken in his belief. True, there are some engineering disciplines that have receded in relative prominence, for example, mechanism design. However, there are still endless opportunities for intellectual advancement in even the oldest disciplines of engineering. There is no reliable way to forecast future developments in engineering, just as in most aspects of human activity. All that one can be sure of is that engineers, in every discipline, will be in demand for the foreseeable future.

What's common between the disciplines?

> Read the rest of this book and find out. Most of what engineers do is common between all of the disciplines. This helps to explain why most engineers are able to change disciplines during their careers, some more than once.

Which disciplines are special, involving something that no other engineers do?

> None of them. That's the fascinating thing about engineering. Everything that is done in any one discipline is also done by some people in other disciplines. There is no such thing as a 'standard' mechanical or civil engineer. Every engineer has unique abilities that are different from every other engineer. Even in a specialised discipline such as fire protection engineering, you will still find ordinary mechanical or electronics engineers working in fire protection without calling themselves fire protection engineers. There are usually many more mechanical, electrical, and instrumentation engineers working in the oil and gas industries than there are petroleum or oil and gas engineers. The same applies in the aerospace discipline, where mechanical, mechatronic, and electrical engineers typically outnumber aeronautical and aerospace engineers.

Which industries provide employment opportunities and where?

> There are obviously links between certain industries and specific disciplines. For example, a road engineer is less likely to be working in aerospace than a materials engineer. However, a road engineer that acquires high-level commercial and project delivery skills could successfully migrate into mine development, and from there into mineral processing systems. By recognising that there are as many, if not more, similarities as differences between engineering disciplines, and also learning these common engineering practice skills, you will have far more opportunities in a wide range of industries.
>
> Also, many industries employ engineers simply for their analytical and mathematical abilities, as well as for their ability to work with abstract ideas. The most common destination where these skills are directly applicable is probably finance. For example, almost all the engineering graduates from one of the prominent London universities end up working in financial institutions.

In summary, then, my recommendation is to always follow your interests, knowing full well that your interests will change and, at the same time, keep watching for opportunities in areas that you might not have considered. In the end, it does not matter which discipline you choose. All the engineering disciplines offer exciting and rewarding careers and each of them offers the chance to become an expert engineer. Every choice on the diverse spectrum of specialties will give you the chance to improve the world, while also making a reasonable income to support your family at the same time. The most essential things you need to learn in order to become an expert engineer do not depend on which discipline you happen to find yourself in at any one time.

Table 2.1 Engineering disciplines and the main disciplines from which they originate.

Discipline	Civil	Electrical	Mechatronics	Mechanical	Chemical	Environmental	Computer systems	Aviation
abattoir engineer			*	*				
acceptance test engineer		*	*	*				
acoustics engineer			|	*				
aeronautical engineer		*	|	*				*
aerospace engineer			*	*				
agricultural engineer	*		*	*				
air-conditioning engineer			|	*				
aircraft engineer			|	*				
aircraft structure engineer				*				
airframe engineer				*				
airport engineer	*	*	*					
analogue electronics engineer		*	|				*	
application engineer			*				*	
application support engineer			*					
architect	*							
army engineer	*	*	*	*			*	
asset engineer		*	*	*				
asset management engineer	*	*	*	*				
asset manager	*	*	*	*				
audio engineer		*	*					
automotive engineer			*	*				
avionics engineer		*	*	*				*
ballistics engineer			|	*				
bearing engineer				*				
biochemical engineer						*		
bio-engineer				|	*			
biomedical engineer		*	*	*	*			
bionics engineer		*	*					
boiler engineer				*				
bridge design engineer	*	*						
broadcasting engineer	*							
builder	*							

Job title	1	2	3	4	5	6	7
building services engineer				*	*	*	
canal engineer				*			*
ceramics engineer			*	–			
chemical engineer				–			
chief information officer	*						
chip design engineer		*		*	*	*	*
city engineer						*	*
civil engineer				*	*	*	*
clean room engineer				*	*	*	
coastal engineer							*
code designer	*			–			
commissioning engineer	*		*	*	*	*	
composite structures engineer	*			*	*		
computer engineer	*						*
computerised maintenance management system	*		*	*	–	*	*
configuration engineer							
concrete engineer	*						*
condition monitoring engineer				*	*	*	
configuration management engineer				*	*		
construction engineer				*			*
construction management engineer	*			*			*
construction planner	*		*	*	*	*	*
contracts engineer	*			*	*	*	*
control engineer	*			*	*	*	
cost engineer	*			*	*	*	*
cryogenic engineer					–		
customer support engineer	*			–		*	
database engineer	*			–	–		
design office manager	*			*	*	*	*
designer	*			*	*	*	*

(Continued)

Table 2.1 Continued.

Discipline	Civil	Electrical	Mechatronics	Mechanical	Chemical	Environmental	Computer systems	Aviation
detailed design engineer	*	*	*	*				
die design engineer				*				
disaster relief engineer	*	*	*	*	*	*	*	
discipline lead	*	*						
drainage engineer								
drill support engineer			*	*				
drilling engineer				*				
earthmoving engineer	*							
earthquake design engineer	*							
electrical engineer		*	−					
electronic engineer		*	−					
elevator engineer		*	−	*				
energy safety engineer	*	*	*	*	*	*		
engine design engineer	*			*		*		
engineering academic				*			*	*
environmental engineer				*				
ergonomics engineer			−	*				
erosion engineer	*					*		
estimator	*	*	*	*	*		*	
explosives engineer	*		*	*	*			
facilities engineer	*	*		*	*		*	
fatigue specialist				*				
field support engineer		*	*	*			*	
financial engineer								
finite element analysis engineer	*			*		*		
fire protection engineer		*	*	*				
flight control engineer		*	*	*				
flight engineer				*				
flood control engineer	*		*	*		*		
food-processing engineer			*	*				
forensic engineer	*	*		*				
forging engineer								
foundation engineer	*							
fracture mechanics engineer				*				
furniture design engineer				*				
gear design engineer				*				

geographic information system engineer	*						*		
geotechnical engineer	*			*			*		
geothermal engineer			*	*					
ground support engineer		–	–						*
heating and ventilation and air-conditioning engineer									
heritage engineer	*	*	*	*	*			*	
hotel engineer		*	*	*					
human factors engineer		*	*	*			*	*	
hydraulics engineer	*		–	*			*		
hydrologist	*						–		
hydropower engineer		*	*	*					
industrial design engineer		*	*	*					
industrial engineer			–	*					
information systems engineer								*	
infrastructure engineer	*			*					
inspector	*	*		*	*				*
instrumentation engineer		*	*	*					
integrity engineer				*					
irrigation engineer	*			*			*		
IT systems manager			–					*	
kinematician			*						
lighting engineer				*					
local government engineer	*	*	*	*			*		
locomotive engineer									
logistician			–	–					
logistics engineer		*	*	*					
lubrication engineer		*	*	*					
machine designer	*			*					
maintenance planner		*	*	*					
manufacturing engineer			*	*					
marine engineer			–	*					

(Continued)

Table 2.1 Continued.

Discipline	Civil	Electrical	Mechatronics	Mechanical	Chemical	Environmental	Computer systems	Aviation
marine surveyor				*				
materials engineer				*				
mechanical engineer		*	*	*				
mechatronics engineer		*	*	*				
medical equipment engineer		*	*	*				
medical instrument engineer				*				
metallurgist				*			*	
military engineer	*	*	*	*			*	*
mine manager	*			*				
mineral processing engineer	*			*	*			
mining engineer		*	– *	*			*	
mission specialist			*	*				
mobile equipment engineer				*				
mould design engineer				*				
naval architect			– *	*			*	
network engineer (computer, IT)		*	*					
network engineer (electrical power)								
nuclear engineer			*	*		*		
nuclear weapons specialist				*		*		
ocean engineer	*					*		
oceanographer	*					*		
offshore engineer	*			*		*		
oil and gas engineer				*	*			
operating system engineer		*	– * –	*			*	
operations engineer		*			*			
opto-electronics engineer			–	*	*			
ordnance disposal engineer			*	*	*			
packaging engineer		*	*	*	*			
patents attorney	*		*	*	*		*	
pavement engineer								
petrochemical engineer					*			
petroleum engineer				*	*			
pipeline engineer				*				
piping design engineer			–	*				

Job title	1	2	3	4	5	6	7
plant engineer			*	*	*	*	
plant operator			*	*	*		
plastics engineer				*			
pneumatics engineer				*	*		
port engineer		*		*	*		*
power systems engineer						*	
power transmission engineer				*	*		
powertrain engineer				*			
pressure vessel design engineer				*			
process design engineer							
process engineer			*	*	*	*	
process simulation engineer			*	*	*	*	
procurement engineer			*	*	*	*	
product design engineer			*	*	*	*	
product engineer	*		*	*	*	*	
product manager	*		*	*	*	*	
production manager			*	*			
production superintendent			*	*	|		
project engineer	*		*	*	*		*
project manager	*			*	*		*
propulsion engineer							
prosthetics engineer					|		
protection engineer						*	
pump engineer				*	*		
quality engineer				*		*	
quantity surveyor			*	*		*	*
quarrying engineer							*
radar engineer					|	*	
radiation protection engineer				*	*	*	
railway engineer				*	*		
reaction engineer			*				
reactor engineer			*	*	*	*	
reactor operator			*				

(Continued)

Table 2.1 Continued.

Discipline	Civil	Electrical	Mechatronics	Mechanical	Chemical	Environmental	Computer systems	Aviation
recording engineer		*						
refinery manager		—	—	*	*			
refractory engineer			*	*	*			
refrigeration engineer	*	*	*	*				
regulator	*	*	*	*	*		*	
reliability engineer		*	*	*				
reliability modelling engineer	*		*	*				
remote sensing engineer	*	*	—	*				
renewables engineer	*	*	*	*	*			
research and development engineer	*	*	*	*	*	*	*	
road engineer	*							
robotics engineer		*	*	*			*	
rocket engineer			—	*				
rotating equipment engineer		*	*	*				
safety engineer	*	*	*	*	*		*	
sales engineer	*	*	*	*	*		*	
sanitation engineer				*				
satellite engineer		*	*	*			*	
scheduler	*	*	*	*	*			
seismic engineer	*	*		*				
semiconductor engineer		*	—	*				
ship engineer			*	*				
signalling engineer		*	*					
simulation engineer	*	*	*	*	*		*	*
site engineer	*	*	*	*	*			
site remediation engineer					*	*		
software architect			—				*	
software configuration engineer			—				*	
software engineer			—				*	
software support engineer			—				*	
sound engineer		*	—	*			*	
space engineer		*	*	*			*	
structural engineer	*	*		—				
studio engineer		*		*				
submarine engineer		*	*	*			*	

subsea engineer

substation engineer

surveyor

switchgear engineer

systems engineer

tailings engineer

technical standards engineer

technical writer

telecommunications engineer

test equipment engineer

textile processing engineer

thermodynamics engineer

tool design engineer

traffic control engineer

transmission engineer

transport engineer

tribologist

tunnelling engineer

turbine engineer

underwater acoustics engineer

underwater weapons engineer

value engineer

vibration engineer

video engineer

water engineer

water supply engineer

weapon systems specialist

weapons engineer

welding engineer

wind tunnel engineer

wireline logging engineer

yacht design engineer

Table 2.2 Comparison between main disciplines: workplaces, thinking, and safety.

	Where, apart from the office	What you can see, feel, smell, and touch	What you cannot perceive, where abstract thinking is required	3-D spatial thinking	Predicting hazards and anticipating failure	Public health and social risks	Is this something I can show my children?
Civil – structural	Outdoors	Surface of structure only	Load paths, stress, construction sequence, planning	yes	Mostly invisible, some signs of failure (e.g. cracking)	yes, severe	yes
Geomechanics	Outdoors	Surface only	Load paths, stress, construction sequence, planning, soil-water interactions	yes	Mostly invisible, some signs of failure (e.g. cracking)	yes, severe	Hard to see
Mining	Underground, outdoors for open pit mining	Bore samples, rock faces	Load paths, stress, mining sequence, planning, soil-water interactions	yes	Mostly invisible, some signs of failure (e.g. cracking)	yes (in mining area)	Hard to see
Hydraulics	Outdoors	Surface only	Ground-water interactions, water flow, planning, forecasting	some	Mostly invisible (out of reach)	yes, severe	Hard to see
Materials	Factory	Surface only	Atomic structure, load paths, stress, faults, fracture, thermal effects	yes	Many obvious, but some invisible (e.g. fatigue)	Indirect	yes
Mechanical	Factory, chemical plant, outdoors	Equipment and machines	Heat transfer, fluid flow, stress, load paths, fract ure & fatigue, vibration, construction sequence, planning	yes	Mostly visible, but some not (e.g. stress, pressure)	yes, severe	yes
Manufacturing	Factory	Equipment and machines	Heat transfer, fluid flow, stress, load paths, assembly sequence, planning	yes	Mostly visible, but some not (e.g. stress, pressure)	yes (esp. to factory workers)	Products, rest hard to see
Mechatronics	Factory, outdoors	Most, except electronics and data	Heat transfer, stress, load paths, assembly sequence, planning, electric current, voltage, electro-magnetics, control, data transfer and manipulation, synchronisation	yes	Mostly visible, but some not (e.g. stress, pressure, software crash)	yes, severe (transport vehicles)	yes

Discipline	Location	Accessible/visible	Processes involved		Visibility	Consequences of failure	Visibility to eye
Software		Only through visualisation	Data transfer and manipulation, throughput, synchronisation, planning	no	Invisible	yes, mainly in telecommunications & banking	Often hard to see
Aeronautical	Factory	Surface of structure only	Heat transfer, fluid flow, stress, load paths, assembly sequence, planning	some	Mostly invisible, some signs of failure (e.g. cracking)	yes, severe	yes
Electrical (power)	Power station, outdoors	Equipment and machines	Electric current, voltage, electro-magnetics, transients, planning, scheduling	no	Mostly invisible (current, voltage, fields)	yes, severe	Often hard to see
Electrical (instrumentation)	Factory, chemical plant, outdoors	Equipment and machines	Electric current, voltage, electro-magnetics, transients, planning, scheduling, control	no	Mostly invisible (current, voltage, fields)	yes, severe	Hard to see
Electronic	Factory	Equipment and machines, components	Electric current, voltage, electro-magnetic fields, transients, atomic structure, planning, scheduling, control	no	Mostly invisible (current, voltage, fields)	yes, mostly indirect	Hard to see
Telecommunications		Equipment, rest only through visualisation	Electromagnetic fields, data transfer and manipulation, throughput, planning	no	Invisible	yes, severe	Cannot be seen
Chemical	Refinery, chemical plant	External containers only, some solids and liquids	Chemical reactions, solid-fluid interactions, fluid flow, heat transfer	no	Mostly invisible	yes, severe	Cannot be seen
Environmental	Outdoors	Pollution, plants, water flow, soil condition	Ground-water interactions, water flow, bio- and chemical interactions, planning, forecasting	yes	Mostly invisible, slow changes with time	yes, long term severe	Often hard to see
Petroleum	Offshore, remote locations	Bore samples	Ground-fluid-gas interactions, fluid flow, bio- and chemical interactions, planning, forecasting	some	Invisible and mostly inaccessible	some	Invisible

Chapter 3

Flying start, no wings, wrong direction

AVOIDING A HARD LANDING

Completing any engineering degree course is tough: at most universities, students consider it one of the most difficult areas of study to pursue. However, it provides an essential foundation for an engineering career, even though two of the most valuable attributes you develop are not in the formal curriculum specifications: the ability to work very hard on short notice and the ability to prioritise. One of the reasons why engineering courses are so challenging is that it is usually impossible to do everything required before the deadlines for submitting assignments and projects. As a student, you had to learn to choose the most important parts, which were usually those that made the best contributions to your grades.[1]

As many engineers soon find out, an engineering degree course provides a great foundation and starting point, but it may not lead to a great engineering career. Many engineers find themselves in dead-end careers with little chance of being promoted; they may even see non-engineers being promoted faster. Often, they see few ways to apply their technical skills in the work they end up performing, which makes them begin to question what all the hard work they put in to become an engineer was actually for.[2]

As a student, you may have found it hard to see the relevance of your courses.

Although you studied under the guidance of dedicated teachers, people who have devoted most of their lives to help you and hundreds of other students succeed as engineers, most teachers have had limited opportunities to experience the world of engineering work that you will spend most of your career performing. How could they? Engineering is inherently dynamic, changing year-by-year, decade-by-decade, and your career lies in the future. Your instructors' experience of engineering may have occurred two or three decades ago. As academics and researchers, their careers were most likely in technical specialist roles, meaning that they had very little experience with ordinary, everyday engineering.

While they have done their best to get you off to a flying start, they have no easy way to be sure that they've sent you in the right direction.

For some graduates, landing their first engineering job can be much more difficult than it seems for others. They quickly find out that employers are not exactly lining up to employ them. They send off countless applications, but receive only one or two replies (and no interviews). In some countries, it can take a long time to find an engineering job, which can be understandably discouraging.

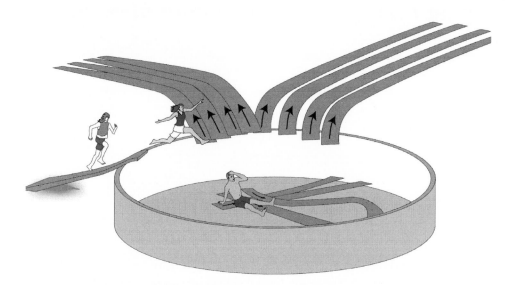

Figure 3.1   Off to a flying start? The steep inclined ramp on the left side represents your undergraduate engineering education. Getting an engineering degree is just one obstacle that lies between you and the wonderful career that lies ahead. It is a steep climb, but it gets you over the wall and into the ring; however, many engineers experience a hard landing like the fellow in the picture.

This chapter explains why many engineers experience this sort of frustrating, hard, and painful landing in the real world, as well as what can be done about it.[3]

The question is, why does this happen so easily? Why do so many young engineers with so much potential end up leaving engineering jobs in frustration?

The main reason is the lack of written knowledge about real world engineering practice, a problem that this book has set out to correct. The research for this book has identified more than 80 concepts that are fundamental to engineering practice that are rarely, if ever, mentioned in engineering texts and courses. The same research identified 17 significant misconceptions about engineering practice that develop due to the absence of written engineering practice knowledge.[4] Therefore, engineering education can actually *prevent* you from learning the capabilities that you will need to become an expert engineer. These misconceptions work just like the water-repellent layer that forms on the surface of soils containing oily vegetable matter when they are baked in strong sunlight. You can spray on water, but it remains in droplets or pools on the surface and does not soak in, sometimes for many hours. Misconceptions can create a 'learning-repellent mindset', one that resists learning opportunities. These misconceptions are also like blinkers, also known as blinders, fitted on the sides of horses' eyes to prevent them from seeing sideways or to the rear. Certain myths about engineering, reinforced by endless unquestioning repetition, effectively impede learning: they help the mind resist new knowledge that can ultimately be very helpful.

The structure of undergraduate university courses, the teaching methods, the classroom environments, and widespread community misconceptions about engineering all contribute to the development of these mistaken ideas about actual engineering practice, no matter which discipline you choose to pursue.[5]

If these misconceptions are not corrected, then it might be much harder for you to become an expert engineer. As explained in Chapter 1 page 8, they can even prevent you from recognising your activity as 'engineering work', even when you have an engineering job.

An engineering degree is not the only pathway to an engineering career, but it is quicker than other routes. You have built an intellectual foundation that helps you learn much of what you will need in the future. You will need that foundation as you work through this book. However, an engineering degree course at university only strengthens certain intellectual capabilities; unless you work hard to develop other abilities your formal education can lead to further misconceptions.

An awareness of these misconceptions has only recently emerged from our research. In a generation or two, engineering schools around the world may very well have new curricula that include the other skills you need to become an expert engineer. For the time being, however, working with this book is the one of the few ways to help you become a truly expert engineer.

To reach one of the many career pathways that take you out of the ring, over the wall, extend your potential into the distant future, and enable you to become an expert engineer, you will need to learn how to fly over that proverbial wall, as well as how to strengthen all the other intellectual capabilities that you will need. You also need to be aware of the misconceptions that we all started with, to a greater or lesser extent.

Are you ready to learn to fly?

You will need to develop some new skills of perception: listening and seeing, to find out which way to fly.

You will also need stamina and fitness. Your engineering degree studies developed both of these; you have withstood the rigours from years of undergraduate study, assignments, deadlines, and exams.

Determination is essential. Sometimes it will take several attempts to find the path that suits you best; remember, there is no single unique career path that suits everyone. You need to be prepared for setbacks, followed by picking yourself up, turning around, and trying once again. You will eventually succeed, but it takes determination and resilience.

You will need two intellectual wings, which can be difficult to grow during your university studies without a deep understanding of engineering practice.

The first is the realisation that you can only become an expert engineer with the help of many other people. You need to learn how to perform technical work with the help of others in order to gain their support, as well as their willing and conscientious collaboration.

The second 'wing' is the ability to value, acquire, develop, and use tacit ingenuity, which is compiled in a vast library in your mind composed of 'how-to' fragments of unwritten technical and other knowledge. Your progress as a student depended on knowledge that you could write down in examinations, tests, quizzes, etc. In engineering, your progress depends much more on knowledge that is mostly unwritten, the kind that is carried in your mind and the minds of other people. To acquire this knowledge, you may need to strengthen your ability to listen, read, and see accurately.

You can't fly in the right direction without a tail. For this, you need to understand what engineering is, how it works, and why it is valuable. Value is a multidimensional concept: economic value, namely making money for yourself and others, is just one dimension. Others include caring for other people, social justice, sustainability, safety, social change, protecting the environment, security, and defence, as we learnt about in Chapter 1.

Finally, you won't be able to find your way without knowing the point from which you are taking off. You will need the ability to understand yourself and where you are today. Otherwise, you won't be able to work out which way to fly.

Your test scores from Chapter 1 will give you some measure of your starting point. With the understanding of yourself that comes from this chapter, you can confidently start to develop your wings and then learn to fly by working through the remaining chapters of this book.

Here's the good news.

You don't have to wait until you have understood everything. It may take years for you to fully appreciate all the concepts in this book, but that doesn't need to delay your take-off. As once said, 'We are constantly leaping off cliffs and building our wings on the way down.'

As long as you accept the reality that there is still a great deal to learn, and you're prepared to work hard to grow your wings, then you can soon be off to a flying start.

This book is all about growing your wings, your tail, and your sense of direction, while also understanding where you are. As your career develops, you will continue to learn how to fly higher and faster.

REWORKING OUR NOTIONS OF ENGINEERING

The first steps in your journey require that you develop some new ideas about engineering and the knowledge on which you rely. The rest of this chapter outlines some of these ideas that have emerged through our research, and we will discuss knowledge in more depth in Chapter 5.

Misconception 2: Engineering is a hands-on, practical occupation

Many students often think that ...	Research reveals that ...
Engineering is a hands-on, practical occupation; therefore, it's important to be able to work with your hands on real equipment.	Most engineers rarely (if ever) perform hands-on work and are often required to ensure that any hands-on work is performed by technicians with appropriate qualifications. For an engineer, hobbies provide the main opportunity to perform hands-on work. If this is important for you, then start a hands-on hobby now.

How do you see the hands-on side of engineering? Is it something that you're keen to do? Not everyone is, but some engineers value this immensely. Write your comments or type them in directly at the interactive website for the book.

Scott Adams' well-known cartoon, 'The Knack', portrays one of the stereotypical notions of a child proto-engineer as the 'fixer', a child who can repair and make anything better. Many experienced engineers involved in teaching engineering complain about the small proportion of engineering students who have 'the knack', meaning the experience of having fixed things at home or through construction hobbies. Hands-on experience with construction, manufacturing, or troubleshooting unquestionably provides an expert engineer with a level of understanding that can be very valuable. However, our research demonstrates that this idea of a stereotypical engineer with hands-on skills conflicts with the realities of practice: nearly all engineers spend little, if any, time performing hands-on work in their professional capacity.

Some do, however. For example, here's a structural engineer talking about his own hands-on experience and how it helped him in the first year of his career.

'I worked at a mine. Basically I was just a boilermaker, a welder, I qualified in welding. My firm got their graduate engineers into it, the hands-on experience we never had at university. They told me "there is the welding machine and some electrodes" and I got trained by the guys who have done it for years and years. They said "this is how you do it". I wasn't welding all the time. I can't do overhead welding, I can't do pipe welding – anything that requires NDT (non-destructive testing). But anything that didn't require NDT was okay for me to weld. Some of the time I thought this was ridiculous – I was lying in mud, welding and grinding, and I was thinking there must be better things I can be doing. But since doing it I don't regret it at all. It was a great experience – now I know what you can do and what you can't do, what a bad weld looks like and what a good weld looks like. I'm smart enough not to ever have to do it again.'

Even more important, this engineer has learnt the language or 'lingo' of welders and other tradespeople who build engineers' creations. Even when engineers are speaking in English (many engineers use other languages as well), their language is different from the ways that other people use English. Knowing the language and words used by tradespeople can give you an enormous advantage. This is an important concept to which we will return later.

One engineer that we interviewed who was particularly keen to keep up his hands-on, practical skills told us how he has equipped a shipping container with machine tools and racks for his materials. As he moves from one engineering project to another around the world, his shipping container travels with him and sits at the back of his house or rented accommodation. Other engineers take up certain hobbies in order to maintain their hands-on skills.

Not all engineers remain hands-off; some will find time to participate in hands-on activities from time to time.

Usually, however, hands-on technical work is performed to a much higher standard with greater proficiency and quality by experienced technicians.

Later in this chapter, I will explain why so few engineers perform any real hands-on work and later chapters will explain how expert engineers without much hands-on skill or knowledge can collaborate very effectively with people who have those skills.

Misconception 3: Engineers are naturally concise and logical

Engineers (and engineering students) often think that …	However, research demonstrates that …
Engineers think and express themselves concisely and logically (in comparison to others).	Engineers are often neither concise nor logical. Engineers are not as good with logical arguments as many other people.

The ability to write and explain a convincing, logical argument commonly falls under the category of 'critical thinking'. It is the disciplined reasoning that lies behind a persuasive argument, the ability to see gaps and inconsistencies in another person's argument, and the ability to produce a valid counterargument.

This is another difficult obstacle to surmount for many engineers. Engineering courses provide few opportunities for students to develop their critical thinking skills, so many engineers struggle to express themselves concisely and logically by writing English text. In fact, a recent study found that engineering academics tend to be unsure about what 'critical thinking' even means.[6] Most of the time, engineering students work with mathematical derivations and equations. Consequently, students develop the skills needed to write mathematically consistent equations and expressions. However, the logic is embodied in the mathematical language itself. For example, given that F is force, m is mass, and v is velocity, any engineer can see that the following equation cannot be valid: force (F) is a different quantity from momentum (mv).

$$F = mv$$

Mathematics provides us with a beautifully concise language in which to express the ideas of physical and engineering science.

Writing and speaking concisely and logically in English, or any other language used for daily human communication, is an entirely different skill. We cannot rely on the intrinsic logic of mathematical expressions when we discuss or write about concepts in the real world. Instead, the logic has to be explained in words, which is a difficult skill to master. Our research evidence derived from extensive interviews and field observations helps to demonstrate that most engineers are often no more logical or concise in their expression of ideas than the majority of other people.[7]

One way to consider this issue is by thinking back to your experience of building with Lego blocks or putting together jigsaw puzzles. Those blocks only fit together in certain ways that were determined by their shape, just like pieces of a jigsaw puzzle. Critical thinking, on the other hand, is like building with smooth wooden blocks that need to be glued together to make an interesting structure. In this analogy, the blocks are logical propositions, such as:

The slope of the pyramid's side is 45 degrees.
We have five elderly people in our tour group.

To make a convincing argument, we need to join the propositions with explanations that link them together, similar to the glue that joins the smooth wooden blocks.

We have five elderly people in our group <u>and because</u> the slope of the pyramid side is 45 degrees, <u>it could be difficult for them to reach the top safely</u>.

Many expert engineers have mastered the ability to reason concisely and logically in the languages of their listeners. The key step here is to recognise that university engineering courses seldom develop this capacity in students. Studying philosophy at university is one of the best possible ways to develop this capacity, along with working hard to develop listening skills. This may be why students educated in the humanities often have better reasoning skills than engineers.

The next step is to accept that other people can be very logical and concise, but their thinking can follow quite different pathways of reasoning from those of an engineer. However, it is no less valid. The challenge for an engineer is to listen, understand, and learn how other people think. Most engineers find this difficult: they often think that people from non-engineering backgrounds cannot think logically.

Misconception 4: Engineers work with objective facts

Students and novice engineers often think that …	Expert engineers have learnt that …
Engineers prefer objective facts: numbers can convey more precise and objective facts than words. Facts explained with words tend to be fuzzy and subjective.	Reasoning with numbers and mathematical equations is only as good as the assumptions behind the data. Measurements often depend on the skill of the operators who are using or installing instrumentation. Numbers taken out of context can be quite misleading.

Engineers often feel challenged when it comes to words: many tend to think that words can be misleading, meaning that facts can easily be twisted by using words. Numbers, on the other hand, present objective facts and cannot be contested.

Here's an example to illustrate the fallacy of this common perception.

An engineer has expressed great confidence in the reliability of a complex chemical process plant at a particular site. He reports that there have only been three failures in two years of operation. Sounds good, right? What if I were to add that each of the three failures took four months to rectify, meaning that in two years, the plant has only been operating for 50% of the time. Does that still sound like a reliable process plant?

What we learn from this example is that numbers are merely symbols, like words, and the same numbers can convey completely different meanings in different contexts.

In other words, human language is context-dependent, whether that language comes in the form of numbers or words. We will revisit this concept several times in this book.

Many engineers feel more comfortable with numbers than with words: that much is true. However, when engineers attempt to present logical arguments to groups of decision-makers, which often include people who are very quick with words, they can often find themselves at a disadvantage. Mostly, this disadvantage arises because too many of us engineers cling to misconceptions about the nature of language and facts, once again exposing weaknesses in critical thinking skills.

Misconception 5: Engineers are problem solvers

This idea is one of the most deeply embedded misconceptions about engineering.

Our research tells us that engineers see themselves, more than anything else, as technical problem-solvers and designers. However, once we properly understand the context of engineering practice, we can understand that the act of problem solving needs to be seen very differently from the perspective shaped by our experience of solving several thousand 'textbook' problems from which young engineers learn their fundamental engineering science.

Students and novice engineers often think that …	Research shows that …
Solitary work on technical problem solving and design form the main components of engineering practice. Other aspects are either irrelevant or of lesser importance.	Solitary design and problem solving usually make up less than 10% of working time; analysis and modelling also make up less than 10% of your time (except for a few engineers). Expert engineers know how to *avoid problems*, rather than having to solve them. Problems without well-known, tried, and tested solutions only add to uncertainty.

As we shall see in the next section, there is much more to engineering than design and problem solving.

There is no question that many engineers see the intellectual challenge of solving difficult technical problems as one of the most satisfying aspects of their work. In our research interviews, many engineers talked about particular problems that they had managed to solve as though they were milestones in their career. However, a more interesting finding was that most engineers could recount a relatively small number of these instances. Engineering is much more about routine process than solving difficult technical problems. Much of the routine has evolved as a way of avoiding the need to find unique solutions for problems that have been effectively solved many times before.

Studies of engineering problem solving show that engineers seek solutions from other people as often as they try to set about solving the problems for themselves. Knowing who to ask is one of the best ways to find a solution for a technical problem in the workplace.[8]

Some engineers might counter this by claiming that nearly all their work amounts to seeking solutions to technical problems: it is just a matter of the particular approach that is used, while most problems can be solved using routine processes. However, once we expand the idea of problem solving to embrace most of what an engineer does, problem-solving itself becomes much less 'special'. In fact, every human being is a problem solver in that sense. Life is a succession of problems; we all solve problems every day of our lives. How do we drive to work and avoid the worst areas of traffic congestion? How do we find a parking place? Which is the best train or bus to catch to match my schedule today? What should I do first when I start work this morning? We all deal with these problems in a routine way without even thinking about them.

So, yes, engineers are problem solvers ... but so is everyone else. The fact that engineers are problem solvers doesn't get us very far in understanding the engineering practice.

Design and technical problem solving are certainly significant aspects in the work of most engineers, but as we shall soon see, other aspects take up much more time and attention from engineers.

Practice concept 5: Engineering is much more than design and problem solving

A research study on our own graduates[9] provided strong evidence that social interactions and communication (mostly on technical issues) take up much more time than solitary technical work, such as designing and problem solving. Separate research studies on engineering problem solving have revealed that it is actually social interactions that provide access to the necessary expertise for expediting solutions.[10] The following tables show perceptions from our own graduates on how they spent their working time after 6–9 months in their first engineering jobs.

126 respondents (graduate engineers in their first 9 months of employment) estimated the time spent in a previous week on each of the aspects of their work that was listed in the tables. They could choose from the following: none, <2 hours, 2–5 hours, 5–15 hours, or >15 hours. Responses were interpreted as median time values: 0, 1, 3, 7, or 15 hours, respectively. The total for different respondents varied (both as an artefact of the discrete choices and also possible overlaps, e.g. calculating while interacting with people face-to-face). Therefore, the time fraction for each category was calculated, and the resulting fraction was averaged across groups of respondents in each discipline shown in Table 3.1. Some respondents may have interpreted <2 hours as indicating a nil response; therefore, percentages at the low end should be treated as negligible amounts of time.

Most engineering students expect that solitary technical work will be a big part of their working life and hope that they will enjoy the challenge of working with difficult technical issues in the context of advanced technology. The results of our study, particularly the relatively small proportion of time devoted to solitary technical work, have helped to explain some of the frustrations I have so frequently encountered among engineers. Many felt frustrated because they did not think that their jobs provided them with enough technical challenges. Others felt frustrated because they thought that a different career choice might have led to a job that would enable them to make more use of the advanced technical subjects they had studied in their university courses. Many of them were actually planning to leave their career in engineering. In our research, we found that more experienced engineers, those who had stuck with it for a decade

Table 3.1 Average perceived percentage of working time for different disciplines.

	Civil	Mechanical	Electrical Electronic	Mechatronics	Others	Pilot Study	Ave.
Face-to-face informal	**13**	13	10	11	12	11	11.6
Writing documents	9	10	12	11	**13**	10	11.0
Calculation and simulation	12	9	3	7	**18**	7	9.2
Searching for information	6	9	**12**	8	7	8	8.2
E-mail	8	8	8	**9**	7	8	8.1
Reading, checking documents	4	6	**8**	8	7	9	7.1
Meetings	7	6	7	5	6	**9**	6.6
Design, coding	6	7	6	**9**	4	7	6.5
With people on site	**9**	4	4	5	4	7	5.4
Training sessions	6	6	**7**	4	7	2	5.2
Survey, inspection, observation	**7**	4	5	3	3	3	4.0
Phone	3	4	2	**4**	3	3	3.3
Operating, testing	3	3	**4**	3	2	5	3.3
IT, filing maintenance	1	2	4	**5**	2	2	2.6
Hands-on work	3	**4**	2	2	1	3	2.5
Debugging	1	2	2	**4**	2	4	2.3
Text messages	2	2	2	**3**	2	2	2.1
Searching for lost items	1	1	1	1	1	2	1.1

Table 3.2 Average perceived percentage of working time.

	Ave.	Cum.
Face-to-face informal	11.6	11.6
With people on site	5.4	17.0
Meetings	6.6	23.6
Training sessions	5.2	28.8
Phone	3.3	32.1
Text messages	2.1	34.3
E-mail	8.1	42.4
Reading, checking documents	7.1	49.5
Writing documents	11.0	60.5
Searching for information	8.2	68.7
Calculation, simulation	9.2	77.9
Design, coding	6.5	84.4
Debugging	2.3	86.7
Operating, testing	3.3	89.9
Survey, inspection, observation	4.0	93.9
IT, filing maintenance	2.6	96.5
Hands-on work	2.5	99.0
Searching for lost items	1.1	100.0

Note: These tables report perceptions of time spent on different aspects of engineering work nine months after commencing the first job. The highest figures in each row in table 3.1 have been highlighted. The pilot study group was predominantly involved in mechatronics. 'Others' included environmental and petroleum engineers and both these disciplines involved more extensive computer modelling work than their counterparts. Table 3.2 reveals that about 60% of the time was spent interacting with other people, either directly or through documents.

Design, coding

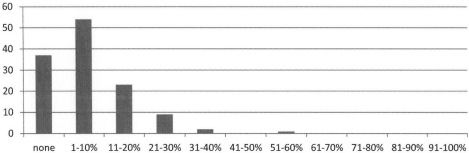

Figure 3.2 Perceptions of working time spent on design and software coding by 126 graduates in their first nine months of engineering employment.

Calculating, modelling

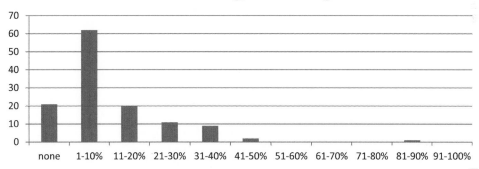

Figure 3.3 Perceptions of working time spent on calculations and computer modelling.

or more, had mostly realised that the real intellectual challenges in engineering involve people and technical issues simultaneously. Most had found working with these challenges far more satisfying than remaining entirely in the technical domain of objects.

The data in the tables consists of averages. The following charts reveal more about the variability in the data: the time spent on different aspects of engineering varies from week to week, as well as from person to person, and between disciplines and settings. The first two graphs demonstrate how most novice engineers in our sample spent less than 10% of their time working on designs and calculations.

The next set of graphs demonstrates how social interactions dominate engineering practice.[11] Qualitative analysis and field observations demonstrate that social interactions cannot be classified as 'non-technical'. Most of the social interactions, whether face-to-face, by telephone, e-mail, or written documents, concern technical issues. The graphs show how there is significant variation for different engineers in any given week. However, notice how the extent of the variation between individuals, expressed as a proportion of the average, is much less than the variation in the time spent on solitary technical work that is shown in the graphs above.

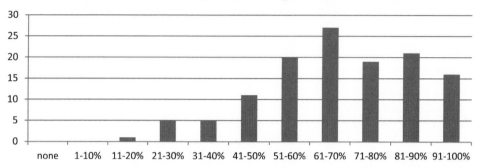

Figure 3.4 Perceptions of total working time spent on social interactions and searching for information prepared by other people, mostly on technical issues.

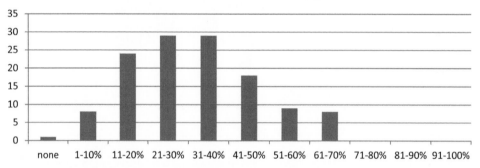

Figure 3.5 Perceptions of total working time spent on face-to-face and telephone social interactions.

In other words, all novice engineers in our sample consistently spent a lot of their time interacting with other people.

What's going on here? How can we explain the apparent fact that engineers spend a much greater proportion of their time talking with and writing to each other (and other people) than they do on the solitary technical work for which their university education has prepared them?

I was legitimately surprised with these results. A study performed in 2010 in Portugal and Spain provided similar data. Therefore, these results do not seem to depend on where the engineering is being performed or in which industry the engineers are working.[12] Several other studies on more experienced engineers over the last 50 years have provided similar data: engineers spend about 60% of their time on communication activities and socio-technical work.[13] What was particularly surprising to me was that novice engineers, most in notionally technical roles, with little or no management responsibilities, were spending exactly the same proportion of their time (on average) on social interactions as more senior engineers. The proportion of time

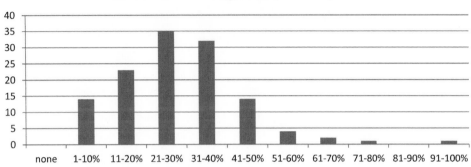

Social interaction through text, e-mail, documents

Figure 3.6 Perceptions of total working time spent on social interactions performed by preparing, exchanging, and reading documents and text messages.

spent on social interactions seems to be independent of discipline, experience level, and geographic location.

Another study using data obtained from interviews and field observations provided similar data for engineers with three or more years of experience. The qualitative data revealed much more about what they were doing in these extensive social interactions.

For a majority of engineers, between 25% and 30% of the time is spent conducting informal leadership in the form of technical coordination: influencing other people so that they conscientiously and willingly perform some form of technical work to contribute to a combined result within an agreed time schedule. Examples of this include asking other people to provide information and arranging for people with appropriate expertise to perform part of the work. This is an informal activity: this coordination work is only documented in brief notes in personal diaries, e-mail messages, and reminder lists.

These studies have contributed to a more defined picture of engineering practice that is illustrated in the following diagrams.

The first diagram portrays engineering as an activity with two distinct threads. The upper thread is comprised of three phases. It starts with discerning, comprehending, and negotiating client and societal needs, and then discussing engineering possibilities with clients and the broader society through various means. The second step is conceiving achievable and economic engineering solutions that will meet those needs. The last phase is performance prediction: working out the technical performance and the cost of those proposed solutions. Engineering education focuses almost entirely on analytical tools to support technical performance prediction, along with some basic design skills, which are shown in the dashed rectangles.

The lower thread is rarely mentioned in engineering education: delivering solutions that meet the needs and requirements on time, on budget, safely, with the predicted performance, and with an acceptable environmental and social impact. Despite how commonly it is overlooked, this is where most of the work for an engineer lies. Without this second thread, engineering would not yield anything useful. Solutions on paper,

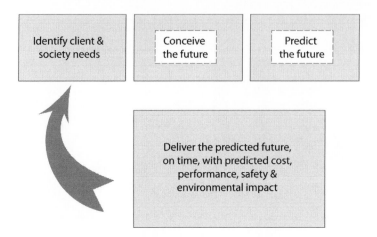

Figure 3.7 A simplified model of engineering practice.

no matter how elegant, provide little value until they can be translated into reality. Effective engineers are able to deliver; they get things done.[14]

The large arrow represents the idea that experience in the delivery aspects of engineering in the lower thread enables an engineer to develop expertise that helps with discerning client requirements, creating new ideas, and predicting performance in the upper thread. This is like a feedback loop: engineers progressively develop more and more expertise as they learn from experience.

What we can take away from the data is that the socio-technical aspects of engineering practice that are required for effective technical collaboration dominate the time and attention of engineers. That's why the core of this book is devoted to technical collaboration. Aspects that tend to dominate engineering education curricula, namely solitary technical work such as calculations, modelling, and design, as well as the aspects that engineers identify as 'real engineering work', are a small, yet critical, part of actual practice.

Practice concept 6: A sequential project life cycle model

Figure 3.8 illustrates an engineering project as a vertical sequence of stages (starting from the bottom: each stage builds on the results from the lower stages).

At the bottom, every project starts with discerning needs and negotiating possibilities, along with securing funding approval and regulatory approval. In practice, funding and regulatory approval precedes each of the main stages through a 'stage gate' decision-making process that is usually unique to a particular enterprise. I will describe this in much greater detail in later chapters.[15] Different engineering ventures naturally place different emphasis on each of the stages. For example, in the context of an engineering consultancy the 'product' is usually information. In construction, plans and specification documents may be the product, while construction supervision may be an optional service. Plans, procedures, contracts, specifications, and estimates would still be a critical stage.

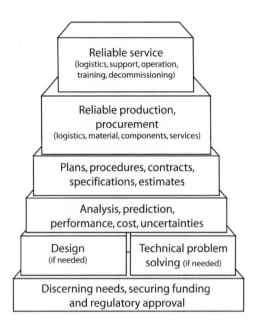

Figure 3.8 Sequential stages in an engineering project, starting from the bottom, explained in the text below. Not all phases have been included in the diagram.

Design and technical problem solving come next, to the extent that they are needed. Again, in practice, design and problem solving are often interwoven with analysis and prediction processes in a series of iterative refinements. Even the requirements may be renegotiated in the early stages.

Detailed planning and preparation of all project documentation is usually the last step before the critical financial investment decision is made to proceed with the rest of the project. Typically, by the end of this stage, between 5% and 10% of the project budget will have already been spent. From this point, everything gets much more expensive.

Procurement and logistics, manufacturing, installation, commissioning, and acceptance testing use up the bulk of the capital investment in any project. Predictability is the key in this stage: engineers are under a great deal of pressure to make sure that everything is completed safely and on time, within the planned budget, and delivered with the required technical performance.

Only in the last step does all this investment start to provide some useful value for people through the tangible products or services provided. This is when payments start to flow back to the investors who provided the funding for the venture. The product has to work reliably for its expected service life, and often much longer. Maintenance plays an important part in achieving this, which means keeping everything working so that the predicted performance is achieved throughout the service life of the product or process. Ultimately, there will be a decommissioning step when the artefacts are removed for reuse or recycling. In the case of a manufactured consumer product, this can happen throughout the life of the product.

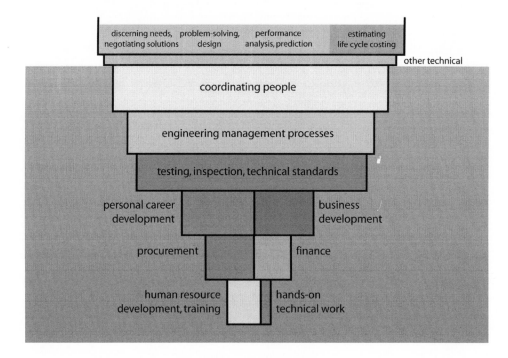

Figure 3.9 Invisible engineering.

Few engineers work on the early feasibility study stages of a project, before the final investment decision. However, lots of engineering projects never pass the investment decision point. Many engineers spend much of their time working out which of hundreds of possible projects is likely to be sufficiently valuable enough to be worth presenting to investors. Most of the projects they work on never go ahead; these engineers create value by providing information for investors that will help prevent financial losses by reducing information uncertainty, resulting in better financial decision making.

Many aspects of the work performed by engineers remain invisible and cannot be directly related to the blocks shown in Figure 3.8. Many of these constitute work that engineers don't regard as 'real engineering'.[16] Figure 3.9 represents similar activity to Figure 3.8, but this time we are looking at a cross-section to expose what is not visible in Figure 3.8. What we see in Figure 3.8 is only the 'top deck' of engineering practice; Figure 3.9 shows the other decks that provide the 'structure' without which the top deck would never remain intact. The proportional sized of the different 'decks' correspond to the relative significance of each aspect, according to what we have derived from our research so far. From this we can see that, like an iceberg, the invisible engineering aspects are much more significant than the top deck. Technical engineering work that you learn about in your engineering degree course represents, at best, only two 'planks' of the top deck. Perhaps you can now begin to appreciate just how limiting it is to see engineering practice only in terms design and problem solving.

These 'invisible' aspects of engineering practice have evolved over time to manage all the uncertainties and unpredictable elements of engineering practice that arise from the reality that engineering is a social system that depends on countless individuals and, to some degree, unpredictable human performance.

Many engineers find it hard to even predict their own work. Between scheduled meetings, they react to problems as they occur, so the results from their work will be inherently unpredictable.[17] Engineers report that they can have 60 or more separate, simultaneous, ongoing issues for which they are personally responsible. Many do not seem to have a systematic way to choose which ones to work on each day, or in what order these issues should be handled.

Most of these hidden aspects of engineering practice provide a measure of predictability for the end results of individually unpredictable performances by the participants. One of these hidden aspects is helping engineers comply with appropriate technical standards to reduce the chance that mistakes will be made that otherwise would not be picked up in time. Technical standards have been created through the experience of other engineers and are carefully negotiated within each specialised engineering discipline, striking a balance between restrictions to promote safety and ease of use, while also avoiding constraints that would inhibit innovation and design freedom.[18]

The social process by which this expertise is shared contributes a significant portion of the hidden layers in Figure 3.9 and helps to explain why building relationships with experienced engineers is so critical for engineers that are just beginning their career.[19]

Figure 3.10 combines the invisible elements of Figure 3.9 with the sequential life cycle model shown in Figure 3.8. The stack of blocks representing the life cycle of an engineering project is enclosed within a coordination ring that continually guides the implementation steps towards the intended objectives. A web of social relationships provides necessary support and functionality to coordinate the technical processes that depend on human activity. The coordination ring helps to produce predictable results, even though each of the human contributions is unpredictable. The coordination ring consists of informal (and often invisible) processes grouped on the left and their formal equivalents on the right. In a way, the formal coordination processes are also invisible: they are often regarded as 'non-engineering' activities.[20]

It is important to note how design and technical problem solving are relatively insignificant aspects in these diagrams. Technical problem solving is actually avoided as much as possible: it is usually much more preferable to use solutions that have been tried and tested in the past, rather than devising new ones with uncertain efficacy. This might seem to contradict the emphasis on innovation and invention that underpins many of the ideas we learn about in our engineering education. We need to understand much more about the perceptions of investors who provide the money for us to have fun engaging in engineering; then we will understand why we need to avoid uncertainty as much as possible.

In many projects, engineers reuse designs from previous projects as frequently as possible. Once again, this saves time and reduces uncertainty.

The diagrams should help you understand how much there is to learn about engineering practice beyond the limited coverage provided in university engineering courses.

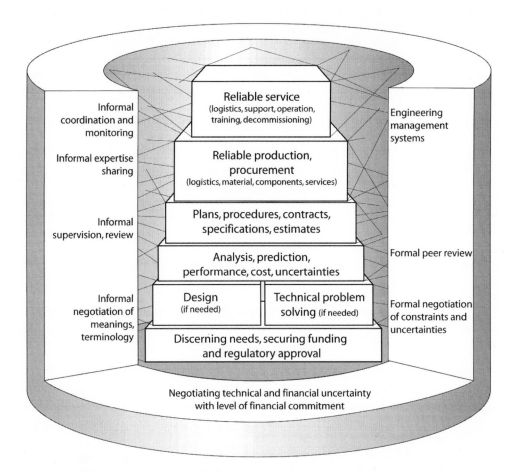

Figure 3.10 Visual representation of engineering practice: the stack of project steps is surrounded and supported by a coordination ring that guides the project and manages all the uncertainties introduced by human performance. Coordination links with the world outside of a given project also occupy much of an engineer's effort; these links are not shown on the diagram, but are equally significant.

Notice how the coordination ring also acts as the foundation for the whole endeavour: this is intentional.

The coordination ring involves continual interaction between all the participants, including the client(s), financiers, engineers, contractors, suppliers, production and service delivery workers, technicians, regulators, government agencies, the local community, and special interest groups.

In the coordination ring base, work starts with negotiations on constraints, even before funds have been committed. Constraints include:

- capabilities of suppliers, production capacity
- technical requirements

- schedule
- regulatory requirements
- health & safety requirements
- environmental impact, emissions
- reliability requirement, client's maintenance capacity
- client's financial capacity
- external financier(s) requirements
- tolerance for uncertainty
- intellectual property

These negotiations provide the decision parameters for committing funds at each stage of the project.

On the formal side of the coordination ring, we find engineering management systems, including project management, configuration management, environmental management (e.g. ISO 14000 series), health and safety management (e.g. ISO 18000 series), quality management (e.g. ISO 9000 series), asset management (or sustainment), document management, change management, and asset management (e.g. ISO 55000 series, formerly PAS-55).

At the informal level, there is continuous negotiation concerning meaning. Different participants initially attach their own meanings to the terms used to describe every aspect of the project, but as the project proceeds, these differences have to be resolved, or at least understood and acknowledged. For example, there may be differences in the way that specifications are interpreted. Many people think this means a non-negotiable statement of requirements: components cannot be accepted unless they pass all tests at the required level of performance. However, others may think this only applies to *production* items. Pre-production versions of certain components would have lower performance. Some people may understand a specification to be 'elastic', meaning that as long as the essential requirements are met, other non-compliances could be negotiated away in the form of a price discount, if and when they became apparent. In one study, we found that many engineers referred to 'reliability' issues that manufacturers considered as 'quality' problems. Different individuals involved in the coordination ring construct their own knowledge and understanding in different ways, which can make the process of sharing that knowledge a lengthy and difficult one at times.

We will learn much more about all of this throughout the rest of this book. Don't worry if it all sounds very complicated: that is intentional. What I want you to take away from this chapter is the realisation that engineering practice involves a huge amount of understanding and know-how of which you can only scratch the surface in this brief introduction.

I am not going to teach you how to do all this in this book, either – that would be quite impossible. The book would be far too heavy to carry and you would become extremely bored while reading it.

Instead, I want to show you how you can learn about this for yourself as you get started in your engineering career. The purpose of this chapter is to explain some of the misconceptions that can prevent you from learning: that is all.

Before we leave this discussion on problem solving, there are three more important concepts that need to be mentioned.

Practice concept 7: Engineering problems are rarely presented with complete information in writing

Students often think that ...	Expert engineers know that ...
Engineering problems will be completely specified in practice, in writing, with 'problem statements'. Leaving out important data or providing ambiguous information is 'unfair'. Problems need to be accompanied by all the relevant information, in writing, to be solvable. Verbal statements are 'subjective' and unreliable.	Engineering problems are rarely, if ever, written down: engineers have to elucidate the issues through conversation and dialogue to comprehend the problem; even getting people to accept that there is a problem at all can be a challenge. The solution is often defined by the problem description that the engineer has to write. Ambiguity is normal, and a complete account of relevant information is almost never available.

Solving between two and three thousand textbook problems over four or five years is a great way to learn the ins and outs of engineering science. The practical necessity of distributing these practice problems to large groups of students lends itself to the use of written problem descriptions. Any information omission from a problem description generates questions from students that can be time-consuming to answer. As a result, both students and instructors prefer problem descriptions that are complete and can stand alone without supplemental information.

In real life, problems mostly present themselves through a gradual process of discovery. A significant part of an engineer's job is to figure out the true nature of problems and reconceptualise them in ways that allow them to be solved or avoided.[21] It turns out that getting many different participants to align their views on a problem is critical for success, but this can take considerable skill and effort.[22] For example, recurrent failures to achieve production targets from an offshore gas platform might lead to enlightening discussions among engineers, especially if they miss out on production bonuses. Eventually, they might start analysing production constraints and then begin to find opportunities to bypass some of those constraints. Eventually, they might be able to even write down a coherent description of a particular production constraint and the description itself will often suggest various solutions.

Research demonstrates that engineers who can take the time to discuss issues with clients are usually much more able to discern the real issues faced by a client. By doing so, an engineer may be able see an opportunity for a simple and effective solution that might not have been obvious from the original comments made by the client. It might be as simple as staggering payments for an expensive item of equipment rather than making a single large payment.

The practical difficulties faced by an engineer in that situation is that the discussions needed to clarify problems take time, and an engineer's time can be expensive, as we have seen in the previous chapter. It can be very tempting for an engineer to avoid the discussion entirely and provide a straightforward solution based on the original statement. The difficulty then is that the client will probably be dissatisfied and may not come back for any more business.

Expert engineers develop the necessary skills for discussing issues with clients (or their representatives), gaining alignment among stakeholders, and listening carefully so they can quickly understand the client's motivation, which may not be evident simply from the words that they have used. Background knowledge and taking the time to prepare in advance can be very valuable. Unfortunately, students in university and college courses rarely get the opportunity to practice these skills. This is something that you will need to focus on, so we will return to this issue shortly, and again in more detail, in Chapter 6.

Practice concept 8: Logical explanations are essential and can be challenging to write

Students and novice engineers often think that ...	Expert engineers know that ...
Only a brief explanation of assumptions is needed, if any, since the lecturer (or senior engineer) will already know more than they do, and mathematical derivations are self-evident. Verbal reasoning is 'loose, qualitative, and imprecise' compared with mathematical reasoning backed up with numbers.	Solutions and conclusions require logical explanations so that others can check them. The most frequent serious errors stem from inappropriate assumptions, critical thinking mistakes, or even conceptual misunderstanding. Engineers save costs and provide better value if they can quickly spot reasoning errors and craft concise, logical explanations that can withstand critical analysis.

Earlier, we learnt that critical thinking and presenting a logical argument can be more challenging for engineers than people with an educational background in the humanities. Here we find a related, but different, issue.

Many people become engineers because their schoolteachers suggested that they build on their strong maths and science performances at school. Logic and reasoning in science is deceptively easy because the language of mathematics provides a robust system for working with logical reasoning. This language provides the basis for engineering studies in universities and colleges.

The most important reason for studying maths as part of an engineering programme is to develop proficiency with the mathematical language. Very few engineers ever have to solve differential equations during their career, yet one cannot become an engineer without having demonstrated proficiency in solving differential equations.

The following mathematical statement should be immediately obvious to you:

$$y' = 0|_{y=\max(y)}$$

It is a very simple and compact way of expressing the idea that the maximum value of a function presented as a curved line on a graph occurs at a point where the slope of a line drawn as a tangent to the graph is horizontal. We know from our mathematics training that y usually represents a function of one or more variables, as well as all that that statement entails. The | is used to represent a particular condition; in this case, it is the particular condition at which y is at a maximum value. y' represents the slope

A 90 kg person, whose centre of gravity is located at point G, is standing on an inclined ladder. The horizontal distance from the bottom of the ladder to the person's centre of gravity is 2.5 m. The ladder is 5 m long and has a mass of 20 kg. You can neglect friction.

(a) Draw the free-body diagram of the ladder.

(b) Write the equilibrium equations for the ladder.

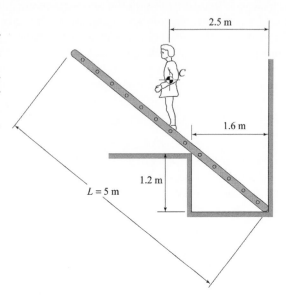

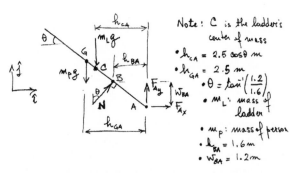

Figure 3.11 Textbook problem and a solution by an expert presented as an example in a research study concerning engineering education (reproduced with permission).[23]

of the curve. There is no need to write all of this because mathematics uses an agreed upon set of symbols to represent these ideas, which require several years of patient study in maths courses at school and university to master.

Years of practice finding mathematically based solutions for thousands of engineering problems in order to develop your knowledge of engineering science can gradually erode your ability to apply similar logical reasoning in situations that are more difficult to represent in the mathematical language.

The idea that the logical basis for a mathematical solution requires some non-mathematical explanation seems to have been neglected, even in educational literature.

Figure 3.11 shows an example of a simple engineering analysis problem with a solution provided by an expert.

Notice how the expert solution consists of an unlabelled diagram and a few lines of mathematics. There is no attempt to include statements about the assumptions needed to transform the original problem statement comprising text and a diagram into the mathematical derivation. Similar explanations are often absent from lecture presentations given in engineering courses: many instructors and students would regard it as tedious and repetitive to include them. As a result of your own studies and practice, you could probably provide a comprehensive statement that would outline all of the necessary assumptions.

The point I am attempting to make is that we, as engineers, through developing years of proficiency with the elegant and compact language of mathematics, have lost much of our ability to write and present a logical explanation in words. As Figure 3.11 shows, we actually learnt from our lecturers to *leave out* explanations and assumptions.

I personally ran into this difficulty throughout the course of the research that led to this book. When it came to writing about this research and presenting the ideas, I found it to be extremely difficult. Time after time, my writing was criticised for not including simple, logical connections between different ideas. The connections were obvious to me, but not to other people, unless I included further explanation. I have found that this same difficulty has affected all of my students who originally trained as engineers and were then required to write logical arguments to justify their research studies. Many of these students helped me to 'discover' aspects of engineering practice and write about them.

Recently, we have started to discover how documents used by engineers in everyday practice, particularly when constructed as PowerPoint presentations, display the same omissions. These omissions usually remain unnoticed.[24]

When we researched the ways that engineers review engineering documents in order to identify and correct mistakes, we found that engineers reported finding very few errors in mathematical reasoning or computation. Instead, most of the mistakes involved incorrect or missing assumptions. In some of the documents that we examined, the assumptions and logical arguments linking the ideas presented in the documents were completely missing. Interestingly, these were all documents that had been checked by one or more other engineers: the checkers had not noticed the missing assumptions and logical arguments; the documents had seemed correct to them.

Mistakes in logical reasoning are much easier to spot if they have been written down in the first place.

What makes this even more difficult is the nature of language itself. We will return to this issue in Chapter 7.

Engineers need to recognise mistakes in documents that provide designs, present project plans, or justify spending investors' money. However, engineers find it hard to write the explanations that reveal the assumptions and reasoning behind mathematical and computer analysis, and may also find it hard to notice when these essential components are missing.

Practice concept 9: Engineers have to work with missing and uncertain information

A few pages back, I noted how engineers rarely start with a complete written description of the client's requirements. In that common situation, there is a further complication that engineers have to work with: uncertain or incomplete information without which one cannot specify either the problem or a solution. Many people characterise engineering in terms of precision and certainty, but this is a major misconception.

Recently, a banker explained some of the difficulties of economic forecasting to my students: '*All that you know about the answer that you get is that it is wrong. You just don't know how wrong it is. It is very unlike a mathematical equation or an engineering solution when you know that the answer is right. You have to take account of that in your thinking.*'

This statement embodies a common misconception about engineering, which is that engineering problems have known solutions for which one can 'know that the answer is right.' In engineering practice, however, it is rare (and often considered a trivial case) that one can know that the answer is right.

Students and novice engineers often think that . . .	Expert engineers know that . . .
Engineers solve well-defined problems that have more or less complete specifications with a single, best solution. Problems require complete information to be solvable.	Most engineering situations are imprecise: engineers have to develop solutions without clearly defined input parameters. The cost of obtaining more precise information may make it infeasible to prepare more accurate estimates. There are often many solutions that can be made to work for approximately the same cost, but the total costs cannot be distinguished because of information gaps, so there is no clear choice.

Unfortunately, textbooks seldom mention the inevitable uncertainties and gaps in the information that engineers use in their work. For example, an engineer can seldom define the precise loading (external forces acting) on a structure in advance. The in-service loading will depend, for example, on how the structure is used and environmental factors that cannot easily be predicted in advance. In many instances, it can be difficult to predict installation and construction loads accurately. Because this issue receives hardly any mention in textbooks, there is consequently little or no guidance for students on ways to choose an appropriate loading for design and analysis.

Instead, almost every sample problem presented in the textbooks has precisely defined parameters, reinforcing the notion that engineering is based on precisely known information and objective certainty. Figure 3.12 illustrates a typical problem that students might be asked to solve from a statics textbook. Students might be asked to predict the reaction forces at the two support points in response to the forces shown acting on the beam.

What if I were to replace the force values above the arrows with question marks? What if one or two of the length dimensions were replaced with question marks? How can you calculate the reaction forces in that case?

Obviously, to deal with this real and practical situation, an engineer needs to estimate appropriate loads and likely dimensions, given a qualitative description that explains the intended use of the beam.

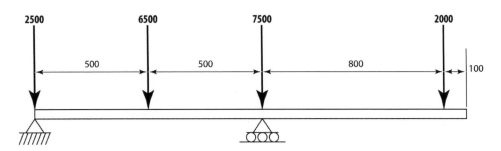

Figure 3.12 A typical student problem in a foundation course on statics.

How would you estimate the loads and likely dimensions? Would it be easier if I were to let you know that this beam will be part of a device to carry heavy oil drilling components?

Initially, you might find this kind of problem unsolvable. By the end of this chapter, however, you should be able to suggest two or three ways to proceed with this problem in order to find possible solutions.

More precise information can always be obtained: you could consider foreseeable loading cases, including multiple failure situations, or you could conduct more detailed analysis of each foreseeable situation, perhaps coupled with a statistical analysis, including probability distributions for each of the forces and loading points. However, in practice, all this work comes at the cost of effort, and possible instrumentation and testing equipment, as well. Engineers have to consider whether the cost of this effort is going to be worthwhile. That's an important issue that will be covered in later chapters. In the meantime, there are several other ways to provide the necessary information. Remember this problem as we work through the rest of the chapter.

Research has also shown that great care needs to be taken when making judgements based on incomplete or insufficient information, due to the simple fact that human thinking can easily be biased.[25]

REWORKING NOTIONS OF DESIGN

Truly creative design is time-consuming and expensive. While it certainly has its place, one only has to remember the high cost of employing an engineer to realise that by using elements from previous designs, a lot of time and money can be saved. Sometimes, an original creative design is preferable, but this can introduce sizeable uncertainties.

At the same time, as engineers, innovation and invention are two of the fundamental values that we respect and cherish. The cost of creativity and the risks of an untried and untested design in practice run up against the commercial imperative to quickly provide a trouble-free solution in the form of something that has been used before and is known to work. One engineer succinctly summarised this inner conflict when he told me:

> 'The only two times I really enjoyed myself in my career were when the clients forgot to ask me "has this been done before?"'

Practice concept 10: Most design is based on a precedent

Students and novice engineers often think that ...	Expert engineers know that ...
Design exercises start with a blank sheet of paper, from scratch, with total freedom to design what you want without any major constraints. Creativity is associated with free, unconstrained sketching. If it's based on an existing design, then it's just a modification exercise.	Design mostly involves the modification of existing artefacts or designs. Design without precedent is rare and can introduce unnecessary risks.

We need to understand the intrinsic difference between a designer and an engineer. As we have seen in the previous chapter, engineering work involves a wide variety of activities, very few of which one would associate with creative designing based on drawings or three-dimensional models. A specialist designer, on the other hand, spends most of the time working with design, and may have started as a drafter, architect, or even an artist. A large proportion of engineering design work is done by skilled and experienced drafters who developed their skills over the course of many years. Naturally, most drafters and designers today work primarily with computer software tools, although sketching remains an important way for many designers to develop their ideas.

While students are introduced to creative design largely without constraints, most engineers are faced with the need to make improvements or modifications to an existing design. If the firm is involved in manufacturing consumer products, for example, much of the firm's capital will be tied up in specialised manufacturing equipment such as jigs, fixtures, dies, moulds, patterns (both physical and virtual in software), and established production systems and procedures. In a sense, this is all manufacturing 'infrastructure'. Therefore, it will be faster and cheaper to adopt a new product that uses as much of the existing manufacturing infrastructure as possible, and is therefore based on existing product designs to a large degree. If the firm is hired to design a large-scale processing plant, it can save a lot of time and money by starting with an existing design and modifying it to meet the requirements for the new project. Not only will this save design time, but there will be construction firms and contractors with the experience of constructing a very similar plant in the past and, if they are well organised, who will have learnt many useful lessons that lead to design improvements.[26]

A large proportion of the effort required for design is frequently devoted to safety issues, often occupying much more effort than the initial creation of the design concept itself. The general public now expects very high standards of safety from engineering enterprises. The financial and reputation risks associated with accidents are so great that companies go to great lengths to ensure safety. Once again, by using a design that is already known to be safe, a great deal of time and money can be saved.

Consider an electrical power supply for a small appliance as an example. Often, this will be a self-contained 'black box' that plugs into a power socket. The appliance may only require a very small current to keep it operating. However, the power supply has to withstand large surge voltages and currents that may be caused by the effects of nearby lightning strikes or faults in the electricity supply network. The appliance needs to work when exposed to full sunlight in even the hottest weather and should be safe even with water splashed on it. It needs to continue to operate even though the output is connected to a short circuit. In no circumstances can it overheat and start a fire. Furthermore, it must not radiate significant electromagnetic interference. Designing a reliable power supply, therefore, involves much more than knowing how to transform main voltage alternating current into a low DC voltage for the appliance.

It might seem like much less of an intellectual challenge to start with an existing design that was possibly created by other people. It might be seen as little more than making some modifications to keep the client happy, rather than starting from basic principles and ending up with a technically superior design in all respects. However, it all depends on your perspective of an 'intellectual challenge'. It can be just as challenging to achieve the necessary improvements while simultaneously retaining almost the entire intellectual content of the existing design.

Practice concept 11: Precedents provide some of the best guidance

Novice engineers often don't even think of looking for precedents . . .	Expert engineers know that . . .
The copying of previously achieved problem solutions or designs is a form of cheating.	Precedents provide some of the best guidance for engineering solutions.

Every now and again, it will be necessary to design something new without being able to adapt an existing design. However, in doing so, there will be very few design challenges that have not been solved before, at least in principle. In most university courses, students are encouraged to devise their own designs instead of searching for existing designs that can be used or adapted. Precedents can be extraordinarily valuable and are applicable in many different ways.

A mechanical engineer once described an interesting example of this concept. Imagine that you need to understand the loading on the rear axle of a car for a particular application. An expensive way to find out would be to fit electronic instrumentation to an existing vehicle being used for a similar application in order to measure the actual forces experienced by the rear axle. Another expensive way to find out would be to conduct extensive computer modelling studies using validated simulation software, backed up by experiments to confirm the simulation calculations. A much cheaper, but equally effective, technique would be to examine bearings from the rear axles of used vehicles with a similar performance requirement. Wreckers' yards are often a good place to look. The bearing manufacturer's catalogue would provide the load rating for the bearings. Close examination of the condition of the bearings and knowledge of the length of time for which the vehicles had been in service would provide additional information from which one could estimate the likely loadings that had been experienced by the rear axles.[27]

Products returned for warranty repairs represent another potentially valuable source of design data. Similarly, older products returned or discarded because they are not worth repairing can also provide useful data. A design engineer can learn a lot by examining these items to understand why they failed and how they were used in practice.

Practice concept 12: Codes and standards enable fast and efficient design

Novice engineers often think that ...	Expert engineers know that ...
Codes and standards constrain designers: you have to comply with their requirements, thereby inhibiting creativity.	Codes (requirements) and standards (recommendations and guidance) represent the accumulated experience that other engineers have learnt from precedents. Following codes and standards saves time and reduces uncertainty, cost, and risk perceptions. Creativity lies in knowing how to work within the constraints imposed by codes and standards to achieve outstanding performance.

Most engineering faculty staff members would like their students to be able to design from 'first principles'. This means that they would like their students to have a sufficiently deep conceptual understanding of what they've studied to develop a design solution on the basis of engineering science principles, such as Newton's Laws of Motion. Students should be able to develop a design without having to rely on 'recipes' or step-by-step instructions that can be followed without deep understanding.

Standards and design guides in component manufacturers' catalogues can be perceived as recipes: they provide proven, simplified design methods.

For example, Australian Standard AS4055 is a simple method for calculating the maximum forces due to wind on a small building, such as a house or shed.[28] For a given terrain type (e.g. flat, undulating, hilly), with or without nearby equivalently sized trees to provide shelter, in a given area of Australia, the standard provides a measurement of maximum wind pressure, which might be given as 1.3 kPa, for example. The maximum force on a section of the roof can then be calculated by multiplying the maximum wind pressure by the area of the roof section. Different pressures are provided, both positive and negative, depending on the relative wind direction, for different parts of the structure of typical small buildings. The maximum wind pressures in areas of Australia affected by cyclones (hurricanes) are obviously much greater.

While helping students to develop deep understanding is an admirable objective, the result is that very little time is devoted to helping students develop the ability to work with standards and design guides. This may partly be because they are seen as 'trivially easy' in comparison to 'first principles' approaches based on mathematical analysis.

Perhaps it is difficult for engineering faculty staff to understand the value of these useful engineering tools because they are not accustomed to using codes and standards in a commercial setting.

Research academics often have to work with technically challenging issues that lie well beyond contemporary standards. For example, the sheep shearing robots that I helped to develop could not have been designed from established standards. They had to be designed by working from fundamental engineering science principles. It is experiences like these that reinforce the value and power of working from first principles in the minds of many engineering faculty staff. Most staff with industry experience have been exposed to the same perspective-altering issue: they were employed as technical specialists in large companies, working with technical issues that could not be resolved using routine methods based on standards. In these situations, one has to work well outside the situations conceived by the engineers who write those standards. One can easily come to see standards as a constraint, or a barrier to be bypassed, rather than as a valuable resource to access prior experience and knowledge. Many times in research, one is trying to break out of ways of thinking that have become entrenched and accepted as a result of experience.

However, everyday engineering practice in the commercial world is very different, of course. Engineering often depends on people who are tired, bored, prone to forgetfulness, and anxious to get home.

Codes and standards enable engineers to routinely produce safe designs quickly and economically, with a much lower chance of making mistakes than if they were working from first principles. In Chapter 1, I explained how expensive it can be to employ an engineer. This makes it easier to understand the commercial value of using codes and standards. They do more than directly reduce costs; there are much greater benefits that occur because investors perceive less risk when engineers follow design codes and standards. That factor alone can greatly increase the value of an engineering project, which will be explained further in Chapter 11. These aspects are not (at the moment) part of the undergraduate engineering curriculum, partly because they lie well outside the knowledge and experience of the vast majority of engineering faculty staff.

The result is that, as a student, you may have developed certain misconceptions about codes and standards.

Most organisations that issue codes and standards act independently of governments and have to recover their costs from subscribers, despite the fact that they may carry an implied connection with governments due to their name. For example, Australian standards are provided by 'Standards Australia', trading under the name 'SAI Global Ltd'.[29] Standards can also be expensive to purchase, even online.

Usually, you would have access to standards during your studies at a university: some of the most valuable documents you can retain from your studies are copies of standards relevant to your intended area of engineering practice. Familiarity with these standards is one way to significantly increase your employability as a novice engineer.

Expert engineers can list nearly all of the standards relevant in their area of practice from memory and will be familiar with most of them.

Practice concept 13: Engineers program with spreadsheets

Students and novice engineers think that …	Expert engineers know that …
Engineers use C or MATLAB for programming at work. Other packages like MAPLE, MathCAD, and Mathematica are valuable, although not as well understood. Programming is seen by many engineering students to be something that you get someone else to do in India or Eastern Europe.	Most engineers write software code for Excel spreadsheets and macros with Visual Basic. MATLAB is normally only used by technical specialist firms.

Many introductory engineering texts provide substantial introductory chapters on programming because computing has become a universal tool for engineers. Until the 1970s, mathematics was just an abstract thinking tool for engineers, a discipline for the mind, and an intellectual language framework for solving limited engineering problems. Practical applications in the workplace were few and far between because symbolic methods were limited by the human inability to perform algebraic and calculus manipulations; computations had to be performed graphically or by using calculators. The original word 'computor' came from the job description of people (mostly women) who were hired to operate mechanical calculators until the 1960s. Now, computers provide the means to transform mathematics into a highly effective tool that can be used in the workplace.

In reality, though, most engineers still only need to do relatively simple calculations. Excel spreadsheets have practically become the universal tool of choice for engineers. However, it is surprising that most engineering students seem to miss out on formal training in the use of Excel as a programming tool. Pressing Alt-F11 opens the Visual Basic programming window for Excel, which provides a means of extending the program's basic spreadsheet capabilities almost indefinitely. The Internet provides plenty of learning resources on Visual Basic for spreadsheet programming.

Excel offers many practical advantages over other software packages. Perhaps most importantly, Excel is almost universally available in a corporate business environment. Many companies impose strict restrictions on the software that engineers are allowed to have on their personal computers, and are thus unable to provide software like MATLAB and Mathematica except for a select few specialists. Providing tools such as these can be very expensive, as the expertise necessary to support scientific programming tools is often hard to find in a business environment, unlike at universities.

Another issue is that most engineers only have a relatively small amount of time to spend on solitary technical work. It is usually more effective for a specialist to handle any computational work that requires more than a few basic calculations that can be entered in a spreadsheet. A person who uses a particular programming tool most of the time is much more likely to efficiently and effectively use it compared to an engineer who might only use it for an hour or so every few weeks.

Practice concept 14: There is never enough time to investigate everything

Students and novice engineers think that ...	Experienced engineers know that ...
Engineers in professional practice do not have the pressure of assignments to complete and lectures to attend, nor part-time work to earn enough money. They can take all day to do calculations and check the results.	Engineers have to make decisions without being afforded the time to find complete information or handle every issue. Engineers have to quickly decide which important issues they will devote their attention to and which others will have to be ignored.

Engineering, like most human performance, is time-, information-, and resource-constrained. In engineering practice, therefore, people have to allocate time and attention to satisfy a wide range of diverse demands. Seldom is complete information available, and the information that is provided always comes with some level of uncertainty. One cannot predict nature completely, although we are getting better at it. Rarely, if ever, did the engineers who participated in our research have extensive, uninterrupted time to think and reflect. Engineering performance requires rapid and difficult choices related to the use of personal time and material resources.

Engineers are often under pressure to keep the cost of their investigations as low as possible, particularly when other people are paying them high fees for every hour that they work. Sometimes, it can be difficult to explain to clients that spending more time (and money) would result in a solution that would ultimately save money. Most clients find that idea a difficult to one to accept.

Fortunately, most engineers learnt to work with this reality during their university courses when there was never enough time to study everything needed to get a perfect set of grades (and enjoy life at the same time). Successful prioritising is something that most students learn through informal methods.

Finding the beam loadings

By now, you should be able to suggest three ways to estimate the load values and the spacing distance represented by question marks in figure 3.13 that will be needed to calculate the reaction forces. Use the interactive website to enter your responses if you have an Internet connection (the self-assessment guide is in the online appendix).

1. _____

_____

2. _____

_____

3. _____

_____

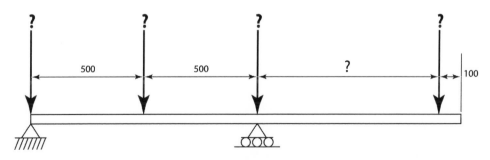

Figure 3.13 Remember the statics problem presented in Figure 3.12? The beam is to be used for carrying heavy oil drilling components in a maintenance facility.

ENGINEERING – A WONDERFUL CAREER

All the engineers in our research studies were happy with their career choice. The only unhappy engineers we encountered were a minority involved in design, much of it prospective, consisting of preparing proposals for projects, most of which would never be built. Plenty of them had frustrations, certainly, but none of them regretted their career choice in any way.

Keep going with the next few chapters: you have made the right decision in pursuing an engineering career. Learn what you can and the path to becoming an expert engineer will be that much easier and rewarding for you.

The next chapter explains what is required to become an expert and how anyone with enough determination to succeed can achieve that.

Chapter 5 explains what engineers know and how they come to know it. It turns out that engineers use several different kinds of knowledge in their work. However, your formal education has emphasised only one kind of knowledge: the type of knowledge that you can write down to help you pass an exam. Not only do you need to be aware of the varieties of knowledge, but you also need to understand how you can learn these kinds of knowledge that you will need for engineering practice.

Chapter 6 revisits some basic perception skills that are often neglected during a university education: listening, reading, and seeing. However, as Chapter 5 explains, you will need to develop these skills to acquire the types of knowledge necessary to become an expert engineer. The rate at which you can learn will depend on your perception skills.

Chapter 7 explains how collaboration, working with other people, is the most important aspect of engineering practice. However, collaboration skills are very different from the communication skills that you learn in your university education. These two areas often get confused, which is why employers tend to think that engineering graduates have very poor communication skills. In this chapter, I will also explain some fundamental concepts that might be quite difficult to grasp: concepts about language and the way we use it in different cultures. However, the rest of the book will be much more difficult to comprehend without being aware of these concepts.

Chapters 8, 9, and 10 will explain the most important collaboration skills for engineers: teaching, technical coordination, and project management.

Although we discussed the value of engineering in Chapter 1, that was only an introduction. In Chapter 11, we will take a much closer look at investment decisions. Investment decisions are critical for engineering because it is investors who provide the money that enables engineers to be active and engaged in such a wonderful profession. You get to spend a lot of that money, even though it belongs to other people.

Chapter 12 explains sustainability issues and the ethical context in which engineers must take sustainability into account. We will learn that sustainability is not merely a matter of building solar power stations. In fact, sustainability can only be achieved if we understand the intimate connections between engineering and human behaviour. In this chapter, I will also introduce another important collaboration skill: negotiation. It is through negotiation that most sustainability issues can be solved, thereby allowing technical engineering to proceed.

Chapter 13 introduces the opportunities presented by engineering in the developing world. It was actually the observations of engineering practice in the developing world that led to this book in the first place. Throughout the course of the research that led to this book, I have come to understand the enormous potential for engineers in the developing world. Expert engineers in countries like India and Pakistan can earn more than their counterparts in industrialised countries such as Germany, France, Japan, and the USA. This book came about because I wanted to understand what distinguished these expert engineers from so many others in the field.

Chapter 14 will help you find satisfying engineering employment if you have not already done so. This chapter explains why most engineering jobs are never advertised, as well as the techniques you need to access them. It also outlines how you can make productive use of your time while seeking an engineering job and also improve your chances of obtaining a job at the same time.

Reading the whole book will be demanding on both your time and patience, if nothing else. I have recommended and included certain lengthy practice exercises to develop the most important skills. I'm confident that you will stick with it because I personally understand the enormous satisfaction that you will gain once you master the skills I describe in this book. However, you need to remember that reading the book is not enough: this is only a guide to help you learn what you need to know. You have to put in the work for yourself, but I am confident that this book will point you in the right direction. Furthermore, by absorbing and implementing the techniques, knowledge, and guidance from this text as you move forward, I believe that you will gain just as much satisfaction from your engineering career as I have gained by writing this book.

NOTES

1. Walther and his colleagues (2011) discuss the accidental formation of a competency, to work hard against imposed deadlines, often at the last moment.
2. Bailyn & Lynch provided these quotations from engineers who thought their career had reached a standstill: 'Professionally my career is stagnant and going nowhere … Whether I can get a decent position after 12 non-promoted years is questionable.' 'The job bores me to death … I need challenge in my work and will not be given that opportunity by [my] employer due to my age bracket.' These were engineers who had not engaged with engineering beyond the technical problem solving aspects (Bailyn & Lynch, 1983, p. 263).
3. Most of this book is based on the analysis of about 150 interviews with engineers in Australia, India, and Pakistan, several participant observation studies in Australia, and one in India. Data from several other research reports from the USA, France, Japan, and the UK, and analysis of major contemporary introductory texts that introduce engineering practice to students have helped. Understanding of students' attitudes comes from extensive teaching experience over 40 years. The appendix (online) provides further details.
4. Practice concepts and misconceptions are introduced progressively through the book and are listed in the appendix (online). The practice concepts resemble threshold concepts (Cousin, 2006; Meyer, Land, & Baillie, 2010).
5. Trevelyan (2011). Begel and Simon recognised similar misconceptions among student software engineers (2008).
6. A study of academics across all main disciplines (Ahern, O'Connor, McRuairc, McNamara, & O'Donnell, 2012), reported a humanities academic describing critical

thinking: 'Always thinking is that true? What is the evidence? What are the counter possibilities? If you are looking for a way of substantiating it, how would you do it? If you are looking for a way of disproving it, how would you go about it? Examining if somebody says something, finding somebody who criticises, and evaluating the counter positions and so on.' The study concluded that critical thinking is 'a dynamic concept that requires academics to guide students through engagement with the context-bound knowledge and the empirical on the one hand, and knowledge that is abstract and reflective on the other. It is the movement backward and forward between these two states that represents critical thinking.' The study provided evidence that engineering academics have the least developed ideas on critical thinking compared with other academic disciplines.

7. More than half a century ago, Becker wrote that 'The ideology tells them that anyone called "engineer" has learnt to reason so rationally and effectively that, even though this has been learnt only with reference to technical problems, it operates in any line of endeavour, so that the engineer is equipped to solve any kind of problem in any area quickly and efficiently' (1956).

8. Jonassen, Strobel, & Lee (2006); R. F. Korte, Sheppard, & Jordan (2008).

9. A three-year study of a single cohort of 160 engineering graduates over the first three years of their (mostly) engineering careers (Trevelyan & Tilli, 2008).

10. Pioneering research by Louis Bucciarelli and many others has demonstrated the overwhelming influence of social interactions on technical thinking and creation in engineering (e.g. Bucciarelli, 1994; Faulkner, 2007; Itabashi-Campbell & Gluesing, 2014; Jonassen et al., 2006; Kidder, 1981; R. F. Korte et al., 2008; Minneman, 1991).

11. Further support comes from surveys of desirable engineering competencies that tend to rate communication skills as the most important (Male, Bush, & Chapman, 2011).

12. B. Williams & Figueiredo (2014).

13. Australian manufacturing engineers were found spending over 70% of their time on documentation-related activity (H. McGregor & McGregor, 1998). Youngman and colleagues quoted a 1975 study on 1300 engineers keeping detailed time records, revealing 60% of their time was spent on direct communication activity (1978). Several other studies have reached similar conclusions with the total reported time on direct communication activity ranging from 45% to 75% (e.g. Kilduff, Funk, & Mehra, 1997; Tenopir & King, 2004).

14. Newport & Elms (1997, p. 325).

15. Cooper (1993); Huet, Culley, McMahon, & Fortin (2007).

16. Perlow & Bailyn (1997, p. 232–235).

17. Trevelyan (2007, p. 194).

18. Shapiro (1997).

19. G. L. Lee & Smith (1992), this issue is the focus of Chapter 5 to come.

20. Informal leadership, technical coordination, is described in Trevelyan (2007). The significance of technical coordination in US practice is explained by Anderson, Courter, McGlamery, Nathans-Kelly, & Nicometo (2010). Invisible aspects of engineering work are also described by (Fletcher, 1999).

21. Tan found that one of the factors explaining the generally low service quality perception by clients of consulting engineers was that engineers often do not devote enough effort to discover the nature of the problems faced by clients (Tan & Trevelyan, 2011).

22. Itabashi-Campbell & Gluesing (2014).

23. Litzinger et al. (2010) performed a research study comparing expert and novice solutions for problems in engineering statics.

24. The report of the Columbia Accident Investigation Board describes how PowerPoint presentations without logical explanations enabled the logical steps in earlier analysis to

become lost, leading to the decision to allow Columbia to re-enter the atmosphere (NASA 2003, p. 7).

25. Tversky & Kahneman (1974); Kahneman (2011).
26. Leveson & Turner (1993).
27. Personal communication from Alan Kerr (2011).
28. Most countries have a comparable standard to help building designers.
29. See http://www.saiglobal.com/our-company/ for details.

Chapter 4

Becoming an expert

By the time you finished Chapter 3 you may have been a little apprehensive, especially if your quiz scores from Chapter 1 indicated you had a lot to learn. Chapter 3 brought more revelations about how much you still have to learn. With several common misconceptions now apparent, you might be questioning the value of all those years studying at university. Now it is time for some better news. This chapter is where things start to get better and there is a glimpse of the light at the end of the tunnel.

There was a time when most people believed that experts were people born with natural, innate, even God-given gifts – special abilities that enabled them to reach a level of performance acknowledged to be superior to the rest of the population. Many people still think this is true while others think that these special abilities are largely inherited through a fortunate choice of parents with the right genes.

In this chapter, you will learn that you *can* become an expert engineer. Indeed, anyone can become an expert engineer. It's quicker and easier if you have graduated from a university engineering course, but even if you worked your way through trade qualifications or as an apprentice, you can still become an expert engineer. This chapter explains the process that enables anyone to become an expert, and the chapter ends with the story of one such expert engineer.

WHAT WE KNOW ABOUT EXPERTISE

Famous performers like Yehudi Menuhin, Nusrat Fateh Ali Khan, Robert de Castella, Elvis Presley, Michael Jackson, and Cathy Freeman, famous composers like Mozart, Bach, Beethoven, Adams, Shostakovich, and so many others, we are accustomed to calling these people 'gifted'. They were lucky enough to have these 'gifts' and the rest of us admire and enjoy watching and listening but few of us think that we could be like them. Whether they are leaders like Nelson Mandela, businessmen like Richard Branson, scientists like Einstein, engineers, mathematicians, or biologists, we have come to speak of them as special people and not like the rest of us.

In the early 20th century, psychologists such as Alfred Binet, Theodore Simon, Lewis Terman, and David Wechsler and many others developed and refined psychometric tests for intelligence and mental ability (such as the IQ, intelligence quotient, test). They were trying to find ways to predict which people would become outstanding performers. These tests continue to be developed today and those of you who have applied for jobs with large corporations may have encountered them. Many human resources experts and companies use these tests to try and select the best performing future leaders.

Throughout the 20th century, there were endless debates about intelligence: is it the result of natural selection, a genetic inheritance? Or is intelligence the result of careful nurturing, attention from parents and teachers? Or does the place in which a child is reared or the people the child grows up with and the home environment determine it?

Towards the end of the 20th century it was widely accepted that intelligence, whether it is measured by tests or academic grades, is the best determinant of future performance. It was assumed that the ability to solve classic mathematical problems, such as the Tower of Hanoi, was a predictor of expert performance at much more complex problems.

But 21st century research tells us a different story.[1]

Education psychologists gradually realised that expectation plays a much greater part in human learning than we previously imagined. Children who are told that they are talented outperform children who are told that they are less able than the rest, even though they have the same intrinsic ability.

Other education psychologists realised that practice and tutoring play a much larger role than had previously been imagined. The psychologists who had developed intelligence tests had carefully removed tasks from their tests that involved perception, memory, and judgement precisely because they found that the results of these were affected by practice. The psychologists were interested only in measuring 'innate' mental ability because they thought that human beings were differentiated by the measure of the abilities that they were born with.

Towards the end of the 20th century, psychologists and cognitive scientists started to study outstanding performers to try and understand what differentiates experts from the rest of us. Ericsson and his colleagues started with musicians while others studied chess players and typists. As early as 1946, Adrian de Groot started to realise that chess grandmasters were somehow able to choose better moves in a game of chess than others who were not quite as good. He gradually came to the conclusion that it was thousands of hours of practice that enabled grandmasters to better predict the course of the game as well as provide them with a wider choice of moves at any given stage.

Later, Chase and Simon found that expert chess players were better able to recall the positions of pieces on a chessboard that they had been shown for a few seconds, but only if the chess pieces were in positions that had resulted from a real game of chess. If the pieces were positioned randomly, there was much less difference in their ability to remember. The experts' memory came from practicing chess games, time after time after time. Memory improves with practice.

Ericsson and his colleagues found that the largest factor predicting the performance of musicians was a certain kind of practice that they called 'deliberate practice'. First, they found that a combination of motivation and practice alone did not necessarily lead to the best level of performance. The motivation was needed because practice is not easy: it takes effort and determination to work systematically on improving the skills that are needed for high levels of performance. They found that achieving the highest level of performance required not only motivation and practice, but also a search for techniques that resulted in higher levels of performance both in the underlying skills and their effective combination.

Expert tutors could make a large contribution as well; these were people who could watch and listen to a performance and identify the need to improve certain techniques. They found that experts actively seek these tutors for one-on-one feedback and guidance.

Finally, the performers themselves had to have the ability to evaluate their own performance. As they were practising they were constantly observing their own performance and trying out alternative techniques in a search for improvement. Sometimes this resulted in frustration and temporary setbacks. However, with enough determination, an expert performer could overcome these difficulties to reach still higher levels of performance. Most experts required at least ten years to reach their ultimate level of performance, at which point they achieved a state of 'effortless mastery' (sometimes called a state of 'flow' by other researchers).

By 2003, Ericsson and his colleagues had extended their studies to physical performance, involving gymnasts and basketball players, and cognitive tasks such as mathematics and problem solving. They found that even when experts perform at a routine level (i.e. not the highest that they can achieve) their performances are more consistent and reliable than those who are considered less of an expert. They found that experts develop special perceptual abilities that enable them to be aware of very subtle differences. This gives them the ability to remember better and instantly adapt or select an appropriate response that enables them to look ahead and anticipate better than others. The ability to look ahead allows experts to perform faster than others. Experts still make mistakes, but they are able to hide them by effective improvisation so that the observer does not notice and the expert avoids making them again through subsequent practice.

Ericsson and his colleagues concluded that the acquisition of expert performance relies fundamentally on deliberate practice. In essence, this is a continual process of problem solving in order to seek techniques that lead to even higher levels of performance. They have estimated that it usually takes around 10,000 hours of such practice. Given the effort needed, especially in the early stages, learners can only tolerate between 20 and 40 minutes of such practice every day. With the help of tutors, encouragement from parents, friends, and peers, and in many cases, inspiration from a chance meeting with a role model, gradually the learner can increase the amount of deliberate practice. For most people it takes at least ten years to achieve an expert level of performance. Normal routine work rarely provides the opportunity for deliberate practice: work performance has to meet employer expectations and can't easily be adapted to the needs of practice. For most people, deliberate practice has to be a spare time, after-hours devotion.

Although researchers are still far from certain, the evidence accumulated so far has failed to find any definite link between the acquisition of expert performance and genetic inheritance, innate ability, or intellectual ability as measured by intelligence tests and other similar instruments. However, there are some exceptions. For example, expert basketball players tend to be taller people and expert gymnasts tend to be shorter people: inherited physical characteristics do confer certain advantages.

BECOMING AN EXPERT ENGINEER

What does this mean for engineers, especially for aspiring expert engineers?

First, we need to put aside the idea that only a few individuals can become truly expert. The evidence from psychology research over the last few decades shows that expert status is within the reach of anybody with the required determination, energy, and encouragement.

The good news from all of this is that any engineering graduate can become an expert engineer; your grades in your university course need not make much difference to your long-term prospects. Even if you are not a graduate, it is still possible to become an expert engineer: it is just harder.

Next it might be helpful to have some idea of what is meant by 'an expert engineer'. There is no absolute measure but I can describe what it feels and looks like to others. First, you will have that quiet confidence that you can overcome any engineering challenge that is presented to you, and do it well. While it is not effortless, it will seem that way to others, and you will be able to do this without losing sleep or having to work until 3 AM in the morning and return to the office at 7 AM.

You will be able to deliver on your promises, on time or earlier, without feeling exhausted when you have finished. You won't necessarily be the world's best engineer, but you won't be much less proficient in your performance within your own field. You will be able to accomplish market-leading performance without undue effort and stress. On a few occasions when you need to do better, perhaps the best in the world, you will have the reserve capacity that will enable you to do that.

Other expert engineers will agree with your decisions, possibly most of the time, but when they disagree they will acknowledge that your decisions are equally good, just different from theirs.

So how do you reach that level?

Graduating from a university course means that you have been exposed to, and to a certain extent mastered, the language of mathematics and science. You will have also demonstrated a capacity for abstract thinking. Most of the engineers interviewed for this research complained that they thought that their university education had been far too theoretical without sufficient exposure to the practical realities of engineering; it did not matter which university they graduated from, whether in the First World or the Third World. The capabilities you developed to work with theoretical ideas, describing invisible phenomena – electromagnetic fields or stress distributions – helped you acquire the ability for abstract thinking that is essential in engineering practice. Theoretical knowledge also enabled you to more quickly make sense of the unexpected situations that you encounter in practice. You gained a tacit understanding through studying the theory.[2] Of course, direct applications of advanced mathematics and engineering science theories are rare, and this can be misleading.

At the same time, no matter how well you performed at university, most of the hard work is still in front of you. If not yet, then certainly by the end of the next chapter you will realise that most of what you will need to learn still lies in the future, and that you will still be learning decades from now. You will need to keep on learning just as hard as you did during your university courses, only harder, because this time the results really count. At the same time, the learning is more rewarding because you can immediately appreciate the practical benefits.

As you will appreciate from the previous chapter, much of the learning that lies ahead will be related to working with other people. We learnt that an ability to work with and through other people is an inescapable aspect of engineering expertise. When I graduated, I was shy and regarded by most others as being quirky and inept in my relationships with other people. In many ways I was a nerd. Through a combination of circumstances I learnt to work effectively with other people, not as fast as some of my peers, but I have improved.

The last lesson from the research is that there is no pre-determined pathway to becoming an expert engineer: there are millions of different pathways and one of them can be yours.

Let's first look at the aspects of acquiring expert performance that we can see are important from reading the most recent research.

Motivation, inspiration, a role model, drive, and determination

The one thing that you cannot avoid on your path to becoming an expert engineer is hard work. In the words of Professor K. Anders Ericsson, 'deliberate practice is effortful'. You will need drive and determination, especially when things get difficult and when your performance decreases after practice or when the new techniques that you try result in failure. You need sufficient drive, determination, and motivation to keep going in the knowledge that you will ultimately succeed. Believe me, I have been there.

There are many different ways to find the motivation. Some of us are fortunate enough to meet role models, people who we have admired from a distance and ultimately get the chance to meet. I was lucky enough to meet several such people when I was young, sometimes with surprising results.

One such person was Professor David Allen-Williams, the first head of the school of mechanical engineering at the University of Western Australia. He started as a civil engineer in the UK, measuring the effects of coal dust explosions. The instrumentation he needed for this led him to electronics and he helped with electronic warfare between 1940 and 1945. After the war, he worked on diesel engine control systems and then came to Australia in 1958. I remember how much he emphasised the importance of people in engineering, though I learnt more about control systems and technical writing from him than working with people. Above all I learnt that an engineering career can take you anywhere and disciplinary boundaries can be crossed at any time.

Another opportunity was when I met a famous aircraft designer: Sir Barnes Wallis. Before World War II, he had devised a lightweight geodetic framework for early fabric-covered aircraft. When I met him, he was working on a supersonic passenger aircraft. 'A long range supersonic aircraft is the only way to guarantee the future of the empire', he told me. 'It has to have enough range to fly from London to Sydney non-stop, avoiding the need to land in those untrustworthy Middle East countries'. When I explained I had flown non-stop from Singapore to London, he seemed not to hear me and continued, 'A single swing wing, it can cruise at Mach 4.0 for five hours'. Then he complained for several minutes about how no one would support his ideas.

He was seated at his drawing board in a vast room full of mementoes of past designs. His ill-fated Swallow aircraft, a huge radio-controlled model, was there. There were parts of missiles, pieces of airframes, each mounted on a wooden plinth. There was no one else evident, not even another desk in the large hall that was his personal design office.

I left, thinking of this person who had been described to me as an irascible genius, hard to work with. I learnt that technical brilliance only leads to frustration without the ability to inspire other people to collaborate and help.

When I first became interested in robotics, I visited one of the world's leading research institutes where there were two eminent leaders, both widely regarded as having the most inspiring and advanced ideas of their time. When I arrived, a receptionist directed me to one of the professor's offices: she told me that he was expecting me and I should just knock on the door and go straight in. I did as I was directed, only to find him on the floor with a woman behind a filing cabinet. This was my introduction to the height of robotics research and the lows of behaviour! I retreated quietly, and waited until the woman walked out from the office without looking at me, and then knocked at the door.

From a technical standpoint my visit was very inspiring, although I never thought I would ever have access to the amount of equipment and the number of assistants that they had at the time.

Shortly after my visit I heard that the two professors separated with a considerable degree of acrimony. I was disappointed but not surprised. Their work as individuals never regained the vitality they created during their partnership. I followed their achievements, hoping for a resurgence, but it never came.

Sources of ongoing encouragement

One of the best ways to maintain motivation for the hard work of deliberate practice is to keep in contact with friends and supporters. For engineers, it can be inspiring when the people who ultimately use the service or products to which you contribute, talk of their hopes for the results of your work. These sources of inspiration and encouragement were evident in all the research interviews with engineers that we recognised as experts in their field.

For me, working for months writing this book, my motivation has come from young engineers who express so much frustration in trying to find a job, and then trying to work with people. Here is an example provided by a software engineer:

'I could see an obvious way to improve the company's information systems. However, the engineers and managers were very satisfied with what they already had and were not eager to change. I prepared a detailed proposal and provided examples from other companies that had already adopted better software tools with improved performance as a result. I became so frustrated trying different strategies to persuade the engineers and managers to support the change. I spent so much energy and time gathering a team of researchers to develop ideas to improve the existing software tools at hand, but it seemed to make no difference to the resulting decision. This proved to me that being technically strong in one dimension is not sufficient: it's even more important to be an excellent communicator,

listener, and a good negotiator. The ability to explain ideas and facts and convince people to follow them is one of the keys to success in any industry.'

These accounts as well as many others have continued to motivate me to understand these difficulties and how expert engineers managed to overcome them.

Another motivation has been the equally frustrated accounts from Pakistan, India, and many other developing countries, like this one:

> *'I learnt that technical ability is not very helpful here. Of course it is necessary, but the boss is very defensive when I raise technical issues, which he does not understand. He only listens to people who have some hold on him, like family members, or influential clients. It is very frustrating: there are so many improvements I could make, but he always withholds the money at the last moment making some excuse or other.'*

As you start on a difficult and challenging journey, think of the people around you who can keep on encouraging you. Your family might do this, or close friends.

Practice concept 15: Ability to learn from others

Most engineers seeking help first look on the web and then in textbooks. However, if you are to become an expert, you will need to seek out much more valuable reference sources: experienced people who can help guide you and provide constructive feedback on your performances. Some of them may be working for the organisation employing you. However, you will also need to find them in other places. You also need to build the skills you need to learn from others: accurate listening and attention to visual detail. Chapter 6 provides pointers for these.

Here are some suggestions.

Senior engineers

In most countries, for reasons that lie beyond the scope of this text, there are relatively few experienced and senior engineers. In developing countries, there are almost none. Most of them are very busy, and it can be quite difficult to persuade them to spend time helping you and providing you with advice.

Many engineers are very economic when it comes to providing guidance and information to help younger people. First, many engineers have acquired the habit of expressing complex ideas in just a few words. Consider the following example from a telecommunications planning engineer, explaining the work he performs every day:

> *'A typical case will be ... say a new estate with a couple of thousand lots proposed over five to six years, we got to first work out the dimensioning, what is the network required to get to that, and then dimensioning also includes elements of the network all the way back to CBD (central business district), major exchanges where all the high level network switches are.'*

In this excerpt, we see this descriptive economy. This engineer is talking about the communication requirements to ensure that people who will live in a new residential estate will have access to the telephone and data communication services that they will need.

'Dimensioning' means working out, in advance, all the details required to install the conduits (underground pipes that contain the cables), cabling, optical fibres, junction boxes, and exchange equipment. The engineer has to be fully aware of the constraints and performance limitations of the current telecommunications system serving several million people living in the city with tens of thousands of major components, and perhaps hundreds of thousands of kilometres of cable and optical fibres. He has to predict the telephone use and data communication capacity required to serve the people who will live in the new estate. Then he has to decide where new equipment needs to be installed and where existing equipment needs to be upgraded to handle the additional capacity that will be needed. He has to take compatibility issues into account: some of the existing equipment may no longer be compatible with the new equipment needed to expand capacity.

Like the business activities of the city, the telecommunications network is centred on the central business district. Just as road engineers are involved in a constant process to improve the capacity of the existing road network in moving people to and from the centre of the city and other destinations, telecommunications engineers are engaged in a similar process, gradually extending the capacity of the less visible, largely underground telecommunications network carrying the voices of city residents, TV pictures, and other electronic data.

Just a few words spoken by an engineer can hide an extraordinarily detailed and complicated intellectual task.

While Chapter 6 explores listening in more detail, you will need to develop the ability to use lots of clarifying prompts such as:

'Could you explain more about what you mean by dimensioning?'

Here's a young engineer talking about his interactions with his immediate supervisor: a senior and very experienced structural designer. My questions are between square brackets.

'There are only two of us working on this job: my boss who is a civil structural design engineer and myself. We share the design work and we check each other's calculations.

'[How do you decide what to do each day?] I decide through liaison with my boss each morning or whenever it is necessary. Whenever I have questions I will go and discuss them with him. He is very approachable.

'[When do you talk to your boss?] That would be in the design part I mentioned. Probably an hour and a half a day. It's because we do everything together, though it depends on which problem we're looking at. I don't just go in there and talk to him in his office all day, it is more that we would be working on the design together.'

This young engineer clearly has a very good relationship with his boss as it is very unusual for a senior engineer to have two hours every day with someone more junior. However, research shows young engineers need at least an hour a day of face-to-face interaction with more experienced people.[3]

In coming chapters, we will talk more about building helpful relationships to access the knowledge and expertise that other people can contribute.

Suppliers

Another invaluable source of knowledge for young engineers lies with engineering component and materials suppliers. While they have an interest in selling their products, most technically knowledgeable sales representatives know that the best way to secure business is to be as helpful as possible with prospective clients. They also relish the opportunity to engage with some of the technical issues involved in applying their products. Most will treasure any opportunity you can provide them to solve a technical problem using the products they are selling, and they can be very helpful.[4]

Foremen, tradespeople, artisans, technicians, and craftspeople

Many young engineers spend their first few months assigned to the care of an experienced foreman, someone with a practical trade or technician background who supervises tradespeople and site or production workers. Here we need to remember that nearly all engineers have neither the skills nor the time to build, construct, produce, assemble, or manufacture artefacts and systems. This means that the ideas conceived by engineers have to be made a reality by other people. Learning how to work with these people, about their language and the ideas they work with, how to convey ideas accurately and quickly and how to watch what they are doing and learn from them lies on your pathway to becoming an expert engineer. Without this experience, it will be much more difficult for you to successfully translate your ideas into reality. In the next chapter we will learn more about the knowledge that these people carry with them.

Financiers, accountants, and lawyers

After reading Chapter 1 you should be in no doubt about the critical importance of finance and money in engineering, even if only to have the enjoyable experience of spending money that belongs to other people. However, it can also be an acutely painful experience if you end up spending too much money at the wrong time. Just as you need foremen and tradespeople to help you understand ways to translate your ideas into practical reality, you also need to learn from financiers, accountants, and lawyers. It is through these people that you will receive the money that enables you to have fun doing engineering. Learning the language and ideas that financiers, accountants, and lawyers work with is a considerably greater challenge than the ideas of tradespeople and foremen. The online appendix provides a guide that deals with the language of accountants, just to give you a head start.

Ordinary people in your community

The people who live around you are one of the most helpful sources for you to learn about engineering, in particular the way that the public view what engineers do and provide for them. Maybe they don't think about engineering at all. You might talk to them and learn why that might be so. Given that the engineering profession seems to have forgotten the reason for its existence, one can understand why the community doesn't understand much about engineering or what it is for.[5]

At the same time, engineering can have a greater effect on the lives of ordinary people than any other single profession. Think for a moment about the supporting infrastructure of human civilisation today: water supply, sanitation, construction, roads, transport, and communications. No matter which branch of engineering you work in, you have the opportunity to bring huge benefits to the people in your community. There is also the risk that some of the things you do can have negative consequences for people in the community. Sometimes these can't be avoided. However, by learning to talk about engineering issues with ordinary people, you will be able to help maintain the bridge that links the engineering community with the much larger social system that engineering helps to support.

As explained in the previous chapter, engineers need to be able to predict the unpredictable and to anticipate uncertainties that will affect the outcomes of the work we do. Studies have shown, however, that engineers are not good at this.[6] Perhaps the skills we developed for abstract rational thinking can actually make it more difficult for us to account for the unpredictable. Ordinary people can be a wonderful source of ideas for some of the obvious things that can go right or wrong, unpredictably. In this respect at least, ordinary people can be a wonderful source of help for engineers.

Practice concept 16: Ability to evaluate your own performance

The research on expertise reveals how important it is to be able to evaluate your own performance in order to acquire the attributes of an expert engineer.

In one respect, engineers enjoy a unique advantage compared with some other occupations. Most of the time, whether the engineering succeeds or fails, the ultimate results are obvious. Sometimes the result is all too obvious, especially with a failure. A lawyer, for example, may prepare extraordinarily detailed contract agreements but may never find out if the words written into those agreements actually worked in practice. That is, did they provide a clear resolution to any difficulty arising in the course of the work specified by the contract?

However, as you will begin to understand, engineering is an extraordinarily complex activity involving lots of different people so it can be difficult sometimes for an engineer to know whether he or she has contributed anything of value; and if so, how to evaluate and reflect on one's personal performance.[7]

Many engineering employers look for the ability for engineers to evaluate their own performance. Here's a quote from one interviewed recently:

> '*I want someone who can reflect on what they've done, like this: "in doing that, I learnt this and this and this, and when I'm an engineer, I will make sure that I never design something like that." That's more the kind of person I'm looking for, somebody with ability to learn and reflect on what they have done, with an understanding about the context in which the work is being performed and why it is relevant.*'

The most important idea here is 'performance'. At university, students are evaluated on what they can write but in practice what counts is how one performs. As explained through much of the last chapter, engineering is a 'human performance'.

Just as with a musical performance, engineering performance comprises many subsidiary parts, each of which contributes to the combined performance. Though it may be difficult to assess the impact of any one contribution, the quality of that contribution can be assessed as a performance (especially by experienced colleagues). As such you should seek out their assessment of your performance whenever you can.

Our research has provided strong evidence that this is the single most time-consuming aspect of practice.[8] One's own listening performance is relatively easy to evaluate, and this is explained in Chapter 6. Other aspects of communication performance can also be evaluated relatively easily, particularly with the help of other people.

Each of the remaining chapters will provide techniques from which you can develop your own methods for evaluating your performance.

Practice concept 17: Deliberate practice, persistence, and resilience

Ordinary practice involves the repeated performance of a task: intellectual, physical, or a more complex combination. Repeated performance strengthens neural connections involved in the task and can result in performance improvement, but only to a limited extent.

One of the main difficulties with ordinary practice is that it is just as easy to reinforce and strengthen counterproductive aspects of performance (bad habits) as it is to reinforce desired aspects (good habits).

Deliberate practice, according to Ericsson,[9] is an effortful problem-solving activity in which the performer constantly struggles to achieve a higher level of performance by evaluating his or her current level of performance and then adopting deliberate changes in technique to see whether they open up the possibility of higher levels of performance. To do this, the performer has to work at the highest levels of skill.

This can be difficult to do on your own without prior experience and insight into different performance methods and evaluation techniques. For this reason, regular contact with a mentor or a coach can be very helpful. The coach can provide external feedback and suggestions on technique, as well as being a source of encouragement.

Even so the rate at which skill improves is slow. It can also be erratic and it is not unusual for a performance to get worse rather than better, particularly when trying new techniques. Sometimes it can even be impossible to go back to an old technique that provided a better level of performance. Once neural pathways have changed, it can be difficult or impossible to reverse the changes!

Deliberate practice can be exhausting, particularly in the early stages and few individuals manage more than 20 to 40 minutes daily. As performance skill improves, it is possible to do more practice because part of the skill improvement results in less intellectual and physical effort being needed.

Remember it takes at least ten years, and around 10,000 hours of deliberate practice to reach levels of performance that would be regarded as being near to an expert. This emphasises the need for patience and persistence: do not expect rapid and sudden performance improvements. However, by engaging in deliberate practice to the extent that it is feasible, you can be assured of gradual improvement. Eventually, and with sufficient deliberate practice, you will reach a performance level at which at least part of your performance becomes enjoyable and you can perform well relatively effortlessly. However, continual performance at the ultimate levels of skill, even for an expert, is never effortless: it is always demanding.

The sheer effort involved in deliberate practice means that you have to choose, over a period of time, where to devote your effort. You cannot be expert at every aspect of engineering. However, you can achieve a high level of performance in many aspects and reserve your energy for improving your performance in the most important ones to you.

Here are some examples of tasks that you can use for deliberate practice.

- Learn to teach other people. Evaluate your own performance from the performance improvements that other people display in response to your teaching.
- Predict the future development of a project, particularly costs, completion schedules, and defects that have to be rectified; note your predictions in a private diary. Evaluate your own performance by comparing your personal predictions with later observations on what actually happened and project performance records.
- Reverse engineer all the artefacts and systems that you can: a dead hard disk drive or DVD player is a good place to start. Wander around engineering works and construction sites, taking photographs to record what you see, and then prepare sketches from the photographs (see Chapter 6) to develop your ability to notice important details. Write down the rationale for the design and construction, manufacturing or assembly methods. Evaluate your performance by contacting the engineers responsible for the work and ask them why they decided to do it the way they did. If you cannot contact the engineers, try other experts who might be able to comment on your ideas. Compare your notes with their explanations.
- Write anything and find someone who can critically review and proofread the text. Evaluate your performance from their reactions to your writing and the mistakes they find.[10]
- Catalogue your own performance slips and mistakes. Evaluate your performance by tracking the number of times you repeat similar errors.
- Draw anything and evaluate your performance by carefully comparing your sketch with a photograph of the object or person you were drawing.
- Chapter 6 provides ways to improve your listening performance.

Along with hours of deliberate practice, you will also need energy, persistence, and resilience.

Developing new skills and abilities takes a lot of intellectual energy, something that has been well documented by psychologists. The human brain consumes around one third of our total energy supply, even when resting. It consumes more when it is engaged in conscious thinking and learning a new task.[11]

Keeping fit, sleeping well, and eating healthy food are all necessary as well.

Above all, you need to realise that your performance will not increase steadily. You will encounter times when your performance hardly improves, or even gets worse. This is when you most need a combination of persistence and resilience. Assistance is needed so that you keep going and continue with deliberate practice, trying to find better ways to accomplish the tasks that are part of your daily work. Resilience is needed so that you recognise that when your performance declines, it is not terminal. It is just part of the process of finding a new level.

As you experiment with different ways to accomplish your engineering performance, not all of them will work the way you expected. Resilience helps you see that and realise that you need to keep going in the confidence that you will eventually be able to improve. Sometimes, your performance will improve remarkably and it can be very hard to figure out what led to that improvement. Other times it will remain static for perhaps months.

By being resilient, you can anticipate the difficult times and know that eventually your performance will improve and reach new levels that you never imagined to be possible.

ROLE MODEL: C. Y. O'CONNOR

This chapter would not be complete without at least one example of an expert engineer. I have had the privilege to know and work with several expert engineers in my career. However, it is easier for you to appreciate the contributions of an engineer whose work has been extensively described in literature. The insights into practice from this book might help you gain more by reading detailed accounts of his life and work.

Western Australia is my home and no engineer from Western Australia can afford not to know about the achievements of Charles Yelverton O'Connor.[12]

Born in 1843, O'Connor spent the first 22 years of his life in Ireland. His family could not afford to send him to Dublin University so instead, his family paid for him to work as a pupil, an apprentice under the guidance of the chief engineer of a railway company. At that time, around 1860, there was a dramatic expansion of railways across Britain and Europe. O'Connor was exposed to a wide range of civil engineering technical issues: drainage, gradients, speed, power, and haulage limitations of different engines, designing the track alignment to keep within these constraints, building bridges, tunnelling, and rock cutting.

O'Connor developed his engineering expertise further in New Zealand. Although he was based in Christchurch, he spent much of his career on the rough and wild west coast of New Zealand's South Island designing, organising, and supervising the construction of ports, railways, and other infrastructure to support the rapidly growing mining industry, with its tenuous and unreliable supply lines, and located far from industrial centres.

He was recruited by the visionary premier John Forrest to lead a huge expansion in government-funded infrastructure engineering in Western Australia. Over the next 11 years, he designed, organised the construction of, and managed near-simultaneous operations of over 1000 km of railway lines, Fremantle harbour, and the world's longest and most ambitious water pipeline to Kalgoorlie. At the same time he was also responsible for countless other smaller public infrastructure projects.

Of all of these, the pipeline was his greatest achievement. Today it is hard to imagine the conditions under which this amazing engineering feat was undertaken. At the time, the population of Perth was around 25,000 people. Just ten years later, the population had doubled with a huge surge of immigrants attracted by the gold discoveries further inland.

The world's greatest water pipeline at the time had just been constructed to carry water from a reservoir in England's Lake District to Manchester. The pipe was about 250 mm in diameter and about 100 km long. O'Connor designed a pipeline that was to be 785 mm in diameter and about 530 km long, with a total hydraulic lift of

Figure 4.1 C. Y. O'Connor (W. A. Water Corporation Library).

480 metres using a completely new method for manufacturing pipes and joining them. Whereas Manchester was one of the main centres of engineering capacity in the world, a concentrated hub of engineering expertise, O'Connor wanted to construct a pipeline in Western Australia, several weeks by ship from the world's manufacturing centres, across terrain for which only sketchy maps were available.

Engineers today have a term for this: step-out. It means the degree to which an engineering undertaking uses untried technology or goes beyond the limits of what has been done before. In today's engineering terms, O'Connor's pipeline was not just a step-out; it was an unimaginably huge step-out in a place with virtually no established engineering capacity, a virtually unpopulated backwater at a distant extremity of the British Empire.

In order to get the project accepted, O'Connor cleverly argued that he was simply constructing a series of 10 to 14 separate pipelines, each running from a pumping station to a holding reservoir, which was well within the realms of possibility at that time in terms of diameter, length, and hydraulic lift.[13] What was new was that each of these pipelines would be placed end-to-end in order to achieve the required objective to bring water from the 'well watered' Darling Range near Perth to the inland goldfields located far from the coast in arid semi-desert country.

In today's terms, O'Connor was responsible for several billion-dollar-projects, undertaken near simultaneously, supported from a town of a mere 25,000 people.

With no detailed maps O'Connor had estimated the cost at £2.45 m, before a detailed survey of the route and the main storage reservoir at Mundaring could be carried out. The project was completed for £2.66 m, a budget overrun of less than 10%, and the first water was delivered to Kalgoorlie in January 1903.

The construction of Fremantle Harbour and the Kalgoorlie pipeline were both highly controversial projects, particularly because of the financial commitment and the size of the loans that would have to be repaid. Both projects had a significant influence on property values in the surrounding communities. In a town of the size of Perth, practically every contractor would have been related by blood or marriage to influential members of Parliament and the owners of the principal news media: *The Inquirer*, *The Sunday Times*, *The Morning Herald*, and the West Australian newspaper. Only the last maintained a balanced reporting style: the other three were happy to represent the uninhibited views of contractors.

O'Connor could only achieve the results he did by being very tough with contractors and he interpreted contracts strictly, being very reluctant to compromise. He believed that contractors were firmly bound by their original estimates and that these could not be altered unless exceptional circumstances warranted a revision. A typical letter to a supplicant contractor expressed his views, which were often repeated:[14]

> 'Having again carefully investigated the question, I have come to the conclusion that contractors, either in law or in equity, but especially in equity, are not entitled to any further payment, beyond what they already got in the contract. I cannot therefore entertain the claim which you advance.'

Many times he argued that contractors had to be made to work within their estimates. Otherwise the contractors would submit lower estimates to secure the contracts knowing that they could revise their estimates upwards at a later date.

The Premier, John Forrest, had recruited O'Connor not only for his technical expertise as an engineer, but also for his ability to manage contractors in small communities. He saw the need to change the local contracting culture and recognised O'Connor's ability to make this happen. These huge engineering projects could only succeed if contractors met their promises and O'Connor made sure that they did.

As work finally got under way on the pipeline, three years after the government initially approved it, O'Connor imposed exacting requirements on the contractors and refused to pay until the work had been done to his satisfaction. It is not surprising that there were many complaints. Contractors also complained about work being performed by public works labour when, they alleged, they could do it more cheaply

Figure 4.2 Work underway on Goldfields Water Pipeline in 1900. In the foreground is an electric caulking machine, one of many technologies devised specifically for this project (W. A. Water Corporation Library).

and effectively. However, O'Connor knew that some aspects of the work could not be entrusted to contractors. For example, he insisted that public works labour install the rock foundations for the harbour because he knew that the contractors had a clear incentive to place less rock than was required and it would be almost impossible to verify that the required amount had been placed because it was underwater.

> 'A contractor who is making a large profit by tipping stone into the sea and has a personal and pecuniary interest in it might have a very vital and strenuous objection to tipping more stone into the sea than was actually necessary.'[15]

Pressure on the contractors emerged in the form of critical comments in Parliament and the press, culminating with accusations of corruption. There was enormous pressure on O'Connor who was supervising the many aspects of the pipeline work, distributed over 500 km of the route, in addition to the other projects all nearing completion at the same time. He was working 16 hours a day, seven days a week for much of the time.

In addition to all this he had to reply to critical letters appearing in the West Australian newspaper and statements made under parliamentary privilege by politicians.[16] John Forrest, his staunch political supporter, had switched from state to national politics, leaving a vacuum filled by a series of unstable and short-lived state governments.[17] A series of different ministers oversaw the project for just a few weeks each, none with the time to understand any of its detailed aspects. Construction of the storage reservoir dam at Mundaring had been seriously delayed because a fault line in the rocks under the dam itself had been missed in preliminary site surveys. There were technical difficulties in maintaining a sufficiently high quality for the pipe section joints. Then, after returning from a three-week visit to Adelaide by steamship, O'Connor learnt that his lead engineer and one of the main contractors had engaged in questionable commercial transactions. Having placed absolute trust in his subordinates, leaving almost all the day-to-day detailed decision-making to them, this was a huge setback for him.[18] On top of this, there was a run of extraordinarily hot weather: there was no air conditioning in those days.

Early in March 1902, O'Connor took his horse for an early morning ride along the beach, carrying a revolver. A little south of Fremantle itself, he stood in the water, placed the muzzle of his revolver in his mouth and pulled the trigger.

Just eight months later, water was flowing into the Mount Charlotte reservoir in Kalgoorlie.

What marks O'Connor as a truly expert engineer?

First the Kalgoorlie pipeline, his most ambitious project, has been an outstanding success. To this day it still provides water, not only for the mining industry in Kalgoorlie, but also for a large rural population between Perth and Kalgoorlie. In 1968 I travelled the route of the pipeline by train and some of the timber sections of the pipe used for low-pressure sections (manufactured using similar methods to wine barrels) were still lying beside the track. They had only recently been replaced with steel pipes.

Like the pipeline, Fremantle Harbour has been expanded to enlarge its capacity but remains in very much the same form as O'Connor designed it. Few engineers create artefacts that are still in use in their original form more than a century later.

Remember how engineering practice has two threads (Fig. 3.7), the intellectual part consisting of understanding needs and making predictions, and the practical part, delivering results by winning the conscientious collaboration of other people? Consider these observations made by close colleagues:

'A genius, possessing extraordinary foresight'.

'He was neither dogmatic nor arrogant [and] possessed an extraordinary capacity to win the interest and cooperation of men whose experience enabled them to throw light upon his plans. Never did he fail to acknowledge information offered.'

He had the foresight to recommend the pipeline over alternative solutions for the Kalgoorlie water supply, mainly based on dams and storage reservoirs. He had just one year of reliable rainfall records and two other years of incomplete records. He knew that the rain that accumulates in a gauge does not reliably indicate the amount of water that runs into creeks that can be channelled into storage reservoirs. He relied on personal accounts from local people and his understanding of hydrology. His decision to build a pipeline has been questioned in the years since, but subsequent sharp declines in rainfall run-off are a testament to his foresight.

Delivering the pipeline with less than 10% budget overrun would be an extraordinary achievement today, even though pipeline technology is well established around the world. That achievement with the simultaneous necessity to completely reform the engineering contracting culture of Western Australia at the time is truly remarkable.

Figure 4.3 Deep excavations were needed under the site of the storage dam near Perth to enable a rock fault to be grouted with cement. The near vertical fault had not been detected by exploratory rock drilling prior to construction. This was the only major delay in the project and accounted for the cost overrun (W. A. Water Corporation Library).

Lastly, O'Connor was intimately involved in all aspects of his projects, even though he had the support of many other engineers who could handle most of the technical day-to-day details. He assessed the requirements and worked with the government to develop possible solutions. He prepared the technical designs (with assistance from many others), wrote detailed proposals, and had them reviewed by leading international expert engineers at the time. He helped to convince investors to lend the money, then planned, organised, and directed the projects. He supervised the work in person and checked the contractor's claims for payment. As a result of this gargantuan effort, the railway extensions and Fremantle Harbour were in service by the time of his death

Figure 4.4 Pipeline being built over a salt lake near Coolgardie where special roofing was included to protect the pipe from the highly corrosive wind-blown dust.

and he succeeded in building an effective organisation that delivered the pipeline after his death, keeping to the revised schedule and budget.

NOTES

1. Research reports that provide readable accounts include: Cianciolo, Matthew, Sternberg, & Wagner (2006), Ericsson (2003, 2006), Ericsson, Krampe, & Tesch-Römer (1993), Schwartz & Skurnik (2003), Stanovich (2003), and Wenke & Frensch (2003).
2. Goold (2014).
3. Bailey & Barley (2010); Gainsburg, Rodriguez-Lluesma, & Bailey (2010).
4. Darr has written two highly informative descriptions of technical sales engineering (Darr, 2000, 2002).
5. Explained further in Trevelyan (2012b) and Chapter 1.
6. See for example: Busby (2006), Busby, Chung, & Wen (2004), Busby & Strutt (2001), Cross, Bunker, Grantham, Connell, & Winder (2000), Leveson & Turner (1993), Mehravari (2007), Nicol (2001), and Waring & Glendon (2000).
7. Schön (1983)wrote about reflection as an essential component of professional practice.
8. Tables 3.1 and 3.2 in Chapter 3 provide data. One can estimate the proportion of time spent listening for novices, including training, as about 25%.
9. Ericsson (2003, p. 64).
10. I used to dread receiving feedback from my professors: every page would be marked with up to 50 or 60 corrections. Eventually, after 15 years of practice, I learnt to improve my writing and the number of corrections dropped to just a few, even none on some pages.
11. For a review of some relevant research, read Kahneman (2011, pp. 40–50).
12. I have relied extensively on Tony Evans' biography (2001).

13. Webb (1995).
14. A. G. T. Evans (2001, p. 72).
15. A. G. T. Evans (2001, p. 210).
16. Parliamentary privilege allows politicians to make statements in parliament for which they could be sued for libel or defamation if the same statement were made outside in public.
17. Webb (1995).
18. Webb (1995).

Chapter 5

What engineers know

The key to understanding engineering work lies in understanding the role that technical knowledge plays. After all, specialised knowledge, much of it technical, is the main factor that distinguishes engineers from other professions, and it also distinguishes the engineering disciplines from each other. For example, the technical knowledge of most electronic engineers is very different from the technical knowledge of most civil engineers.[1] However, understanding specialised knowledge is not that simple: most of it is not what was learnt at university. Much of it exists invisibly in the minds of engineers who are entirely unaware of it. Much of it, learnt informally on the job, can easily be taken for granted.

In the last few chapters, you have learnt a little about some aspects of engineering practice, such as how collaboration and finance are critical aspects. By the end of this chapter, you will appreciate how much more there is to learn than you might have thought while you were at university. You will always be learning something new every day, which means that *you will never stop learning*. Engineering knowledge is limitless in its scope and detail and there is no way of knowing what you are going to need around the next corner of your career.

This chapter will help you to map this vital knowledge domain – what engineers know and how we get to know it. By the end of the chapter, you will see that certain special learning skills are very useful: accurate listening, accurate visual perception, and accurate reading. These skills can easily be forgotten in higher education. In most university courses, everything you need is provided for you. You are told what to learn in order to get high enough marks to pass the exams. It is presented to you in so many different ways – lectures, recordings, PowerPoint presentations, notes, text-books, websites, past exam papers, discussion forums, learning management systems, friends, and peers. However, workplace learning requires a different approach. What you need to learn is all around you, but there are few, if any, helpful teachers to point out what you need to learn, and there is no process of formal assessment. For many of the things you need to learn, you only hear or see them once: they won't be repeated or recorded for you, nor will there be notes and handouts. You have to make your own notes and diagrams, as they won't be available on the Internet, waiting for you to copy and paste.

What, then, is this specialised knowledge, know-how, capability, or competence, this invisible 'currency' of engineering? Here, a young mechanical engineer tells us about some of the things she found that she had to learn, which provides a revealing insight about the nature of workplace learning:

> *'Practicality, how it all works, what a valve looks like, how you pull a pump apart. The next most important thing is where you fit in. Like others when I first started I didn't think that I fitted in anywhere. I didn't have a job that no one else could do. As the scheduler I had no idea how long it took to put in foundations. I had to ask somebody and figure out who to ask. I had a terrible lack of knowledge of actual "stuff".'*

Let's begin with a common misconception.

Misconception 6: We need to know it all

Students and many engineers think that ...	However, research demonstrates, and expert engineers will tell you that ...
Technical expertise (individual skill) is what distinguishes an engineer. You have to know your technical 'stuff' or you will lose the respect of your peers and boss. Asking for help is a sign of giving up, a kind of cheating – a last resort. If you don't know it, look it up in texts, online, at the library, Wikipedia (if you can get away with it), and never admit that you don't know.	Good engineering works because knowledge is distributed in the minds of different people: it is not necessary to know everything, but you do need to know how to find someone you can ask and how to get them to help you.[2] Above all, when you don't know, say so.

> What is your anxiety level about your own technical knowledge? How comfortable do you feel about your technical knowledge that you had to learn at university? Write your own comments on this. If you have a word processor, write your comments and then have them ready to copy and paste into The University of Western Australia learning management system (UWA LMS) later.

If you lack confidence in the technical knowledge that you had to learn at university, or if you feel that you don't really remember it accurately, that you would be hard pressed to explain entropy in the context of the second law of thermodynamics, then you have probably made a realistic assessment.

Don't worry. One of the strengths of a university education is that you learnt at least some of the maths and the other technical stuff once, which means that you could learn it again if you need it. It also helps, without you necessarily being aware of it, when you find yourself in an unfamiliar situation. Exposure to mathematics and engineering science theory actually helps you make sense of these situations more quickly if the theories provide useful insights.

It takes courage, especially when you are young, to admit that you don't know something that might seem very basic. That's why many young (and not-so-young) engineers choose the web, even Wikipedia, as their first point of reference. There is nothing to be ashamed of: everyone else has experienced similar feelings of utter ignorance and helplessness.

With over 260 different engineering disciplines, how would anyone know where to start? Since no one can immediately predict which discipline they're going to end up in, how can any engineer know what they need to learn next?

This chapter can help answer that question. As we shall see by the end of this chapter, much of the working knowledge needed by engineers lies in the minds of the people who work in the engineering enterprise. Accessing that knowledge effectively is the key to success in engineering. Learning and building new knowledge is also a vital part of the picture that we will return to in later chapters.

DEFINITIONS

We will need to understand more about what we mean by 'knowledge', but first, we need to understand the notion of an engineering enterprise.

Practice concept 18: An engineering enterprise

In order to understand engineering knowledge, it is essential to explain the idea of an *engineering enterprise* and, in doing so, define engineering differently from the traditional notions that emphasise only the practical application of science.[3] An engineering enterprise is an organisation that depends on the application of specialised knowledge that emerges from engineering schools and allied communities of practice. An enterprise is not the same as a firm or company. The people involved can come from several different firms, all collaborating with the same purpose – delivering a product, a service, or information. Even the customers, clients, and end users are part of the enterprise: without their knowledge of how to use the products and services, as well as their willingness to pay for it, there would be no enterprise. The investors who provide finance also play an integral role. The people in the enterprise are not all engineers; in fact, there may be very few engineers and they may represent a very small minority of the people involved. Engineering relies on many different people collaborating, including the financiers, clients, end users, consultants, and even government regulators.

Engineering, therefore, is much more than what engineers do.

Engineering enterprises use tools, machinery, energy, information, and components to transform materials and information into deliverable products, information, or services by using special knowledge contributed by engineers. The aim is the predictable delivery of reliable products, services, or information, all provided through the most economic use of financial capital, human effort, and material resources possible.

The products, information, and services are often provided as inputs to other engineering enterprises in a complex web of connections linking companies and people in a network that, these days, extends across the planet.

Figure 5.2 helps to illustrate these connections. It shows six engineering enterprises linked to each other in what is called a *value chain*. The outputs of each enterprise, collectively, have greater value than the inputs. The diagram is highly simplified:

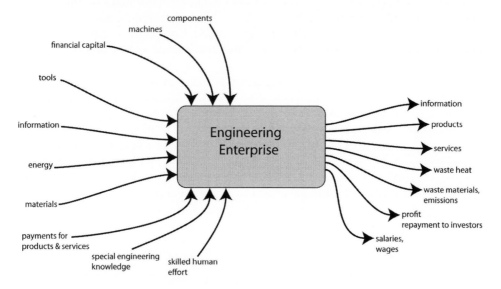

Figure 5.1 An engineering enterprise.

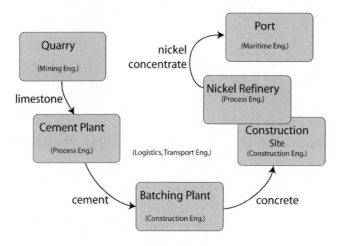

Figure 5.2 Simplified diagram of an engineering value chain: refer to text for explanation

only the principal flow of products has been shown. With the help of machinery and energy, limestone rock at a quarry is transformed into crushed limestone powder and transported to a cement plant. The main product of this plant, cement powder, is combined with granite chips, sand, water, and special additives and transformed into wet concrete at a batching plant. The wet concrete is carried by special cement trucks to a nickel refinery that is under construction. Here, the concrete is combined with steel, pipework, reactor vessels, and instrumentation to construct the nickel refinery. The refinery, when completed, will process crushed nickel-bearing ore from a mine into nickel concentrate: a powder containing a much higher proportion of nickel. The powder is transported to a port where it is loaded into ships that transport it to a faraway city (not shown), where electric arc furnaces convert the powder into metallic nickel, which is supplied to . . . and so on.

Figure 5.2 shows an extreme simplification: each enterprise provides just one product to the next. In reality, every enterprise has most, if not all, of the other inputs shown in Figure 5.1. The quarry uses mining, crushing, conveying, and loading machinery that, in turn, require fuel, electricity, and spare parts. The cement plant also requires water and energy in the form of fuel (oil, gas, or powdered coal) and electricity. Ash and other minerals are roasted in a large rotating kiln with powdered limestone, producing cement clinker. Crushing, conveying, and processing machinery had to be installed at the cement plant, along with instrumentation, controls, and product storage silos. Furthermore, all these elements require ongoing maintenance and spare parts. Transport and logistics[4] engineering knowledge, the ability to transport and store materials and components, is an integral part of this network of enterprises.

Each *enterprise* provides *information, products,* or *services* that are *more valuable* than the inputs. Each relies on special knowledge, mostly focusing on just the operations performed in the enterprise. Engineers work in many of the firms that are involved, but not necessarily all of them: many engineers are consultants who come to provide advice only when needed. Engineers also work in banks and investment firms that provide financial support for these enterprises. They help insurance companies assess the risk of financial losses from fires, breakdowns, and accidental damage. Engineers also work with consultancy firms that provide technical advice to the enterprises and also for government agencies that regulate their operations.

A consultancy, for example, collects a mass of information. Consultants with special expertise[5] know how to find, extract, and combine only the particular parts of that information that their clients need, and they also know how to present it in a form that is meaningful for their clients. A manufacturing firm employs people that know how to transform information, components, and materials into products using energy. A construction firm transforms materials and components into buildings, which requires the significant use of human effort, information, and energy. All of them require financial capital, as well, representing the confidence of investors to provide money long before it is returned, with the expectation of making a profit in the end. A small firm that is just starting up, particularly in engineering consultancy, needs human capital – accumulated know-how and expertise, as well as a list of prospective clients. Eventually, the consultants will earn higher fees, once their reputation has been established.

How does the special technical knowledge of engineers influence the engineering enterprise that depends on it?

Remember the aims of the enterprise? Predictable delivery of reliable products, services, or information, all provided through the most economic use of financial capital, human effort, and material resources possible. First, engineers provide the planning and organisation needed for predictable delivery. Second, engineers can quickly produce cost-effective designs and make accurate predictions about the technical performance, cost, safety, and environmental impact of proposed solutions. Third, engineers provide the special knowledge needed to ensure that products, services, and information are reliable enough to meet the needs of clients and end users long after they are provided. The expectation that products will meet their needs, a human perception, is vital, because it results in the willingness to pay for the products. In other words, the products must represent a sufficiently greater value than the combined inputs to economically sustain the enterprise. Fourth, engineers apply their ingenuity to minimise the use of financial capital, human effort, and material resources needed to sustain the enterprise. Fifth, engineering knowledge provides reassurance for investors, reducing the perceived risks and uncertainties, increasing the perceived value of the enterprise, and stimulating the willingness for investors to build and sustain the enterprise in the first place. As we work through this book, we will learn more about how engineers do all this.

Sometimes, when you look at some of the engineers' activities individually, it can be difficult to understand their value. Take, for example, an engineer who checks specifications and incoming components to ensure that they comply with safety standards. Considered by itself, the work seems to have contributed no real value. The engineer has not contributed any design or performance improvement, planning, organisation, reliability improvements, or cost reduction. However, in the context of an enterprise, the engineer has reduced the perceived risk of failure simply because the checks have been made, even if no non-compliance was found. In doing so, he has increased the economic value of the enterprise. We will learn more about this in later chapters. For the time being, we need to understand the vital significance of technical knowledge and how it helps an engineering enterprise.

Knowledge and information

Within this engineering enterprise framework, we can begin to build a map of the types of knowledge that are needed by the people who work in it.

However, we need to review what we mean by *knowledge*. Defining knowledge is something that philosophers have debated and focused on for thousands of years. Some philosophers have even specialised in discussing the kinds of philosophy that might be helpful in engineering.

What do we mean by knowledge? Some philosophers have focused on knowledge as a product of rational thought in the form of written statements of truth that are independent of any particular individual. Others have explained knowledge in terms of beliefs held by an individual as the result of experiencing the physical world.[6]

In this discussion, we need to focus on the latter view, because we are interested in the knowledge held in the minds of individual engineers that enables them to influence the operations of an engineering enterprise.[7]

A useful starting point is to consider that knowledge is 'justified true belief'. It is 'justified' in the sense that the person has taken personal responsibility to establish the truthfulness or validity of the belief.[8] The belief, therefore, has some basis in the experience of the person, and perhaps other people who the person believes to be reliable informants.

Information, on the other hand, is *data*: the content of *messages* exchanged between people, machines, and systems. Information and knowledge are not synonymous. We receive an enormous amount of information every day, but a relatively small percentage becomes knowledge that we retain and most of it is discarded. Try and imagine, for a moment, quantifying the informational content of every conversation you have ever had, everything your eyes have ever seen, your ears have heard, your skin has touched, your nose has smelt, your tongue has tasted, the books you have read, the movies you have seen, and the courses you have taken.[9] Now, how much of that is still part of your knowledge? Perhaps a lot of it, but certainly not all of it.

In creating new knowledge, we *interpret* information in the light of our *prior knowledge*. Philosophers, education psychologists, and learning scientists have studied this process extensively.

One of the most important kinds of prior knowledge that we rely on is our *language*. As we shall see in coming chapters, the idea of language as a convenient set of symbols with agreed upon meanings is a bit too simplistic to explain human communication. However, it will do for the moment, provided we understand that we are continually learning about new meanings. We cannot take it for granted that the listener has the same understanding of a word or symbol as the speaker. Therefore, as we interpret new information in the light of existing knowledge, we have to understand that our prior knowledge base is continually evolving.

Knowledge develops in our own minds as a result of interpreting information that we receive every day. Much of our knowledge develops as a result of *social interactions* with other people, including parents, teachers, friends, and peers. As we discuss our own views and beliefs with other people, new ideas and perspectives gradually emerge as a result of these interactions. This is particularly important within an organisation like an engineering enterprise, where the quality of the knowledge being developed and applied by people in the organisation is a critical factor in its overall success.

Practice concept 19: Types of knowledge

Philosophers, psychologists, and many other writers have helped by describing several different kinds of knowledge.[10]

Explicit, codified, propositional knowledge

Most of the knowledge that you learnt in engineering school falls into the category known as explicit or propositional knowledge, also more broadly defined as 'codified knowledge'. A proposition is a simple statement that can be verified as being either true or false. These are examples of formal propositions:

> *'There are 242 pages in the geopolymer application field guide.'*
> *'Young's modulus, E, is the ratio of applied stress divided by the resulting strain in an elastic material.'*

Explicit knowledge is relatively easy to propagate and distribute. It can be written down in a language that appropriately educated people will understand, with a fair chance that their interpretation, based on their own prior knowledge, will align fairly closely with the intentions of the author. Explicit knowledge can be transmitted by using symbols, such as words. In written form, explicit knowledge can be transmitted without any meaning being lost. However, as soon as someone has to listen to or read, and then interpret the words and symbols, some of the knowledge is inevitably lost or changed because of the variations in prior knowledge between individuals. As Collins explained, all human language use is, in effect, translation.[11]

Explicit knowledge can be taught to young people who have less advanced or mature prior knowledge. This is not easy: considerable effort on the part of both pupil and teacher is needed, as you probably appreciate.

Your experience at university should have been enough for you to realise that acquiring explicit knowledge from another person, the Internet, or a book, can be unreliable and time-consuming. I am sure you can remember long hours of study for examinations, as well as the ache in your stomach upon finding exam questions that you still could not answer. Transferring knowledge, even explicit, codified knowledge, is not easy and is prone to errors and misunderstandings.

Other kinds of knowledge are even more difficult to transfer.

Procedural knowledge

As you would appreciate from your studies, it is one thing to acquire explicit knowledge, but quite another thing to effectively use it. We can refer to the latter as procedural knowledge: knowledge that is needed in order to effectively make use of explicit knowledge. Sometimes, procedural knowledge can be conveyed using explicit knowledge, in the form of a list of instructions. However, it is not really useful until one has practised the sequence of instructions and reached a stage when one no longer needs to refer to them.

You can study a textbook on mathematical statistics and acquire explicit knowledge that is relevant for the analysis of data resulting from experiments. Once you reach that stage of understanding, you can probably pass an exam.

However, you need to solve many practice problems in order to acquire the procedural knowledge that would enable you to competently apply statistical techniques to analyse your data so that you can confidently draw statistical conclusions from it.

In your final year or capstone project at university, you may have learnt this for yourself. While you may have passed exams in maths and other engineering science subjects, applying that same knowledge for yourself in a project may have required a lot more trial and error, learning, and maybe a good deal of frustration. In other words, developing procedural knowledge is not easy.

Implicit knowledge

Implicit knowledge, on the other hand, is knowledge that has not been made explicit. Philosophers disagree on how much implicit knowledge can actually be made explicit, but we don't need to concern ourselves with this difficulty.

From a practical point of view, implicit knowledge includes things like knowing where the nearest bathroom is located. It might be written down on a plan of the building somewhere, but we don't usually have building plans lying around in order to figure out where the nearest bathroom is. Instead, we ask someone else, 'Could you show me to the nearest bathroom, please?' Most of us can remember where we need to go after the first visit, so we don't need to write it down. We could if we wanted to, but showing newcomers the location of the bathroom is part of the welcoming ritual in any organisation in almost every human society. How would you feel, starting your first job, if someone told you to get a plan of the building to figure out the location of the nearest bathroom? Most of us would ask ourselves, 'What kind of strange people work here?'

The important thing to understand is that, unlike explicit knowledge, implicit knowledge can only be learnt by experience, mostly through the help of other people. A few of us might be inclined to explore the entire building, find every bathroom, and then decide which is the most convenient to use. However, most of us simply ask someone to show us where it is or how to find it.

Tacit knowledge

The next kind of knowledge is slightly more difficult to appreciate. The term tacit knowledge was devised by Michael Polanyi, who wrote, 'we can know more than we can tell'.[12] An example is the knowledge you need to ride a bicycle or to tie a bow with your shoelaces. This is knowledge that we acquire by practice, often frustrating practice, until one day we get it right. From then on, it seems to come very naturally. Just for a moment, think of trying to describe how to ride a bike with words alone.

There are several different kinds of tacit knowledge. Riding a bike could be described as 'psycho-motor' knowledge or 'sensory-motor' knowledge. This is knowledge that enables us to respond very quickly with movements that are appropriate for certain sensed conditions. Being able to walk without falling over is another good example of sensory-motor knowledge.

Tacit knowledge becomes embedded in our minds and bodies so deeply that we forget it even exists. We simply use it when we need it. Being able to communicate in a particular language, to speak and understand other people speaking, as well as to read and write also relies on a vast amount of tacit knowledge.[13]

Tacit knowledge is a kind of implicit knowledge. It can only be learnt by practice and experience, often with the help of someone else. However, unlike much of the implicit knowledge that we hang on to, tacit knowledge is nearly impossible to verbally describe in a way that would be meaningful to someone trying to learn it.

Another aspect of tacit knowledge can be summarised as 'social knowledge', which is knowledge about how to behave and interact with other people. Some of this is contextual: there are particular ways that you interact with people in a particular enterprise that will be different in other enterprises. This type of knowledge includes social customs, acceptable clothing, ways of talking with people, ways of negotiating with people, body language, and ways of telling if someone is being honest with you.

Tacit knowledge also confers an ability to recognise artefacts, materials, defects, failure symptoms, noises, smells, material feel, vibration, and kinaesthetic cues. Once you develop this knowledge, it will enable you to acquire valuable insights into what is happening all around you. For example, the ability to recognise the difference between mild steel and stainless steel is something that you probably had to learn. Being able to recognise the likely causes of electromagnetic interference that affect a circuit can be extremely valuable for an electronics engineer when trying to diagnose performance deficiencies.

Another kind of tacit knowledge that is particularly valuable for engineers is accurate visual perception, particularly the ability to work with graphical representations, such as circuit diagrams, and the ability to visualise a three-dimensional artefact from a two-dimensional drawing or image. Just like words, this visual understanding is a form of translation involving symbols. Not everyone develops the same level of ability. For example, an electronics engineer can quickly see the relationship between a

Figure 5.3 Unwritten tacit and implicit knowledge in engineering practice. Take a look at this construction site: how much of what you can see in the picture is shown on the drawings? Answer:
not much, because the drawings show details of the *finished* building. They do not show
all the steps needed to make it, such as in a LEGO construction kit directions leaflet. For
example, how did the builders know where to put the crane and how to construct the
scaffolding? Most likely, those elements were not shown on the drawings.

physical electronic circuit and an abstract circuit diagram. While there might be a
one-to-one correspondence between components in the diagram and the circuit, it is
almost certain that the physical layout of the circuit will be completely different from
the circuit diagram. Also, the physical circuit will have many parts that are not even
represented in the circuit diagram, such as power supplies, circuit board mountings,
cooling devices, and connectors. An electronics engineer develops tacit knowledge that
enables intimate understanding of the relationship between the circuit diagram and the
physical artefact that it represents.

An often-unconsidered area of tacit knowledge is the appreciation of intrinsic
beauty, or aesthetic knowledge, particularly the ability to create objects that will be
visually attractive or beautiful to a chosen class of spectator or user.[14] An expert will
be able to create a design that is not only visually attractive but also has appropriate
proportions to make the best use of materials.[15] This is valuable for anyone involved

in consumer product design or the design of large structures that become part of the visual landscape: it is an essential knowledge category for architects. However, even engineers can find it useful: a proposal that includes attractive sketches, diagrams, and object designs may be accepted much more readily by clients or peers. Engineers, as a group, tend to appreciate visual images more than words, so it can be useful to identify features that are instinctively attractive for engineers: aesthetic knowledge has instrumental value. Aesthetic knowledge can be based on other senses as well; for example, sound is the most important aesthetic sense in the realm of music recording.[16]

Still another kind of tacit knowledge exists in our subconscious minds. Our minds can process knowledge without us being aware of it occurring. Some of the best ideas surface from our subconscious mind when we least expect it, maybe even in the half-dreaming, half-waking darkness of dawn. Other times, ideas emerge as we talk to someone else or walk down the street, without any prior warning. Some people call this intuition, others call it creativity, and still others call it inspiration. Often, the best way to access this knowledge is to deliberately 'switch off'. If you're the kind of person who needs to occupy your mind, then think about something completely unrelated for a while. The one thing you cannot predict is when this kind of knowledge will surface in your mind. It very rarely comes exactly when you want it to!

> **Tacit knowledge**
> You don't know you need it.
> You don't know you know it.
> You don't know how to learn it.
> You don't know where to find it.
> You don't notice it even if you are looking at it.
> You don't notice that you have learnt it when you have.

The use of mathematics in engineering is also, surprisingly, more tacit than explicit. Engineers seldom, if ever, apply the methods taught in university courses. However, engineers frequently use mathematical concepts when they discuss technical issues, and instinctively appreciate, for example, that the highest point on a curve or surface has a horizontal tangent, a concept that originates in calculus. Engineers also make instinctive mathematical decisions when choosing, for instance, a simplified model that will enable them to estimate design parameters faster.[17] The tacit knowledge that engineers use instinctively in the workplace can only be developed through extensive mathematics practice in engineering schools, even though the same engineers may have difficulty in recognising that they are making extensive use of their mathematical knowledge.

We can measure tacit knowledge, even though we are not necessarily aware of it. Recent developments in psychology have demonstrated that tacit and procedural knowledge development can be assessed by observing performances and also through online, multiple-choice questionnaires.[18]

Embodied knowledge

There is another kind of knowledge that is built into the artefacts that make up our world, even in natural objects that have been set up in certain places in order to convey meaning. For example, the arrangement of a supermarket represents knowledge that has been developed over decades that makes life easier for shoppers, while also creating marketing opportunities for sellers. The arrangement of shelves, price tags, product packaging, and labelling, categorisations of products on the shelves, signs that direct you to 'cereals' or 'tea and coffee', the arrangement of checkout desks and cash registers ... these are all manifestations of what we call embodied knowledge.[19] This is knowledge that is embodied in the artefacts that populate our world. Take roads, for example. Look at all the details of road design, line markings, reflectors, speed humps, kerb design, drainage, signs, and traffic lights; they all embody knowledge about helping people make effective use of the roads for safe transportation.

Embodied knowledge exists outside of individual people because it is embodied in artefacts. It can be distributed by distributing the artefacts. Annotated images and sketches alone may be sufficient to capture much of the embodied knowledge in an artefact: these are even easier to distribute.

Embodied knowledge is not necessarily easily accessible in the sense that a person can inspect an artefact and acquire all of its embodied knowledge. On the contrary, many aspects of embodied knowledge may only be apparent to someone who has enough background knowledge to appreciate particular features of the artefact and the knowledge from which they were derived. For example, much of the embodied knowledge in roads may not be apparent to a person with little or no experience of driving motorised transport vehicles. Even a driver would not recognise the more subtle aspects of road design, some of which are concealed underground. It takes an experienced road construction engineer to recognise most of these aspects.

Here, we will spend a moment emphasising one of the fundamental concepts that you need in engineering practice.

Contextual knowledge

Another helpful knowledge classification in engineering is the idea of contextual knowledge, which is knowledge that is specific to a particular context, perhaps a particular organisation or even a specific workplace. Using tacit knowledge, most people can recognise an electrical switch, even if they haven't seen that particular kind of switch before. They might not be absolutely sure that it is an electrical switch, but they can be reasonably confident. However, knowing what a particular switch controls is a form of contextual knowledge. For example, knowing that a switch is on the wall of an American home and seeing that the movable part is in the uppermost position tells you that it is in the 'on' position. In many other countries, it would be in the 'off' position. This is an example of contextual knowledge – knowledge that is specific to a particular setting.

There are many other kinds of knowledge and ways of knowing it, many of them unique to a particular culture or civilisation. However, for the time being, it is sufficient to understand the distinctions between explicit and implicit knowledge, between tacit and embodied knowledge, and between contextual and procedural knowledge.

The following table provides a quick summary. Remember that the different types of knowledge are not all mutually exclusive. For example, it is possible to have knowledge that is both explicit and contextual: the location of the nearest bathroom in a particular building could be provided in the form of direction signs close to the doorway of every office. Some implicit knowledge could also be tacit knowledge, while some implicit knowledge could be made explicit if we took the time and trouble to write it down.

Practice concept 20: Competency depends on knowledge from other people

Most engineers think that …	Research tells us, and expert engineers know that …
Competency is an attribute of an individual. Individual assessment reinforces this notion, as well as statements of engineering competency that are used in accreditation.	Competency comes from an individual's ability to work with objects, tools, materials, artefacts, and documents that embody knowledge from other people. Competency also comes from the ability to collaborate with other people that possess knowledge and skills.

The idea that competency is an attribute of an individual person is remarkably widespread and misleading when applied to engineering practice. It conflicts with the origin of the term 'core competencies', which traditionally referred to collective learning within an organisation.[20]

When misunderstood in this way, the implication is that competency is a measure of an individual's capability. In reality, an individual's capability is determined by many factors other than intrinsic knowledge, skills, and abilities. Some of the most significant factors are the artefacts all around us: the office space in which we work, our desks, our computers and the operating systems, the filing systems that enable us to find information, lights, transportation systems, telephones, e-mail, the Internet, and so on. All these artefacts represent embodied knowledge: they represent the accumulated knowledge and experience of many other people over time. These artefacts include software and databases, processed materials, and even objects arranged in a particular way; they can all represent knowledge passed on from other people.

An engineer can be highly effective when supported by appropriate embodied knowledge, and yet without this, can find it hard to make any useful contributions at all. An expert engineer is distinguished by an ability to quickly access knowledge embodied in tens of thousands of different components, design fragments, and other ways that the knowledge of countless other engineers has been stored so it can be found, recalled, and reused.

Novice and less experienced engineers regard engineering procedures, such as procurement processes, specifications, test plans, quality assurance, document review, change management, configuration management, and regulatory approvals, as a painful administrative burden, imposing what seems to them to be unnecessary, bureaucratic paperwork (or, these days, time spent completing electronic forms). One engineer told me that he was working with a company known as 'the company with 10,000 procedures'.

Artefacts like procedures, checklists, standards, manufacturer's catalogues, design guides, and textbooks all represent ways of sharing the knowledge created by other people who came before us.[21] Expert engineers understand that these artefacts enable engineers to work and collaborate more effectively, with fewer misunderstandings and the consequent delays that they cause. It takes an extraordinary amount of time and effort to develop effective procedures, not the least of which is the time needed to reach an agreement on what would be an effective procedure, and then for engineers and others to learn how to use it effectively. More and more engineers are now gradually coming to appreciate how much intellectual capital in a firm is tied up in the design and training that is associated with these procedures and processes.

Business and organisational software systems, such as SAP,[22] also represent embodied knowledge. Their design reflects experience from countless organisations that have used them, and then requested improvements that were included in later versions.

Practice concept 21: Knowledge is difficult to learn and transfer

One of the most important things we can learn from Table 5.1 is that the only kinds of knowledge that can be represented with objects that are external to our mind are explicit, procedural (only as information, in the form of action lists), and embodied (partly, in the form of artefacts and documents).

Acquiring these types of knowledge, either for oneself or when helping others acquire them, can be time-consuming and troublesome.

Table 5.1 Types of knowledge.

Knowledge type	External representation	How to acquire knowledge
Explicit	Logical propositions, knowledge	Reading or listening, memorising, written writing, and practice.
Implicit		Experience with the help of others, reflection, with help of prior related knowledge and skills.
Procedural*	Written or spoken instructions	Following instructions, practice.
Tacit – social, spatial, graphical, and visual language sensory-motor recognition		Imitation, practice, with help of prior related knowledge and skills.
Embodied	Artefacts	Examine artefacts, or detailed descriptions of them, with help of prior related knowledge needed for interpretation.
Contextual		Experience, conversation with experienced people, with help of prior related knowledge needed for interpretation.

*It is easy to confuse 'procedural knowledge' with a written procedure, which is an ordered list or a written statement of actions. The latter is merely a method of storing and transferring information. The former means knowing how to do something.

Implicit, tacit, and contextual knowledge cannot be represented with external objects and are therefore even more difficult to acquire or transfer from one person to another. This limitation becomes crucially important in engineering practice.

Remember these human limitations – they are critical.

Implicit knowledge, tacit knowledge, and much of the contextual knowledge important in engineering practice exist only in the minds of individuals. Together with embodied knowledge, they form a large component of the knowledge required for engineering practice, all of it in the form of unwritten knowledge. Obviously, unwritten knowledge cannot be transmitted or made available through books or the Internet.

Practice concept 22: Knowledge that influences perception and action

Another closely related concept is that the only knowledge able to influence human perception and action is what is in a person's head at the time. While it is useful to have knowledge in an accessible form, written knowledge needs to be learnt (or readily available, with a person's attention drawn to it) in order for it to guide human behaviour. From Chapter 6 onwards, we will frequently return to this issue, which is a key to understanding collaborative performance in engineering.

Practice concept 23: Mapping engineering knowledge

We now have some understanding of the different kinds of knowledge, and we understand the notion of an engineering enterprise that combines components, materials, energy, information, and knowledge to provide products or services that have a greater value than the inputs.

We can start to draw a map of the knowledge used in engineering work, essentially the knowledge needed by people who influence the operation of engineering enterprises, including those who work in them. A significant part of the knowledge exists in the minds of people other than engineers.

In the early stages of the research that led to this book, I listened to dozens of engineers telling me about the details of their work in lengthy research interviews. I visited them at their workplaces so I could understand the environments in which they worked. With the help of research colleagues and my own twenty years of work experience spanning mechanical, electrical, electronic, software, and some aspects of civil engineering, I identified 80–90 separate activities that characterise the work of engineers. Then, I analysed the interview transcripts in detail, steadily refining the descriptions of these activities by adding some new ones and pruning others from the list. Each engineer that I interviewed mentioned approximately a third of these activities. No one does them all at any particular stage of their career, but some might see them all at least once. These engineers came from all the main engineering disciplines, so I was able to write about the activity descriptions that would be meaningful in any engineering discipline.

I was surprised by the degree of consistency between different disciplines, different work settings, and even different countries and cultures. No matter what discipline or type of engineering, it was surprising to learn how much they all have in common.

In the same way, I worked on a list of different kinds of knowledge and skills that characterise engineers. About half of the list represented technical knowledge, which was related in some way to aspects of engineering, while the other half was composed of a more general kind of knowledge that one needs to work in any organisation. Later, I compared these results with earlier published accounts of 'what engineers need to know' in order to check my findings.

Any attempt at this kind of qualitative classification depends on the starting point and no two researchers would end up with the same list. However, by comparing the results with those devised by other researchers, I could at least check that everything they found was also in my lists.

The classification of engineering activities and specialist knowledge is available in the online appendix.[23]

Mapping knowledge types

One of the first knowledge categories required in an engineering enterprise is an understanding of the different tools, machines, components, and materials that could be used, what they are called, what they look like, what each one does, its function in a particular context, and some understanding of its value or cost. This understanding does not include any detailed knowledge of how it is actually made. You do not need to know how a component or material is made in order to be able to use it. However, some knowledge of its manufacture and design can be helpful in understanding its limitations and possibly identifying faults or failures. For example, it is not difficult to appreciate that a screw made from steel is likely to be much stronger than a screw made from plastic. However, it is more difficult to appreciate that a screw made from heat-treated alloy steel is likely to be much stronger than one made from mild steel without any special treatment, partly because they may look and feel almost identical.

A related or relevant body of knowledge helps an engineer acquire the tools, machines, component parts, or materials: which firms supply them, in what quantities, and with what delivery times? Which firms hold them in stock, locally? How much do they cost in a given quantity? How long do they have to be ordered in advance? Can the supplier deliver them or does transportation need to be arranged? Which firms provide helpful technical information, application knowledge, expert advice, or training? How is the machine, tool, component, or material safely transported and stored without affecting its performance when it is used? How long can the component or material be stored for and in what conditions? How heavy is it? How can it be handled safely?

A third category of knowledge related to tools, machines, component parts, and materials is called 'detailed technical information', usually represented by technical data sheets and other technical information that enables an engineer to decide whether it will adequately perform the required function. This kind of knowledge also enables an engineer to predict the performance of an artefact or service that uses the component or material, how much it will cost, when it is likely to be available, the environmental and safety consequences of using the component or material, what kind of training will be needed by people who are going to work with the component or material, what kind of protective equipment they will need, and what kind of safety precautions have to be taken. Apart from the knowledge of information specific to each item, engineers also need to know about combinative effects. Can tool X be used with material Y?

What happens if fluid F flows through pipes made from P, lined with R and S, controlled by valves made from V and W? A new fuel distribution system at a major Australian airport had to be scrapped and rebuilt after it was discovered that an anti-corrosion coating used in some of the pipelines could contaminate jet fuel. Why, you may ask, was the anticorrosion coating needed, as hydrocarbon fuel cannot oxidise steel?

In cold weather, water can condense from air trapped in fuel pipelines and tanks. The water can become highly corrosive as it absorbs components of additives and contaminants in the fuel. Furthermore, bacteria that grow on the interface between the fuel and the water can produce highly corrosive waste products. Knowledge of components and materials has to include knowledge of how to use them in combination with each other.

This takes us on to the fourth category of component and material knowledge: failure modes. Knowing how to anticipate and recognise failures is critical knowledge for an engineer to possess. An engineer must be able to predict the performance of a product or service before it is built or delivered. To do this, an engineer needs to ensure that the relevant tools, machines, materials, and components will be used well within their capacity. An engineer needs to anticipate the consequences of accidents and emergencies, even the consequences of failures in other parts of the product. Will a certain material have sufficient strength to withstand the forces applied during manufacture and assembly, as well as during transport to its final destination? Furthermore, when tools, machines, components, or materials fail unexpectedly, an engineer can use knowledge of failure modes to diagnose the causes and devise improvements to prevent future failures.

I have represented these four aspects of knowledge in Figure 5.4. Lines with arrows on the map indicate possible dependencies. For example, knowledge of logistics and procurement requires basic knowledge of the tools, machines, components, and materials themselves, as does any detailed knowledge of properties, characteristics, and models. Knowledge of properties, characteristics, and models is essential in order to understand and predict failure modes.

I chose to represent categories of knowledge with clouds, unlike Etienne Wenger,[24] who has proposed related ideas using the notion of mountain peaks to represent expertise. To me, clouds provide a more appropriate metaphor.

First, a cloud is diffuse and has no clear edges or boundaries, just as knowledge classifications have fuzzy, ill-defined boundaries.

Second, it is difficult to capture a cloud, and in a similar way, it is difficult to appreciate or fully summarise a particular area of knowledge.

Third, you can pass right through a cloud at night without even being aware of it. A person who lacks awareness can be surrounded by extremely knowledgeable people and not notice that they possess such valuable knowledge. Many Australian Aboriginal elders have a vocabulary of up to 300,000 words, ten times more than the average professor of English. Yet, for the first 200 years after they arrived in Australia, Europeans thought that Aboriginals were ignorant, uncivilised, primitive people.

The fourth idea captured by the idea of a cloud is that you cannot tell how far into the cloud you are seeing from the outside, nor how thick the cloud might be. Unless you already have the required knowledge, it is difficult to understand how much you don't know.

Fifth, even when you have some knowledge, when you are partly immersed in the cloud, you cannot see how much more knowledge you need to learn, just as you cannot see how much further into the cloud it is possible to penetrate.

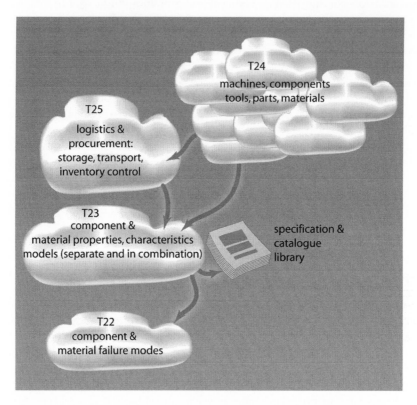

Figure 5.4 Mapping knowledge of tools, machines, components, and materials used in an engineering enterprise. Arrows convey typical associations linking fields of knowledge.

Lastly, once you are fully in the cloud, it is very difficult to see the world outside. In the same way, once you become totally immersed in a particularly specialised aspect of knowledge, it can become more difficult to appreciate what is happening in the world around you. You can only see the outside world within the framework of the knowledge that preoccupies your mind.

Practice exercise 1: Mapping component and tools knowledge

Try to identify the types of different knowledge associated with different tools, machines, components, and materials. To make this easier, choose a particular engineering enterprise that might be familiar to you, even a software engineering firm. Try and identify the tools, machines, components, and materials that are commonly used in the enterprise. Then, write down as many examples as you can of the different kinds of knowledge represented by the map in Figure 5.4. Which aspects of knowledge comprise explicit knowledge? Which aspects relate to procedural knowledge? Which aspects are considered tacit knowledge? What about contextual knowledge? Implicit knowledge? Embodied knowledge?

Not every engineer will have detailed knowledge of all the tools, machines, components, and materials used in a particular firm or enterprise. In fact, there may be no one who knows about every single component and material used. Most engineering enterprises produce a range of products and services: not every product uses every single component or all the different kinds of materials. Therefore, we need to appreciate that the knowledge map represents knowledge pertaining to a particular product or service. Think about a layered map: a map with different layers corresponding to the different products and services provided by an enterprise.

Refer to the classification of specialist knowledge in the online appendix to help evaluate your responses.

So far, our map has included only four aspects of engineering knowledge: there is still much more to come.

How much knowledge do you start with?

A recently graduated engineer has more up-to-date and freshly activated knowledge of abstract science and mathematics than most other people in the enterprise.[25] They should be able to describe different aspects of the behaviour of a product using mathematical or computational models to predict its technical performance in different situations. A recent graduate might also have more recently applied knowledge of testing and measurement methods from laboratory classes.

A recent graduate cannot be expected to know much more at the start. Much of what we learnt at university comprises explicit codified knowledge: there's not much incentive to remember something that you can't write in an exam paper.

Engineering models may be partly based on known science, but use other sources as well. Contextual knowledge based on experience is important, as is experimentation. Engineers' models may not even be strictly valid from a scientific standpoint. As Schön pointed out, engineers map known ideas and relationships onto the unknown in an attempt to resolve their problems, whether the mappings are valid or not.[26] Vincenti took this notion to a greater level of detail.[27] One of the best examples he presented is a case study on propeller testing: an empirical search for an optimal solution where engineering and scientific theory could not provide satisfactory explanations. He concluded that engineers use empirical searches guided by any theoretical understanding that happens to be available to build up knowledge that can be used for design. Most areas of engineering use 'rules of thumb', which are simple rules with little or no logical connection with scientific theories, but are nevertheless based on extensive practical experience. Depending on the courses they experienced, graduate engineers may only have a limited awareness of rules of thumb.

Our models don't need to be completely accurate, nor even correct.

Orr mentioned that photocopier repair technicians work with scientifically incorrect models of machine operating principles, but these models provide 'field tested' methods to find faults, given the symptoms of failure.[28] He also noted the systematic approach adopted by technicians, particularly carefully writing up their work in logbooks, parts, and order forms, ensuring that they keep their tools and parts neat and tidy. Chaotic work is associated with finding and solving problems by luck, rather than by competence, and is frowned upon in the broader community of technicians.

However, our exploration of engineering knowledge is still only beginning.

There are two key misconceptions that are worth mentioning here before we go too much further.

Misconception 7: Self-learning is needed because technology keeps changing

Students and many academics often think that . . .	However, research demonstrates, and expert engineers report that . . .
Lifelong self-learning is needed to keep up with technological changes. You need to be good at reading books for that sort of thing.	Engineers, even experts, need to learn on the job all the time. Most of what engineers need to know is learnt on the job, preferably from more experienced people, such as engineers and site supervisors, but occasionally in formal courses. Expert engineers learn every day of their career and even after retirement. That's why they have become experts.

Research demonstrates that engineers learn almost everything they need on the job. Every day brings a new learning opportunity for most engineers. Few have any idea of how much they still have to learn when they start work.

Technology changes certainly require you to learn to keep up with recent developments. However, as you will appreciate from the length of this book, you also need to acquire practical (implicit, tacit) knowledge and you will also need to learn about engineering practice: a subject that is rarely, if ever, taught in engineering schools. Most engineers have had to learn all this on the job as opportunities arose. You have the advantage of being able to read some of it in a book. By the end of this chapter, you will have a better idea of how much learning lies ahead: certainly far more than you have learnt at school and university.

Most young engineers prefer to ask their peers for help when they find that they need to learn something. It is much more difficult to lose face by being ignorant in front of a senior engineer when you are asking for help. However, it is not particularly wise to ask those people who are least likely to be able to help. Research has helped to demonstrate that the single most important factor that predicts the career success of young engineers is the ability to ask more experienced engineers for help and to learn from them – we will return to this idea later.[29]

Misconception 8: Grades matter

Students and novice engineers often think that . . .	Expert engineers know that . . .
Getting good grades at university makes you a better engineer. A high average grade demonstrates a better work ethic.	Workplace performance is unrelated to course grades, because practice requires so many other skills and abilities that are not learnt at university. The ability to learn these other capabilities determines success, for example, building collaborative relationships with more experienced engineers, suppliers, and site supervisors. Also, course grades depend on many factors other than academic ability and motivation.

Higher course grades can help you get the job you want, but only because employers are often faced with an enormous pile of job applications when they advertise a position, particularly on the Internet. When faced with 500 applications, which is not an unusual number, employers have little choice. Eliminating applications with lower grades is one of the easiest options. Also, some employers think that a student's grades are a measure of their 'work ethic', essentially representing how prepared a student is to work hard.

In reality, there are many factors that influence student grades, including the perceived relevance of the course, emotional pressures from home and family, the necessity (or choice) to spend time earning money in a part-time job, or even a full-time job, and other outside interests that may or may not be highly relevant to engineering. There are other ways to find jobs that don't require formal applications, meaning that they are less influenced by course grades; these are described in Chapter 14.

The more important issue is that, as we shall see throughout this book, an expert engineer requires many capabilities that are not explicitly taught, many which are not even mentioned in university courses. That's precisely the reason why you purchased this book: to learn something about them. While technical understanding and other abilities learnt by completing university courses are definitely valuable in any engineering workplace, they only comprise a small proportion of what an engineer needs to know and be able to do. That's why grades have relatively little, if any, impact on engineering workplace performance.[30]

You might be scared and apprehensive to read this: you may be someone who has worked diligently for high grades throughout all your courses. This revelation may leave you wondering why you bothered and why you worked so hard to achieve those marks. The ability to get high grades is certainly an advantage, but what you need to understand is that there are many other aspects of engineering practice that are just as important as your ability to perform well in examinations. High grades, by themselves, are not enough.

One of the most expert engineers I have been privileged to know actually gave up her university studies in Melbourne after her first year of engineering to work with a US-based multinational engineering firm. Despite that apparent academic failure, even in her late seventies and early eighties, she continues to be highly sought after by international companies anxious to benefit from her skills.

At that time, and it is still true today, the best way to learn about most aspects of engineering practice is from more experienced people around you, including engineers. The ideas in this book will help you more accurately make sense of what these people tell you. The book will provide you with the names of ideas that you can attach as 'labels' to what engineers may describe in seemingly vague or contradictory language.

Listen to what they tell you and emulate their performances by watching what they do and how they do it. Ask people such as site supervisors, technicians, operators, suppliers, and drafters what they need from you. Developing cooperative and helpful relationships with older or more experienced people can be challenging but is the best way to get you started on your quest to become an expert engineer. That being said, you will still experience the difficulty faced by many novices: having to endure the experience of seeming to be dumb or ignorant by asking simple questions. Simply make sure that the person you are asking will give you high-quality answers.

There is one thing that makes you look even more ignorant than having to ask a basic question: finding out later, often the hard way, that you should have asked someone for an answer to that simple, basic question at the start.

Finally, some evidence from the research literature on expertise must be discussed. One of the best predictors of grades in formal education is the traditional kind of IQ test, which measures cognitive intelligence. However, there are many other kinds of intelligence that measure technique. In Chapter 4, we learnt from the research literature on problem solving and expertise that there is no measure of intellectual ability that predicts problem-solving ability and expertise.[31] However, perception skills and knowledge of the task are widely regarded as being highly influential. In short, grades don't matter as much as much as perception and the ability to learn on the job, particularly from other people around you.

TECHNICAL KNOWLEDGE IN THE WORKPLACE

What do you need to learn once you start work as an engineer? Almost everything.

Case studies of engineering work can help us understand more about what you will need to learn. The case studies are based on detailed publications that provide further details.

Case studies also help provide further background for maps that include all the aspects of engineering knowledge that we have identified so far.[32]

I have included case studies from several different engineering disciplines: you can skip the less interesting ones, but make sure that you read the one on technicians, please.

Case study 1: Aircraft engine design

G. L. Wilde, a senior aircraft engine designer with Rolls Royce, argued for improving engineering education in order to help prepare engineering designers for the kinds of work that will be expected of them.[33] In doing so, he provided us with a first-hand outline of some of the technical knowledge that aircraft engine designers need. He advocated creating drawings with freehand sketching, learning to read drawings, and making parts, which consists of 'thinking with a pencil' with economy and completeness of communication – we already mentioned visual perception skills in the tacit knowledge section.

He identified relevant engineering science knowledge across related disciplines:

- mechanical design (form, function, material properties, stress analysis, and fatigue),
- thermodynamics, fluid mechanics, and aerodynamics (fluid flow, pressure, temperature, turbulence, and energy transfer),
- tribology (lubrication, wear, and surface treatments), and
- vibration (noise, acoustics, and vibration suppression).

Wilde also hinted at the contextual knowledge of materials, manufacturing methods, and working with dimensioned drawings.

He described the 'humiliation' felt by graduate engineers starting work in a large design enterprise with little contextual knowledge to complement their theoretical and analytical skills. He proposed a master-apprentice arrangement in which young design engineers would work side-by-side with the most experienced designers before being transferred to detailed product development and enhancement assignments. He reported subsequent discouragement and demotivation, associated with inadequate supervision and guidance, which led to the loss of young engineers and substantial company investment. Pressure on engineering staff to complete demanding technical work competed with the need to spend time with less experienced engineers to help them gain knowledge that, Wilde asserted, could have been more easily learnt earlier. He complained that design 'is a skill not sufficiently well understood by company specialists and mostly undervalued . . .'

This tension between the time pressure on experienced engineers and the need to spend time helping young graduates continues today. Part of this is because young engineers have little idea about what they have to learn: I expect that you will gain a much better idea of what awaits you by reading this chapter.

We could classify the different aspects of technical knowledge that Wilde mentioned in this way:

- Abstract knowledge, both mathematics- and science-based, from formal education to construct an abstract model of a machine, organism, or physical system needed to predict performance.
- Documentation, ways to represent the internal operation of machines, organisms, and physical systems.
- Manufacturing and assembly methods to produce finished assemblies.
- Design for manufacture: ways to design products that result in economic manufacture and assembly in a given context.
- Product definition, including how it works and how each of the components contributes to product performance.
- Knowledge of components and materials.
- Knowledge of component and material properties, both individually and in combinations.

Wilde went on to describe a design office hierarchy: designers, design section leaders, senior project designers, assistant chief designers, and a chief designer, all of whom were supported by specialist groups in stress analysis and engine systems, such as lubrication, fuel handling, and aircraft installation.

He then outlined a selection of the main technical constraints facing the designer, apart from the need to deliver the predicted technical performance within an agreed overall cost and delivery timescale.

Existing designs: the entire design builds on existing practice that is thought to be well understood, as it has provided hundreds of millions of incident-free flying hours (design from precedents was discussed in Chapter 3).

- Turbine blade shrouding, gas leakage, blade vibration, blade cooling, turbine inlet temperature and pressure (Figures 5.7, 5.8).
- Minimising turbine blade tip clearance, location of shaft bearings close to turbine discs to limit shaft deflection, and the use of turbine inlet nozzle vanes as the bearing support structure within the turbine housing.

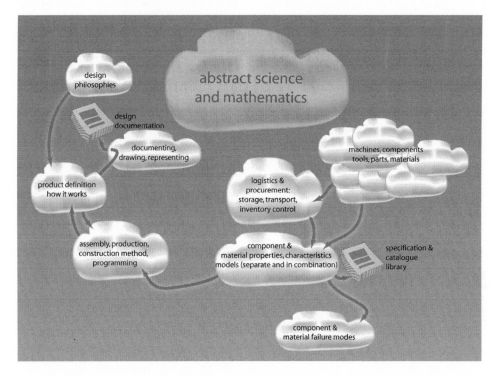

Figure 5.5 Knowledge map extended with fields of knowledge reported by Wilde – arrows convey the main associations linking different fields of knowledge.

- Adjustment of seals and clearances to regulate pressure and flow of cooling air to the internal engine components, such as bearings and turbine discs, handling of engine axial loads, and the effects of cooling air pressure on lubricant flow in the bearings.

Turbine blade-cooling air needs to be arranged to maximise convective cooling within the internal passages of the hollow blades and, by allowing air to pass to the outer surface of the blade through tiny holes, to provide a protective insulating film of cool air around the blade's external surface, ensuring that there is no chance of engine pressure fluctuations that would be strong enough to allow hot combustion gas (up to 700°C above blade material melting temperature) to enter the cooling air passages. While it is easy to describe the knowledge categories, much of the knowledge used in the design is tacit, unwritten knowledge.[34]

Aircraft engines require extremely high-quality blade manufacturing methods to achieve the highly complex, internal cooling air passages that are desired with material properties to prevent high temperature creep (gradual elongation) under high centrifugal and bending loads. This necessitates experimental investigation of different blade designs to find the best cooling airflows and internal passage arrangements. Experimental studies are also needed to investigate transient thermal expansion effects during rapid engine speed changes that cause changes in seal clearances, cooling airflows, and internal air pressure differences.

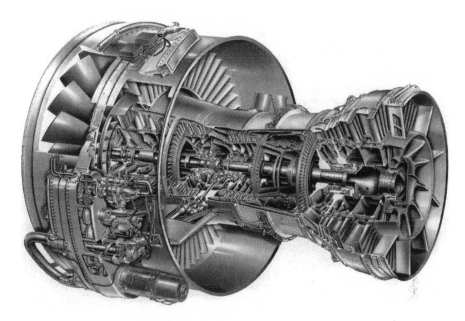

Figure 5.6 Large bypass ratio jet engine, typical on a contemporary passenger jet aircraft. The turbine at the rear of the engine is at the right-hand end of the picture (Rolls Royce, 1973).

Modern aircraft engines are expected to be extremely reliable. Failures are so rare that they often become headline news around the world. Engines will usually run for a year or more of continuous flying with only occasional inspections before they are taken into workshops for maintenance.

Wilde's account shows how the accumulation of all this technical knowledge is a dominant constraint on the capacity of designers. It takes young designers many years to accumulate this knowledge within the firm as they perform their design work: it cannot be obtained from other sources. Younger designers will acquire detailed knowledge in different technical aspects, but the design cannot be completed without taking the interactions between all these aspects into account.

This is not easy when the knowledge of all these different aspects is shared between many different specialist designers and the design process is constrained by limited time.

Wilde's account lies at one extreme of engineering practice: high technology design work where technical performance close to the ultimate limits of our knowledge is the desired outcome. Most engineers work further from ultimate performance limits, but under much tighter time and cost constraints. The accumulation of technical knowledge is still important, but there is much less time to learn, so access to shared knowledge becomes the principal constraint.

Adding to the previous list of technical knowledge aspects, we could classify the following:

- Previous and current designs of assemblies and products, and implicit knowledge of proportion.[35]
- Abstract models of products and assemblies that describe certain aspects, such as heat transfer, vibration, gas leakage, and fluid flow.

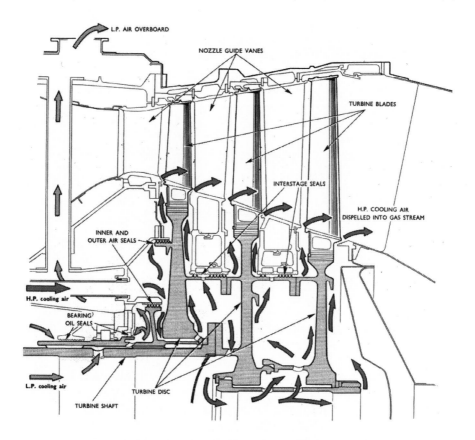

Figure 5.7   Cross section of the top half of a 3-stage turbine in a jet engine showing air cooling around turbine discs. LP (pale arrows) refers to 'low-pressure air' and HP (darker arrows) refers to 'high-pressure air'. The engine centre line is below the bottom of this diagram (Rolls Royce, 1973).

- Ways to control the production work environment to prevent disturbances and contamination, ways to organise work to reduce the chances of error, and how to maintain high-quality standards.
- Measurement, testing, and inspection methods for product and components.
- Diagnosis methods for product.
- Failure symptoms of product.
- Component and material failure modes.
- Failure modes of product.
- Production faults and defects in product and components, symptoms of 'trouble'.

Case study 2: Aluminium extrusion

Ravaille and Vinck described some much more mundane and routine engineering design work – designing aluminium extrusion dies.[36] The special aluminium sections used for window frames, for example, are all manufactured by extrusion from dies.

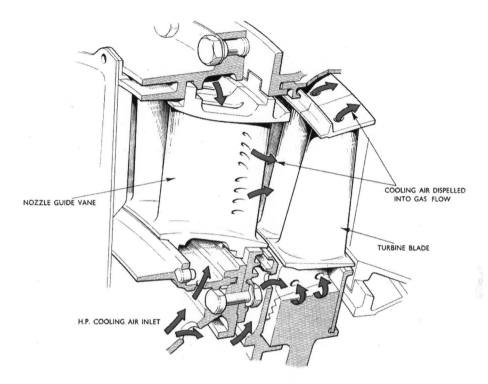

NOZZLE GUIDE VANE

COOLING AIR DISPELLED
INTO GAS FLOW

TURBINE BLADE

H.P. COOLING AIR INLET

Figure 5.8 View of nozzle guide vane and turbine blades showing cooling air flow. Modern turbine blades have many more holes and cooling air passages and operate in gas temperatures well above the melting point of the material (Rolls Royce, 1973).

Extrusion works just like toothpaste, and aluminium is extruded through the small hole at the end of the toothpaste tube. The die defines the shape of the final aluminium extrusion. Many aluminium extrusions have very complex shapes, and some are hollow.

Of course, there is a big difference between extruding aluminium and toothpaste. The special aluminium alloy, chosen to have the best possible properties for extrusion, is heated to become soft and easier to be extruded. It is forced through the die passages at great pressure. The die passages have to be carefully crafted so that the aluminium progressively forms the ultimate shape that emerges from the face of the die on to a conveyor that keeps the soft aluminium perfectly straight until it cools and hardens.

Although the researchers were more interested in the design process, their account reveals interesting aspects of daily work for three distinct groups of people: die designers, die makers, and extrusion machine operators. The designers and makers rely heavily on experience with previous designs. Each has a different knowledge domain with a small amount of overlap. The designers seem to have little appreciation for the wider context in which the machine operators live with all the constraints of a manufacturing operation. Different operators provide conflicting feedback on die performance, so designers tend not to take much notice of their comments. The operators never mention manufacturing (application) details that they think would be obvious to anyone in their position, so the designers never get to hear about them. Some operators change their dies and send simple two-dimensional sketches back to the designer, even though they have made complex three-dimensional changes that the sketches cannot adequately represent. The die makers have to interpolate the sketches and instructions from the designers. The designers provide a precisely detailed final extrusion shape, but fewer details on other aspects of the die. The designers have just enough understanding of the die-making process to be able to generate a viable design: the die makers don't let them know much more because they guard their own expertise and don't think the designers need to know the details anyway. Although the die makers craft a three-dimensional object with complex internal passageways, the authors reported that the designers almost exclusively thought in two dimensions.

Another interesting observation came from comparing two different companies specialising in die manufacturing. One company used more or less fixed design rules that were developed in a separate R&D department. They stored details of previous work in numerical sequence, making it almost impossible to refer back to previous designs, even if the designers wanted to. The other firm relied heavily on looking up and adapting previous successful designs. Designers used tacit rules to choose an earlier similarly shaped extrusion and would adapt the die for the new extrusion from the design of the earlier one.

We can see evidence for technical knowledge in several domains here. First, we should note that the die designers and die makers work for firms that sell dies to aluminium extrusion makers. The machine operators are clients that use the dies. This is common in engineering work: the client is also involved in engineering work. The die designers have some basic knowledge of how the dies are used, but face the problem that they have many clients with very different requirements and constraints. They choose to limit their knowledge of how dies are used in practice by taking little notice of comments and feedback from machine operators, on the grounds that the comments tend to conflict with each other. The machine operators, on the other hand, have much more detailed knowledge of their particular situation and how they actually use the dies purchased from the die-making firms. The die makers have their own knowledge of how to make dies. The designers share some of this, but only just enough to enable them to do the design work in a way that the dies can actually be manufactured. Interestingly, according to the authors, there is no satisfactory theory that enables the die designers to predict die performance. They cannot make accurate predictions, so they rely on knowledge of previous designs. One firm uses actual designs that designers can copy and adapt, while the other uses design rules that emerge from an in-house R&D department. The authors say little about the actual way in which the R&D department operates because their main concern was the die designers.

In this account we can find evidence of the following areas of technical knowledge, but the knowledge is not shared equally. Different people possess different aspects of knowledge, which may not be complete or even consistent with each other.

- Definition of product (designer, die maker, operator)
- Customer needs (technical) (operator, some with designer)
- Applications of the product (mainly operator, some with designer)
- Operating the product (operator)

- Maintaining the product (operator)
- Repairing the product (operator)
- Failure symptoms, signs of 'trouble' with the product (operator, some with designer)
- Failure modes of product (operator, designer)
- Diagnosis methods for product (operator, designer)
- Documentation techniques and standards, representing the product, ways to represent the internal operation of machines and physical systems (designer)
- Properties, models of product assembly for predicting performance and behaviour (note: in this case, this was not feasible because there was no satisfactory theory)
- Manufacturing methods, assembly of product, construction methods, time/cost, and resources needed (mainly die maker, some with designer)
- Knowledge of components, materials (designer, die maker, and some with operator)
- Locating required technical information in large amounts of mostly irrelevant written documentation (some designers)
- Design for manufacture: ways to design products that result in economic manufacture and assembly in a given context (designer)
- Previous designs for similar products and components, neat, well-structured, economic designs (designer, die maker)

Case study 3: Laboratory and photocopier technicians' knowledge

Technicians develop special kinds of knowledge that most engineers seldom think about.

In their 1997 collection of studies, Barley & Orr raised the issue of technical knowledge in attempting to define technical work.[37] They suggested that such work 'requires understanding and utilisation of abstract knowledge', presumably mathematics or science acquired through formal education. They acknowledge that many technical workers, particularly technicians and trade workers, acquire their knowledge and skills through experience and working with more experienced workers (e.g. apprenticeships), rather than through formal education. They also suggest that, 'technical acumen is the result of contextual and tacit knowledge'. In another chapter, Barley and Bechky identified many aspects of technical working knowledge required by biotechnology laboratory technicians. Doron also discussed similar lab technicians:[38]

'Technicians are usually charged with ensuring that machines, organisms, and other physical systems remain in good working order. Caretaking often requires technicians to employ theories, diagnoses, documentation, and other representations drawn from the symbolic realm they support.' (p. 89)

Yet this knowledge is much wider than the obvious specialisation of a technical worker:

'[Technicians] focused more on the instruments than on the cells they analysed, knowledge of optics, lasers, for instance, and computers was considered more crucial than knowledge of cell biology.' (p. 92)

'Highly developed tactile sensory-motor skills also seemed to be important.' (p. 99)

Avoiding sources of disturbance that could lead to mistakes also seemed to be an important issue for lab technicians. Such sources could be visitors, lack of concentration, being preoccupied with something else, being interrupted, doing new things, dirt, or contamination. Methods for preventing trouble included being well organised, almost to the point of obsession, knowing the proper place for everything (tools, instruments, and supplies), and careful scheduling of the work to avoid unnecessary time pressures.

> '*Troubleshooting was, therefore, both a discipline and a way of life among technicians and research support specialists. A panoply of undocumented practices for increasing the odds of a technique's success enveloped every procedure performed in the two labs and constituted most of the labs' routines. Although mentions of documentation, habitual cleanliness, rules of thumb, and strategies for recovering from mistakes or confronting enigmas were conspicuously absent in published discussions of scientific methods, the fortunes of the labs largely rose and fell by their exercise. Such a state of affairs challenges not only standard conceptions of the technicians' role, but prevalent images of the distribution of knowledge in science's division of labour ... To achieve this task, lab personnel constructed and employed a body of knowledge that most scientists lacked – a contextual understanding of materials, instruments, and techniques that was grounded in hands on experience.*' (pp. 115–116)

We could classify the different aspects of technical knowledge that emerge from these accounts like this:

- Abstract knowledge: mathematics- and science-based, from formal education, to construct an abstract model of a machine, organism, or physical system needed to diagnose performance problems.
- Documentation: ways to represent the internal operation of machines, organisms, and physical systems.
- Knowledge of tools and equipment required for a technician to perform work, how to keep the tools in good working order, and how to use the tools.
- Failure symptoms and signs of 'trouble'.
- Ways to control the production work environment to prevent disturbances and contamination, as well as ways to organise work to reduce the chances of error.
- Production scheduling and the effective deployment of productive resources.

Much of the knowledge is contextual knowledge and is drawn from first-hand experience. Much of the knowledge is also tacit knowledge, as the individuals cannot necessarily articulate it. Much of the knowledge, even if articulated, is unwritten, particularly working methods to maximise the chance of success without trouble. Some of the knowledge is readily applicable in another context. For example, computer knowledge and skills can be readily applied in other contexts, except, of course, for specialised software knowledge. Like the die makers, much of the technicians' knowledge is not shared with the scientists, who are the professionals attributed with mastery of the subject.

In reviewing the work of technicians, Orr described the wide range of information that he observed technicians using to help them diagnose problems.[39] They liked to talk to the actual users, not necessarily the customer representative. Users could provide more valid descriptions of fault conditions, even though they themselves did not have a correct understanding of how the machines worked and would use words that directly conflicted with the words that the technicians used. Built-in machine diagnostics and stored electronic logs recording error conditions and standard parameters also provided useful information. The technicians consulted their colleagues and the technical expert or specialist that was attached to each team of technicians. Their own experience was a useful guide, for example, knowing that certain wires could have different colours on different machines, perhaps because the manufacturer simply ran out of pink wire in the middle of a production run. Subtle variation in the appearance of the machine, its operating sounds, and even the feel of the machine and individual parts revealed subtle fault conditions more effectively than documented procedures. Even examining spoilt copies in the trash bin provided useful indicators about likely faults. Examination of the dirt deposits they removed as they cleaned the machine also provided valuable clues about the faults. Technical documentation and procedures seemed to be consulted a lot of the time, but it took a while before technicians knew which parts could be relied upon and which parts contained errors.

His account helps to draw our attention to more aspects of engineering knowledge, and particularly to 'unofficial' knowledge that technicians use in practice. Some of this knowledge, if it was passed on at all, only seemed to be relayed in 'war stories' at social gatherings, and addressed the following aspects of specialised knowledge:

- Human behaviour in design, production work, maintenance, and operation of the product
- Maintaining the product
- Repairing the product
- Disassembly and reassembly of the product
- Failure symptoms of the product
- Failure modes of the product
- Diagnosis methods for the product
- Measurement, testing, and inspection methods for product and components
- Procuring components and materials, storage, logistics, and transport

Case study 4: Planning a construction project

Sometimes, it is hard for outsiders to appreciate that engineers often encounter far too much technical information and knowledge. The challenge is relevance: which knowledge is needed and for what? In our research, we have seen plenty of evidence that engineering involves sifting through huge amounts of information to obtain relevant fragments that need to be reassembled in a way that is meaningful for a particular aspect of the work.

Winch & Kelsey[40] reported that British construction planners, particularly the ones working for trade contractors, tend to be overwhelmed with information, much of it not relevant to their role, yet they spend considerable time sifting through it. They described how planners find mistakes in drawings, for example, not allowing installation space for pipework. They described instances when insufficient space had been allowed for machinery to install piling close to property boundaries. This shows that the planners have to spend considerable time extracting information from drawings, specifications, and planning schedules. As they extract information and reassemble it in the form appropriate for issuing instructions to contractors, they find inconsistencies and mistakes.

When they find mistakes, they have four options. First, they can provide the corrected building details based on their own experience. Second, they can qualify the problem in the tender documents, which means drawing the attention of the architect and building developer to the mistake and letting them know that they will need to cover the cost of making changes to the design during the construction phase. Third, they can adjust the risk premium and consequently increase the price to cover the cost of making changes during construction, and not tell the architect or building developer. The fourth option is to negotiate a variation to the tender, the drawings, specifications, or the scope of work. Some planners quantify risks and present them as 'risk items' in the tender with individual prices attached.

We can classify some further aspects of technical knowledge here:

- Defining complete list of parts and materials required for given assembly and tools, as well as other equipment required for construction, production, and assembly.
- Locating necessary information in large amounts of mostly irrelevant written documentation.
- Finding mistakes and inconsistencies in plans and documents, correcting mistakes, and judging where it is possible to guess the original intent of the authors.
- Commercial negotiations with suppliers and clients.

In these knowledge categories, we see many overlaps with the kinds of knowledge used by foremen, often promoted after long experience as a technician, or engineers with an appreciation for practical 'shop floor' issues.[41]

Case study 5: Software engineering: a Japanese photocopier becomes European

Button and Sharrock illustrated further issues with knowledge in engineering in an interesting account of software engineering of the microcontroller for a complex mechatronic product – a photocopier.[42] The photocopier controller had been originally designed in Japan for Japanese office workers. The controller and displays had to be completely redesigned for the European market, because European office workers have different ways of using a photocopier compared with their Japanese counterparts.

They described how the team started with documentation describing the Japanese microcontroller from the machine designers in Japan, which was written in Japanese. They had this translated into English, only to find out that it was incomplete, with many important details missed, which they needed to design and develop software code for a new microcontroller. The team leaders were initially briefed by managers of the Japanese design team in English, but when they wanted to find out the missing details, they were referred to individual design engineers who, it turned out, could only communicate fluently in Japanese. E-mailed questions were translated into Japanese and, unsurprisingly, the responses from the Japanese engineers, translated into English, were still incomplete. In the end, they gave up asking and had to reverse engineer the photocopier themselves to figure out how it worked, and then set about designing the new microcontroller software.

This account illustrates a common engineering problem: the written documentation is often incomplete, even after requests for missing details that need to be supplied. It also reveals some surprising gaps in technical knowledge. Interestingly, only two of the programmers had previous experience in the chosen programming language (C). Some of the others had experience in a language called Sequel, used in earlier photocopier microcontrollers. Apparently experienced photocopier microcontroller software engineers are few and far between, even in a large photocopier company. The project was also stigmatised because it was referred to as 're-engineering' and, the authors report, this made it more difficult to recruit experienced engineers to the project. None of the experienced engineers had used CASE (Computer Aided Software Engineering) tools before and none had used the chosen design methods. The lack of experience with the C programming language forced engineers into ad hoc methods to learn the language quickly, but progress was slow, which came as no surprise. The slow progress was blamed on the particular choice of systematic software design method using UML, a software modelling tool. The engineers circumvented this difficulty by publically explaining that they were following the chosen design method by preparing documentation in UML, while, in fact, they were not constructing the code using UML at all. Instead, they were writing the code without using any systematic design approach.

Again, we can understand much of this in terms of different types of technical knowledge:

- Programming the product and predicting the time/cost of programming.
- Locating required information in large amounts of mostly irrelevant written documentation.
- Finding mistakes and inconsistencies in plans and documents, correcting mistakes, and judging where it is possible to guess the original intent of the authors.
- Knowledge of the tools and equipment required for technicians to perform their work, how to keep the tools in good working order, and how to use the tools.

In this case study, we can see how important it is to distinguish the technical knowledge in engineering. Labelling knowledge as 'old technology' so that the work becomes a 're-engineering' of someone else's design significantly reduces the prestige of the work. Engineers regard advanced technology and designing something for the first time as having much higher status. Engineers associate generating new knowledge, pioneering, and solving difficult problems with a higher status because they can tell 'war stories' about their heroic efforts to develop new technologies to their friends at social gatherings.

Case study 6: Japanese electronics and consumer product engineering

Knowledge dissemination and sharing practices vary from one country to another. In a study of British and Japanese electronics firms, Alice Lam revealed that Japanese engineers working on a new electronic consumer product spent much more time than their British counterparts on collaborative trial and error approaches, working together, and experimenting with real prototypes on their work benches.[43] Direct participation in the technical experimentation and development seemed to be taken for granted by Japanese project managers and information sharing seems to be more 'organic', with engineers spending up to a third of their time participating in that collaborative behaviour. The Japanese product development team involved a wide diversity of participants, including sales and marketing professionals and production line engineers. In British firms, she observed more solitary work, and an emphasis on written documentation and scientific design based on calculations. She also observed more of a 'silo' mentality, with vertically structured projects split into functional departments, with upper management being the exclusive holder of organisational and marketing knowledge. She found much less information transferral between different elements of project teams. Just like Button and Sharrock reported, she found that Japanese engineers worked informally with much more knowledge that was never written down. British engineers easily became frustrated with the lack of documentation when working with their Japanese counterparts, while Japanese engineers were unaccustomed to reading documentation provided by British engineers.

Case study 7: Waste disposal in food processing

In an interview with an engineer working in a food-processing factory, he explained how he had managed to make a big improvement in waste disposal from the factory with the help of the person who might have seemed to be the least significant person in the entire enterprise. That person was the old man who earned some pocket money sweeping up vegetable scraps from the floor of the transfer shed. That was where fresh produce in crates and palettes was unloaded from trucks and transferred to conveyors at the start of the production line. One day, the old man made a suggestion to the engineer. The next day, he came to work with several goats. At the end of the shift, when all the fresh produce had been unloaded, he brought the goats into the shed. The goats gorged themselves, eating all the fresh vegetable scraps from the floor. Then, he led the goats out onto the playing field adjacent to the factory, tethering them to stakes.[44] After some time, the vegetable scraps were converted into goat dung, a high-quality fertiliser that the goats deposited on the playing fields near the stakes. Every day, he put the stakes in a different position to ensure that goat dung was applied evenly across the entire stretch of playing fields. The local council paid the company for the fertiliser and for spreading it. The old man never had to sweep the floor again.

Case study 8: Impossible engineering

One of the striking findings from research for this book was a detailed historical account of engineering that showed the same patterns of technical thought and social interactions that we have observed in our own studies. This is the story of the *Canal du Midi* (Southern Canal), entitled *Impossible Engineering*.[45] The canal was built between 1660 and 1680 to link the Atlantic Ocean with the Mediterranean Sea, bypassing the British stronghold at Gibraltar. Proposals for its construction were initially treated with disdain, particularly as the main proponent was a prominent local tax collector, Paul Riquet, who had a dubious reputation and very little engineering know-how. Despite that, he persisted and eventually brought together all the different people who were necessary to make it possible. No one knew enough at the time to complete the project, even how to design a viable route through the complex hills and valleys in the foothills

Figure 5.9 Lock on the Canal du Midi.

of the Pyrenees Mountains that divide Spain from France. Furthermore, no one knew how much water would be needed to keep it operating through the dry summer of southwest France. Choosing a route meant finding appropriate elevations for each section of the canal, separated by locks, like the one shown in Figure 5.9. The route had to follow areas of relatively impervious soil and avoid sandy stretches to minimise seepage and water loss. The canal represented a departure from the established practice that restricted canals to relatively short routes between rivers across flat land. To make matters even more complicated, local people spoke a regional dialect and could neither converse with the French educated elite, including the military engineers whose expertise was essential, nor did they seem to have a common cultural background.

As Chandra Mukerji wrote:

'If learning depends on a basis of common culture, this raises questions about the Canal du Midi. Given the social and cultural divides among the participants in the enterprise, how did they find common ground to learn from each other? The answer is surprising. They shared remnants of the classical past; they all knew some elements of Roman engineering. Academics and military engineers studied literature written in ancient times, and scholarship derived from it. Locals, on the other hand, unknowingly reproduced elements of classical culture over generations, thereby making them taken-for-granted elements of daily life. The

Roman presence had been pervasive in the region ... they had built bath towns near hot springs with extensive waterworks, leaving behind a complex tradition of hydraulic engineering. The people did not see themselves as guardians of ancient knowledge, but they also had not abandoned ancient techniques of building or hydraulics just because Rome had fallen; they continued to use the skills pertinent to their lives, simply forgetting their provenance and treating them as common sense.... When they try to reuse stones from old buildings in new ones, it was easier to reproduce the classical forms for which they had been cut than to recut them, reproducing some ancient building techniques over time.'

The excitement engendered by the hugely ambitious project was infectious:

'Even labourers continued to work on the project toward the end when Riquet was broke and they were slow to be paid. The excitement of doing the impossible was intoxicating.'

Riquet drew on some unlikely sources of expert knowledge. Bandits and rebels who harassed or even killed tax collectors knew every centimetre of the landscape, so they knew how to hide and evade state law enforcement agencies. Riquet knew that these people had the knowledge he needed for the canal, so he included them in his enterprise, drawing on their intimate knowledge to help find the route for the canal and sites for the water supply reservoirs. Even local women became engineers to help finish the project.

'These were the people who knew what you could do in this particular country-side, its topography, seasonal rainfall, rock formations, rivers, forests and coastal sands.'

What we see here is a remarkable piece of engineering, far ahead of comparable projects that would take another century or more to come to fruition. However, it was only made possible because of distributed knowledge and expertise. No individual had the required knowledge, or the means to acquire it. The Canal du Midi worked because a large group of people came together and shared the knowledge needed for the project by talking and working with each other, all of them captivated by the size and sheer ambition of the project. They were bound together by excitement, an emotional bond with the project, and the community that they were serving.

This idea of distributed knowledge, so expertly applied more than 300 years ago for the Canal du Midi, is also the key to understanding contemporary engineering practice.

Practice concept 24: Distributed knowledge

After analysing all the interviews and checking the results with other published accounts of engineering practice like the ones above, I had generated a list of 37 aspects of specialist engineering knowledge and 27 aspects of general organisational knowledge. I constructed two maps to represent these, shown in Figures 5.10 and 5.11, found on the next pages. In the online appendix, I have included the full classification list, including descriptions of each aspect of specialist and organisational knowledge.

Note that these maps *do not* distinguish between the knowledge contributed by engineers and all the other human actors, contractors, accountants, clients, and government regulators. As an engineer, you must know about all the kinds of knowledge needed in an engineering enterprise, even though you will rely on other people for much of it.

What became apparent from the analysis of the interviews, maps, and descriptions surprised me in three ways.

First, even though I had extensive work experience as an engineer over three decades, the knowledge map was much more complex than I had imagined at the start. Over time, I had acquired my own knowledge in many aspects, but certainly not in all of them. Also, what knowledge I had was only for very specific classes of engineering enterprises, for certain kinds of products, services, and information. You need to remember that the maps have to be replicated for every single product, service, and information package provided by an engineering enterprise. While there are areas of common knowledge between different products and services, there are also many differences. Think of these maps in layers, partly merged, partly separated, a different layer for every product and service. The layers have to be subdivided yet again for producing the same product or service in different locations, or in different countries. This complexity explains why it is so difficult for any one individual to know everything, even for a single product or service.

I am sure that you can appreciate now that even in a lifetime it is not possible for an engineer to get to know all this 'stuff'.

The next surprise was that the technical knowledge that engineers reported using in their work was predominantly unwritten knowledge, learnt informally in the workplace. I have tried to capture this in the knowledge maps: whenever engineers reported *using* extensive written information (on paper, the Internet, or internal databases), I represented that with a sheaf of papers on the maps. Of course, in every category of technical knowledge, there was a combination of all the different types of knowledge explained earlier. Take, for example, knowledge of the product and how it works, located near the centre of the map in Figure 5.10. While this knowledge can be, and frequently is, written down, engineers reported that they carry this in their heads: they seldom referred to written descriptions. What became apparent was that engineers used unwritten knowledge, often not strictly correct according to scientific principles, far more than written knowledge. Furthermore, the range of knowledge categories in which each engineer had developed technical and organisational knowledge was quite limited.[46]

The third surprise was that, in one way or another, nearly every aspect of technical knowledge is required somewhere along the way in nearly every enterprise.

Remembering the earlier discussion on different kinds of knowledge, and that most of the knowledge that engineers *use* is unwritten knowledge, we can immediately appreciate that most of this knowledge is extremely difficult to transfer from one person to another.

Therefore, engineers are left with no alternative. Since the knowledge cannot be accessed by transferring it from one person to another, it can only be accessed by arranging for the people who have the knowledge to share it through skilled and knowledgeable performances. For example, production supervisors have much of the knowledge about controlling the production environment, as well as managing safety, quality, and minimising errors (just to the right of product function knowledge). Therefore, when it comes to the choice of production methods and the design of production facilities, engineers need to work with production supervisors, as they are much more knowledgeable.

Furthermore, strong evidence for this hypothesis, that knowledge is distributed among people in an engineering enterprise, came from observations by young engineers that they spend most of their time on social interactions with other people in the same enterprise. Most of these interactions concern technical and specialised knowledge. The real surprise was that the amount of time that young engineers spent on social interactions was almost exactly the same as the proportion shown by many previous studies of more experienced engineers. Here was strong evidence that social interactions are the means by which this knowledge is shared, largely by coordinating skilled performances of people who held the knowledge in their heads. That was the factor I had been searching for that explained why engineers spend so much of their time coordinating technical work performed by other people.

All this is apparent even in the few accounts presented above: engineering knowledge is distributed unevenly among the participants in each engineering enterprise. Remember the aluminium extrusion die designers, the tool makers who convert the designers' two-dimensional drawings to the three-dimensional tools made from hardened steel, and the extrusion machine operators who used the dies and modified them on the basis of their experience? They effectively collaborated with each other, yet they had little knowledge of the others' contributions and understanding. Instead, each simply made their individual contributions at the right time.

In just the same way, on a construction site, each kind of specialised tradesman comes in to do their work at the right time, precisely when everything is ready for them. The scaffolders come in and provide working platforms for the formwork erectors to construct the concrete formwork. Then, the reinforcing fixers have to complete their work, assembling the steel reinforcing mesh, before the concrete can be poured. This is also specialised work, often requiring concrete pumps, vibrators, and site workers who understand how to get relatively stiff, wet concrete to flow into every space it has to, under and around the steel reinforcing mesh. Engineers coordinate this work, often starting as site engineers working with an experienced foreman who can help the engineer learn about every aspect of construction. Nevertheless, it is still the engineer's job to ensure that all the materials and people are ready at the right time and place on-site, exactly when they are needed.

In setting up and managing maintenance in a complex facility, factory, or process plant, every building, road, pipeline, and separate item of equipment (e.g. centrifugal pump, valve, or even a mobile crane) has to be listed in a computer database. The computer system[47] contains data on the make and model number, location, size, dimensions, and weight, among others. One vital parameter is the equipment criticality: an indication of the importance of the equipment in supporting safe and reliable operation.

For example, equipment can be considered critical for many different reasons, such as:[48]

i.   failure of the equipment might result in one or more people getting hurt,
ii.   failure of the equipment might result in major environmental damage,
iii. the equipment provides early detection of failure in other safety and environmental critical equipment, allowing time to react and prevent catastrophic events, such as explosions or major oil leaks,

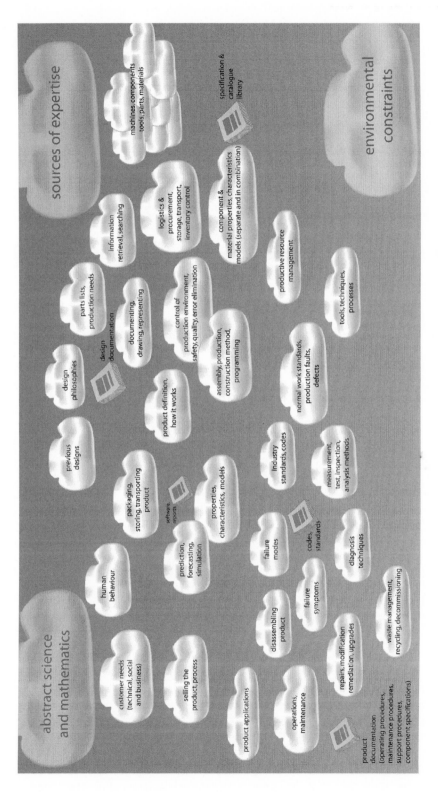

Figure 5.10 Aspects of technical knowledge used in engineering enterprises.

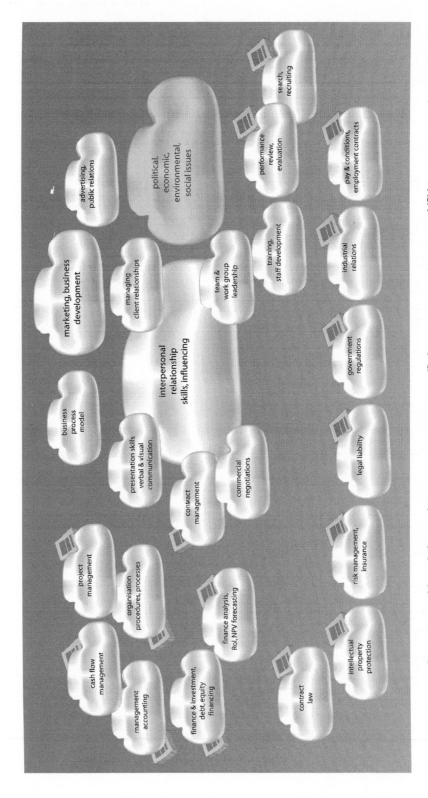

Figure 5.11 Aspects of organisational knowledge used in engineering enterprises (RoI is to return on investment; NPV is net present value, explained in Chapter 5).

iv. the equipment mitigates the consequences of failure of other safety and environmental critical equipment,

v. failure of the equipment will result in a large production loss,

vi. the equipment will be costly to repair,

vii. the spares required to repair the equipment have long lead times, possibly resulting in a lengthy out-of-service period,

viii. the equipment design or location will make repairs complex and potentially dangerous,

ix. the equipment cannot be replaced because of obsolescence – potentially requiring lengthy and costly redesigns of surrounding systems and equipment in order to replace it with a modern equivalent, or,

x. equipment failure will influence the reputation of the organisation.

Equipment criticality is used in many ways by different groups in an organisation. For example, it is used by reliability engineers and maintenance strategists in an analysis to determine what maintenance is required on the equipment, while it is also used by maintainers to indicate the priority of corrective and recurring work orders. Schedulers use equipment criticality to ensure that critical equipment is fixed before non-critical equipment, and that limited resources are prioritised for equipment that contributes towards safe and reliable operation. Spares analysts use equipment criticality to determine the minimum spares holding requirements to ensure that the spare parts needed for critical equipment are available on short notice. Managers use KPIs that track the open safety critical work orders to ensure that the overall plant is safe to operate.[49] Regulators use equipment criticality to ensure that organisations are aware of and can sufficiently manage the equipment that – if it fails – can hurt people or damage the environment.

Checking and inputting all this data is usually the job of junior engineers, typically in their first three years of working in the plant. However, they know very little about the plant, especially when they start. Remember the quote on the second page of this chapter: 'I had to ask somebody and figure out who to ask. I had a terrible lack of knowledge of actual "stuff".' Even the most experienced engineers seldom know enough to consistently assess equipment criticality. Although they know plenty about their own areas of responsibility, it would be unusual to find even one engineer who knows enough details about every aspect of a complex plant or a piece of machinery, such as a military aircraft, to be able to assess all the reasons listed above.

Despite that lack of knowledge, engineers, even junior engineers, manage to do this, albeit with occasional mistakes and inconsistencies. The only way they are able to do this is by working with an extensive social network that brings all the different people, each possessing part of the necessary knowledge, to collaborate together. Even though they are collaborating, most of them may never be in the same room at the same time.

As an engineer, you need to know that all this knowledge exists, that it is essential, and that someone will know it. You don't have to know it all yourself – that's impossible. However, much of your job will be coordinating the work of people who do possess the knowledge you require and making sure that they understand what's needed and when. It's like being the conductor of a symphony orchestra, except that you have to write the music and plan the performance, along with doing the conducting.

Here is another fundamental concept to learn from this discussion.

Practice concept 25: Knowledge is carried by people

Novice engineers often think that …	Expert engineers know that …
Asking someone else is like cheating. You are supposed to know it all and be able to do it for yourself. Even if I don't know something, I should know where to locate the information in a book or an online source.	Engineers rely on others to provide know-how and expert advice when needed. Experts take much less time to figure out what kind of expertise is needed, and it is much quicker and more effective for an expert to perform the work than to learn how to do it from an expert.

Some student misperceptions are soon displaced by the daily reality of engineering work. Young engineers soon learn the impossibility of knowing everything for themselves and the necessity to rely on and seek the help of other people. They soon realise, often with a degree of frustration, that much more time is needed to communicate with other people than they expected and much less time is spent on the solitary technical work for which they were trained. It doesn't take long for them to realise that they have to learn a lot every day, that the logical development of ideas is much more difficult than it seems, and that the boss is much happier when he finds out that you have already done what is needed without having to be told. Many people learn to work effectively, despite being in dysfunctional teams, which can be more frequently encountered in workplaces than in university team project exercises.

One of the greatest misperceptions prevalent in both practice and published literature concerns the idea of 'knowledge management', which suggests that knowledge is something that an enterprise can capture in written form, store, and distribute to where it is needed. In engineering, far more useful knowledge is carried in the heads of people than is ever written down. People remember this knowledge precisely because it is useful for them to remember it: it is needed sufficiently regularly that looking it up, even on the Internet, is unnecessarily time-consuming.

Acquiring explicit codified knowledge from texts, as explained previously, is time-consuming, difficult, tiring, and highly susceptible to errors and misunderstandings.

Also, that's only for explicit knowledge. Unwritten tacit, implicit, procedural, and contextual knowledge cannot be transferred easily at all: those more intangible elements of knowledge can take years of special training to acquire.

That's why it is much better to get people with the knowledge to contribute it directly by performing those parts of the work that require that special knowledge. Much of this is organised informally using technical coordination, which is explained in more detail in Chapter 9. An engineering enterprise, in other words, is a network of people with special knowledge and expertise. While engineers have special technical knowledge in their own right, it is understandably insufficient, given the massive scope of relevant knowledge in an enterprise. The enterprise only works well when everyone can access the knowledge they need by working collaboratively with other people.

There are two ways for other people to contribute their knowledge and skill.

The first is to arrange for people with the knowledge to present and explain it to you when you need it. A typical example is relying on an in-house expert or an external consultant who has the technical expertise. Even though consultants can be expensive, it is almost always cheaper than attempting to acquire the necessary level of expertise yourself.

The second is to arrange for an expert to contribute knowledge through skilled performances.

Let's consider a simple example. A skilled pipeline welder using appropriate equipment can reliably produce high-strength welds between sections of pipe. It is usually much more cost-effective to engage an expert welder than it would be to engage an expert to teach inexperienced welders how to accomplish the same task. This is what I mean by a skilled performance. It can apply equally well in the context of engineering analysis. It makes much more sense to engage an engineer who is highly experienced in designing and analysing bolted joints in a situation where a new design for a bolted pipeline flange joint is needed. It is also beneficial for a novice engineer to watch and learn as he does this: the novice may feel inspired enough to become an expert in their own right.

Practice concept 26: Value of unwritten knowledge

One obstacle that you will need to overcome is remarkably deeply embedded in the ways that we think, particularly in academic circles.

Novice engineers often think that …	Research demonstrates, and expert engineers know that …
Knowledge is only valid if it is written. Other kinds of knowledge are 'subjective' and of less value. Ideas in people's heads don't constitute knowledge until they have been written down.	Engineering relies as much, if not more, on unwritten knowledge and tacit knowledge as it does on explicit, written, codified knowledge.

All through your education, as a student, you achieved nearly all your marks for providing responses on examinations, tests, and quizzes, along with preparing written reports. On rare occasions, you may have been marked on the basis of a performance, such as the quality with which you made a component in a workshop or delivered a spoken presentation. Almost invariably, however, students regard these marks as subjective and often complain that these marks should count less than 'objective' marks based on written answers.

For thousands of years, philosophers have valued rational understanding and abstract reasoning expressed in written logic above contextual knowledge and practical expertise in the minds of experts.[50]

The written knowledge accumulated by Western civilisations has traditionally been viewed as 'science' and 'philosophy', representing a much more advanced civilisation than, for example, tribal cultures based on oral traditions like those found in Australian Aboriginal people. Yet, Aboriginal languages typically include roughly 300 terms for different family relationships, making European languages seem rather primitive in comparison.[51]

As we saw in the maps reproduced in Figures 5.9 and 5.10, engineering relies on around 60 identifiable aspects of specialist knowledge, much of which is contextual, in the sense that it is specific to a particular situation and is mostly unwritten, carried in the minds of people. People acquire knowledge for a large number of different components, processes, places, firms, countries, regions, and geographic environments. Once we realise this, we can begin to appreciate just how much knowledge is carried in the minds of people in an engineering enterprise, more than could ever be feasibly written down or learnt by others.

One disadvantage inherited from our Western philosophy is that we tend to underestimate the value of unwritten knowledge. There is no better illustration of this than the typical engineering hierarchy represented by:

i.   Engineer
ii.   Technologist
iii. Associate or Technician
iv.   Tradesperson

The engineer is always placed at the top of this hierarchy. The longer period of formal education required to become an engineer reflects the greater value placed on explicit written knowledge. Technicians typically attend a college or a technical institute for two or three years at the most and are therefore often regarded as possessing a smaller subset of engineering knowledge and competencies. In reality, skilled technicians take years to accumulate 'hands-on' expertise that is tacit knowledge: knowledge of which most people are unaware, even the technicians themselves.[52] Almost invariably, this knowledge is regarded as being less valuable than written knowledge.

Another way to understand the knowledge carried in the minds of technologists and technicians is to see it as complementary to formal, explicit written knowledge. Engineers need technicians to work with and mediate between them and information systems, process plants, machinery, and other artefacts. It is more appropriate to see the hierarchy laid on its side, representing each group of people collaborating with the others:

Engineer | Technologist | Associate or Technician | Tradesperson

Reviewing each of the brief case studies above, we can see how a great deal of the knowledge those situations require is not written down.

Practice concept 27: Distributed cognition

Novice engineers often think that …	Research tells us, and expert engineers indicate through their actions that …
An engineering hero solves the problem by himself, at the very last moment.	New engineering knowledge emerges from the social interactions between people with complementary knowledge, working on similar issues together.

We have already discussed the idea that expert engineers know that it is more cost-effective to call in an expert than to try and acquire the necessary technical expertise in a hurry.

Here, we take this idea one step further. The notion of distributed cognition implies that new ideas and knowledge can be socially constructed: it results from the social interactions of people with different and complementary expertise, knowledge, and skills.

I am referring to the process by which a group of people with overlapping and complementary knowledge work together to construct new shared understandings that could never have arisen without social interactions. The story of the Canal du Midi is just one of many examples that show how people can achieve the seemingly impossible, simply by working together, building new knowledge through their conversations, and sharing stories with each other. This is remarkably similar to certain modern techniques commonly used in classroom learning.

Ann Brown and her colleagues reported behaviours in collaborative learning that are remarkably similar to everyday behaviours that we observed in engineering workplaces, as well as behaviours that we saw in the Canal du Midi project described above.[53]

This is not easy. It depends on building cooperative relationships between the participants, which are critically important, but take time to build and maintain. Each participant uses different terminology, or even the same terminology but with different meanings. It can take days or even months of patient negotiation to come to a shared understanding that represents the new knowledge shared between the participants.[54]

> *'Children (like young engineers) can participate in an activity that they cannot yet fully understand, and the teacher doesn't need a full understanding of the children's understanding of the situation to start using their actions. The child is exposed to the teacher's understanding without necessarily being directly taught . . .. Learners interact with each other as they develop at different rates.'*

As they do so, the meanings of words is constantly changing, often imperceptibly, as the participants develop new understandings and, through listening to each other, see new associations with the words that were out of their reach before. As Brown wrote:

> *'Meaning is constantly being re-negotiated – conjecture, evidence, and proof become part of the common voice of the community. Conjecture and proof are open to re-negotiation in many ways, as the elements of which they are composed, such as terms and definitions, are re-negotiated continuously. Successful enculturation into the community leads participants to relinquish everyday versions of speech activities, replacing them with "discipline embedded special versions" of otherwise ordinary activities.'*

As we shall see, it takes a special set of skills to make this happen. While many engineers see social interactions as a non-technical aspect of their work, in reality, these social interactions involve highly technical discussions. Much of this occurs outside of working hours in relaxed social settings, and highly technical conversations can often be intermixed with jokes and speculation about upcoming sporting events and 'small talk'.

I would like to try and explain this using a set of diagrams. The first diagram shows part of the map of technical knowledge from Figure 5.10. Remember that this is only part of the story for a single product, and there is a complimentary map of organisational knowledge to go with it. A female engineering graduate has just joined the enterprise. As a result of formal studies, she has a firm grasp of abstract science and mathematics, shown in the top left-hand corner. She can apply this knowledge in performing prediction and analysis calculations using mathematical models and computer software, but she can't do this in isolation. She needs a lot of other knowledge in order to come up with realistic estimates for the parameters that define her mathematical models.

Fortunately, she is not on her own.

There are several other engineers that cover the other aspects of the technical knowledge needed by the enterprise. Our young graduate will need to spend a lot of time working with them and figuring out realistic estimates for the mathematical models at the centre of her work. Of course, this is a hypothetical situation. She might just as well have been placed at a production site, learning to organise basic maintenance tasks or production work under the guidance of a foreman or another experienced engineer. In that kind of situation, she would know even less: hardly any of her university education would be directly applicable. She is completely in the hands of the other people around her, but she is a fast learner. She soon learns that almost everything she needs to know is already known by someone else in the organisation: it is just a matter of finding out who that person is.

There is something else that's very important to understand. What we have represented with the map of technical knowledge that forms the backdrop of Figure 5.12 is the knowledge that might be relevant for one particular product or project. However, engineers are seldom dedicated to one particular project at any one time. Each engineer will have many different tasks, many of them corresponding to e-mails waiting in their inbox that they will need to attend to.

If she is relatively inexperienced, or perhaps a little shy, our young graduate may start off by sending an e-mail to one or two of the engineers asking if they can help her pin down some rough estimates for her modelling work. She may not realise that her e-mail is just one of 50 or 60 matters vying for attention from each of the other people we can see in the diagram. They don't recognise her name immediately, so it goes to the bottom of their priority lists and may even be forgotten in a few days.

On the other hand, if she has already had the courage to find these people and talk to them in person, asking about their interests and some of their stories about working in the organisation, her e-mail has a much better chance of being opened and read. In other words, if she has developed a cooperative relationship with the people from whom she wants information, they are much more likely to respond. If she calls them on the phone or, even better, passes by their desk or cubicle a day or two after sending the e-mail (assuming that she has not received a response from them), she is even more likely to get a response. In fact, it may take some face-to-face discussion to elaborate on the original e-mail message . . . 'Oh, now I understand what you're asking for! I couldn't quite work it out from your e-mail.'

The next diagram attempts to illustrate this situation using a network of pipes with valves as an analogy for relationships between people within the organisation. We could just as well use wires with transistors to represent the relationships. The important aspect is that the valves or transistors modulate the interactions between the individuals at each end of the link. If they are open, meaning that a cooperative relationship has developed, then they will be more likely to give priority to responding to each other's requests over all the other matters that they have to deal with on a day-to-day basis.

Figure 5.12 A graduate (female, centre) with abstract science, mathematics, and modelling knowledge has joined several other engineers in the enterprise that possess different aspects of technical knowledge. Between them all, if they collaborate, they can access all the knowledge they need.

Depending on the valve settings, each pair of people in the network will have a stronger or weaker tendency to collaborate. We could represent this tendency with lines of varying thickness, as I have attempted to portray in the next diagram.

It is important to understand that a high-quality collaborative relationship between two individuals does not necessarily depend on friendship. Granted, friends could more easily develop a productive working relationship, but it is also quite possible to have a cooperative working relationship with people who are not your friends.

It is also important to understand that collaborative relationships don't simply develop by themselves. It takes time, effort, and some basic people skills that you will learn more about later in this book.

Practice concept 28: Social relationships lie at the core of technical engineering

So where does all this leave us? We started this chapter focusing on the technical knowledge that distinguishes engineers from other people. We have now reached the stage of realising that even the most technically oriented engineering work relies on a network of personal relationships. In fact, the more specialised and narrow the field of

Figure 5.13 The same group of engineers. We could represent the quality of the relationship between each of them by pipes with valves. The valves may be open, closed, or somewhere in between. As shown in the lower right-hand corner, the valve position can be changed by the person at either end of the relationship.

technical expertise, the more an engineer will have to rely on other people to provide all the other knowledge and expertise that they need for their ideas to reach fruition.

This is the most important concept to remember from this chapter.

We can now see the fallacy in the very common perception of engineering students and recent graduates that learning about social interactions and building relationships has nothing to do with engineering.

Misconception 9: It's all psychology and management, not engineering

Students and novice engineers often think that …	Research, and the actions of expert engineers, demonstrates that …
Interaction between people has nothing to do with engineering: that's psychology or management. If it's not a technical issue, then it's a management issue, not an engineering issue.	Engineering practice relies heavily on people: engineers need help from other people and only create value through the work of other people. Interactions between you and other people will define your success, just as the strength of steel constrains the height of a skyscraper.

Figure 5.14 A hypothetical map of the social network at any one time.[55] Thick lines represent strong relationships with a strong tendency to collaborate. Thinner lines represent weaker relationships: this is where there could be problems within the organisation. People who don't easily work together and bounce ideas off each other could represent blockages that impede the enactment of distributed cognition within the network.[56] We can see that our young graduate at the centre of the diagram has developed strong relationships with two of her peers and weaker relationships with four others. There are also people in the network with whom she doesn't have any relationship with yet: for the time being, she will have to work with them indirectly, with the assistance of some of the others. Large enterprises, therefore, require coordination tools and IT systems to enable newcomers to work beyond their immediate work group and find knowledgeable people within the larger enterprise.[57]

Getting used to the idea that you will have to rely on other people is the most difficult obstacle to overcome in this chapter. As an engineer, you will hardly ever lay your hands on the artefacts that your work helps to create: other people will do that. You may not even see them. You and they may be on opposite sides of the globe, and you may have moved on to many other challenges before they are even built.

Engineers rarely perform any hands-on work, as we have already seen in Chapter 3.[58]

I often repeat this advice: if you ever see engineers with tools in their hands, doing something, particularly if they are working on a bridge, then move away to a safe distance ... quickly. Something unusual is happening.[59]

Engineers rely on highly skilled people: technicians, drafters,[60] artisans, contractors, operators, maintainers, programmers, designers, and tradespeople. These people *mediate* between engineers and the systems that engineers create and work with. These can be machines, information systems, process plants, factories, materials, specialised tools, or vehicles: in skilled hands, they all become human extensions of the self. As an engineer, you will never have time to develop the level of hands-on skill that these people nurture through years of experience. As a civil engineer, you will rely on skilled surveyors to provide the data from which you can begin your design work, skilled drafters and designers to work with CAD systems to produce working drawings, and skilled operators to control machines and vehicles to make your vision actually happen in reality. You may have heard people say, 'Engineers construct bridges, roads and buildings . . .', but in real life, engineers almost never actually construct anything for themselves, except perhaps a backyard barbeque.

As an engineer, you will rely on those other people to turn your ideas into reality. Your ideas can only become useful through the minds, hands, and actions of other people.

Unless you are very experienced, and have legitimately become an expert engineer, you won't even be able to tell these mediators what you want them to do. Instead, you will be relying on those other people to already know what can be done and, equally importantly, what cannot be done.

The way you interact with other people will determine how effective your engineering contributions can be.

While theories of psychology, management, and other areas of academic knowledge can help you, they are insufficient in the context of engineering. I challenge you, if you want to contest this point, to demonstrate that any more than a few pages of this book can be learnt from psychology or management courses.

Why is this?

Management and psychology theories provide useful guidance for working with other people, but they do not take engineering technical expertise and skill into account. While some theories apply equally well for seamen and software engineers, engineering practice does not fit well with most of them.[61] In this chapter, we have just begun to see where and why this difference exists. Engineering technical knowledge is something profoundly different that management researchers and psychologists have found difficult to understand.

We are almost ready to move on and delve even deeper into the special nature of engineering knowledge.

In the next chapter, we will revisit some fundamental skills for acquiring technical knowledge that tend to be overlooked in university courses. By the time you graduate from an engineering course, you should have developed effective learning skills for explicit written knowledge. However, as we have seen in this chapter, much of what you need to learn in the future will be unwritten, implicit, and tacit knowledge, often embodied in artefacts in a way that requires prior understanding to recognise it. For this kind of learning, you need a new set of skills, different from the ones that you used at university. You may think that you are a 'good listener', that you have good visual skills, and that you have top-notch reading skills. However, have you ever thought to test these skills? We'll do that in the next chapter.

Practice exercise 2: Mapping your knowledge network

In the following pages, there is an abbreviated list of some different aspects of engineering technical knowledge.

For this exercise, you need to select an engineering context in which you have first-hand experience. If you are already working, choose one of the main products or services provided by your firm, one on which you are working at the moment. Otherwise, think of a context that might be related to your past experience or perhaps a project based on your capstone project in the final year of your university education.

For each aspect of technical knowledge listed in the table, try and identify individual people that you could go to for expert knowledge of each aspect. These people may be in the organisation that you are or were a part of, or they may be outsiders.

When you have completed drawing up your list, try to sketch a network similar to Figure 5.14 that would illustrate the quality of the relationships you had with each of these people, as well as the relationships that the people had with each other.

You can assess the quality of the relationships between each of these people by trying to measure the following aspects of their interactions.

Classify them on the scale of face-to-face meeting time and frequency that follows.

1. More than 2 hours daily
2. More than 30 minutes/day, or more than 2–3 times per week
3. More than 1 hour/week, meet at least weekly, catch up by phone if necessary
4. More than 10 minutes/week, at least once every three weeks, by phone if not present
5. Meet for 30 minutes–1 hour, every three months or so, by phone if not present
6. Meet on odd occasions, more by chance than design, know face and name
7. People are physically distant (e.g. different cities, countries) but would get together for at least a meal if they knew they were in the same place at the same time. Typically meet at least once every 6–12 months, but for extended time and conversation. Can be confident that they will reply to each other's e-mail requests

Now rate the strength of their relationships on this scale:

a) Highest degree of trust and confidence; each would give the other access to their bank account if needed.
b) High trust and confidence; each would recommend them to other people on their contacts list. If they ask them to do something, they could be confident they would help.
c) Medium trust level; each would have some reluctance if the other one asked for contacts, but would provide information, if needed. Each would not be completely confident that the other would help when asked, but they probably would do something, although they may need reminders.
d) Low trust level; each would feel reluctant to pass on names of contacts, might help the other if asked, but neither could not be confident about that.

Now draw a map using these ratings. Draw lines between each person, with the thickness inversely proportional to the number scale 1–7 (i.e. frequent contacts have thick lines). If you can, colour the lines according to the scale: a, b, c, and d.

Survey of Expertise.

Context (productor service)

Technical Expertise

Category	Description	People with this expertise in the firm	People outside firm
Definition of product	Knowledge of product details and components, function and location of each component, and how each component contributes to the function of the whole assembly.		
Customer needs (technical, social, business)	Understanding of individual client needs, including technical requirements for performance, life, maintainability, compatibility, industry standards etc., understanding client social needs, both within the local community and as individual people, understanding client business needs, such as price, financing method, delivery timing, client organisation logistics, business rules, installation, maintenance, service and support, repairs etc.		
Selling the product	Knowledge of appropriate techniques to represent the product to a client so that decisions can be influenced; knowledge of aspects of the product that are important to a client and how to describe and present these (also see business development aspects).		
Applications of the product	Knowledge of how the product is applied and used by the client's personnel. The client is more likely to have detailed knowledge than the engineers developing and making the product.		
Operating the product	Knowing how to operate the product safely and correctly to obtain the desired performance and avoid damage, how to recognise operational problems, how to start the product, how to stop the product and shut it down, how to 'mothball' the product for extended storage, how to protect the product from external environment or potential damage, and how to care for the product (other than maintenance).		
Maintaining the product	Knowledge of maintenance procedures for individual components and for the finished product, including the labour skills and knowledge required, appropriate maintenance planning and scheduling techniques to be used, spare parts management, procurement, and logistics.		

(Continued)

Survey of Expertise. (Continued)

Context (productor service)

Technical Expertise

Category	Description	People with this expertise in the firm	People outside firm
Repairing, remediation, modifying the product	Knowledge of repair methods to restore a damaged product to working condition, including repair methods for damaged materials and components. Modification of the product to obtain improved, even satisfactory, performance in a given application (knowledge of modification methods typically lies with users and may not be known to the makers or producers).		
Disassembly, reassembly of product	Disassembly and reassembly knowledge required for inspection, installation of upgraded components, and repairs.		
Failure symptoms, signs of 'trouble' with the product	Problem indicators, cues and symptoms can include unusual noises, smells, heat, and visible fragments (chips, smoke, rubbings, flakes, fragments, dust, particles – particles may be attracted or retained by secondary effects, such as wet or oily surfaces). Evidence of heating can include discolouration, burn marks, heat haze, and shimmering, as well as direct tactile contact or temperature indications. Visible damage can include cracks, dents, scratches, marks, rubbing, bends, crazing, discolouration, wrinkling, distortion in light reflections, and burn marks. With electronic equipment signs of trouble can include intermittent failures, heating, and even audible noises. Problems can also be evident from operator comments, even the contents of trash bins, for example, discarded copies from a photocopier.		
Failure modes of product	Symptoms suggest specific failure modes, given an understanding of the product and how it operates. For example, a sticking valve in a process plant can cause chemicals to spill from a tank. The failure symptom is spilled chemicals. The failure mode is a valve failing to operate correctly.		
Diagnosis methods for product	Technical methods for collecting data and analysing data to determine the cause of performance loss or failure. This also relies on having a model of how the product functions and relevant physical principles.		

NOTES

1. I use the word 'most' because there are some engineers who have remarkable in-depth knowledge of more than one major engineering discipline.
2. Various studies (R. F. Korte et al., 2008; Larsson, 2007) and our own (Trevelyan & Tilli, 2008) demonstrated just how much is learnt from others, and how important it is to seek the help of others in the workplace. Eraut et al. (2007) demonstrated this across several professions in a study of engineers, accountants, and health workers.
3. In an attempt to build a theory of engineering practice that provides an intellectual foundation for this book, I have argued that 'engineering' itself only makes sense in the context of an engineering enterprise (Trevelyan, 2014b). I have departed from earlier definitions, such as Layton's and Treadgold's, based on the notion of making practical use of applied science and optimisation (Layton, 1991; Rogers, 1983) or of productive activity to meet human needs (Mitcham, 1991). My observations of engineers at work demonstrate the overwhelming significance of collaborating with and influencing others in the context of a formal or informal organisation. To exclude these aspects denies the validity of most of what I and others have observed engineers engaged in. Even though many engineers see these other aspects as 'not real engineering', they also concede these aspects are essential and require engineering insight: they mostly cannot be handled by non-engineers. This strengthens the case to adopt a revised definition of engineering based on observed engineers' workplace performances.
4. Logistics – activities associated with procurement (buying) of components and materials, transporting and storing them, and monitoring their use and consumption to ensure that there are adequate quantities ready to use whenever they are needed.
5. 'Expertise' here denotes knowledge and skills that characterise an expert. The fact that consultants are paid for their knowledge input is a form of recognition that marks them as experts. However, there are many other expert engineers in other kinds of employment.
6. Rescher has provided a discussion on knowledge that is easier to read than most philosophical writing (2001).
7. There are many discussions on classifying technical knowledge in the literature (e.g. Gorman, 2002). Others, like Gainsburg and her colleagues, have classified technical knowledge and also when they observed engineers using it (Gainsburg et al., 2010; Henriksen, 2001). This section of the chapter primarily draws on my own research, but I have used excerpts of published accounts to illustrate in different ways how my own ideas play out in practice.
8. This discussion is adapted from an influential paper by Nonaka (1994, p. 15).
9. To do this, try and estimate the amount of computer storage needed for this. Remember that a full-length high definition video movie requires around 10–20 Gigabytes, say 1.5×10^{11} bits of information.
10. To learn more, read about different approaches to epistemology, the nature and scope of knowledge, and ontology, the nature of reality, what actually exists, and how we classify different approaches to these ideas. Both are traditionally concerned with propositional or explicit knowledge that can be explained in written language. However, for this discussion, we need also to understand knowledge that influences human action and perception, forms of knowledge that are more difficult to represent with written language.
11. Collins (2010).
12. Polanyi (1966, p. 4).
13. Collins (2010).
14. Ewenstein and Whyte have discussed this in the context of building design (2007) and Ferguson has drawn attention to a similar appreciation for form in mechanical design (1992).

15. The modulus concept in which a design for a particular application is obtained by adapting the proportions used in earlier designs has been widely used since the early 19th century, and can be traced back several thousand years in architecture (Guzzomi, Maraldi & Molari 2012).
16. Susan Horning has described this in the context of the work of sound engineers (2004).
17. Gainsburg (2006; Goold, 2014).
18. Razali & Trevelyan (2012; Sternberg, Wagner, Williams, & Horvath, 1995).
19. Latour (2005, pp. 204–209).
20. Shippmann et al. (2000, p. 712).
21. e.g. Bucciarelli (1994, pp. 133–135).
22. SAP: One of the most widely used enterprise resource management software systems, commonly used in large enterprises.
23. http://www.mech.uwa.edu.au/jpt/pes.html.
24. Known for the idea of communities of practice (Wenger, 2005).
25. Prior learning needs to be reactivated in order for it to be useful, and connected with the procedural knowledge needed to apply it (Ambrose, Bridges, Lovett, DiPietro, & Norman, 2010).
26. Schön (1983, pp. 184–187).
27. Vincenti (1990, Ch. 5).
28. Orr (1996).
29. Dahlgren, Hult, Dahlgren, Segerstad, & Johansson (2006); Eraut (2007); Eraut, Alderton, Cole, & Senker (2000); D. M. S. Lee (1994); Martin, Maytham, Case, & Fraser (2005); Moore (1959).
30. Gibbs and Simpson (2004, p. 7) reported that exam assessment is almost unrelated to subsequent recall. Newport and Elms (1997) reported that engineering work effectiveness is unrelated to academic assessment. Lee (1986) reported that academic performance was unrelated to engineering graduates' performance in their first job as assessed by supervisors.
31. For a review of these studies, read Wenke & Frensch (2003).
32. We are still working on this research.
33. Wilde (1983).
34. Wong & Radcliffe (2000).
35. Guzzomi, Maraldi & Molari (2012).
36. Ravaille & Vinck (2003).
37. S. Barley & Orr (1997).
38. Doron & Marco (1999).
39. Orr has described technicians troubleshooting (1996, pp. 98, 105, 114) with remarkable parallels in the work of sound engineers (Horning, 2004). We can clearly distinguish engineers whose work leads them to become highly engaged with machinery and systems in performing their work, often acting as intermediaries or mediators for other engineers who work with more elaborate social networks.
40. Winch & Kelsey (2005).
41. Mason (2000); Prais & Wagner (1988).
42. Button & Sharrock (1994).
43. Lam (1996, 1997, 2000).
44. Tethering – the goats had loops of rope around their necks. The other end of each rope had another loop, which was attached to a stake in the ground. Each goat was free to walk around the stake but could not go further than the length of the rope allowed.
45. Mukerji (2009).
46. Rachel Itabashi-Campbell and her colleagues revealed that earlier studies identifying 'bounded rationality' applied equally well to engineers in problem-solving situations. Engineers had great difficulty in expanding their thinking boundaries beyond their immediate

day-to-day job concerns. Successful problem solving, it turned out, only happened when engineers were able to accomplish this, and 'see the bigger picture'. This requires considerable learning and effort, and the opportunity to set aside normal organisational constraints (Itabashi-Campbell & Gluesing, 2014).

47. CMMS – computerized maintenance management system.
48. Leonie Gouws, personal communication (2012).
49. KPI is Key performance indicator, a quantifiable measure of personal or organisation performance.
50. For a detailed explanation in the context of engineering philosophy, see Goldman (2004).
51. European languages typically contain only about 15–20 terms for brother, sister, father, mother, uncle, aunt, cousin, grandfather, grandmother, great uncle, great aunt, brother-in-law, sister-in-law, father-in-law, and mother-in-law.
52. Collins (2010) and Polanyi (1966) provide interesting and thought-provoking discussions on tacit knowledge and its manifestations.
53. Brown et al. (1993).
54. Distributed cognition, distributed creation of knowledge, or what Gibbons and his colleagues have classified as 'Type 2' knowledge production (Nowotny, Scott, & Gibbons, 2003) has been discussed in the context of technical work since the 1980s and possibly earlier than that (Hutchins, 1995). Unlike Navy crews, however, engineers don't necessarily know what others know. For a discussion in the context of design, read Larsson (2007) and see also Eraut (2000). Rachel Itabashi-Campbell and her colleagues explored this in a study of engineering problem solving in the automotive industry (Itabashi-Campbell & Gluesing, 2014).

Hutchins' model of distributed cognition has limited applicability in the context of engineering practice. His case study, an aircraft carrier entering port, describes the roles of different crewmembers contributing to the navigation of the ship.

Distributed cognition in the context of engineering departs from this model in several important aspects.

First, in the case of the navy or a ship's crew, individual crewmembers have clearly designated duties performed with respect to clearly defined procedures that are learnt in training and developed over many years. Each crewmember has a specific duty to perform. In the case of engineers, they collaborate and coordinate their work, drawing on each other's knowledge, but there is no clearly defined script or set of procedures by which they do this in most cases. Most interactions are informal. Some interactions, however, follow clearly defined organisational procedures, many of which are designed more to promote the opportunities for informal interaction rather than to transfer information from one person to another.

The second aspect in which engineers differ from the ship's crew is that engineers have numerous other responsibilities to perform at the same time; for example, they will have at any one time 50 or more issues to attend to. Many of these correspond to e-mails arriving in their inboxes every day. Engineers are relatively autonomous and choose when and how to perform on each of these issues. Furthermore, their interactions are relatively informal and may happen as a result of chance encounters rather than planned meetings.

Technical coordination described by Trevelyan (2007) is one of the ways that engineers collaborate and coordinate with others in a network of distributed cognition.

The third aspect in which engineers differ from the case study described by Hutchins is the degree of overlap of technical knowledge. Engineers collaborate because they need to rely on each other's expertise: they may or may not know very much in the area of expertise of their colleagues. Usually, the reason why they collaborate is precisely because they need skilled contributions from other people. Therefore, while an engineer may have detailed knowledge about what the other person contributes, it is equally possible that they do not; this makes errors and misunderstandings more difficult to detect and correct.

55. Social networks can be analysed using quantitative methods (Knoke & Yang, 2008).

56. In our research, we discovered many types of social interference with distributed cognition, including relationship issues, language difficulties, and social hierarchies.

57. E.g. Davenport (1997).

58. Note, however, that some technicians use 'engineer' to denote their occupational status, such as locomotive drivers in the USA, and aircraft maintenance 'engineers'. They spend much of their time performing hands-on work.

59. Sadly, this actually happened when the Westgate Bridge collapsed in Melbourne in 1971. Engineers were working inside the bridge at the time of the collapse and many of them lost their lives, along with many other workers.

60. Draftsmen and draftswomen: people who are highly skilled at operating computer-aided design and drafting systems.

61. As Mintzberg noted, business strategy can be learnt in an afternoon, but not the engineering needed to manufacture a better motorcycle engine (2004).

Chapter 6

Three neglected skills: Listening, seeing and reading

Communication skills are widely regarded as being the most important competency for engineers.[1] When it comes to learning materials and courses, however, the focus is almost always on one-way communication, namely writing and giving a confident technical presentation or sales pitch. The focus of this chapter is on the almost entirely neglected area of *perception* skills: listening, reading, and seeing.[2]

Why are these so important?

The problem is not simply that these skills are widely neglected and taken for granted in education. Nor is it that many people, not just engineers, are poor listeners, inaccurate readers, and failures at noticing things that they can clearly see. By attempting the exercises later in the chapter, you will most likely find that, as a result of taking these skills for granted, there is plenty of room for improvement in your own abilities.

These three skills are fundamentally important for two reasons.

First of all, the main *transmission* aspects of communication, writing, speaking, and drawing (or some other use of graphic images or visual content), all depend on your ability to read, listen, and see accurately. Further explanation for this will have to wait: the following chapters address this issue comprehensively. In the meantime, I ask that you accept this assertion and devote as much time and energy as you can to developing your neglected *perception* communication skills.

The second and even more compelling reason that these skills are so vital is that becoming an expert engineer is a long-term journey of learning: it will take up to 10 years or more before you can count on being reasonably expert in some aspects of engineering.

The rate at which you can learn and acquire engineering expertise is almost entirely dependent on these three *perception* skills: listening, seeing, and reading. Your ability to learn depends entirely on your ability to perceive reality, which can be much more difficult than it might appear.

Perception skills are not only demonstrated by accurately perceiving the world around you; there is also the need to notice important details. Throughout your formal education, your teachers have guided and helped you learn about yourself and the world. You did not have to rely very much on your own ability to perceive the world: instead, for the majority of the time, you were experiencing the world through words and recorded images prepared by other people. Your teachers pointed out what was important to learn and you accepted that as the scope of necessary knowledge. However, in the world outside formal education, you have to rely almost entirely on your own senses.

Therefore, the first and most fundamental capabilities you need to cultivate for your learning journey, and for you to become a competent and articulate engineer that is capable of influencing your engineering enterprise, are these three perception skills. This chapter addresses each one separately, but first we need to understand some recent results from psychology research to understand just how difficult it is to master accurate perception skills.

It is useful to specifically understand how formal education can allow your perception skills to become neglected: one might even argue that formal education can result in serious damage.

As a graduate emerging from tertiary education, your ability to listen and see for yourself has probably not been seriously practiced for several years. You have spent a significant part of your recent life studying texts and notes provided by your teachers and lecturers. During most of that time, you relied on their eyes to point out what was important and what could be safely ignored. If you graduated recently, your lectures were probably recorded and supplemented by PowerPoint presentations and lecture notes. Furthermore, you probably had access to other notes and past exam papers from previous students that they had kindly placed on Internet discussion boards to help you pass your course with the least amount of work.

The world of engineering practice is a very different place.

Most of the people that you can learn from throughout your career will not provide you with notes when they teach or tell you things that are really important, nor will you be able to record what they say, at least not most of the time. Even if you can make a recording, you probably won't have time to go back and listen to it for a second time. You need to quickly and accurately understand what is being said, often with heavily accented, grammatically incorrect English, using words in atypical and unexpected ways.

When you go on site, you will be expected to be able to see for yourself. You will almost certainly have a phone camera with you, but you need to know what to point the camera at and how to take photographs that show what's important. No one will be pointing to the essential components that need to be documented or recorded; those decisions will be left up to you.

In other words, you need to reacquire the ability to use your own ears and eyes, a process that takes a decent amount of time and practice.

PRIOR KNOWLEDGE AND PERCEPTION

Practice concept 29: Perceiving reality requires us to understand how prior knowledge can help us and deceive us

The world is often different than what we expect. One can argue that learning only occurs when we discover that our expectations don't match with reality. Most of the time, it is very difficult to make sense of the world without some prior belief that provides us with an expectation about what we are experiencing. However, the same expectations and beliefs that help us make sense of the world also prevent us from seeing reality and learning where our expectations are wrong. That is why learning can be very difficult. To learn effectively, we need to understand how our normal senses are influenced by prior beliefs that can actually prevent us from learning, and then, temporarily 'disable' that influence.

Our human senses – sight, hearing, touch, smell, and taste – are the means by which we experience the world around us. Proprioception concerns another set of senses within our body by which we sense the state of our own body – the positions of our limbs and fingers, our temperature, our orientation and balance, whether we are turning, and so on.

For many years, psychologists have struggled to understand how we manage to make sense of the constant flow of information that arrives through our eyes, ears, and other sensory organs. Artificial intelligence and computer science researchers have continued to struggle to replicate our senses using computer hardware and software.

For many years, artificial intelligence researchers and cognitive scientists held that perception was a layered process that started with low-level analysis of visual images or sounds. For example, sharp changes in brightness in an image often indicate where the edges of an object are located. Image segmentation, the locations of edges, colours, corners, and other features, was presumed to be the most fundamental step in visual perception. Everything else, it was thought, depended on information derived from the segmentation step.

Now we understand that human perception is much more complex. The ways we reliably make sense of what we see and hear depend on our prior knowledge of *what we expect to see and hear*.

I can illustrate how powerful this concept can be by explaining a small practical contribution in image analysis that embodied some understandings about human perception.[3]

We wanted a means to locate certain features in images of sheep, lying upside down on a cradle, ready to have the wool shorn off by a robot. By accurately locating these features, our software could estimate the likely shape of the sheep's body underneath the layer of wool, which is up to 12 cm thick.

It was impossible to analyse these images using conventional computer vision techniques. The light and dark shading caused by the natural profile of the sheep's fleece, as well as the dirt that characterises sheep's wool grown in the dusty conditions of the Australian outback causes extreme variation in the images. We needed a technique that could recognise the major features, obvious to the human eye, but elusive for computer image analysis algorithms, which typically start by identifying the edges between light and dark areas of an image. Figure 6.1 shows several examples.

We adapted a technique known as 'active contours' or 'snakes', which was originally developed by Terzopoulos, Kass, and Witkin at the Schlumberger Palo Alto laboratories.[4] We imposed a prior belief about the shape of the feature we were trying to locate in the sheep image and used a pre-shaped active contour, or snake, that could deflect elastically, to a certain extent, in order to locate nearby features in a particular sheep image. The software analysed the image intensity along 30–50 short lines crossing the snake at right angles to each of the snake's line segments to determine which way the snake needed to be moved and whether it needed to be bent to match that particular sheep. Not only were we able to reliably locate subtle image features, but our method was also about 500 times faster than conventional image analysis methods. It had to be. At the time, we were relying on computers that were approximately 100 times slower than a typical smart phone today.

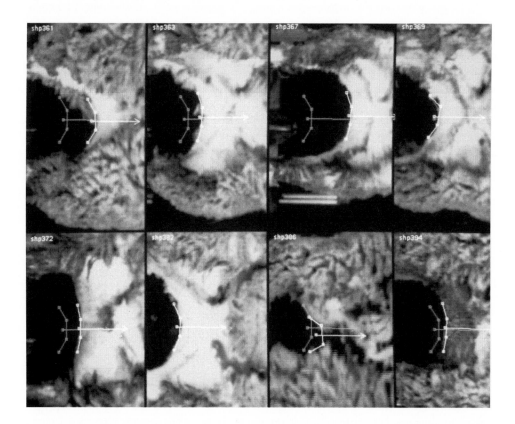

Figure 6.1 Fast computer vision measurements with 'messy' sheep images. Each image (approx. 150 × 150 pixels) shows the crutch, the rear end of a sheep between the legs, seen from above. The sheep was lying on its back, prepared for shearing. A pre-shaped 'snake' line starts at a fixed location in the black area between the rear legs on the left side of the image. It slides to the right, stopping at the end of the sheep. The snake can adapt to the shape of the crutch, as the software provides the snake with flexibility to bend and slip sideways if needed. By combining our prior expectation of what the crutch looks like with image data encoded into the initial shape, position, and flexibility of the snake, we obtained far higher speed and much greater reliability of image analysis. The pronounced striping at the bottom of the seventh image was caused by sheep movement at the instant the image was captured using an interlace scan camera, yet the image analysis still produced a correct result. (You may need an enlarged image to see this.)

In doing this, we were replicating a fundamental aspect of human perception. We are able to quickly discern important aspects of the world around us because our mind already knows roughly what to expect; this prior knowledge or belief framework helps us make sense of the signals that we receive through our eyes, ears, and other senses. Without knowing what to expect, it would take us far too long to avoid being run over by a cyclist heading directly towards us. Our ancestors would have been eaten by tigers had it not been for their ability to make sense of the world by using their prior knowledge.

Figure 6.2 Demonstration of the 'Cornsweet Illusion', in which the flat surface of the upper grey tile appears darker than the shaded surface of the lower white tile, whereas in fact they are the same shade of grey, as shown in the rectangles sampled from the image above. Covering the white and dark shading between the tiles with a finger or pen will help to expose this. Our beliefs about the tile colours influence our lower level perception of the shades of grey. Image reproduced by permission from Purves, Shimpi & Lotto (1999). This illusion and others by the same authors have been made available in colour in the online appendix.

We are not alone in this. Professor M. Srinivasan and his team at the Australian National University discovered that bees and other insects have similar visual perception abilities, despite having a tiny visual cortex compared to the one found in the human brain.[5]

At the same time, however, our powerful ability to use expectations to rapidly perceive what is happening around us can also deceive us, as shown in Figure 6.2. Our beliefs shape our perception. The BBC video 'Is Seeing Believing?' provides many entertaining examples that demonstrate this relationship.

Later in this chapter, some simple challenges in sketching will help demonstrate this for you in a very practical manner.

The challenge in perceiving reality, therefore, is to disconnect at least part of the influence of prior expectations in order to reduce their influence on what we see or hear. This means much more work, as it is much more difficult to perceive reality without prior expectations, partly through lack of practice. Most of the work that you need to do for the exercises in this chapter reflects this difficulty. It takes time and patience. This is the first application of the concept of 'deliberate practice' that was explained in Chapter 4. Don't expect it to be easy.

Gradually, through doing these exercises, you will come to learn how prior knowledge influences what we see and hear, and how you can compensate for this influence.

Once you can compensate effectively for prior knowledge, and more accurately perceive reality with less effort, your own learning process will be much faster. As long as you continue to allow prior knowledge to get in the way of accurate perception, your learning will essentially be confused by the inaccurate perception driving it.

In later chapters, when we take up the challenges involved in influencing other people, we will return to this discussion of prior knowledge. It is helpful to understand the way that prior knowledge can influence perception. Only then can you appreciate how difficult it is for other people to learn something new so that you can begin to influence what they do. That is why it is so vital that you take the time and trouble to work through the practice exercises prescribed in this chapter. This is an integral foundation for the topics covered in the later chapters of this book.

Practice concept 30: Engineering is based on accurate perception skills

When it comes to visual communication skills, students and novice engineers often think that ...	Expert engineers know that ...
Visual communication is all about preparing fancy images and graphics by copying and pasting them from the Internet whenever possible.	Engineers need highly developed perception skills. For example, visual skills are needed in order to notice subtle features in artefacts (e.g. cracks, crazing, distortion, and slight defects) and unusual features in graphical information displays (e.g. unusual data trends on a graph). The ability to read drawings, which demonstrates a capacity for spatial imagination and visualisation abilities, is also critically important. Expert engineers also know that sketches, graphics, and drawings are often useful ways to help convey intent and meaning to others, but the ability to prepare sketches depends on seeing accurately.

This brief discussion also highlights the importance of visual skills.[6] An expert engineer has to be able to read and interpret displays showing sensor readings and information.[7] He or she also has to interpret engineering drawings, and then be able to watch how the designers' intentions in the drawings are translated into reality in order to detect possible slips and mistakes. At the same time, an expert engineer can quickly check materials and components to make sure that there are no obvious flaws before they are installed and then hidden out of sight by other construction work. A visual imagination helps an engineer anticipate the processes of translation into reality in advance, once again in order to avoid problems before they actually occur, while also effectively allowing them to plan the work in a logical sequence.

We depend on prior knowledge for interpreting drawings and photographs, just as much as we do to interpret speech, and words on paper. Sometimes this can lead us astray. For example, a mechanical engineer explained how a large and very expensive reactor vessel made from exotic material was specified with drawings using 3rd angle projections. The 3rd angle projection is a drawing convention for describing three dimensional objects using two dimensional views (projections) of the object from the top (plan view) and sides (elevation views). While the 3rd angle convention is common in the Americas, the mirror image 1st angle convention is commonly used in Britain and former British dominions. In this instance, the reactor vessel was fabricated by technicians who mistakenly interpreted the drawings using the 1st angle convention, even though the 3rd angle symbol was included on all the drawings. They made a mirror image of the specified reaction vessel, and by the time the mistake was discovered, it was too late.

Much of what you need to learn to become an expert engineer will be based on what you see with your eyes. The more accurate your visual perception skills, the more accurately and quickly you can learn.

Are you ready for this? How good are your visual skills?

The third part of this chapter, we will put your visual skills to the test and help you to develop much more accurate visual perception skills. First, however, we will focus on listening.

PERCEPTION SKILL 1: LISTENING

This will be a brief introduction to the most neglected skill in communication. Even if you look at nothing else in this book, the single most effective way to improve your ability as an engineer is to practice listening and writing notes that enable you to accurately recall what was said by others. As we shall see later, the aim is to capture not only the essence of what was said, but also the language and the precise words used by the speaker.

If you look around, you will find that nearly every communication skills book or course misses this seemingly fundamental ability.

Listening is the single most important skill for engineers.

Our research has demonstrated that engineers spend between 20% and 25% of their time listening, on average: that's far more than any other single activity.

In my engineering career, many of the most valuable things I learnt came from listening to other people: there were no notes, handouts, or recordings. Often, the truly useful information is *only* conveyed through speaking. People may even be reluctant to write it in an e-mail. I was very fortunate to receive some superb training that developed my listening skills early in my career.

Listening, and subsequent note taking, does not come naturally. These are acquired skills that need to be practiced. Developing these skills requires hard work, just like any other aspect of expertise.

Your clients will only give you money to spend if you respond to their needs. Most clients with money tend to be verbal people: they don't typically express their needs in writing or drawings. That's the first reason why listening is so important for engineers: you need to listen in order to thoroughly understand your clients' needs.

The second reason why listening is so important is that engineers cannot achieve much without a great deal of help from other people. You need to be sure that they are listening to you, and you need to listen to them when they tell you about problems that may arise. Accurate communication is one of the best ways to avoid nasty engineering problems. The major contributing factors to most engineering disasters have involved communication failures. The Transocean crew operating the BP Macondo well in the Gulf of Mexico tried to communicate their concerns about the condition of the drilling operation to their superiors, but the risks of a catastrophe were never fully appreciated. Eleven of them died in the subsequent fire and explosions.[7] Knowing when someone else has properly listened and understood is as important as listening itself.

Most people think that being able to hear means that they can listen. However, just a few minutes of observation will tell you that those two skills are very different things; many people could definitely benefit by improving their listening skills. You may be no exception.

When I have asked my students in the past about which aspect of communication skills they would most like to improve, their most common request is often phrased like this:

> 'I would like to be able to get my point across more often. I find that other people don't listen to my ideas. I'd like them to listen more carefully, because I get frustrated when they seem to misunderstand what I am saying or miss the point completely.'

Knowing more about listening skills can really help you in that sort of situation.

The chances are that the 'other people' are not listening particularly well, but you have not noticed. Also, you have probably not listened carefully enough to them to realise that they have, after all, understood more about what you were telling them than you thought. Communication is a two-way street, requiring the effort and attention of all parties involved.

If you can recognise the level of other people's listening skills and notice when they are not listening, then you can save yourself a lot of trouble. Once you can tell when another person is not listening, you should simply stop talking. Continuing to talk is a waste of your time and effort. Instead, you need to regain their attention and work out how to retain it so the information you are offering can be effectively absorbed.

The suggestions in this chapter have been developed with the help of an invaluable reference text: *People Skills* by Robert Bolton (1986). It is a valuable companion for developing many of the skills discussed in this book.

Practice exercise 3: Observing listening lapses

Join a group of people talking about something: it could be a project meeting, a casual conversation, or just a group of people trying to organise a social activity. Even better, if you're living with two or more other people at home, simply observe a routine conversation around the dinner table.

Watch and listen carefully.

See if you can notice when someone starts speaking before the other person has finished talking.

When this happens, the interrupting person most likely switched their mental focus a few seconds before they opened their mouth to figure out what they were going to say. They have not only missed what the other person said once they started to speak, but also what was said for the last few seconds before that.

Now, think of three other indicators that could alert you to a listening failure.

1. _____
2. _____
3. _____

(See the later section on listening road blocks for hints, if you need them.)

Repeat this exercise whenever you can.

The main trick for good listening is to keep your focus on what the other person is saying, right through to the end.

It's hard, and sometimes genuinely tiring, until you have practiced this skill and it has become natural.

If you're like me, you will often find your attention drifting ... that strange way the person said a certain word ... the piece of food stuck to their lower lip ... there are countless things that can divert your attention. How many times have you been listening to someone, perhaps in a meeting, and found yourself thinking about something completely irrelevant?

Fortunately, since most people start with poor listening skills, it's not that difficult to significantly improve.

Practice concept 31: Listening starts with engaging the other person

There is not much point in listening unless the other person is talking. That sounds obvious, right? In fact, it is not that easy to keep the other person talking unless you convey to them that you are actually interested in what they're saying.

Let me warn you, this can be quite challenging with many engineers, because engineers often like to express very complex ideas using minimal amounts of language. Therefore, when you are listening to someone who is economic with their words, you need to employ some special techniques to keep them talking and explaining their initial comments in greater detail.

There are several techniques you can learn that help keep other people talking, which can then provide you with valuable learning opportunities.

Building rapport with the speaker, developing a relationship

We will return to this theme a number of times throughout this book. People are much more helpful if you take the time to build a relationship with them. This means extending the conversation to include anything of mutual interest: it might be sports in some contexts, political gossip in others. Talk about anything other than the information or advice that you need, at least for a few minutes, so that the other person sees that you value them as an individual, not just as a source of help when you need them. Whenever you pass near their workplace, take a moment to call in and say hello, even if you don't need anything in particular at the time.

When you meet people for the first time, ask them to talk about their interests other than work for a few minutes. This helps to set most people at ease in your presence and starts the foundation of good rapport.

Posture

The best place to start is with your posture.

Notice that when people are really listening carefully, their posture often clearly shows that; you don't even have to watch or listen carefully to understand whether they are paying attention and listening.

However, a person who is leaning back in their chair, with their hands folded behind their head, staring at the ceiling ... chances are that they are not listening very carefully. Compare this to someone sitting forward in the chair, with head and eyes fixed on the person who is speaking. This person is probably concentrating on everything that the speaker is saying.

Be careful, however, when observing others and judging their level of listening. Posture can be deceptive. A person who does not look at the speaker may actually be listening very attentively, even though it might seem to put you off as a speaker.

Yes, that's right. The posture of the person listening can actually communicate a great deal to the speaker. Therefore, as a listener, sitting up attentively, leaning forwards, keeping still, and focusing on the person to whom you are listening to can help hold the speaker's attention and urge them to continue talking, because you are sending a strong, non-verbal message to them that you are listening carefully.

Appropriate body motion

Once again, observe people while they are communicating. What do they do when other people are talking? Do they fiddle with their papers or books? Do they shuffle their feet?

Someone who is really listening hard will probably keep very still because their mind is focused on the speaker.

Try to develop an awareness of your own body movements when you're listening: keep as still as possible.

Eye contact

Like posture and body motion, eye contact is very important.

Keeping your eyes focused on the speaker does two things.

First, it helps you to avoid distractions and increases your focus on what is being said.

Second, like posture, it sends a strong non-verbal message to the speaker that you are focusing on what they are saying. Eye contact is a very powerful tool for the listener.

However, in many cultures, it is considered impolite to look directly at a person of higher social status, or even at a young person of the opposite sex. These sorts of cultural cues will often come through experience, but this means that you need other methods of listening well or assessing whether others are listening to you.

In other words, in certain situations, eye contact is a useful indicator, but you need other techniques to make sure that listening is working.

Environment

When you want to listen to someone carefully, it helps to choose a good environment, one without too many distractions. You cannot always choose your environment, but it helps to be aware of how different environments can affect listening.

Listening is all about focusing on the speaker, so any interruption is likely to interfere with concentration.

Switching your mobile phone to silent mode before starting is a good way to prevent interruptions, particularly in our modern world where hyper-connectivity seems to be the 'new normal'.

If you can, avoid meeting at times when other people are likely to come looking either for you or the person to whom you are listening. If you can't go somewhere private, see if you can put up a 'Meeting in Progress' sign to minimise interruptions, but remember to take it down afterwards.

Make sure that both participants in the conversation (you and the speaker) are comfortable. Choose a room with comfortable furniture, but not so comfortable that one of you may fall asleep.

Sit lower, or at the same level as the speaker: avoid a situation where you are sitting on a higher chair than the person you are listening to.[8]

Reduce glare: avoid a situation where one of you has to look at the other person against a bright light or a window.

Sometimes it is best to choose an open space, either outside or in a busy coffee shop. A person who does not know you that well may not feel comfortable sitting by themselves in a room with you. A busy coffee shop has enough background noise to keep the conversation private, but will still put them at ease due to the informality of the locale. However, don't choose a place where either of you are 'well-known' or it will be difficult to avoid interruptions.

The only problem with meeting in public is that background noise makes listening more difficult, particularly in the form of other people talking. Be sure to choose a place with an undisruptive level of background noise.

Emotion and fatigue

Anyone in a heightened emotional state tends to find listening much more difficult. This is precisely why intense relationships can be so challenging after the initial stage of infatuation has passed. Even positive emotions can disrupt or decrease listening abilities!

If you are tired or upset, even if you are full of overwhelming joy and excitement, try and defer the conversation for another time: your listening skills will be reduced by your emotional and physical state.

Fatigue is the other killer of your listening abilities! Alcohol is also usually fatal for your listening and retention.

Unless you have professional training as a counsellor, avoid any situation requiring accurate communication when the other person is in a heightened emotional state, particularly when the emotions are negative, including anger, frustration, loss or bereavement, insecurity, or anxiety. Wait for a day or two: there will be other opportunities to communicate more effectively.

On the other hand, don't suddenly break off a conversation if you realise that you're not listening well, or that the other person has switched off: that can kill a relationship quickly! Wait for a suitable moment to pause the stream of conversation, and then take your leave, or simply apologise for being too tired or distracted to continue, even if you think that the other person is the one who is too tired or emotional and isn't listening!

If you need to calm down quickly to improve your own listening ability, there are several useful techniques that you can practice. Moderate exercise for 30 minutes or more (e.g. walking), deep breathing, sleep, concentrating on something completely different (but not too demanding), or even sitting down for a relaxing meal can help. If you are religious, prayer can also be very helpful.

Remember, also, that listening can be tiring because it demands your full concentration, particularly if you are not used to doing it for long periods of time. It is even more exhausting in a noisy environment or if the speaker has an unfamiliar accent. Take time out for a rest after 45 minutes or so and let your ears, brain, and patience recharge.

Active listening

Active listening is a special type of interactive conversation, one in which the listener carefully uses occasional participation to help the speaker along. This is a skill that you can easily master, and it can make listening more fun and enjoyable. It is also very useful in meetings: your active listening will help other people in the meeting to understand what someone is saying and promote additional engagement.

Active listening takes practice and can feel embarrassing at first. However, it sends a powerful message to the speaker that you are really paying attention and genuinely respecting what they are telling you. When they see your level of interest, the speaker will probably tell you much more than they otherwise would have done.

The main technique in active listening is to engage the speaker in a dialogue. Active listening is a two-way conversation. It builds on the idea of building rapport, as mentioned earlier.

Respond non-verbally to the speaker

While you are listening, respond to the person who is talking. Smile when there is a hint of humour, put on a slight expression of concern when they look concerned or begin talking about a difficult situation.

Emulate their posture: if they are sitting up and leaning towards you, try to do the same. If they are sitting back with their legs crossed, try adopting a similar pose without making yourself uncomfortable.

Maintain eye contact, but not necessarily with a fixed stare that might make the speaker uncomfortable. If the speaker's eyes wander occasionally, let yours wander too, but keep a close watch at the same time, as many people change their posture and body language over the course of a conversation.

Minimal encouragement

Insert brief 'minimal encouragement' phrases into the conversation when the speaker pauses, like 'Oh yes?', 'Really . . .', or 'Cool . . .', etc., without actually interrupting the speaker. This will encourage them to continue and will demonstrate that you are fully engaged.

Attentive silence

If the speaker pauses, don't respond immediately; they may be thinking about what to say next. Try 'attentive silence'. This simply means waiting for the person to continue talking. Silence is nearly always more effective than phrases such as 'Tell me more' or 'What else did they say?'

Taking notes

Take notes if the conversation is important. However, in sensitive situations, always ask if it is okay to take notes first!

You will learn about more efficient ways to take notes in a later section.

If the speaker informs you that what they are about to say is confidential or 'off the record', stop taking notes. Then, try and listen especially carefully and when you have an opportunity later, write notes to help you remember what was said in confidence. When writing notes, always anticipate that someone may stumble across your notes one day: use code words to protect anyone who might be vulnerable if the notes ever become public. To minimise the chances of this ever happening, remember to keep your notes secure and password-protected, if possible.

If you feel the need to use a recorder, always ask permission first, and use a discrete recording device to avoid unnecessary questions from other people.

Infrequent, open questions, probing, or clarification

Ask occasional, infrequent questions, and always use 'open-ended questions'.

A 'closed' question is one that will evoke a 'Yes', 'No', or 'I don't know' answer, or at least one variation of a very small range of possible answers.

An 'open' question invites a longer, more informative response. Instead of asking, 'Was Jane there?', you can get a lot more information by saying, 'Tell me about the other people who were there.'

It's often important to ask clarifying questions. Sometimes you will feel afraid to admit that you don't know what a word means or that you've forgotten about something very important that the speaker just told you. Ask politely for an explanation, perhaps with a confession that you're not quite sure what they meant. Here are some examples of 'probes' which are exploratory, open-ended questions that seek further clarification.

'Could you tell me what you mean by that?'

'Tell me what you mean when you use the word _____?'

'I'm sorry, I think I missed that. My mind must have switched off for a moment; could you tell me again please?'

It is very important to clarify the meaning of technical terms. As engineers, we assume that all technical terms have a well-defined meaning that is understood the same way by everyone. This is not necessarily the case.

Door openers

Sometimes, the speaker will be stuck, perhaps trying to think of words to describe what they want to convey to you. If attentive silence does not work, try a gentle 'door opener' in the form of an open question, possibly about something not entirely relevant. The effect is simply to get the speaker back into talking, telling you things that are on their mind. Even if you take them off the track slightly, they will soon come back.

This might happen at the start of a conversation, try asking about people you both know. 'How's Sam these days? What have you heard about him recently?'

Midway through, you can take the speaker back to something they may have mentioned briefly in passing, such as 'Could you tell me more about that instrument you said you found really useful for measuring surface temperatures? Can you remember where you got it from?'

Paraphrasing

This is probably the most important 'active listening' skill to master, and it takes a significant amount of practice to truly perfect. At first, it seems embarrassing, but never mind that. Keep working on this skill, even though you may feel rather embarrassed.

After the speaker has said something that's important to accurately understand, ask them to listen to your own understanding of what they just said, in your own words, and tell you if it's right:

'If I heard all of that correctly, what you have just told me is ____ ____ ____ ____ ____. Is that what you meant?'

or

'Did I hear you correctly when you said ____ ____ ____ ____ ____ ____ ___?'

This is particularly helpful in meetings, especially if you are the meeting chair. If you are unsure whether you really understood what the speaker was saying, the chances are good that the other people present have possibly also misunderstood what was said. By asking for clarification, or better yet, paraphrasing what you think the speaker just said, you will help other people understand more accurately.

You might think it takes extra time and trouble, or you might feel that you will annoy the speaker by doing this. However, you will actually make the speaker feel more reassured that you are sincerely trying to understand what they have said.

Roadblocks

Just as active listening helps to improve listening accuracy and is an important skill to practice, there are also 'roadblocks', which are actions that inhibit effective listening.

There are some actions that really disrupt face-to-face communication. If you notice a person using them, they can provide strong evidence that the person is not listening well.

Train yourself to notice when other people deploy these roadblocks and remind yourself that we all make the same mistakes more often than we should!

Criticising, name calling

You are listening to someone and they tell you about a difficult situation or conversation. You say, 'I don't think it was a good idea to say that.' In a more extreme version, it might come out as, 'You idiot! Why did you say that?'

The speaker will feel embarrassed and will be reluctant to tell you what really happened, because no one likes being told that they did something wrong, also known as direct criticism.

If you really think they could have handled a situation better, try this: 'That must have been tough for you! I'm not sure I would have been able to handle that.'

If asked directly for advice and criticism, it is beneficial to provide it, but do it gently.

Diagnosing or praising evaluatively

You are listening to someone and they tell you about a difficult situation or conversation. You say, 'You must have been tired or pissed out of your mind to say something like that!'

Your statement is quite possibly true, but once again, your response is unlikely to make the speaker comfortable enough to tell you the remainder of what they still have not told you; perhaps they will be too reluctant to talk about it now!

Praising someone for doing 'the right thing' can also block communication, because the speaker might be silently thinking, 'I must have been a dumb idiot to have said that!'

Diagnosing

It is easy, especially as an engineer, to diagnose the situation the speaker has been trying to describe to you and give your analysis of what the speaker has just said.

This diverts attention from what the speaker was talking about and can weaken the speaker's confidence in telling you everything. Also, the intellectual effort needed to diagnose the situation will inhibit your listening capacity, making the speaker feel like you aren't really engaging in what they are trying to express.

Thinking about something else while the speaker is talking

I constantly find myself doing this and I know that it is my main listening weakness. Thinking blocks your ability to listen, so learn to switch off your own thoughts. That is contrary to what we do for the vast majority of our lives, so it takes practice.

Ordering

For example, 'You should tell him about ...'

Once again, this shifts the speaker's attention from the situation they are describing to possible responses, even though they may not have finished describing everything that you need to know.

Threatening

For example 'If you don't (X), then (Z) will happen.'

As with ordering, this diverts the speaker's attention from the situation they are describing to possible consequences, despite the fact that they may not have finished describing everything you need to know.

Moralising, advising

For example, 'It's not right to say that.'

> or

'I think you should have told them to ...'

The speaker needs to feel completely comfortable that you will accept their point of view in order for them to fully describe the situation. We all make mistakes and take actions that we would prefer not to have done in that way. However, most people are very reluctant to talk about their own mistakes and need constant reassurance that they will be supported and protected as much as possible. Moralising can undermine this confidence very quickly and put you at odds with the speaker.

This does not mean that you have to agree with the speaker's point of view and accept that what they did is correct. However, it is necessary to defer your own response in order to ensure that the speaker feels free and comfortable to tell you, so you can listen accurately and get the whole story.

Logical argument

For example, 'If it's true that you said ... then it follows that you must have ...'

Most people, even engineers, cannot necessarily remember past events in a perfectly logical sequence. Our memory is selective, as is our ability to recall, so we often reconstruct stories of past events in a way that makes sense to us, but not necessarily to other people. Trying to explore the logic of what somebody has just said is more than likely to disrupt their account and make them doubt their own memory.

If you can see a gap or an inconsistency in what someone has just explained to you, try posing an exploratory (probing) question such as, 'That's really interesting, could you please tell me more about what you just explained ...?' or 'I'm not sure I quite understood that; could you please explain that last part again ...?'

Excessive questioning

Too much questioning can be just as inhibiting. A skilled listener will recognise when a speaker feels uncomfortable talking about a sensitive issue. The speaker may change the topic, pause for a long time, start talking really quietly, or even show visible signs of distress, such as sweating or turning pale.

In these circumstances, it is best to move on to a different topic. Once the speaker has enough confidence, they may return to the more sensitive issue of their own accord if they think that it is important for you to know. Alternatively, it may be possible to try again on another occasion. Sensitive issues need to be handled with care, not questioning.

Reassuring

For example, 'There's no need to worry about that, it will all blow over very soon.'

While you might say this with the best of intentions, trying to help the speaker feel better, the effect is usually the opposite. No one wants personal problems to be diminished or treated as though they aren't important. Instead, it is much better to reflect the feeling expressed by the speaker in a way that enables them to understand that you have heard what they just said. For example:

'I can see that you would be very worried about that. It must be tough for you.'

Just remember that the aim here is to help the speaker tell you everything they can, and for you to understand the speaker fully, including how the speaker is feeling about the issues that they are telling you about.

Practice concept 32: Listening accurately and taking notes

So far, we have dealt with the important issue of keeping the other person engaged and talking: the more they talk, the greater the opportunity you have to learn. The other aspect is to be able to listen accurately and take notes in order to capture as much detail about the words actually used by the speaker and the overall meaning of what they said. There is always great temptation to paraphrase and substitute different words from the ones actually used by the speaker. Try and avoid doing this as much as possible. This is very important in certain situations but the value of this practice will not become clear until we reach Chapters 8 and 12.

Unlike almost every other aspect of engineering practice, everyday life provides endless opportunities for you to practice and improve your listening skills, provided that you can evaluate the accuracy of your listening.

Try listening to lectures and taking detailed notes: see the section below for guidance on taking accurate notes with minimal effort. Compare your notes with written handouts and notes taken by other people attending the lecture. Try to understand why you missed some things or misunderstood the big picture concepts. Perhaps try sitting closer to the lecturer so you can hear better and with fewer distractions.

When appropriate, use paraphrasing more often during conversations, explaining that you are trying to improve your listening skills.

The best aspect to this kind of practice is that any improvement can be very rewarding. Relationships will improve and you will learn more easily, with less effort and fewer mistakes.

Writing accurate notes

It is never too late to start learning how to take accurate notes. I learnt this technique midway through my career and it has been immensely helpful to me ever since.

Being able to write comprehensive notes that capture the essence of what was said is a vital aspect of listening skills. Surprisingly, many students never acquire this skill, even though it can enable almost complete recall of a lecture. I have seen some students furiously typing lecture notes on laptops, but there are much easier and more relaxed ways to write notes that will enable you to achieve near 100% recall, even months later. Education research has helped to show that students who take effective notes score up to 15% higher in course assessments.

Here, I have shown you a method that personally works well for me. There is a multitude of different ways, however, and you can also develop your own. Another way is to learn how to write in shorthand, a skill that many journalists develop for speed, without sacrificing comprehension or accuracy.

The method I have used since I was a student is based on the same ideas that Tony Buzan used in his books on 'mind mapping' (e.g. Buzan & Buzan, 1993). When I first started taking notes, I used to write key phrases in more or less the same sequence as my lecturers, something like this:

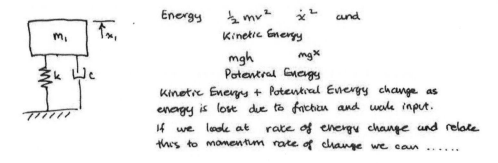

If the lecturer was well organised, so were my notes, but I still had difficulty keeping up. Furthermore, real problems never seemed to follow the lecturer's notes! Disorganised lecturers seriously taxed my memory! They told us extremely useful stories and anecdotes, but I could never notate all of it in the same manner as I demonstrated above.

Then, I found a much more useful approach that used branching diagrams, also known as mind maps, and I found that this enabled me to take clearer notes much faster. Also, I could recall everything that was said in a much more organised way. I could recall it using my own organisational techniques, rather than having to rely on the speaker's system of organisation, if there even was one. I was no longer stuck or exhausted, simply because the speaker was disorganised.

The method is very simple.

Also, it works for everything from casual conversations and phone calls to meetings, seminars, and lectures.

While the speaker talks, write down the key ideas and concepts as single words or brief phrases on the page. It does not matter where you start writing, but the middle of the page is often a good place to start. Leave a little space between each of the phrases or words. As the speaker continues, write more key words or phrases near the ones to which they are related, and draw lines or arrows showing the relationships. As you notice connections that even the speaker has not referred to, you can add more arrows and connections, even adding a note or two of your own alongside the lines to explain your mental deductions.

You may end up with what seems like a disorganised maze of words and lines all over the page, but the initial form does not matter at all.

As soon as you can, underline the most important concepts and phrases.

After the speaker has finished, perhaps after the meeting or lecture, add additional notes of your own to clarify the concepts and connections based on what you understood from the speaker. Add notes for items that you did not have time to write during the meeting. When I use this technique, I enclose any thoughts or side observations of my own in square brackets to distinguish them from what the speaker said.

When you need to recall what was explained, start with the underlined words and phrases. Then, follow the arrows and lines to recall what the speaker talked about; your memory will fill in the gaps between the words on the paper. After a little practice, you will be surprised at how much you can remember, concepts and ideas that go far beyond the words actually written on the page.

Look at the diagram below. It is my representation of what I have just told you, yet it only requires around 30 words, some lines and a handful of symbols to represent around 350 words of text. That's around 10 times less than the number of words in the explanation above. Given that most people speak at around 100 words per minute, you only need to write approximately 10 words per minute to keep up with what they are saying and still be able to recall everything important.

This technique not only enables you to write notes while keeping up with the speaker, but it also provides you with an excellent way to organise your thinking before you start to write something. We will come to that application of this skill later.

Practice exercise 4: Listening and note taking

1. At any meeting, or even during a casual conversation, ask permission to take notes. Write your notes as clearly as you can. Later, reconstruct what was said in

bullet point form. Ask the speaker(s) to review your notes so they can point out any mistakes, misinterpretations, significant omissions, etc.

2.   Watch other people in conversation or sitting in a meeting. Try using the listening skills worksheet (available in the online appendix) to take notes about what you observed. Make sure you do this discretely, or ask permission first, as many people can be quite offended if they think you are trying to watch them too closely.

3.   The Australian Broadcasting Corporation (ABC) Radio National website provides an enormously valuable resource to practice listening and note taking. The website provides podcasts of many programmes, as well as full transcripts. (http://www.abc.net.au/rn/)

  i)   Each day, listen to any recorded programme that interests you, for which ABC also provides a transcript. Listen to no more than 7–10 minutes while taking notes. *Do not press pause: let the recording play at normal speed.*

  ii) After completing the notes, reconstruct what was said as best you can from your notes. This should take you about 20 minutes. You don't have to reproduce exactly what each person said: a set of bullet points will be enough.

  iii) Use a word processor to open the transcript from the ABC website, or print out a copy of the transcript.

  iv) Compare your reconstructed bullet points with the short section of the transcript recording to which you listened.

  v)   Highlight all the text in the relevant section of the transcript that was accurately conveyed by your bullet points.

  vi) Do not highlight any important words or ideas that you missed, or where the word you wrote does not correspond to the word in the transcript. For example, if you wrote 'specification' (or even abbreviated it as spec'n), but the word in the transcript is 'requirements', do not highlight the transcript, even though the words have similar meanings.

  vii) Estimate the percentage of the transcript text that you have highlighted to calculate your score (only the section of the broadcast transcript corresponding to the section that you listened to).

Many of my students have started with a score of 5–10%. However, after practicing a few times, they have improved to 50% or better. Others who started at a level of about 50% have improved to about 80%, some even as high as 90%. Individuals interpret the evaluation criteria differently, so comparison with others is not meaningful. However, anyone can monitor their own progress using this technique.

Remember that like any other aspect of developing expertise, deliberate practice is required (Chapter 4). It takes time and effort to improve your note taking and recall. As it can be tiring, initially limit your deliberate practice to no more than 30 minutes per day, gradually increasing as you get used to it.

Practice concept 33: Contextual listening

So far, we have covered two important aspects of listening – maintaining the engagement of the other person and accurate listening (capturing the exact words that were used by the speaker as best we can). The third aspect of listening requires that you develop sensitivity to particular word choices by the speaker. Let's start with a simple example of a quotation by an engineer:

'They stuffed that one up really badly; it took weeks to recover the drill stem they had dropped down the hole.'

Suppose this engineer had started with the word 'we' instead of 'they'. Would that have made any difference? You might argue from a logical point of view that the aim of the sentence is to convey the seriousness of an incident in which the drill stem, possibly thousands of metres of steel piping with an expensive diamond drilling bit on the end, had been dropped down a hole in the rock. That word choice would not make much difference in terms of the subject: the essence of the incident remains unchanged. However, there is a big difference in terms of the speaker's perspective.

By using the word 'they', the engineer has implied that he or she is not associated with the people responsible for the incident. Using 'we' would have conveyed the idea that the engineer is part of the group responsible, and is therefore sharing some of the responsibility. Social scientists refer to this notion using the word 'identity', which is how we as people tend to associate ourselves with particular groups. When we discuss a local football team and say, 'We thrashed them last Saturday', it means that we identify ourselves as being associated with the team. We have many different simultaneous identities: with our workgroup, company, profession, occupation, sporting teams, community, graduating class, and even our nationality.[9] We often change identities very quickly depending on the circumstances, or even combine various identities at the same time. For example, if our football team lost, we might have said, 'They got thrashed last Saturday', temporarily de-emphasising our normal association with the team.

Noticing the words that people use as an indication of their identity enables you to perceive important social relationships that have a significant influence on the way people act.

This is what Juan Lucena has referred to as contextual listening:[10]

'a multidimensional integrated understanding of the listening process wherein listening facilitates meaning making, enhances human potential, and helps foster community-supported change. In this form of listening, information such as cost, weight, technical specifications, desirable functions, and timeline acquires meaning only when the context of the persons making the requirements (their history, political agendas, desires, forms of knowledge, etc.) is fully understood.'

We will return to this idea in later chapters of the book.

Practice concept 34: Helping others to listen

You can apply your understanding of listening skills to help others listen more accurately. Most of the points listed above can be helpful in this situation as well.

For example, eye contact helps immensely. When I am speaking to a class of students, even when it is a large group, I scan my eyes across the entire audience all the time while I am speaking. I look into the eyes of the audience constantly. This way, I am not only able to hold their attention longer, but I can also tell when I start to lose the audience, because some people might start to move their eyes around the room, losing focus on me. Soon after that initial sign, if I don't regain their attention, I will start to hear shuffling feet and papers being moved around. Essentially, I will know that I have lost the attention of my audience!

Once again, be careful … some people feel uncomfortable being stared at, particularly in a small group or a one-on-one situation, and will actually look away while you are talking, even though they are still listening carefully. Other people may have a natural squint: their eyes seem to be looking somewhere else, even though they are actually looking straight at you.

When it comes to lectures and presentations, remember that a PowerPoint presentation is a great attention vacuum! People in the audience will look at the screen in order to look at and possibly read what's there, which means that they may soon stop listening to you. It is difficult for many students to read and listen at the same time.

If you present more than about ten words on the screen, people who naturally prefer reading to listening will stop paying attention to what you are saying while they read the text. In an engineering environment, it is not uncommon to find that most people will read the text and stop listening.

The same distracting quality applies if you distribute a handout or a report before or during your presentation. If you need to have this tangible element to your presentation, wait until after you have finished speaking to distribute the material.

Sometimes, a picture can tell the whole story, with minimal commentary. If this is the case, stop speaking for a few moments so that your listeners can devote their attention to the picture. When you want to regain your audience's attention, you can simply press the '.' key: the screen will go blank. (Press it again to get the picture back.) Alternatively, you can insert a black slide into your presentation, which will be a clear signal for the audience to shift skills from picture analysis back to listening.

Summary: Listening is an imperfect, interactive, interpretation performance

While this chapter deals with perception skills, it should now be apparent that listening perception is not a one-way process that starts with hearing and ends with making sense of what we hear. Instead, prior knowledge is needed to accomplish effective listening perception: we cannot make sense of what we hear without some prior understanding of language, for example. That being said, this prior knowledge can also interfere with what we think we have heard at the same time. Partly to try and resolve this potential weakness, and partly to evaluate our listening performance, we resort to expanding a conversation with the speaker to help with apparent ambiguity or misunderstandings. Therefore, listening is a truly interactive performance: not a one-way input process.

Perfect listening is rarely possible in real situations. Even with the best interaction, the closest relationship, extended conversation, and explicit clarification, misunderstandings still persist as a result of the prior knowledge that we need to be able to make sense of the spoken words in the first place. Because of this, listening is always an act of reinterpretation: the listener is reconstructing the ideas being explained by the speaker. That reconstruction can never be the same as the speaker's original ideas.

The challenge for engineers, therefore, is to learn how to ensure that this reinterpretation by the listener still allows technical ideas to be faithfully used or reproduced, even though there will always be a certain degree of misunderstanding.

We shall learn in the next two sections that reading and seeing are also imperfect, interactive, interpretation performances, although there are distinct differences, of course. Part of the interactive element will be a conversation with ourselves, while what we see may have been created in a previous time by people who are unable to be with us, thereby making direct interaction with them impossible.

PERCEPTION SKILL 2: READING

We all have preferred channels for receiving information about the world around us. The fact that you have read this far suggests that your preferred channel may be reading. You may be an above average reader, but that does not mean that you are necessarily a comprehensive reader: a reader who can engage with the text in detail in order to see different meanings and interpretations.

This section is the shortest in this chapter because reading this book and completing the practice exercises will help improve your reading skills.

Only about 10% of the population are comfortable with reading extended texts. That percentage seems to apply to engineers just as much as any other segment of the population. As a reader, you may be sought after by other engineers who have lower reading skills, as you will be seen as a source of information and advice. Therefore, you need to develop the skill to read comprehensively so that you don't lead others astray.

In essence, reading is no different to the other principal modes of perception – seeing and listening. Comprehensive reading depends not only on perceiving the actual words that are written on the page, but also on the prior beliefs that we have to begin with. Our perception depends on these prior beliefs, but at the same time, they can deceive us.

Comprehensive reading does require that you read every word. When we read silently, we often skip words; sometimes we bypass entire sentences or paragraphs without necessarily being aware that we missed part of the text.

One of the best ways to make sure that you read every word on the page is to practice reading out loud to the wall of your room. This is also valuable practice for writing: reading out loud helps to build tacit knowledge in your head about the patterns in which words regularly occur. Eventually, this will make it easier for you to write better.

Just as with listening, try and be aware of where your attention lies as you are reading. If your attention starts to wander as you read and you cannot remember the content of the last few paragraphs or sentences, stop yourself from moving forward. Take a break and rest. Reading accuracy is significantly affected by fatigue, both mental and physical.

You can assess the accuracy of your reading from the quizzes that you can access through the online appendix. Some of the quizzes are designed to test the accuracy of your reading.

Comprehensive reading can be tedious and slow, but it doesn't make it any less essential.

You can also find a wide variety of helpful texts on how to read quickly (speed-reading). This technique depends on the conventional way of writing in Western countries, however: the main idea of paragraph will be conveyed by the first sentence. Therefore, you can read quickly by skipping from paragraph to paragraph: you don't have to necessarily read the rest of each paragraph, because you know that it simply elaborates on the idea conveyed in the first sentence. Speed-reading can be a great way to build up a map or a framework of the content of the book and its main ideas so that you can return to the parts that are most informative for you and read them carefully at a later time.

The ideas that you absorb through reading will soon vanish from your head unless you take some action to reinforce those memories before they fade. There are several ways to do this.

It is a good idea to take notes as you read. These days, I use the iAnnotate App on my iPad whenever I can. It gives me the freedom to mark up the text: by highlighting or underlining, the program will automatically extract those sections of the text into a summary document for me, giving me the page number where they are located in the original PDF document. If I can, I make a record of the location of a quotation on the page: 35a.4 means page 35, left hand column (a), 40% of the way down from the top of the page.

Another way to reinforce and explore the ideas that you have just read is to imagine that you have another person who is unable to read sitting in the room beside you. Explain the sense of what you have just read so that they can understand the content of the book. You may also have certain questions on your mind. Ask the imaginary person, 'I don't quite get this part, when it says (X). What do you make of that?' You can even imagine what the other person might say in response, like 'Why don't you look it up on the Internet and see what Wikipedia says about it?' Can you remember a time when something that happened at work resembled what you have just read? Explain to your imaginary companion what actually happened and explain to them how reading the book has helped you understand that episode differently.

Much of what you read in this book will help your understanding of yourself and others around you. After all, the biggest parts of engineering practice that were lacking in your formal education are ideas that can help you understand how people behave and perform engineering work. Try and think of questions that you can ask other people around you to explore these ideas. Go back to some of the ideas about different kinds of knowledge and how you can access that knowledge through social networks, as explained in Chapter 5. Ask other people about this, such as what they have noticed about the different ways that people build on the knowledge of others.

How can you evaluate the extent of your reading, especially when the reading material is not supported by quizzes and other learning activities?

One trick you can use is to choose any paragraph at random after you have completed reading. Make sure you only reread the first sentence of the paragraph. Using your notes and memory, see if you can write a summary of the rest of the paragraph as accurately as possible. When you have done this, read the rest of the paragraph and compare your summary with the actual text in the paragraph.

Remember that prior understanding and knowledge can fundamentally alter the meaning of words for you. It is impossible for you to have the same prior understanding and knowledge as the person who wrote the text. This means that you will almost certainly interpret the text in a different way from what the author intended. Just like listening, reading is an act of reinterpretation. As it is very important that you remember this, we will return to this theme in the next chapter.

Practice exercise 5: Reading for learning

Like active listening, comprehensive reading is a two-way conversation. However, the original author is no longer present, so the conversation has to be with oneself or another person. Here is a practice exercise that illustrates some of the ways to conduct such a conversation in order to help with learning.

I provide similar instructions to my students; in fact, I tell them that I don't believe they have read anything until they have written a 'conversation piece' responding to what they have read.

Read a section of text, for example, part of this book, which is between 5 and 20 pages long.

Write a brief description of the content, including the main ideas presented in the text (beyond what is already given in a summary or abstract, if available).

Describe your assessment of the reliability of the research and the strength of the evidence presented to support the text. Sometimes, you will find no evidence presented at all. Try and distinguish between material based on research or other systematic collected evidence and 'recipes' that are based on opinion. Note any references or further reading that you think you need to follow up on. Often, we may be tempted to classify opinions without stronger evidence as 'subjective' and unreliable. Yet, we often find that opinions can offer insights that more systematic research can miss. A comprehensive reader can make sense of diverse contributions, reconstructing ideas that may not have even been apparent to the writer of the text.

Next, explain what the writing contributes towards the questions and issues relevant to your own aims that motivated you to read the text in the first place. It can be very helpful to quote sentences or even full paragraphs from the text, but you must explain how each quote is particularly meaningful for your goals and interests.

Include a precise page reference for each of these comments.

Finally, write down your own impressions and issues arising from the text, as well as questions that remain in your mind that can stimulate an ongoing conversation with yourself and a later search for meaning.

If I were able to converse with you, the reader, as you work your way through this book, we would both learn a great deal. As I cannot be with the vast majority of you, I suggest that you write a reflective journal as you read this book, section by section. The following practice exercise will give you some pointers.

Ultimately, the test of your reading of this book lies in one or more of the following:

a)   insights that you develop into the world of engineering in which you immerse yourself every day,

b)   a demonstrable performance improvement that brings personal rewards for you, the reader, in terms of job satisfaction and (preferably) improved remuneration in the long term, or,

c)   accumulation of evidence to demonstrate that the ideas in this book need revision or restating, in which case I hope that you will send your accumulated evidence to me so I can take it into account for a subsequent edition.

Practice exercise 6: Written requirements

Even though many clients find it difficult to provide engineers with written requirements, there are some that do provide detailed written statements of requirements. Furthermore, engineering work often has to be performed so that it complies with contract conditions, recommendations contained in standards, or explicit requirements stated in codes. Engineers need to carefully read these statements to ensure that they allow time for any work that will be needed to meet the requirements and demonstrate compliance. In these circumstances, comprehensive reading is essential.

Here is an exercise that will help you develop comprehensive reading skills. Two or more people read the same text and compare notes. If you have not tried this before, you may be quite surprised to find how much you miss when you read a complex document. Some engineering companies formalise this process: in those companies, it is normal practice for at least two people to read a statement of requirements and exchange notes on what they found to reduce the chance that something important has been missed.

When two or more people read the same text, it is likely that they will find more ways of interpreting the text than a single reader would, often through interacting with each other. To see how this works, try it for yourself.

In reading the statement of requirements, pick out each of the requirements that call for some action or a response, and give each one a priority rating score as shown below:

4 = critical, cannot be missed;
3 = important, compliance will enhance value;
2 = necessary, but not important. Compliance will provide minor additional value; and
1 = optional, compliance will not influence the value of results.

To evaluate your reading performance, calculate the following:

N = total number of requirements identified by all readers in combination
M_i = number of requirements identified by each reader 'i' as a proportion of N
P_i = for each item where all the readers agreed on the same priority rating score, add the given priority rating. Divide the resulting sum by the sum of all the priority ratings assigned by this reader.

Here is a small worked example, with the scores given to a series of clauses in the following table.

Clause #	Priority given by A	Priority given by B
1	4	4
2	1	2
3		
4	2	2
5		3
6	3	
7	3	3
8	4	
9	3	4

$N = 8$ (Total number of clauses for which a response was seen as necessary by at least one reader)

$M_a = 7/8 = 0.88$ (Proportion of clauses identified by reader A)

$M_b = 6/8 = 0.75$ (Proportion of clauses identified by reader B)

$P_a = (4+2+3)/(4+1+2+3+3+4+3) = 0.45$
(Proportion of clauses identified by reader A with agreement on priority)

$P_b = (4+2+3)/(4+2+2+3+3+4) = 0.5$
(Proportion of clauses identified by reader B with agreement on priority)

The closer your M and P values are to 1, the more consistent your reading is in comparison to the other readers. However, a score of 1, apparent perfect consistency, probably indicates that you have a problem.

As we shall see in later chapters, human language can never be understood in the same way by every reader. That is why arranging for two or more people to read a single set of requirements is more likely to expose different meanings and interpretations. These interpretation differences either need to be clarified with the client or allowed for in planning the work needed to fulfil the requirements.

Therefore, a score of 1 indicates that you have not discovered interpretation differences: this should be a signal that you need to enlist the help of a reader with a different background to your own.

Does this seem like hard work?

Like any aspect of expertise, developing accurate reading skills takes deliberate practice, which inherently requires effort (Chapter 4). It takes time and persistence and can be much more fun if you have the support of friends and colleagues who can help you on your journey.

PERCEPTION SKILL 3: SEEING AND CREATIVITY

Earlier, we came to understand that any listening performance is an interactive, two-way conversation in which the listener responds to the speaker, sometimes almost entirely with non-verbal cues, other times with minimal responses, and often with questions to elicit more detailed explanations.

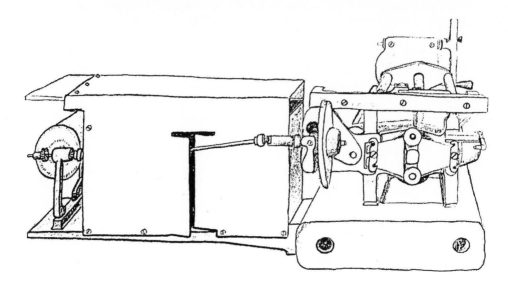

Figure 6.3 Sketch of a tensile testing machine by a student.

In the same way, seeing is a performance in which the viewer can respond in different ways. One of the most telling responses is freehand sketching, because it provides evidence about what the viewer has actually seen. Other responses can be annotated photographs, verbal or written field observation notes, or engaging a companion in conversation about the visual experience. Just as in the case of reading, where the text author is probably unable to be present, a seeing response can also require a conversation with oneself, such as annotated sketches or photographs. However, it is only when we sketch what we have seen that we can be sure that something has been noticed, in the same way that a paraphrased response to a speaker can help confirm that we have listened to and understood the intentions of the speaker. For this reason, accurate freehand sketches with appropriate annotations offer the best ways to evaluate the development of our seeing skills.

Therefore, this section of the chapter is just that: a series of simple sketching exercises to help evaluate your seeing skills. If you think that they could do with improvement, an online supplement to the book provides a graded set of exercises that you can work on independently over several weeks. These have been 'road tested' by generations of my students over the last two decades.

Most engineers see themselves, in some ways, as creative designers, or at least as creative people that can innovate and devise original engineering solutions that meet the needs of their clients.

Like any other aspect of expertise, creativity is something that anyone can develop with deliberate practice, as explained in Chapter 4. Creativity is the ability to come up with original and innovative ideas when needed. Good ideas don't simply emerge from nothing. Creativity depends on accumulating a vast memory of ideas and observations in your mind: that internal library is composed of tacit knowledge of which you are normally never even aware. You can't write it down, since you often can't remember that you possess it. However, when you need it, these memories emerge ... but not always precisely when you want them, of course.

Design expertise, in particular, relies on a vast memory of design ideas. Every time you see an engineering artefact, whether it is a culvert under a road, a bridge, a machine, an optical waveguide, a special connector, or any one of hundreds of other artefacts, you can potentially store this in your mind. However, this only happens if you learn to see the details, which depends on accurate seeing skills.

Except for engineers with significantly impaired vision, most learn more with their eyes than any other sense, but it is a mistake to think that just because you can see with your eyes, that you can see accurately (remember, just because you can hear doesn't mean you are listening). As we will soon see, your eyes and brain, when working together, can easily deceive you.

Practice concept 35: Developing seeing skills by sketching

Can you write your name? Can you draw reasonably straight lines on a piece of paper? Can you draw a square or a circle? Most people can do these simple tasks reasonably well. However, drawing the person sitting next to you or even your own hand can seem like an impossible challenge.

The reason for this stark increase in difficulty is simple. You can write your name or draw a square because your brain is more than able to control the movement of the pencil on a piece of paper. The only reason why drawing the person sitting next to you seems impossibly difficult is because your brain substitutes the image perceived by your eyes with a preconceived idea of how to draw a person. This preconceived idea takes the place of what your eyes actually see. Accurate seeing, therefore, requires that we master the mental discipline needed to suppress this automatic response by our brain.[11] Learning to see is synonymous with learning how to stop your brain from taking over your mind, successfully putting aside preconceived ideas, and allowing your pencil to reflect what your eyes are actually seeing, just like a photocopier.

'Seeing' is an active process that you can improve with training and practice. It is a state of mind that requires discipline and the ability to put aside distractions and preconceived ideas. The quality of our sketching, therefore, reflects our ability to see.

Learning to draw accurately is like learning a sport: you need to build your tacit knowledge, which is knowledge that connects your eyes with your fingers and enables you to *accurately* reproduce what your eyes see.

Training is essential, as well as deliberate practice: remember that deliberate practice can be very tiring, so build up your stamina in stages, starting with no more than 30 minutes of training each day.

Practice exercise 7: Evaluate your seeing skills

You may be surprised when you attempt these exercises. If you have learnt technical drawing skills or the use of CAD software like AutoCAD, Pro-Engineer, or SolidWorks, you may think these exercises will be easy for you. Many of my students who had good technical drawing skills have been surprised to learn how difficult they actually are.

For this evaluation exercise, you will need:

- Several A3 sheets of paper, preferably with a cardboard base (size $42\,cm \times 30\,cm$)
- A clutch pencil (0.5 mm, 2B lead), or 2B wood pencil
- Eraser

If you have a digitiser tablet or touch screen and a sketching app or basic version of Photoshop, you can do the following exercises without using paper or pencils.

Reduce distractions as much as possible. You need a quiet room with good diffuse lighting, preferably free from sharp-edged shadows. You also need a comfortable chair. Furthermore, music may help you concentrate. Some psychologists have found that certain classical music can increase concentration – works by Mozart and J. S. Bach are often recommended.

Before you start, take a few minutes to relax and let go of the tensions and distractions of the day that have built up thus far.

You will do four evaluation drawings. Take no more than 30 minutes for each and be sure to write your initials and the date on each of them.

Drawing 1:

- Draw a square, a circle, and a rectangle, side by side. Write your name immediately below.

Drawing 2:

- Draw a person sitting in front of you: either the whole figure or just a head. If you're in class, draw a fellow student. If you are alone, use a mirror to draw a self-portrait. If you can't get access to a live figure, draw a portrait by working from a photograph of a person.

Drawing 3:

- Draw your own hand.
  Place your left hand (your right hand if you are left-handed) on one side of an A3 sheet of paper, and prepare a drawing of your hand on the opposite side of the paper.
  If you think that you might be interrupted, lightly trace around just the tips of your fingers before starting so that you can return your hand to the same position on the paper.

Drawing 4:

- Place a shiny metal spoon on a book and draw them together.

  Take a short break.

Now, judge the results. You are probably much more satisfied with your square, rectangle, and circle than you are with your portrait or the drawing of your hand.

You managed to write your name? What this shows is that your ability to move the pencil is not an issue. The only reason why the other drawings were not as good is that your eye and brain combination is not yet allowing you to move the pencil in appropriate ways.

How well do the shape and shading of your portrait correspond to the likeness of the person? Would you recognise the sketch as your hand? Can you see the reflections in the shiny metal parts of your drawing of the spoon? Can you see the texture of the binding of the book?

The shortcomings in your sketches demonstrate that your ability to see can always improve. When you can see accurately, you will be able to draw accurately too!

Practice exercise 8: Evaluate your potential for improvement

In her book, *Drawing on the Artist Within*, Betty Edwards described how she stumbled on the idea of drawing upside down as a way of tricking your brain to help disable your powerful habitual recognition process. This is what Daniel Kahneman refers to as 'thinking fast'.[12] It is an essential tool for survival, but not necessarily for seeing. As you walk across the road, thinking fast is the process that tells you that there's a car heading in your direction and the approximation of how fast it's travelling. It helps you quickly pick out the face of a friend or a loved one in the crowd at a football match or on a city street. Your brain is constantly working with the visual images coming from your eyes, recognising complex patterns almost instantaneously.

While this instinctive capacity for recognition is essential for survival, you have to learn to disconnect it when you need to accurately observe all the relevant features of an object or a person – the subject of your drawing. Learning to draw is one way to do that. It is the equivalent of what Daniel Kahneman calls 'thinking slow'.

One way to disconnect your recognition engine is simply to copy an upside-down drawing. You are going to test this technique for yourself in this next exercise.

Resist the temptation to turn the drawing or your copy the right way up until you have finished. This is very important for the process.

Choose the drawing of the woman with a shawl or the wheeled robot, depending on your preference.

Draw a frame with similar proportions to the one around your chosen reference drawing.

Copy the drawing, starting anywhere you like. Most people prefer to start at the top of the upside-down drawing and work downwards.

Draw smaller, complete pieces, one at a time. Do not draw around the complete outline first and then fill in the middle.

Even if you manage to recognise parts of the picture, try and suppress these thoughts as you work. Just focus on the actual shape of each of the lines, and copy that shape to the paper: pretend that you are a photocopier.

Try not to think of anything except how the lines fit together. Ask yourself, 'What angle does this line make with the horizontal or vertical? Where is this corner in relation to other parts of the drawing?'

Use your pencil as a gauge to see where a particular part of the drawing is in relation to other parts. Observe which points lie vertically with respect to each other, and which lie horizontally.

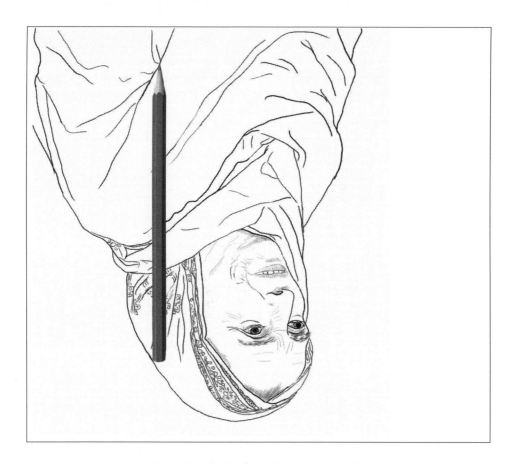

Figure 6.4 Upside-down drawing exercise 1.

Finish the body of the woman before you start on her head. Before drawing her head, construct a faint rectangle to serve as a guide for size. Then, faithfully copy the lines and dots that make up the head. Focus on each line and copy it, without trying to work out whether it is part of an eye, the hairline, or an article of clothing.

Be careful not to draw missing lines that you think should be there. The point of this exercise is to draw without recognition; to be a photocopier, not an artist.

When you have finished, turn your drawing the right way up. Most people are quite surprised at the result, which demonstrates, once again, that you can always improve your seeing skills.

Write your initials, and the date, and keep the results as another record of your work. If you have time, go on to try another upside-down drawing. Images of people or animals in motion provide wonderful practice and are easy to find on the Internet.

The rest is up to you: it normally takes 20–30 hours of deliberate practice (see Chapter 4) to make a difference for most people, but you may respond faster than that.

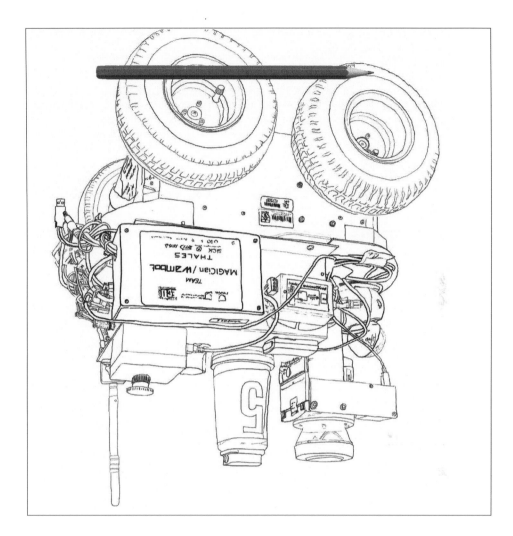

Figure 6.5 Upside-down drawing exercise 2.

NOTES

1. For example, in a study of 300 Australian engineers, of the 63 competencies rated by participants, six of the seven most important competencies concerned social interaction with other people (Male, Bush, & Chapman, 2009). These were written communication; managing communication by following up with people and keeping up to date; verbal communication such as listening and giving instructions; teamwork such as building trust and resolving conflict and disagreement; speaking and writing fluent English; and interacting with people in diverse disciplines, professions, and trades. The one other item in the top seven was managing oneself, time management, and motivation.
2. Lucena & Leydens (2009).

3. This was an element of the sheep-shearing robot's computer vision system for measuring the size and shape of a sheep in order to estimate the shape by statistical modelling (Trevelyan & Murphy, 1996).
4. Kass, Witkin, & Terzopoulos (1988).
5. Srinivasan, Zhang, & Bidwell (1997).
6. So far, we have not encountered any blind engineers in our research. However, there must undoubtedly be some. We would love to hear about them and how they work without the sense of vision that the rest of us take for granted. There are some Internet accounts. For example: http://www.nasa.gov/centers/goddard/news/topstory/2008/soyuz_reentry. html http://thehottestgadgets.com/2009/01/blind-google-engineer-creating-a-touch-screen-phone-for-blind-people-002022.
7. It is curious that BP chose to name their well after the name of the cursed town in the famous novel, *100 Years of Solitude* by the Nobel Prize-winning author Gabriel Garcia Marquez.
8. Bolton (1986).
9. For an extended discussion of this concept, refer to Ashforth & Mael (1989).
10. Lucena, Schneider, & Leydens (2010).
11. Many of these ideas were derived from Betty Edwards' books on drawing skills (Edwards, 2012a, 2012b).
12. Kahneman (2011).

Chapter 7

Collaboration in engineering

Collaboration is the central activity in engineering.

In Chapter 3, we learnt that engineers rarely spend a significant amount of time performing hands-on technical work. Most of the time, other people do that: technologists, technicians, operators, maintainers, artisans, tradespeople, and labourers. Engineers simply don't have time to develop the required tacit knowledge, which consists of the highly practiced manual skills needed to perform hand-on work efficiently with a consistent level of quality. Along with hands-on technical work, there's another aspect that is rarely performed by most engineers: drafting and detailed design. Once again, specialist drafters and technologists have the special skills and abilities necessary to work with complex computer systems and databases, skills that engineers rarely develop to a sufficient level for themselves.

Instead, as we have seen, engineers spend much of their time influencing and collaborating with other people across the engineering enterprise. Our research provides strong evidence that between 60% and 80% of the work performed by engineers is collaborative work in some form.

As examples, engineers *educate* clients and others about engineering possibilities and do their best to *understand* and *negotiate* technical requirements. They collaborate with clients to conceive of ways to satisfy requirements in an economic and socially responsible way. Often, that involves original designs using commercial off-the-shelf (COTS) components and materials, and it nearly always includes reusing a substantial amount of material and ideas from previous designs. Engineers collaborate with each other to find the most economic ways to do this, gathering information from a network of peers, suppliers, end users, and other relevant individuals. Engineers *predict* the performance of products and systems, along with organising and *coordinating* technical work performed by other people in order to deliver results, both informally and by using extensive project management systems and techniques.

In Chapter 5, we saw how engineers rely on an extended network of people, particularly other engineers, to access much of the special expertise they need to do all this. Sixty percent or more of their time is spent interacting with other engineers, clients, supervisors, specialists, technicians, drafters, technologists, and many other people who have special knowledge, experience, and understanding. They interact face-to-face, on the phone, via e-mail, by writing and reading documents, and working with information systems. The core technical 'stuff' that defines engineering is distributed unevenly in the minds of all the people who contribute. No one knows it all or can do it all. In other words, collaboration with other people forms the greater part of an engineer's work.

Engineering is, above all, a productive activity: something real and valuable happens as a result of engineering.[1] This activity requires a collaborative effort by people in an engineering enterprise. In other words, an engineering enterprise *only works* because people work together and collaborate effectively. The quality of the activity depends on the quality of social interactions between the people in the enterprise to ensure that technical ideas, concepts, and details are *accurately* exchanged. Engineers need to utilise special collaboration techniques to plan and organise predictable socio-technical human performances that rely on a high level of technical knowledge, even though every individual performance is, to a certain extent, unpredictable, and many of the people involved are forgetful, bored, tired, prone to making mistakes, disinterested, and anxious to get home.

Some would cite software engineering, specifically programming, as an obvious exception, because software engineers are directly involved in producing software code; in other words, it is inherently 'hands-on'. How can you be a competent software engineer if you don't know how to program, and how can you be good at programming without practice? Our research shows that some software engineers are engaged in programming for *part* of their time; however, they also have to collaborate with many other people *most* of the time, just like other engineers.

When I was starting my career, I wanted to be an expert technical engineer. I had set my heart on aerospace engineering and saw myself as a specialist, working with new displays and controls that would be used by aircraft pilots. Yet, the more I developed my technical skills and knowledge, the more I found that I was dependent on many of the other engineers working around me. Becoming a technical expert *increased* my dependence on collaborating with others, so being skilled at collaborating was essential to my success.

The very large amount of time that engineers spend interacting with other people involves communication; communication can provide evidence that collaboration is happening. Notice how much effort is needed for this: collaboration does not happen without effort from everyone involved.

What, exactly, do we mean by communication?

For people to be able to collaborate, they have to be able to communicate with each other. To put it another way, communication is the means by which people collaborate. Rather than reading from the last chapter and thinking about different communication skills, such as listening and observing, it is more helpful to learn about the different engineering activities that require people to collaborate (and sometimes experience conflict, as well). Developing your ability to become an expert engineer will depend on your ability to master all the different ways that engineers collaborate with other people.

Therefore, this chapter is all about collaborating effectively with other people. The most important aspects of this book for you to master are the coming chapters on collaboration skills for engineers. Your collaboration skills will determine how far you go in your engineering career.

Let's start by examining, once again, some of the misleading assumptions that we tend to bring into engineering practice from our formal education. Here is one of the most common ones.

Misconception 10: I'll need communication skills when I'm a manager

Students and novice engineers often think that ...	However, research demonstrates that ...
Engineers work in technical roles for the first few years and don't need to develop communication skills any further. Later, if an engineer goes into management, better communication skills will be needed.	Engineers can only achieve results with the help of others. Engineers need to work with and influence other people from the very start of their careers. The ability to communicate dictates the ability to collaborate.

Throughout your educational career, your grades were the tangible indicator of success. Almost all your grades have been the result of your personal efforts, except for various group assignments when you might have shared a mark with two or three other students. Basically, you have a deeply embedded understanding that your personal effort is what counts most. Furthermore, nearly all your grades were earned by writing something, usually in an exam or perhaps a computer test.

As an engineer, you have to face a very different reality. The value and quality of the results from your work *depend on the quality of work performed by other people*. Devoting more effort to your own work is unlikely to help. Instead, spending more time helping others do their work well is much more likely to improve the overall results. You will be judged by what others achieve while acting under your guidance. Some of my students become very resentful if I award them grades that depend on work done by other students, but this is the reality of engineering practice.

That being said, due to the fact that you have been rewarded according to your individual efforts for so long, it will be hard for you to adjust. You need to learn completely new skills. Your ability to influence and support the work of other people will enable you to become a truly expert engineer.

Many young engineers see themselves heading for a career as a technical specialist, partly because they are shy, just like I was at their age. I felt that other people were much more socially competent than I was. I tended to stay quiet and just get on with things that I could do for myself. These days we call people like this 'task-focused'. However, I soon learnt that the more technically specialised I became, the more I depended on other people to translate my ideas into practical reality. I found that I could not escape from the need to influence other people to adopt and build on my ideas. Given my personality, I found that skill set to be intimidating and challenging.

I have seen very talented engineers, both young and old, who spend most of their time writing, designing, coding software, developing elaborate computer models, or even building hardware. They often feel ignored and frustrated when confronted with organisational obstacles, mainly encountered when other people don't seem to listen. I know that frustration because I have been there myself.

Gradually, I learnt that as an engineer, you're completely stuck without other people supporting your efforts. Keeping them on your side, retaining their support, is time-consuming, but eventually becomes the most enjoyable and satisfying part of your working life. I soon found that I could improve my social skills with a moderate amount of guidance in practice. Before long, I was just as competent as most other people, and I had soon lost my shyness and gained the necessary confidence. Like anything else, as Chapter 4 explained, anyone can learn these skills with sufficient, deliberate practice.

At first, the intellectual challenges of doing this can be daunting, but if you're like me, it has been precisely these intellectual challenges that have been the most satisfying aspects of my engineering career.

While different aspects of collaboration take up the majority of time for most engineers, it is vital to remember that it is your technical knowledge and your proficiency in using it that define your identity as an engineer. However, without other people helping you, your technical proficiency counts for little.

You might see this as the 'non-technical' side of engineering. However, this is a misleading label. Most engineers talk about technical issues with colleagues a large portion of the time. These social interactions are a critical part of getting the technical work done.

This is what we call 'socio-technical' work. Gaining the support and willing collaboration of other people is the most rewarding path to success. As we saw in Chapter 5, it is also the way that you get access to much of the technical knowledge on which you will build your own expertise.

Through our research, we have learnt that most undergraduate engineering students think that communication is important when they start their studies, but by the time they graduate, solitary technical capabilities like design, calculations, and computer modelling seem much more important.

Another surprising result from our research was that engineers spend just as much time interacting with other people right at the start of their careers as much more senior engineers. The idea that communication becomes a bigger part of your life once you reach a management position is not supported by the research data.

Many young engineers soon find themselves immersed in analysis calculation studies. However, the quality of design, calculations, and computer modelling reflects the quality of the information and input data. Obtaining input data with the help of other people and assessing its quality and reliability is usually very time-consuming and takes much more time than actually performing the computer modelling work. Since the advent of high-speed computers, the time needed to perform calculations once the data is available can be very short.

However, calculations only become valuable when people act on the results, and getting this to happen can be much more difficult that it seems. Young engineers find, often to their surprise and frustration, that it can be difficult to influence engineering decisions based on the results of their work. Gaining that 'influence' is where collaboration skills are so essential.

Learning how to collaborate effectively with other people is critical to becoming an expert engineer.

How good do you consider your collaboration skills? You may be in for a surprise: here's another misconception in the minds of many young engineers....

Misconception 11: I already have good communication skills

Students and many novice engineers often think that ...	In reality ...
They have good communication skills. They gained reasonably high marks for written assignments, particularly project reports, and also for technical presentations.	Employers frequently complain about the poor quality of graduates' and novice engineers' communication skills.[2]

Both perceptions can be correct. The mismatch can be understood by realising that the communication skills that lead to high marks in the academic environment are quite different from the *collaboration* skills required on the job. Effective collaboration skills and techniques are very different from the communication skills that boost marks in higher education. We will revisit this theme many times.

Misconception 12: Communication skills and teamwork cannot be taught

Students and novice engineers often think that ...	Research reveals that ...
Teamwork and communication skills are learnt by practice: they cannot be taught.	Novice engineers do not seem to know much about collaborative work practices, nor do they seem to practice them. Engineers who have been taught communication skills have greater confidence in their communication abilities.

Surveys indicate that engineering graduates, particularly recent graduates, rate their team skills highly and appreciate the importance of team skills.[3] Therefore, it has been surprising to observe engineers in real world workplaces with little or no awareness of, and very little practice in, genuine collaboration skills. For example, we have observed critical weaknesses in peer review processes in all the engineering workplaces that we have observed in detail. While we have observed engineers attending meetings with clients and other key stakeholders, rarely did we observe them collaborating effectively. Furthermore, in our interviews with them, they rarely related anything to us about how they collaborate effectively, for example, to maximise the accuracy of their records of those client meetings they had attended. Few of the engineers that we interviewed demonstrated extensive use of close collaboration in their work, except as a means to share the work that had to be done to achieve a given objective.

As we shall see later, most engineering students have few, if any, opportunities to learn about effective team skills: they are seldom taught.[4] Instead, it is widely believed in engineering schools that team skills develop through practice. The difficulty is that practice without feedback reinforces inappropriate and destructive skills as much as it reinforces constructive team skills. In the absence of instruction and constructive feedback from teachers, it is easy to believe you are doing the right thing just by practising it over and over, when you may actually be doing the wrong thing without realising it.

There is a further factor at work here. Collaboration is often confused with cheating in formal education. Our formal education systems value individual learning performance. Your degree transcript will show the marks awarded to *you* in each of the subjects that *you* studied: it will not reveal anything about your ability to collaborate with other students. In fact, helping another student might help to build a friendship, but it can get you into trouble if your friend simply copies your assignment submission, since both of you might be accused of plagiarism. In some schools, marks are normalised such that the average mark and pass rate falls within acceptable limits. In these circumstances, helping weaker students succeed can make it harder for everyone.

Naturally, the majority of students help each other informally. Most students enjoy socialising with others, and often work on assignments in groups of friends. Students often collaborate and help each other strategically, as well, particularly when working on group project assignments. For example, if one student is stronger at mathematics and another is stronger with written explanations, it seems natural for the former to do most of the work on the maths-based elements of a group assignment while the latter writes the reports, even across different subjects in which both students are enrolled.[5] This logical breakdown of responsibilities means that neither student gets the chance to improve their weaker skills. Deception starts when both parties claim the same marks, even though they have not contributed equally to all parts. This kind of collaboration, therefore, is a form of cheating, and is often associated with questionable behaviour.

For that reason, collaboration is practiced in higher education but is less frequently valued and acknowledged.

Therefore, it is no surprise to find that novice engineers have a limited understanding about collaborative work practices; this weakness could limit their capacity to be an effective engineer.

One possible exception may concern students who find themselves struggling to pass courses. They rely on other students to help, often getting them to do part of their assignment work for them. While this might be questionable from an ethical standpoint if it is not openly acknowledged, the skills that the students develop by engaging the willing support and cooperation of their peers may eventually help them get ahead in the workplace.

If you are still studying, learn to collaborate with other students. They can help you see the same assignment problem in a different way, which helps you develop your ability to anticipate all the different ways that the same problem can be interpreted by different people. Once you reach open-ended and ill-defined problems, or if there is insufficient time to complete all the assigned work, take those opportunities to practice making your own judgements on the best trade-off between completing the scope (doing everything demanded of you) and working to an acceptable level of quality and rigour.

Misconception 13: My boss will tell me what to do

Students often think that ...	However, in reality ...
Their boss will tell them what to do.	Engineers are expected to know what to do.

For most of the time, you will know much more about technical issues in your own area of responsibility than anyone else. Even your boss will be relying on you to acquire detailed technical knowledge that he has neither the time, the inclination, nor even the necessary prior understanding of the subject to learn for himself. In the words of one engineer:

'My boss wants me to bring solutions, rather than problems.'

Acting on your own initiative by anticipating what your boss will need, without waiting to be told what to do, is a great way to enhance your reputation. At the same time, doing the wrong thing can be equally foolhardy. The smart way to proceed is to discuss your intentions to ensure that your boss knows what you are planning to do. He or she will then provide guidance and direction for you, if you need it.

It is important to find an appropriate balance between acting on your own initiative and knowing when to ask others for advice.

Misconception 14: In the real world, slackers will be fired

Students often think that . . .	However, in reality . . .
Teams work well in the real world because freeloaders or social loafers will be fired; everyone contributes equally in the real world.	Team members contribute unequally in practice. Teams are often dysfunctional in the workplace and freeloaders are occasionally rewarded. Firing people is often not an option because of the inherent delay and cost of finding and training a replacement person.

In engineering, teams are usually formed from selected specialists with complementary technical skills, expertise, and experience levels. This is quite different to the typical university experience in which teams consist of students, all with more or less the same (learner) level of capability and experience.

In normal workplace situations, it is entirely unrealistic to expect equal contributions from all team members. The extent of the contribution from each specialist will depend on the needs of each particular project.

In a university course, most students will have a similar set of assignments to complete, even though students differ in ability, personality, and the time needed to complete a given task.[6] In workplace teams, however, different engineers can have very different demands on their time. Practically everyone experiences periods of frantic activity in which everything seems to be urgently needed 'by yesterday', followed by more relaxed times with very few urgent demands. This means that the ability of any individual to contribute to teamwork can depend on many other unrelated project commitments, which can make teamwork very difficult at times.

Dysfunctional teams and freeloaders are not uncommon, so an expert engineer learns to achieve the desired results despite those obstacles. It is quite possible that a freeloader is developing a closer relationship with your boss instead of contributing to the project and may actually be promoted ahead of you. Key team members often need to be transferred to other projects partway through, which means that you may have to help the replacements catch up. Be ready for those unpredictable challenges!

Firing people who are not contributing to a project team is often impractical. It can take months to find and train someone who can effectively replace an engineer: by then, the project may well have been completed. Furthermore, a person often has to be compensated if they are made redundant.

The notion held by most students that 'all members contribute equally in an ideal team' is probably developed from the common assessment practice in formal education of awarding equal marks to all team members. We will return to this topic later when we discuss diversity in the context of teamwork.

Practice concept 36: Communication is all about people collaborating and coordinating their work

Students and novice engineers often think that ...	Research demonstrates that ...
Communication is information transfer from the engineer to other people. Communication skills explicitly assessed in university courses normally include technical writing and graphics for presentations.	Communication is the means by which people interact in order to collaborate in a coordinated performance. Human behaviour is guided by perceptions and relationships: communication enables us to modify perceptions, build relationships, and consequently influence human actions.

Our research shows that there are more misconceptions on communication in engineering than on any other aspect of practice.

The idea that communication in engineering practice is *only* about information transfer is remarkably deeply embedded in the literature and practices of engineering education. That is why it may be a struggle for you to overcome this misconception. Even engineers who have successfully run large companies will still explain communication in terms of information transfer, even though their *actions* demonstrate that they understand the basic idea that communication is all about collaboration.[7]

Can you remember the last meal that you shared with one or more of your friends? Did you remember what you talked about? Unless it was an exceptional meal, you probably can't remember all the details. You probably exchanged a few comments about current events, the latest sporting results, the weather, and maybe other family members. Was there any transfer of information going on that was important enough to write down? Almost certainly not.

Now, I would like you to think what would happen to your friendship if you never talked with your friends (or in the case of complete deafness, you did not communicate in any way). Except in very exceptional circumstances, your friendship would probably not last very long if you did not chat with other people. Friendship requires social interaction.[8]

The point I'm making here is that there are many different styles of communication. Linguists call each communication style a 'genre'.[9] Informal chatting over a meal is often meaningless by itself: people repeat stories that others have already heard, they share gossip based on half-truths and misunderstandings, and they exchange a few jokes, some of which might be in very poor taste. What is said is not important. The importance lies in the act of enjoying a meal together, in order to strengthen a relationship based on countless similar shared experiences, to bolster a friendship, and to build trust and confidence in each other. This kind of socialising can be critical for engineers who often share 'war stories' about critical learning experiences at informal social gatherings instead of in the workplace.[10]

In just the same way, engineers and other staff members chat with each other around the coffee machine at the office and seem to be just passing time, doing nothing in particular. It would be easy to think that they were just wasting time. They may do the same thing over a few beers after work at the pub. They don't talk about anything of importance for much of the time. Yet, this is all about building and maintaining relationships and trust. Those social bonds can transfer back into their working life, especially in gaining willing and conscientious cooperation.

Another reason for communication among engineers is to maintain or modify perceptions. People often have to make decisions very quickly. Here is an example of a scenario that emerged several times in our research interviews:

A lead engineer[11] may not have time to read and understand a 200-page technical analysis of several equipment purchase options that may have taken you several weeks of quiet 'back-office' research to prepare. Three hours after receiving your analysis, you hear from someone else that your lead engineer has completely ignored your recommendations and decided to purchase equipment that was not even analysed in your report. You might be angry and disappointed that you spent so much time on seemingly pointless analysis that has been ignored: the lead engineer has obviously been 'swayed' or influenced by someone else in the organisation. Many engineers call this type of situation 'office politics': the term is used to describe seemingly irrational behaviour and decisions that don't seem to follow any kind of systematic logic and do not appear to be based on the 'objective facts'.

This is an example that demonstrates how many decisions seem to be based on 'irrational perceptions' rather than 'scientific facts', at least to 'us engineers'.

There is another possible explanation. The lead engineer may have spent a lot of time with other people who had carefully shaped the perception that they were well informed and had lots of experience with similar types of purchasing decisions. By the time your report landed on his desk, the lead engineer had already made up his mind. He may have briefly glanced through the report, selectively looking for evidence that would confirm the decision he had already made. He would have typed a brief e-mail to his manager stating his conclusions, possibly including excerpts from your report to justify his decision, even though the decision was quite different from the conclusions in your report.

Even if you were to query this decision at a subsequent meeting, the lead engineer might simply have replied, 'Your report was very useful; in fact, the final decision was largely based on the evidence you collected.'

It would be a great mistake to assume that senior engineers and managers are the only people who make decisions based on perceptions. We all do this. In Chapter 6, I explained how our perceptions are based on the beliefs and preconceived notions that we have already developed in our minds.

This is an example that shows just how easy it is to let perceptions influence the kinds of decisions we make every moment of every day.

As a student, you almost certainly had to perform at least one technical presentation during your studies, probably as a summary of the work you performed in your final year, in a capstone technical investigation or a design project. The presentation was probably brief, only 15 minutes or so: it was quite impractical to present a fully detailed explanation about a technical undertaking that took you many weeks to complete. You probably found that there was insufficient time to try and explain all the technical details. Therefore, the main aim of your technical presentation was to create a perception in the minds of your assessors that you had competently performed the investigation.

Thus, what is commonly referred to in the engineering education literature as simply 'transferring technical information from the student to the audience', is therefore much more about creating a desired perception in the minds of particular members of your audience. Your aim, therefore, is to build a perception in the minds of your assessors that you, the presenter, understand the technical details sufficiently well: there is no need to present all of them.

Our research has shown, over and over again, how engineers experience immense frustrations in their work that can ultimately be traced to misconceptions about communicating with other people. Nearly all engineering disasters can be attributed to communication failures among the people responsible. It is all too easy to blame communication failures on other people: 'they had no excuse … I sent them the information in an e-mail nine months before!'

We need to understand that the narrow idea of communication as information transfer has become deeply embedded in the ways that engineering educators (if not many others) discuss communication. Few researchers have broadened this idea significantly.[12] Even though you will gain a wider appreciation of communication and subsequent collaboration from reading this book, you may continue to find that most other engineers still think about communication as information transfer.

One complaint that I have often heard in research interviews and conversations with middle level engineers relates directly from their interactions with accountants. For example, 'This company is run by f—ing accountants! They don't understand even the simplest ideas in engineering.' These days, when I hear this, I ask 'What do you think an accountant means when they mention accrual accounting?' I have yet to meet an engineer who can provide an explanation. Yet accrual accounting is an even more basic concept for an accountant than conservation principles are for engineers.

The lesson here lies in language. If you wanted to take out an attractive French partner for a romantic dinner for two, and the partner could not speak any English, you would most likely take the time and trouble to learn basic French (like j'aime tu, 'I love you') and you would probably take a phrasebook or electronic translator with you. Dinner would not be much fun without being able to share some basic and simple conversation.

Therefore, it follows that engineers need to understand basic concepts and language that accountants use in order to have anything close to a productive conversation. It is not very helpful or productive to simply blame them for a lack of understanding about engineering fundamentals.

The expert engineers whom I have interviewed have not necessarily understood accrual accounting either, but they have all had sufficient knowledge about investment, accounting, and finance to be comfortable working with accountants, financiers, and bankers. Few of them had completed an MBA degree; however, all of them had taken the time and trouble to make sure that they could converse using the language of finance and accounting sufficiently well.

Using written communication for collaboration is essential for engineers for many reasons. Firstly, written communication provides a compact record that can be retrieved later. In the present moment, as it is occurring, oral communication is rarely recorded. (Most of us can be thankful for that, having made very unwise comments in casual conversations.) Secondly, written communication is largely independent of time: messages can be exchanged without both the writer and recipients being present (or online) at the same time. Lastly, written communication carries much more weight from a technical legal perspective: the written record will usually be relied on much more than recollections of what was said, at least in an Anglo-European-American culture. This is not necessarily the case in all cultures, however.

While formal written communication is taught in engineering courses to a limited extent, in the context of genres such as technical reports, written communication for coordinating collaborative activities is hardly mentioned. Most engineers find that they need to devote a lot of effort to improving this ability: it is hard to be accurate, concise, and comprehensive at the same time. It is also difficult to anticipate how simple written statements can be interpreted in quite different ways, depending on the context in which the recipient reads them (carefully or otherwise!).

Misconceptions about communication are the main reason why so many engineers end up hitting a brick wall that blocks their career paths, as I tried to portray in Figure 3.1. In a recent survey of several thousand civil engineers in the USA, most complained that they had been passed over for promotions and management responsibility in projects that were instead awarded to people who were not qualified to be engineers.[13] Without the opportunity to learn the concepts in this chapter, these perceptions held by so many engineers are not surprising. These perceptions are also close to the reality for many engineers in Australia and other countries. When it comes to communication, most other professions see engineers as 'challenged'. Overcoming several deeply buried misconceptions could change that image: there is no reason for you to be seen that way if you can master the ideas in this chapter, as well as the concepts in Chapter 8.

Practice concept 37: Communicating technical ideas effectively relies on technical understanding

Students, novices, and even experienced engineers often think that ...	Expert engineers know that ...
Communication is a non-technical subject: it is not central to engineering.	Communication is an essential technical skill: communicating effectively in the context of technical work is highly specialised and difficult to learn to do well.

In the context of engineering practice, communication is nearly always referred to as a 'non-technical' skill. Any expert in communication can provide training and

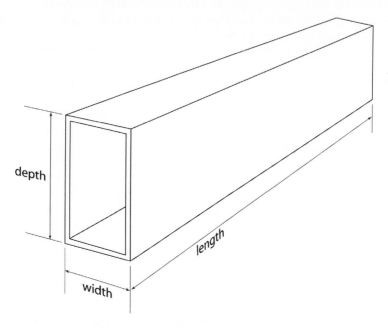

Figure 7.1 A rectangular hollow section beam.

assistance: it is not necessary for the trainer to have any understanding of technical issues.

Once again, this arises from the misconception that communication skills are the same as collaboration skills. There is a further misconception here, as well – that technical collaboration is the same as any other kind. That is not so, as we shall shortly see. Research reveals that effective technical communication is much more complex than it might initially seem to be.

Let's start with an example.

Figure 7.1 shows a sketch of a rectangular hollow beam. Imagine that this beam is required as part of a building renovation and refurbishment: the beam will support part of the floor above a meeting room.

You can assume that the designer has already chosen the width, depth, and section thickness from the standard beam sizes commonly supplied for the construction industry by steel supply companies.

The length has been determined from the spacing between the supporting columns that will support the floor.

The material is to be AS 1163 Grade C350L0 (an Australian structural steel material).

Practice exercise 9: Specification writing

Your task is to write an outline of a technical specification for a supply contractor or fabrication firm to use to ensure that the beam incorporated into the refurbished building will be appropriate for the intended purpose and will function as intended for approximately 200 years, at least. You don't have to write the full details of the specification; you simply need to indicate how the specification could be written and what aspects it would cover.

You can use the interactive online website for this (or save the responses in a word processor to copy and paste later).

1. _____
   _____
   _____
   _____
   _____

2. _____
   _____
   _____
   _____
   _____

3. _____
   _____
   _____
   _____
   _____

4. _____
   _____
   _____
   _____

5. _____
   _____
   _____
   _____

Once you have completed this exercise, carefully read the self-assessment guide in the online appendix and analyse your responses. You may be surprised!

You might also be surprised to learn that few standard industry courses available for young engineers would help you in an exercise such as this. Industry courses on writing specifications spend much more time on how to lay out the text, ensure correct grammar, and so on. They spend very little time helping you understand what you need to include in the specification and why.

A technical specification is one of the several different ways that engineers convey intent: how to try and ensure that other people acquire an accurate understanding of what the originating engineer intends to happen. We will discuss specifications again in Chapter 10.

Practice concept 38: Engineering is a series of collaboration performances

Remember that engineering is a productive, collaborative activity. There are many different ways in which people collaborate.

Contemporary linguists see communication in terms of performance, partly because each style of communication is more than just words or pictures. Each different style of communication performance is called a *genre*. For example, friendly informal face-to-face conversations, formal dialogues, debates, arguments, explanations, novels, letters, plays, and poems are all different genres of communication.

Collaboration is a different concept from communication.[14] However, as with communication, we can conceive of collaboration as an activity that resembles a 'performance', just like acting in a play. It is helpful to understand what the aim of the performance might be. After all, the intention of collaboration influences the nature of the human performances involved. Two people might be collaborating to build an artefact, such as a table or chair. Alternatively, a larger group of people might be collaborating so that they can enjoy themselves during a social occasion like watching a movie. In a still different performance, two people might be collaborating with each other to cultivate mutual trust so that at some future time, one person would happily trust the other person to do something for them that would demand integrity, honesty, and competence. We can refer to each of these different styles of collaboration as different collaboration genres. These performances do not involve communication acts alone; there are many other aspects involved, such as choosing the right clothing and picking appropriate times and places for the performances.

Trust is an essential element of collaboration in engineering.[15] Trust is vital to support the effort needed for accurate listening and respect for other viewpoints, and can also reduce the need for close supervision and monitoring. Without trust, the time and effort involved in monitoring the performance of technical work performance by other parties will likely impede one's own performance, particularly when the monitored work is considered critical for success. Effective engineering relies on the ability to influence others, but that influence must first be founded on building respect and gaining the trust and confidence of others.

Collaboration involves many different social interactions between two or more people. These interactions can take the form of words uttered as sounds or written in symbolic form, as images, drawings, or illustrations. Along with tangible interaction, human interactions involve many other aspects, such as changes in vocal quality when speaking, posture, facial expressions, other sounds, or bodily movements. These have become known as 'non-verbal communication' or simply 'body language'. The postures adopted by people, the clothing they wear, where they choose to interact, the ways that they emphasise different sounds, the way that text is laid out in a document or e-mail: these are all components of a performance, although they can also be considered different forms of non-verbal communication.

As explained before, an engineering enterprise is a large collaboration performance by many people. In fact, as explained by the concept of a 'value chain' in Chapter 5, engineering enterprises collaborate (and often compete at the same time) with each other.[16]

In order for you to understand how we can make use of the idea of a performance, let's consider something that we would more readily recognise as a performance, such as a ballet. Let's set this out in a way that we can see the corresponding ideas that emerge in an engineering performance, shown in table 7.1.

Now that we can understand more about this idea of an engineering performance, let's review all the main aspects of engineering practice, meaning what we have observed

Table 7.1 Comparing a ballet and an engineering performance.

Collaboration	Ballet	Engineering
Discovery of desire or need among people who would appreciate its value.	Promoter discovers an opportunity to stage a ballet performance in a city.	Engineer discovers an opportunity to supply a product or engineering service in a city.
Further discovery through discussion and feedback from people who will pay for the performance or product.	Promoter discusses the ballet performance with friends, acquaintances, and representative patrons in the community.	Engineer discusses the idea for the product or service with clients and end users to understand the level of commercial demand in the city.
Discovery of common interests between promoter and engineer with financiers and major sponsors, agreement to proceed with initial preparations.	Promoter discusses project with financiers and major sponsors to seek financial support.	Engineer discusses project with clients, financiers, bankers, and others to seek financial support.
Promoter and engineer set up their production leadership teams.	Promoter appoints his leading team members: a choreographer, creative director, music conductor, and possibly a composer if new music is required.	Engineer appoints his leading team members in each of the main disciplines including fabrication, design, operations, procurement, sales, and marketing.
Production leadership teams prepare detailed plans and complete everything needed for final investment approval.	Choreographer and creative director work with music conductor, possibly the composer, to adapt the music score and develop the choreography (dance movements) with a deep understanding of what dancers can do. They design the staging arrangement and arrange the music parts to suit the musicians in the orchestra.	Engineers develop the product design and service specifications, implementation plans, gain the necessary approvals, and oversee the production of detailed design documentation, project plans, and budget. All this happens with an intimate understanding about the capacities of local firms to deliver all the engineering capabilities needed.
Promoter and engineer seek final approval from their financiers and sponsors.	Promoter develops a business plan and seeks final approval from sponsors to hire the dancers and musicians, and then go ahead with the performance production.	Engineers present detailed plans and budget to financiers for a final investment decision. Once the go-ahead is given, there is no turning back.
Production leadership teams mobilise their larger workforce, as well as contractors, and arrange production facilities.	Creative director arranges auditions to hire dancers and musicians, books rehearsal studios, appoints a stage designer, and commences stage construction.	Engineers call for expressions of interest from contractors, review bids, select contractors and equipment suppliers, finalise contracts, obtain space to set up production facility, and hire equipment needed for production.

(Continued)

Table 7.1 Continued.

Collaboration	Ballet	Engineering
Production leadership teams organise and oversee full preparations, adjusting plans to suit the limitations and abilities of everyone involved. Sales and marketing organisations move ahead with campaigns to attract buyers.	Creative director works with the choreographer and dancers in daily rehearsals over many months, teaching them the choreography, discovering the limits of their performance abilities, working around those limits, and then adjusting the score and choreography accordingly. The promoter starts a marketing and ticket sales campaign.	Engineers organise and oversee the set-up of production and service delivery facilities, and train and develop production and service delivery workforce capacity and skills. Sales and marketing professionals organise business development and preliminary sales and distribution contracts.
Production teams make their final practice runs in rehearsals to ensure that everything is ready.	The team leaders come together for the final dress rehearsals; the promoter arranges for sponsors, leading journalists, and reviewers to see preview performances.	First products off the production facility go to friendly users who ensure that they meet all the necessary requirements before final adjustments to the production process and service delivery arrangements are made to iron out any last-minute problems.
Production teams commence their full-scale performances.	Production of the performance over the limited season. As the audience excitement develops through the performance, the dancers gain energy from the audience response, pushing themselves to greater efforts to gain even more lift from the audience appreciation. Months of intensive work finally result in what seems to be an effortless performance by the dancers, yet it is anything but effortless to produce the performance.	Full-scale production and service delivery commence. Customer appreciation for the product and service generates enthusiasm among the engineers and production workforce, lifting their desire to achieve production and sales targets, while also giving them a sense of great satisfaction and achievement. What seems like such a simple product to produce has been anything but simple and effortless to bring to the market.

engineers doing in their work, so that we can begin to see all the different collaboration genres playing out.[17]

Engineers use their special knowledge of materials and physical and abstract objects to decide how to rearrange them so they perform some required function with desirable properties, such as working within the engineering capabilities of locally available firms, labour, and suppliers, while also yielding economic or social benefits for people. We can describe this thinking as 'technical'. Thinking is human, however, and we need to recognise that even technical accomplishment is limited by human capabilities.

None of the engineers who we observed in our research worked alone. To a greater or lesser extent, they all relied on interactions with other people in an extended network; their practice was based on distributed knowledge, most of it unwritten, that had been developed through years of practice, making it difficult to transfer to others.[18]

Engineers informally coordinate with other people, which mean that engineers willingly and conscientiously contribute their expert knowledge, mostly through skilled performances. Sometimes engineers start with little or no overlapping understanding, so helping others to learn and learning from others is always a part of practice. Translation[19] and negotiating shared meaning to enable understanding across different areas of expertise is part of this as well.[20]

Engineering, therefore, is a combined performance involving clients, owners, component suppliers, manufacturers, contractors, architects, planners, financiers, lawyers, local regulatory authorities, production supervisors, artisans and craftspeople, drafters, labourers, drivers, operators, maintainers, and end users, among others. In a sense, the engineer's role is both to compose the music and conduct the orchestra, all while working outside the lines of formal authority.

Engineering performance, like most human performance, is time-, information-, and resource-constrained. In engineering practice, therefore, people have to allocate time and attention to satisfy many diverse demands. Complete information is rarely available, and the information always comes with some level of uncertainty. One cannot predict nature completely, although we are getting better at it, so there will always be some unpredictable elements. Rarely, if ever, did the engineers who participated in our studies have extensive, uninterrupted time to think and reflect. Engineering performance requires rapid and difficult choices in terms of using personal time and material resources.

The value that arises from the contributions of engineers is only created through the actions of many other people, often far removed from the setting in which engineers actually perform their work. Therefore, an engineer has to ensure that everyone involved has sufficient understanding of the essential features that will create value to ensure that they are faithfully implemented and reproduced by other people through planning, detailed design, production, delivery, operations, and maintenance.[21] The people who use the products and services also need to understand how to make effective use of them to gain their full value. In other words, engineers have to explain, often at a distance and through intermediaries, how the products of their work need to be designed, built, used, maintained, and disposed of.

Engineers' stories about their work reveal that nearly all engineering projects follow a similar sequence. Many engineers who participated in our study were contributing to several projects at the same time, each one at a different phase of the sequence presented here.

Phase 1: Understanding clients and their needs

At the start, engineers attempt to understand and simultaneously shape clients' perceptions of their needs, and work with clients to clarify requirements. Helping the client to see their objectives in terms of engineering possibilities is part of an engineer's job. Gaining clients' and investors' trust and confidence is essential because a great deal of time and money will be spent before anyone gains the benefits of the project and money is repaid to the investors. At the same time, engineers also develop their own businesses by helping clients anticipate future needs and then helping their clients with the groundwork for future engineering projects.

Phase 2: Conceiving an alternative future

Engineers collaborate with stakeholders to conceive different ways to economically meet requirements, propose solutions using readily available components, and design special-purpose parts when needed. Much of the design work involves rearranging elements drawn from a vast memory of design fragments in the minds of different participants and piecing them together in new ways.[22] Engineers solve technical problems, though many of the engineers observed in this study avoided technical problems as much as possible through a combination of shaping client expectations, foresight, experience, making use of well-understood techniques and known solutions, careful planning, and methods to organise effective collaboration.

Phase 3: Predicting an alternative future

Engineers collect data through their network of peers and others, and create mathematical models based on scientific knowledge and experience to analyse and predict the technical and commercial performance of different solutions so that sensible choices can be made. The level of precision depends on investors' acceptance of risk and uncertainty. Engineers usually describe uncertainty qualitatively, occasionally quantifying it, accounting for incomplete data, uncertainty in the available data, and from external uncontrolled events. They also diagnose perceived performance deficiencies (or failures), conceive and design remediation works, and predict how well the modified system will perform. They also negotiate for the appropriate approvals from regulatory authorities.

Phases 1, 2, and 3 may be repeated with progressively more certainty, particularly in large projects, until prediction uncertainty can be reduced to match the investors' expectations.[23] The term 'engineering problem-solving' is often used in a sense that embodies phases 2 and 3.[24]

Phase 4: Planning and organising for the future

Using the engineers' predictions as a starting point, the client, investors, regulatory authorities, and contractors must decide whether to proceed with the project (the FID or Final Investment Decision). The lead up to this point is often called 'front-end engineering'. Once 'project execution' starts, engineers prepare detailed plans, designs, and specifications for the work to be performed, followed by organising the people and the procurement of materials, components, machinery, and other resources that will be needed for construction, commissioning, operations, and maintenance.

Phase 5: Delivering the future as predicted

Engineers coordinate, monitor, and evaluate the work while it is being performed, adapt plans and organisational techniques to different circumstances, explain what needs to be done, and ensure that the work is performed safely, on an agreed schedule, within an agreed budget, and within negotiated constraints such as regulatory approvals, effects on the local community, and the environment. Although engineers carry these responsibilities, they are reluctant to use formal authority (and it is only rarely available to them). Instead, they rely on informal technical coordination. The aim is to deliver the intended products and utility services with the predicted performance and reliability.

Phase 6: Reuse, recycling, remediation

Engineers conceive, plan, organise, coordinate, monitor, and evaluate decommissioning, removal, reuse, and recycling at the end of a product's life span, as well as the rehabilitation, remediation, and restoration of the site and the local environment.

Since engineering is a human performance, we need to accept that the performers have to deal with unpredictable aspects, like nature. Given that the aim is *predictable* delivery of *reliable* products and services, engineers need to know how to ensure that the *unpredictable* aspects of countless individual performances produce results in a *predictable* way. Assessing risks and uncertainty, checking and review, technical standards, organisation, training and procedures, coordination and monitoring, survey and measurement, teamwork, configuration management, planning, testing, and inspection are all components of an engineer's repertoire for containing human and natural uncertainties.

Engineers are involved in the training and development, not only of other engineers,[25] but also of all the other people who contribute to the process, including end users. Engineers also work on technology improvements and explain technological possibilities to society, businesses, and governments. They help ensure that policy decisions are properly informed and that the costs, risks, consequences, and limitations are clearly understood.

COLLABORATION GENRES

By closely examining all the observations of engineers at work, it is possible to identify a small set of different performance styles that engineers use when they collaborate with others. Often, they are used together in a combined performance. Naturally, each one plays out in a variety of ways in different situations. A given performance can sometimes be classified differently, depending on the circumstances, particularly if it is part of a combined performance.

A script

Engineers usually define a script that specifies the sequence of actions to be performed with or without verification actions as a series of written documents. Creation of the script usually requires collaboration. While not a collaborative performance in itself, a script lies at the heart of any collaborative performance in engineering.

Note that the existence of a script does not necessarily lead to the actions being performed. It is necessary for people to learn that they need to perform the script as a whole, and then learn to perform the actions listed in the script. Not only does that require a good deal of learning, but it also possibly requires teaching as well.

Examples:
i) a construction plan,
ii) a cable installation diagram,
iii) a drawing showing the sequence of assembly,
iv) an organisational procedure,
v) a negotiated agreement,
vi) a specification, and
vii) a testing and acceptance plan.

Teaching

Engineers facilitate learning for others, including explanations, instructions, monitoring their performances, providing them with feedback to improve their performances, and assessing their performance quality.

In addition to the brief introduction in a section later in this chapter, teaching is so important that it justifies an entire chapter on its own. Engineers have no option: success depends on them being able to teach so they can help other people to learn about their ideas and work with them effectively.

Examples:
i) educating clients about engineering possibilities and regulatory requirements,
ii) conveying details of the intent behind the design of an artefact or process so that others can manufacture it with the required attributes or operate the process successfully,
iii) reassuring a client that the project will be completed on time and within budget, with the required technical performance,
iv) helping another person learn how to perform a task in order to achieve the desired objectives,
v) informing others, enabling them, or empowering them to make use of information, or
vi) warning someone about a technical issue and the need for pre-emptive action to avoid undesirable consequences.

Teaching is a complex combined performance involving the other collaboration genres; Chapter 8 provides detailed suggestions and guidance for the development of this essential skill.

Learning

Most engineers devote a lot of their time to learning: learning requires that engineers first appreciate the need for it and the value derived from it. They find their own learning resources, ask others to suggest learning resources or to provide demonstrations and explanations, read, watch, or listen, practice certain actions, assess their personal performance, or ask others to assess their performance and provide feedback. The motivation for learning comes from the perceived need to master new knowledge or skills, a perceived need that all expert engineers maintain throughout their career.

Examples:
  i)   arranging for a senior engineer to check and review their work in order to improve their standard for technical work,
  ii)   seeking clarification on a technical issue,
  iii) personal research, reading,
  iv)   attending courses or seminars,
  v)   watching and emulating others' performances, attempting to make one's performance indistinguishable from an expert,
  vi)   practicing engineering activities under supervision,
  vii) learning from the people who deliver the results of your work, such as technicians and tradesmen who build the artefacts that you help to design, the people who sell or use the products that result from your work, or analysts and drafters who work with computer models to help you make performance predictions,
  viii) monitoring ongoing activity to assess progress, and
  ix)   learning how to perform a task requiring special expertise from another person.

Relationship building

Engineers know that relationships matter. They often engage in conversations or other engineering activities with another person with the aim of strengthening a personal relationship and developing trust, manifested as confidence on the part of the other person about one's personal competence and integrity.

Relationship building is very important for informal technical coordination, as described below and in more detail in a later chapter.

Examples:
  i)   asking another person to review or check one's own work: admitting to vulnerability or weakness is a powerful way to build a trusting relationship,
  ii)   exposing one's own lack of knowledge to another person,
  iii) engaging another person in casual conversations, inviting them to eat together,
  iv)   deliberately entrusting another person with a special responsibility requiring a conscientious performance by the other person, for example, by asking them to look after one's own money or equipment that requires special care,
  v)   engaging in social activities with other people both during and outside normal working hours, such as sports, games, walking, driving to work, etc.,
  vi)   entertaining, providing a performance that engages the audience's interest, humour, or admiration, and
  vii) sharing confidential information with another person.

Discovery learning

A discovery learning performance is one in which all the participants are unsure about what they know or don't know; unlike normal learning, there is no clear idea about what is to be learnt.

Discovery learning involves engaging others in conversation to elicit insights that might not be provided in response to a simple question. It can involve shared learning, negotiating meaning, explanations, discovering ideas, discovering unexpected influences, and solving problems together.

Discovery learning is related to a self-learning performance, but neither oneself nor the other people have clear answers or prior understanding, or participants may have a completely different understanding of the issues. Distributed cognition is often an example of discovery: no one has a clear idea initially, but through one or more extended conversations, everyone can learn and get a clearer idea, and possibly emerge with a shared insight that no one could have come up with on their own.

Examples:

i)   working with clients to shape clients' perceptions of their needs and to articulate requirements,

ii)   ascertaining what a person already knows, particularly the level of their understanding and capabilities in an area of specialised technical knowledge,

iii) discussing a script, such as plans or drawings, with skilled people who will perform technical work to expose practical issues known to them that will make particular actions more difficult than they need to be, discovering information that is missing or incorrect in the script, and seeking their suggestions on easier ways to achieve the desired performance objectives,

iv)   comprehending a problem or technical issue, discerning the problem that needs to be solved,[26]

v)   negotiating shared meaning or a common understanding, taking diverse perspectives into account, and aligning stakeholders, which is shown to predict successful collaborative problem solving,[27]

vi)   arranging for technical work to be reviewed by peers, possibly external to the enterprise, in order for both the author and reviewers to learn something,

vii) ascertaining possible misunderstandings or differences in understanding requirements, methods, or objectives, thereby requiring conversation about objectives and methods,

viii) seeking to understand more about other people or stakeholders by learning about their opinions, viewpoints, languages, and interests, or

ix)   working with a patent attorney to prepare a patent specification: a claim for the limited monopoly rights of an inventor for a particular innovation; the engineer understands the invention, but not how to write a patent, whereas the patent attorney understands how to write a patent but needs to learn about the invention from the engineer.

Asserting

Engineers, like all people, occasionally need to take the lead and assert their physical, emotional, or intellectual territory so others can learn behavioural boundaries.[28]

Assertion is different from teaching and helps to establish boundary markers: behaviours that are acceptable or not to an individual. For example, an electrical engineer who takes responsibility for designing all the cable arrangements and connections required for a project can reasonably assert that any proposals for changes affecting electrical cabling must be presented to enable comments to be written on the implications and consequences of the changes, and possibly for granting direct approval.

Examples:

i) stake out or assert your technical territory and competence: helping other people understand your competence, expertise, and ability to take responsibility for a given aspect of a technical undertaking,

ii) assert critical constraints (technical, commercial, safety, environmental, and social) to ensure that anticipated performance requirements will be met, often against opposition from others who may interpret the situation differently or have conflicting priorities,

iii) preparing your CV or résumé, and

iv) applying for a job (often involves a negotiation of pay and conditions as well, could also be classified under seeking approval, as explained below).

Seeking approval

When explaining intended actions, the aim is for the approver to develop confidence that one can be trusted to use given resources to perform intended actions while complying with regulatory limits or other requirements, as well as to gain permission from the approver to proceed with the proposed actions without direct supervision. This is similar to a teaching performance, although with the specific aim of gaining approval. The learner in this performance is the approver: the aim is to change their understanding of your proposal and gain their consent to proceed.

Examples:

i) gaining approval to proceed and implement the agreed upon plans and budget for a project,

ii) gaining environmental or other regulatory approvals from government agencies,

iii) gaining approval for a design or other configuration change,

iv) being awarded a contract,

v) winning a job, and

vi) gaining a promotion.

COMBINED PERFORMANCES

Teaching, negotiation, coordination, and project management are four examples of performances that are very common in engineering practice. I will introduce the first three briefly in later sections of this chapter, but all four are sufficiently important to justify dedicated chapters later in the book.

There are many examples of combined engineering performances: engineering itself is a large symphony of combined collaboration performances. It is important to be aware of them and recognise them when they occur.

Practice concept 39: Engineers have to be informal leaders: they perform technical coordination by influencing and teaching others

Novice engineers often think that ...	Research shows that ...
Coordinating the work of other people is management, and 'that admin kind of stuff' – not 'real' engineering.	Coordination and leadership, as well as gaining alignment, willing and conscientious collaboration, and the support/engagement, particularly of senior staff, dominates engineering practice, taking up 25–30% of an engineer's time and effort right from the start of an engineering career.

This may come as a surprise to you: the idea that right from the start of your career you will have to be coordinating the work of other people, along with influencing and leading them, without having any authority over them.

Technical coordination seems to be the predominant aspect of professional engineering practice, taking up between 25% and 30% of the time of engineers in Australia. While this observation emerged from studies of Australian engineers, subsequent studies in India, the USA, France, and Portugal have provided strong supporting evidence.[29]

The idea of technical coordination came out of our research after analysing hundreds of isolated comments made by engineers in research interviews.[30] None of the engineers that we interviewed provided a coherent understanding of this aspect of their work. Instead, the idea emerged from the analysis of all these fragments of conversation put together.

It is well known that project management is a large part of the work of many engineers. However, unlike technical coordination, this is a formal process relying on documentation, including plans, activity schedules, budgets, project cash flow forecasts, contracts, and so on. Technical coordination is mostly undocumented. Some engineers make occasional notes in diaries and technical notebooks, and most keep records of telephone conversations, while e-mails exist until they are deleted.

One of the important characteristics of technical coordination is that it involves influencing and coordinating technical work performed by other people outside the lines of formal authority. However, you cannot just tell people what to do: it is far subtler than that. Some engineers have referred to this as 'managing upwards' or 'managing sideways'. Sometimes, as an engineer, you literally have to manage your manager.

During your final year capstone project, you probably had to 'manage' your supervisor, an academic with countless other demands on their time. You had to find ways to provide gentle reminders to get the things that you needed – forms signed, feedback on chapter drafts, etc. This is another instance of informal leadership.

We will explore the details of technical coordination in a later chapter of the book because it is such an important aspect of engineering practice. However, for the time being, it is sufficient to understand how it is contingent on your ability to influence other people. Inevitably, this means building relationships with those people, something that is not included in any official curriculum for engineering schools . . . yet.

Technical coordination also builds on the ability to convey the intent that we talked about in the previous section of this chapter. Technical coordination is a collaboration performance that requires teaching, discovering, and learning, explained in Chapter 9. Influencing a person to perform some form of technical work implies being able to convey what needs to be done sufficiently well to avoid misunderstanding, at least most of the time. As we have seen, this is not a simple process.

Technical coordination has many similarities with teaching, and teaching is an important aspect of coordination. It can be seen as a three-step process. First, the engineer explains what needs to be done, why it is needed and when, and informally negotiates a mutually agreeable arrangement with the other people who will be contributing their skills and expertise. Next, while the work is being performed, the engineer keeps in contact with the people doing the work to review their progress and spot misunderstandings or differences of interpretation. The engineer will engage in discovery performances and discussions to expose unexpected issues that arise; the engineer may even need to compromise on the original requirements. Third, when the work has been completed, the engineer will carefully review the results and check that no further work or rectification is needed. Technical coordination is an undocumented, informal process that relies on personal influence rather than lines of formal authority. This corresponds closely with pedagogy: first, setting the task for the students, secondly, monitoring the students as they perform the task, offering help and guidance when needed, and elaborating on the requirements when the students misunderstand, and finally, checking the students' work and assessing it against the criteria that define relative levels of performance. Technical coordination also describes the interactions in which engineers seek information from other people. Engineers seeking information normally need to rely on other people, such as equipment suppliers: even if the information is available in an archive or library, specialised know-how is needed to locate it and provide the specific information that the engineer needs. This means that an engineer needs to explain exactly what it is that they are looking for so that when someone else provides the information it fits the desired purpose.

Practice concept 40: Learning requires much more than logical explanations

Novice engineers often think that ...	Expert engineers know that ...
Teaching other people requires a concise, simple, and logical explanation, or a PowerPoint presentation. Presenting information is sufficient for people to learn it.	Making sure that other people have learnt something new requires much more than an explanation. Consistent monitoring of their subsequent performance is essential to make sure that they have understood and are acting in accordance with expectations. Misunderstandings are common and have to be patiently overcome.

Our studies of engineering practice have revealed that many of the situations in which engineers collaborate with other people, including other engineers, closely resemble teaching performances. These are different from formal classroom or seminar teaching, which has been observed before.[31] Most engineers spend some of their time assisting less experienced engineers.[32]

Many other aspects of engineering practice, while not seen as teaching, can be better understood in terms of teaching and learning interactions. Engineers base their work on special knowledge and insights gained by predicting how materials, objects, and abstract systems can be arranged to achieve a desirable outcome. In order to secure the collaboration of other people, they need to explain some of these insights to ensure that enough care is taken with all the work that is needed, particularly in terms of managing all the aspects of uncertainty that can affect outcomes. Effective risk management, for example, relies extensively on making sure that everyone involved understands how uncertainty is to be managed.[33]

The people who produce and deliver the products or services provided by an enterprise need to know what to do and how to do it so that the results predicted by the engineers will actually be achieved.

We have observed many other interactions that resemble teaching. Engineers seeking to clarify their clients' requirements need to educate clients about engineering possibilities and limitations so that they can begin to articulate their clients' needs in terms of practical engineering possibilities. Engineers often find that they need to work hard at convincing other people about the merits of a particular way of achieving a desired result. Engineers find that it is not easy for other people to appreciate all the factors that the engineer sees as relevant in choosing an appropriate course of action. Engineers also need to explain to all the other people involved in producing the ultimate information, product, or service how their contributions will create value for end users. Moreover, engineers need to explain how end users can obtain value from products and services, often at a distance or through intermediaries.[34] Last, but not least, engineers need to educate both their peers and younger or less experienced engineers on how to pass on lessons learnt from their own experience.

Bailey and Barley observed engineers in a structural design consultancy and a computer-integrated circuit design office. They observed between 30 minutes and an hour of explicit face-to-face learning interactions every day for younger engineers. A senior engineer will therefore, be more than fully occupied with five or six novices to look after.[35]

Engineers need to be able to listen accurately to other people to learn their language and understand the ways in which they think about relevant ideas. Only then can an engineer start to explain technical issues in a way that is meaningful for the intended listener by building on pre-existing knowledge and understanding.[36]

What is teaching? Ultimately, teaching is leading another person to a different understanding such that their subsequent actions are significantly altered. Since engineers do this all the time, through relying on other people to translate their ideas and plans into reality, an engineer's job has numerous aspects similar to teaching. In addition to this, an engineer can only gain the resources needed to translate plans into reality by gaining the confidence and respect of people who can provide those resources. This is also remarkably similar to teaching: a teacher has to gain the respect of a student in order to influence how the student learns.[37]

Practice concept 41: Engineers negotiate for time, space, and resources to perform their work

Negotiation is another combined performance: one that involves discovering, teaching, scripting, learning, and seeking approval. It is a formal collaborative process of discussion, both verbal and written, in which participants find a way to accommodate their common and competing interests in a solution that is acceptable for all of them.

A negotiation starts with outlining the interests of the different stakeholders, people who can influence outcomes or whose interests will be influenced by outcomes. The next phase involves further discovery performances: identifying solutions that would advance the interests of as many stakeholders as possible. Engineers teach other stakeholders about the possibilities and limitations of different solutions. Learning takes place as each stakeholder comes up against certain boundaries; stakeholders gain further understanding about the boundaries that constrain each of the other participants. Once a solution has been identified, it is jointly designed in the form of a script – a sequence of actions to be undertaken by one or more of the parties involved in the negotiation. The final stage is seeking approval from all stakeholders for the intended script – the negotiated agreement. However, engineers are frequently involved in negotiations, often for the technical space in which to perform their work.

'Technical space' is not office space for engineers to have their desks; by technical space, I am referring to spaces within an artefact that can be used for elements of the artefact that confer performance aspects for which a particular engineer has responsibility.

Negotiations often concern physical space. For example, a structural engineer working on the design of a building tries to achieve a satisfactory structural design that meets the requirements of the architect. One aspect of the building structure is a shear wall, the part of the structure that resists horizontal forces imposed by wind or earthquakes, for example. Openings will reduce the ability of the shear wall to resist horizontal forces. Therefore, the engineer negotiates for fewer and smaller openings, whereas the architect may want greater freedom of access for people using the building. The engineer may have to teach the architect to appreciate the range of engineering solutions that could be explored, and the architect may have to teach the engineers about the client's needs and how the building is intended to be used by its occupants.[38]

An electrical engineer may have to negotiate with a mechanical design engineer to obtain sufficient space for cables and connectors to pass through the structure. A software engineer may have to negotiate sufficient memory space, or hard disk capacity, for some performance aspect of a particular piece of software. Maintenance engineers may have to negotiate for space needed to access equipment, and for dismantling or reassembly operations that often need lifting devices to be installed.

Technical space can also mean virtual space or time. An engineer may negotiate for more time to complete assigned work. Another example could be for a software engineer to negotiate sufficient latency or time delay during for the software code to perform the required computations. A communications engineer may request communication frequency bandwidth in a telecommunication system in order to provide a certain signal transmission rate with a guaranteed reliability or error rate. A mechanical engineer may need to negotiate for a certain amount of noise and vibration to be transmitted from a machine in order to achieve a required level of fuel consumption efficiency and weight.

Yet another aspect of technical space is performance margin, which is a similar notion to a safety factor. This is the capacity for additional technical performance beyond the strict requirements to allow for inaccuracies in performance prediction, performance loss over time, and performance losses due to imperfect assembly or components failing to perform in accordance with their manufacturers' performance specifications. The client or lead engineers in the project may be reluctant to allow for the additional cost in their budget.

Engineers are also involved in negotiations for a share of the budget to be spent on their aspects of the project: obtaining a greater share can make their jobs easier or bring greater prestige and authority within the project.

Engineers are often involved in commercial negotiations with suppliers or clients. While they might not be directly responsible for setting a price or accepting a contract, they will often be party to the negotiations because of their special knowledge, as well as their responsibility for coordinating aspects of engineering work to a particular time schedule. Often, the time schedule will be one of the aspects of a negotiated agreement.

SOME NECESSARY COMMUNICATION CONCEPTS

In this chapter, we have clearly distinguished collaboration skills and techniques from communication skills. However, in order to develop your collaboration skills, there are several important concepts that you need to understand. Collaboration relies on communication and communication relies on language: we need to understand the fundamental limitations of language in order to use it effectively.

These important concepts are critical for understanding the material in later chapters.

Practice concept 42: Vocabulary and jargon

The next issue that we need to discuss is vocabulary. As engineers, although we speak in English (in most engineering enterprises around the world), we use specialist technical words like 'transistor' or 'database' that have particular meanings within a given engineering discipline. We learn many of these words during our undergraduate studies at university. All professional disciplines use their own technical 'jargon', which are special words that convey a widely accepted meaning to people who have acquired the necessary technical background to understand them.

Would you know what a tundish is? For air conditioning engineers, it is an important component that helps to drain water from the cold parts of heat exchangers that cool the air in a building. It is like an open-top funnel that allows air to enter the drain, which prevents contamination from entering the air conditioning system.

However, there is a particular difficulty with engineering vocabulary, particularly in English. In England, engineering developed as much as a craft as it did a profession[39] and early practitioners adopted common English words and started using them in ways that were quite different from common English used today. Gradually, more and more words have been adopted into engineering, many of them with meanings in a technical context that can be very different from the meaning used in a normal conversational English context.

Take, for example, the phrase 'sign off'. In common English, this might be used by a presenter of a weekly TV show to tell us that tonight's show is the last that she will present. 'This is Suzannah Carson, and I'm signing off. Goodbye and God bless you all!'

In an engineering context, signing off has almost the opposite meaning. An engineer 'signs off on a drawing or a technical specification'. The engineer signs the document, often with their hand-written initials or signature, and in doing so, accepts ongoing legal responsibility for the consequences. For example, if the technical specification or drawing provides instructions for the erection of a building, and the building subsequently collapses, causing damage and possibly injuries or death, then the engineer will be personally liable and can be prosecuted by a court or a governmental commission of enquiry. In other words, signing off marks the start of legal responsibility, whereas for the TV presenter, it marks the end of her association with the show.

I have compiled a glossary of confusing English terms used in engineering with their meanings in both general and technical contexts (available in the online appendix). This list is not intended to be complete or comprehensive. It mainly lists words that have different meanings in an engineering context as a guide for novice engineers.

What does this mean for novice engineers?

Novices can only learn their specialty-specific vocabulary from more experienced engineers: few engineering organisations provide glossaries. Inevitably, they will use words in different ways to what novices might expect. When this happens, a novice may initially interpret the word with the normal meaning (or no meaning at all, in many instances), but after a few seconds, a novice may realise that misunderstanding a single word has broken the translation; the novice simply loses track of what has been said or intended.

Trying to understand what somebody with much more experience has been talking about can be quite intimidating for a novice, let alone when the novice has lost track of the conversation. However, there is no alternative but to ask for an explanation. In our research, many senior engineers have expressed frustration at the difficulties of explaining something to a novice or a student who is not able to listen accurately. It can be a waste of time for both people. Therefore, it is much more preferable that novices *immediately* ask for clarification, rather than keeping silent because they are afraid to interrupt and in the hope that they will eventually pick up the meaning. Few manage to do this, however, and will have to get an explanation later.

What you need to know from this brief introduction is that many novices initially find it difficult to understand what other engineers are talking about. The message for novices is this: whatever you do, *don't just sit there and hope that the meaning will become clear later!* Ask for an explanation and clarification, and take notes.

A few years ago, I invited a senior project management engineer to talk about her experiences and techniques with senior students. One of my colleagues who happened to be listening complained afterwards that she had found the entire presentation extremely difficult to understand. I was surprised: I thought it had been a clear and lucid presentation that would have been valuable for most of the students. Later, I listened to the presentation recording and made a note of all the vocabulary terms used by the engineer that would not have been familiar to my colleague, nor to the students. There were more than 150 words and phrases that few, if any, of the students would have understood. This was a very valuable observation. Equally notable was that not one of the students asked for any of the terms to be clarified. When I asked some of the students later why they had not done this, the reply was 'they sounded like words that we ought to know about; I guess we will eventually find out what they mean.'

Practice exercise 10: Project vocabulary

What is the likely meaning when a project manager engineer uses the following words? Can you write a sentence to illustrate how the word might be used?

Deliverable: _____

_____

_____

_____

_____

Driver: _____

_____

_____

_____

_____

Execution: _____

_____

_____

_____

_____

Front-end loading: _____

_____

_____

_____

_____

Check the glossary in the on-line appendix to evaluate your responses.

Practice concept 43: Culture – habitual ways in which people interact socially with each other

In every social setting, people interact with each other in more or less predictable ways. As human beings, we establish 'social norms', which are habitual ways of greeting each other, helping each other, influencing each other, meeting strangers, making joint decisions, forming and maintaining friendships, and speaking with one another. Think back to your time as a student: what did you wear when you came to the campus for classes? Would anyone have noticed if you had arrived dressed for a wedding?

In a casual, relaxed, social culture like Australia, many people think that it does not matter what you wear. They imagine that people will react to you the same way, regardless of your clothes and grooming. The example above demonstrates that this is not the case. Even in Australia, with its seemingly relaxed and diverse social culture, if you had arrived dressed for a wedding at your university classes, every person in the room would certainly have noticed you. At least one person would have asked you why you were dressed that way.

At the same time, we are all different from each other. While we tend to follow social norms, we all have our own variations, particularly in the extent to which we follow social norms ourselves. Some people only feel comfortable by strictly complying with social norms, whereas others only feel comfortable if they defy social norms to a certain degree.

Take a look at the list of performance genres presented earlier in the chapter. Several of them involve influencing other people either to perform some technical work or to change their perceptions about an issue or a person. Others involve conversations with certain people, such as senior engineers, managers, or clients, who you would not normally converse with in the casual way that you would with your immediate colleagues.

If you come from a relatively informal culture like Australia, as a junior engineer or student, you may have addressed me by saying something like this:

'James, could I borrow your ASME Pressure Vessel Code for a few days, please?'[40]

In a much more formal culture, such an approach would have been unthinkable. First, you would have to gain my attention: it is quite possible that I would ignore you completely. This is how the conversation might start:

'Professor, how is your health today, sir? I wonder when you might have some time that you could help me? Would it be possible to ask you something now or would you like me to come back later at a time that is more convenient for you?'

It is tempting to imagine that engineering, being a technical discipline, is independent of social culture, which is formed by the ways in which people normally interact with each other. Our research observations, although mostly restricted to Australia, India, Pakistan, and Brunei, provide evidence that casual conversations among engineering peers are much the same in all those locations. At least on a superficial level, the surrounding regional culture seems to make little difference, particularly in an engineering setting. However, as soon as the interaction involves influencing other people, or involves people that have significantly greater influence over others (older people, more experienced people, managers, and senior engineers), then the way in which the social interaction proceeds can be very different from one's own culture.

Engineers in a Western European culture, such as Australia or America, usually spend more than a quarter of their time on technical coordination (influencing other people) without having any formal management authority, as outlined above. In other cultural settings, this informal activity can be much more difficult and challenging for an engineer. Knowing how to navigate the culture in order to influence others is critical to become an expert engineer.

For example, anyone working as an engineer in Brunei would need to know that food is an essential factor in gaining agreement among stakeholders, which is often an important goal in most technical meetings. Even though many of us who live outside the region tend to see all people from Southeast Asian nations as being similar, their individual cultures can be very different from each other.[41] As one Indonesian explained to me, 'Indonesia today is an extraordinarily multicultural society: we have around 5,000 distinctly different cultures among our people, just in one country.' Our research demonstrates how, in certain cultural settings, it can be very difficult for technical knowledge to be effectively exchanged between people. This inability to exchange knowledge can cripple an engineering enterprise. I will return to this later in the book.[42]

Regardless of how different cultural practices are between people from different countries, regions, districts, and even different communities living in one place, the variation between individual people tends to be even greater. Therefore, while most people who live in Asia adopt a strict practice of respect for elders, it would be misleading to assume that all people from Asia always respect their elders.[43]

As a novice engineer, it will be very important for you to be aware of social norms, particularly when you need to influence and coordinate work being performed by other people. Learn to study the behaviour, language, gestures, and appearance of more senior engineers and study the strengths and weaknesses of the different ways in which they influence other people.

Practice exercise 11: Observing culture

One of the most difficult aspects of culture is to perceive it, unless you happen to be an outsider. Therefore, if you want to learn about your own culture, as an engineer, one of the best ways to start is to talk with another engineer who is an outsider, perhaps someone who has recently arrived from a different culture. Ask the other engineer to tell you about some experiences in which cultural expectations resulted in unexpected results or confusion. Remember that culture is something that is seen not just in a particular country or region, but also in particular companies, an individual enterprise within a company, or even a specific workgroup.

Cultural issues are equally relevant for handling technical issues. For example, think for a moment about how you would go about explaining to your boss that he (or she) had made a significant technical mistake, a mistake that, if made obvious to others, might be rather embarrassing. Ask an outsider how they would do this in their own culture. Compare the differences.

Practice concept 44: Language is context-dependent, person-dependent, and time-dependent

The previous section is sufficient to understand how language can be context-dependent. We saw how a simple phrase 'sign-off' means almost entirely the opposite within an engineering context from its meaning in the context of a TV presenter completing a broadcast. The glossary presented in the appendix provides many different examples of other common English words used in quite a different sense within an engineering context. However, there are more complex issues in the way that we use language that create great difficulties, particularly for engineers, but also for others as well.

Many years ago, I was approached by a computer scientist to review the feasibility of a research project that aimed to construct a set of computer-based rules to encode the entire legal system in a database. Users could then ask any questions about our legal system and obtain an automatically generated answer that was correct, based on the entire set of laws and regulations in force. This, it was argued, could greatly reduce the need for the legal profession and the justice system that we now have.

I replied that such a system was impossible to construct because the entire idea was based on a false understanding about the nature of human language. The author of the proposal had assumed that every English word has a finite set of alternative meanings according to its use in different contexts. However, this is not the case: words have an infinite variety of alternative meanings, as we shall see in this brief discussion about language.

Here, we build on the understanding from Chapter 6 that perception involves an imperfect, interactive, interpretation performance. This is this fundamental in understanding uncertainty in engineering: most uncertainty emerges from differences in understanding and interpretation over time and between people. The notion of the fluid nature of meaning is unfamiliar and can be deeply threatening to almost everyone in the engineering community today, even engineering educators.

Almost all engineers think that …	Research shows that …
Words have standardised meanings that are defined in dictionaries and engineering standards. We rely on reference definitions. One needs to know 'the meaning' of a word and 'it must have been my ignorance if I didn't understand'. It is often important to compose written text in a way that makes misunderstandings impossible. 'You can't have a productive conversation if we have different understandings of what the words mean.'	Language is context-, person-, and time-dependent: engineers use a dialect of English with many meanings at variance from common English. It is useful for engineers to understand that language conveys context-, time-, and person-dependent meanings that are based on the prior experience and understanding of both the speaker or writer, and the listener or reader. Ascertaining what the words used by the other person mean is an integral part of the conversation.

The position explained in the right-hand box above is widely associated with post-modernism, a philosophical movement expounded pre-eminently by French philosophers of the 1980s, such as Foucault and Derrida.[44] Most engineers find this incomprehensible and completely illogical, arguing that one cannot have a conversation without common shared understandings about the meaning of the words that we use. How could we possibly organise traffic flow in our streets if we all had a different understanding of the colours red, amber, and green?

Fortunately, most of us have a shared idea about the colour of traffic lights: we readily agree on their colours. We are also adaptable. People who have red-green colour blindness (about 5% of the European male population) learn to read traffic lights by interpreting the position of the light shining most brightly, as well as watching the actions of other drivers.

This is a good analogy for language as a whole. Most of us, most of the time, have similar understandings of the words that we use. The point I am making here is that these understandings can never be identical.

This is not a new idea.

It is instructive to read the words of Plato, who ascribed this quote to Socrates almost 2,500 years ago.[45]

SOCRATES: Then it shows great folly – as well as ignorance of the pronouncement of Ammon – to suppose that one can transmit or acquire clear and certain knowledge of an art through the medium of writing, or that written words can do more than remind the reader of what he already knows on any given subject.

PHAEDRUS: Quite right.

SOCRATES: The fact is, Phaedrus, that writing involves a similar disadvantage to painting. The productions of painting look like living beings, but if you ask them a question they maintain a solemn silence. The same holds true of written words; you might suppose that they understand what they are saying, but if you ask them what they mean by anything they simply return the same answer over and over again. Besides, once a thing is committed to writing it circulates equally among those who understand the subject and those who have no business with it; a writing cannot distinguish between suitable and unsuitable readers. And if it is ill-treated or unfairly abused it always needs its parent to come to its rescue; it is quite incapable of defending or helping itself . . .

Let's think about this for a moment. What, exactly, is a word? What is language? What did he mean by the 'parent coming to its rescue'? What are suitable and unsuitable readers?

The word is either an arbitrary visual symbol or an arbitrary pattern of sound, both of which can be recognised and distinguished from other symbols or sounds, to an extent.

Human beings are remarkably capable of recognising visual symbols and sounds amongst the clutter and noise of our daily environment, as explained in Chapter 6. Research demonstrates that we still rely on context to an extraordinary extent to do this effectively: not knowing what another person is talking about makes word recognition much less reliable; this is the prior knowledge that we rely on for interpretation.[46]

We gradually learn to associate words with ideas in our heads. Herein lies the key to understanding the concept of language. Since we all have different upbringings, and we have all come to understand ideas through different experiences, we all have different stories, memories, images, sounds, feelings, and ideas that we associate with particular words. Through social processes such as playing together, formal education, sharing stories, and common history, literature, and traditions, we increase the degree to which we have a common understanding of the words that we use, at least within a particular culture. However, there will always be differences in the ideas in our minds that we associate with particular words based on our prior knowledge. Social scientists have described this as the 'lens' through which we see and interpret our world: our lens is shaped by all our prior experiences and learning.[47]

What this means is that words and language cannot convey the full meaning of the ideas in our minds that we summarise with a particular word or phrase any more than I can convey all the insights and experience that I have accumulated in 40 years of engineering practice through this book.

This leads us to the idea, so succinctly expressed by Plato in his quote from Socrates that words can only remind us of what we already know.

Written in a book, like this one, words can be misinterpreted just as easily as they can be interpreted as the author intended. This means that the words I am writing now will be interpreted differently from the way that I intended. If I were present and could converse with you, I could help you reduce the extent of your misunderstanding (i.e. come to rescue my own words), but only to a certain degree. For most of you reading this book that would not be possible.[48]

An unsuitable reader, in Plato's perspective, is someone without the prior understanding to make sense of the written words. They misunderstand the text, like many naïve people can misunderstand religious texts without understanding (or even wanting to understand) the context in which the words were written. Suitable readers have sufficient background knowledge and the desire to understand, and so have a better chance of correctly interpreting the text.

Documents describing an engineering project plan, a script, also need suitable readers; the plan will have been prepared by senior engineers, possibly with extensive consultation amongst all the engineers involved. Any of these engineers could be considered to be 'a parent' of the script, someone who can 'come to the rescue' when others misunderstand. As for 'suitable' and 'unsuitable' readers, many people without a technical engineering background can easily misunderstand technical documents that are prepared by engineers. Engineers, therefore, can mostly be classified as 'suitable' readers. People who are not engineers may be 'unsuitable' readers, depending on the level of technical understanding required.

For this reason, you need to be working as an engineer, ultimately, to make sense of this book. That is the reality check that we, as engineers, love to rely on. It is only when you have the opportunity to observe what actually happens between people that you will come to a deeper and more useful understanding of the ideas that I'm trying to describe here.

Practical implications

Let's return to something practical for you, the novice engineer reader. What does all of this mean for you, in practice, as a working engineer?[49]

Here is a real example of this issue. You would think that the words 'can do' would be too simple to misunderstand, right?

You would be quite wrong.

Computerised maintenance management systems (CMMS) have become essential for organising maintenance for large engineering systems, such as process plants, factories, fleets of ships and aircraft, or even buildings. These are large software systems, often part of enterprise resource management systems (ERM systems) that cost many millions of dollars to set up, often tens of millions of dollars. SAP is one such system. A research study on maintenance has shown how many, if not most, CMMS fail to achieve their objectives: they don't work as the company maintenance engineers intended.[50] But why?

These complex computer software systems require consultants, often called SAP configuration specialists, to install them and write the software needed to make SAP behave in the particular way necessary to meet a firm's needs. Here's what one maintenance manager told our researcher:

> *'When we started, those consultants told us that SAP can do what we wanted. Yet, it seems they didn't bother to even talk to our people, because when the system was installed, it didn't do anything like what we wanted. Then they said we would have to spend millions of dollars to fix it.'*

During the negotiations on the contract to install the CMMS, when the SAP configuration specialists said, 'SAP can do what you need', what they meant was that SAP *can provide the necessary functionality, provided suitable software is written to configure the reporting modules to generate the data in the exact format needed by the customer.* Being software specialists, this is common knowledge: anyone would understand that, when installed 'out of the box', no major software system can do much at all. The software can be made to do almost anything, but it has to be configured, and writing the configuration code is costly. It takes considerable time and effort to figure out precisely what the customer needs, because most people in the customer's organisation have their own variations on what is needed. Then, it takes time to write the configuration software, test it, load representative data, and check that it works for the customer and that the customer is satisfied that it is performing the way they intended. Additionally, it takes even more time and effort to train everyone in the customer's organisation to use the software effectively in order to achieve the intended benefits.

On the other side of the negotiating table, the customer's maintenance engineers, knowing little about complex software systems, understood the words 'can do' in a very literal sense. They imagined that the SAP software could perform the necessary functions now, in the present tense, not that it 'might' be able to perform them sometime in the future, only after spending a lot of money on software specialists and training. In their view, the meaning of the words 'can do' is common knowledge and they never expected that there could be an argument about such a simple phrase.

These interpretation differences can end up in costly legal battles. Even two such simple words, 'can do', can have radically different interpretations depending on which side of the negotiating table you happen to be sitting on.

Here is another example from my own experience. In the early part of my career, I was working for an aerospace firm designing the electronic flight control and other systems for a military aircraft – the avionics side of the design. The firm was responsible for the system design rather than the detailed design of the major subsystem components, such as the forward-looking radar. Instead, specialist companies supplied those components. The aircraft was a joint venture between three European nations: Britain, Germany, and Italy. A special company, Panavia, had been set up by a consortium representing the leading aircraft manufacturers from each of the three countries, British Aerospace, Messerschmitt-Bölkow-Blohm (MBB), and Aeritalia. Companies within the three countries supplied most of the major subsystem components. However, the three defence ministries had jointly agreed to purchase the forward-looking radar from an American corporation, largely because no European company had the capacity to meet the demanding technical specifications for the radar at that time.

My firm was responsible for integrating all the subsystem components and testing them to make sure that they would operate together as a combined system. Engineers in the firm had designed the overall avionics system and had written the procurement specifications. These documents defined the essential technical performance characteristics for each of the subsystem components. It was a challenging task just to prepare the documents. In practice exercise 9 you would have learnt this for yourself when you wrote an outline specification for a simple beam. The engineers had to be confident that when the entire system was incorporated within the finished aircraft, it would achieve the required performance to satisfy the needs of the three defence forces that would be using it.

Gradually, each of the major subsystem components arrived in the integration test facility at one end of our building. Technicians had constructed a mock-up of the front part of the aircraft, complete with the cockpit and all the equipment bays where the electronics boxes comprising the major subsystem components would be installed in racks. Fat wiring harnesses with multiple connectors, some with over 50 connecting pins, linked all the bays with the control panel racks in the cockpit. Flexible tubing that carried cooling air to each of the major equipment boxes had also been installed: the heat generated by the 1970s-era electronics had to be removed by forcing high-pressure air through the tubing. Alongside the mock-up, there were computers with multiple display panels for the engineers to be able to monitor the performance of the overall system.

When the forward-looking radar units arrived, they were also installed in the mock-up for testing. This required special precautions: the radiation from the main radar antenna was capable of delivering a fatal amount of heat to a person, like a high-power microwave oven. High-pressure hydraulic controls moved the antenna, almost a metre in diameter, and they also required special safety precautions.

It soon became evident that the radar was not capable of the performance defined in the specification documents. This was a potentially catastrophic development for the entire aircraft development program. There was no component more critical to the performance of the aircraft as a whole than the forward-looking radar.

An urgent meeting was convened, which was attended by representatives from the three defence ministries, specialist engineers from the three aircraft companies, leading avionics consultants from Britain, Germany, and Italy, engineers from our own firm, and a delegation from the American radar supplier. The engineers from our firm presented the evidence demonstrating that the radar performance would not comply with the specification documents.

At the meeting, it soon became evident that there were differences in the way that engineers from the four countries interpreted specification documents. The British engineers interpreted a specification in terms of function. The radar would comply with the specification if it operated in a way that achieved the performance defined by the specification. The radar did not have to be constructed following all the requirements of the specification as long as the performance requirements were satisfied. The German engineers, on the other hand, considered that the radar could not comply with the specification since it was not constructed in a way that met all the detailed requirements. There were many details in the arrangement of the waveguides carrying the microwave signals to and from the antenna, there were smaller clearances between components, and there were limitations in data transmission between the radar and the flight control computer that did not comply with the specification requirements. The Italian engineers regarded the specification as a negotiable agreement between the supplier and the Panavia consortium building the aircraft. As work progressed, the specification could be changed if agreed to by all parties concerned. The American engineers, with yet another different interpretation, argued that the specification could be seen as a performance requirement that would be met by production radar units. The first units off the production line would not necessarily meet the specification requirements. However, once the technical performance limitations were overcome during the course of production, the earlier units would be upgraded or replaced so that all the units in service achieved the required specification performance. This explained why the first units delivered did not comply with the specification. They were intended merely to help the engineers integrate the radar subsystem into the rest of the avionics system.

From this case study, we can see how the misunderstanding occurred because of the different ways that engineers viewed the specification documents.

Differences in interpretation and understanding emerge in every engineering project, particularly between engineers with different types of specialist expertise. With the benefit of understanding how these differences originate, based in the fundamentals of how humans construct and use language,[51] we can see that what many regard as an exceptional circumstance is actually completely normal.

Not only are differences of understanding and interpretation normal, they can also provide very useful advantages.

When two or more people work together, if they can collaborate and talk about their different understandings of the issues in a way that accepts and regards the differences as beneficial, then the collective understanding that results from this collaboration will be deeper and more complete than either could have achieved on their own. In the same way, this book has emerged from collaborative research and embraces many contributions beyond those provided by my own research students, each of whom has contributed a different perspective. Together, we have reached a level of understanding that was beyond any of us working on our own.

This is the essence of good teamwork, and we will return to this theme in coming sections.

There is a further critically important aspect of language that we need to take into consideration.

Goethe, the great German scientist, philosopher, writer, and poet of the 18th century, described this when he wrote about the difficulties of capturing descriptions of objects in words. He remarked how a symbolic language never describes objects immediately, but only 'in reflection', and how the act of thinking about an object changes our conception of the object, and hence the way we would describe it in words. Derived from that same logic, he suggests that the act of describing an object in words changes the way we think about an object.[53]

As engineers, we spend a lot of time thinking about objects, which inevitably changes the ideas that we have in our minds about these objects. Goethe's observation predicts the inevitable difficulty that we have when we subsequently attempt to describe our findings to others, only to find that they no longer seem to understand what we are talking about. Our language has changed because the ideas in our minds have changed: the word associations remain the same, but the ideas change and therefore the language changes without us being aware of it. The act of writing this book has inevitably

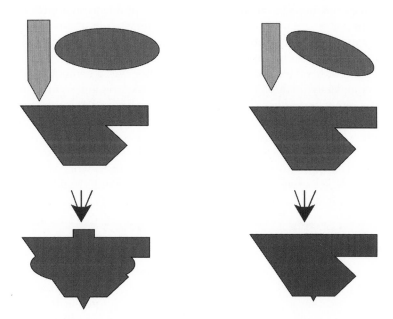

Figure 7.2 This diagram illustrates how using a language involves translation for all of us, and why there is always a degree of misunderstanding.[52] The cluster of three shapes with different colours at the top on each side of the diagram are superimposed and combined together at the bottom with a merged colour. The three shapes on the top left side are very similar to the three shapes on the top right side. However, the resulting combination on the right side is different in shape and colour to the one on the left. Now, imagine that the three shapes correspond to three words. We share similar but not identical ideas, images, and experiences that we associate with each word. However, the same sentence, spoken with the same words, recognised perfectly, still results in a different meaning for both of us, just as the combined shapes at bottom of the diagram are different. Realising this, we can collaborate through conversation to clarify and understand differences of interpretation to achieve a depth of understanding on an issue that would elude both of us when working separately.

changed my ideas about engineering practice: I can only hope that the language is still sufficiently meaningful that you obtain some benefit from reading it.

Visual language is equally affected by context

The ideas in this section apply equally to visual language: images, illustrations, and diagrams. Just as words are symbols that we use when attempting to share ideas with others, diagrams are also symbols that we use to represent an infinitely more complex reality. Just as words connote similar but different ideas, diagrams, images, other illustrations, videos, and even more elaborate representations connote similar but different ideas of reality. All are subject to our differing, complementary interpretation and understanding. By sharing and discussing our different interpretations, we can draw great power from this intrinsic human use of symbolic language.

Simple diagrams used in engineering mechanics problems and electrical circuits are assumed to convey unambiguous meaning for students. However, unless students have played with a pulley, for example, they may completely misunderstand a diagram that represents a pulley with a simple circle.

Neither the words we use, nor the illustrations, nor even a combination of the two can ever be completely unambiguous. Without conversation and mutual understanding, as well as receiving feedback, the chances of mistakes and misunderstandings are greatly increased.

Engineers have developed many different techniques that recognise this difficulty and help to ensure that accurate meaning can be shared and acted on in any engineering enterprise. The challenge for engineers, therefore, is for us to work with human language and learn different ways to reduce the chance of misunderstandings or mistakes being made that could be catastrophic or costly in other ways.

Practice exercise 12: Meanings and contexts

Pay attention to the language and concepts used by different people in the work environment around you. For example, most managers talk about different things than engineers. What do they mention most often? By taking notes in briefings held by managers, you can identify these ideas, as well as the words that they use in different ways from engineers. How do these words relate to the concepts of value that we discussed in Chapter 1?

Observe people using words with different meanings, particularly when one meaning has an application in engineering terminology. Look out for the common English words that have radically different meanings from the sense in which they are used among engineers. Are these listed in the glossary (online appendix)? If not, please send them in as examples that I can include in a later edition.

Watch out for the ways in which the meaning of words changes you, perhaps as you examine something in detail and become immersed in thinking about it. To do this, compare something that you wrote before you started this phase of your work with something that you have written since, perhaps an e-mail or a segment of the technical report. See if you can spot the ways that your language has shifted.

Practice concept 45: Conveying intent accurately is critically important, challenging, and difficult to master

As we have seen, engineers almost always have to rely on other people to translate their ideas and plans into reality. An engineer's efforts could be completely wasted unless the ideas are faithfully translated so that the ultimate end users can experience the real value of the engineer's ideas and investors can begin to be repaid from money contributed by those end users. However, as we have now learnt, each time a new person becomes involved, they naturally reinterpret the ideas in a way that is different from the original.

After the last three sections, you might begin to think that this is impossible: how can an engineer possibly convey ideas to other people reliably, given the differences in the way we all understand both text and diagrams? Understanding how we might be able to do this successfully is one of the most critical lessons to be learnt in this chapter.

Students and novice engineers often think that ...	Expert engineers instinctively know that ...
Something that has been explained concisely and logically needs no further explanation. Presentation (transmission) is all that is needed. Another manifestation is describing the importance of writing text so that misunderstandings are impossible.	Conveying intent accurately is a time-intensive, step-by-step, iterative, interactive process with close analogies to teaching. It is difficult and challenging.

The previous two sections should be sufficient for you to now understand that it is impossible to write a piece of text that is completely fixed in its meaning, entirely unambiguous, and impossible to misunderstand (or understand with alternative unintended interpretations). Here is an example that my children pointed out to me with great amusement. It is a very common sign, and you will find this exact phrase or a similar variant in almost every lift in a high building.

DO NOT USE LIFT IN CASE OF FIRE

We understand this to mean that we should not use the lift if there is a fire burning in the building: the lift system may malfunction and we might be trapped in a burning building and not be able to escape. My children pointed out one day that we should not use the lift in case it might start a fire by malfunctioning. This is an equally valid interpretation of the text, and I suspect that the original authors did not anticipate this. It also gave my children the opportunity to race up and down the stairs and get lots of exercise: hardly a disadvantage.

Fortunately, engineers are always devising effective ways to deal with the limitations of human symbolic language in the form of both words and diagrams. Engineers have managed to overcome these limitations, more by instinct and trial and error than on the basis of a detailed understanding of the nature of language.

As explained in the box above, many engineers will argue that something that has been explained concisely and logically, particularly if it is in writing and drawings, requires no further explanation. Information only has to be 'logically and concisely presented' for it to be conveyed accurately to anyone who needs to use it.

However, what people say they do is not necessarily what they actually do in practice.

We started to discover how expert engineers reliably convey the intent of their plans and designs by simply asking them how they do it. 'When you needed to be absolutely sure that other people have completely understood the requirements that you have specified, with 100% accuracy, with no mistakes, could you please tell us how you managed to achieve this in practice?'

Here are a few quotes from those interviews. The first quote is from a control systems engineer working for a multinational firm. He is talking about a design process for upgraded control and instrumentation systems to be fitted for a chemical process plant. Once the design has been completed, it needs to be checked very carefully by the client's engineers to make absolutely sure that it will meet their requirements. The firm does not rely on the client's engineers to sit down and read the documents and check them by themselves. Instead, they provide the documents for them to read prior to the formal review meetings, which will take place on-site at the plant itself, along with the engineers who will be responsible for operating and maintaining the plant.

> 'We have very formalised document review systems. For the critical design phases, we have actual site meetings and we have checklists of what has to be done at those meetings. We have found that they are very effective. In this project we have probably gone further than just about anyone else has in the world in terms of formalising what previously has been a very informal process. We have check lists for all the major meetings, we have a formal checklist for document review, we have templates for all the major documents, so I think we are very structured in that regard. The client has to check all our conceptual design documents.'

The firm's engineers would work through the documents, sitting with the plant engineers, line-by-line and page-by-page, following checklists to make sure that every aspect of every detail is discussed and checked with the engineers on-site. This is time-consuming, tedious work. However, the results pay off. The firm is able to charge a much higher rate per hour of work for its engineers because the way they work with such meticulous detail results in fewer mistakes being made and a much more reliable implementation of the control system that has been designed for the plant. In the end, the client is much happier, and the plant is much safer to operate, more reliable, and more profitable for the owner.

We could see this as a discovery performance, scripted by a checklist in a procedure.

Of course, it doesn't stop there. Here is another quote from a different interview, this time with a much younger engineer in his first two years of employment. He was working for another multinational firm installing signalling equipment to coordinate the movement of buses in a large city. He was explaining how he had to work with electricians who were installing the cables linking the control equipment, sensors, and indicators in each of the bus stations:

> 'For example, I had to show them how particular wiring is done. I had to make sure that things are done the way we specified in the cable schedules . . . and in the design diagrams and carried out in a timely manner.
>
> '[How did you know how the cabling was supposed to be done?] That was easy. You just follow the cable diagrams. Actually, one of my first tasks was to look at the cable schedules and make some modifications. That sort of got me to know how the cabling is done. It wasn't that difficult to pick up.'

This engineer started by checking the cabling diagrams himself so that he understood the original intent of the designers. He did not design the arrangements himself;

therefore, since he had a different way of interpreting the drawings than the original designer, he was able to spot errors that had been missed by the designers and the people who checked the designs. He also had to make modifications, because the arrangement of the bus stations was not quite the way that had been intended at the time the designs were prepared. He went on:

> '[And how did you know how long it should take, how could you tell when they were going slowly?] Okay, I was doing a very hands-on job when I started, I was working alongside the electricians, I was working as an electrician like a tradesman. I know how long it takes to commission a bus station. In a bus station we have PA systems, the communication system, we have 32 stations along the busways and they are all similar. So once you have done one you know how long each one should take. You can also know by the quality of their work. They always make mistakes because some of them just cannot be bothered to follow the schedules. Some of them don't mark the cables correctly, others do not terminate the cables correctly. You can just go in and find out the quality of their work when you just start turning things on. Sometimes you can spend just as much time fixing up their mistakes. But that was all very valuable experience on how you handle this kind of commissioning work because you learn from their mistakes.'

In this second part of the quote, he explained how he had gained experience by working alongside the electricians himself. The electricians had taught him much of what he knew about the installation task that he had to supervise. He would sit down and work through the drawings in the documents with the technicians (described elsewhere in the interview), and then he would watch them as they performed the installation work on-site, checking to make sure that they performed the work correctly. Notice how he reported the inevitable interpretation errors: just because something was correctly represented on a drawing does not mean that the person performing the work would interpret the drawing correctly. However, by monitoring what happened, he could pick up most of the mistakes at the time they were made. Later in the interview, he talked about the commissioning process in which they tested the entire system, where they were able to find other wiring mistakes.

What we are seeing here is a complex discovery performance that leads to reliable and repeatable replication of the intent of the original designers. However, it is not as simple as it sounds. First of all, the designers cannot anticipate every possible eventuality when their designs are translated into reality. Nor can they produce completely error-free designs in the first place: there will always be slips that remain undetected, despite careful checking at the design stage. Some of the slips can be detected by the site engineer responsible for supervising the construction and installation when he works through the drawings and documents in order to understand what has to be built at the site.

Then, the engineer sits down with the people who will actually perform the construction and installation, or at least a foreman, who is an experienced technician, probably with much more extensive practical experience than the engineer himself. Site engineers tend to be young and relatively inexperienced: they are placed in site engineering jobs in order to learn the practical side of translating engineering ideas into reality. A well-organised firm will arrange for a young and inexperienced site engineer to work with a highly experienced foreman. Part of the foreman's job is to help site engineers learn the practical realities of working on-site.

By working through the documents and drawings, detail by detail, the engineer and the foreman will jointly pick up more details that have been missed and will make minor changes to deal with the practical realities of the site that may not have been completely anticipated by the designers. Here, the experience of the foreman will be invaluable: extensive experience will help the foreman detect slips and work around problems before they actually occur.

As the work is performed by technicians, the foreman and the engineer regularly check to catch mistakes and rectify them as quickly as possible. Finally, a comprehensive set of tests will be performed on the finished installation to make sure that it operates as the designers intended, although slightly influenced by the local interpretation of the site engineer and the foreman.

Notice how much learning is taking place: this is also a learning performance. The young site engineer is learning about the practical aspects of translating engineering designs into reality. He is also learning about the intentions of the design engineers, along with discovering slips made by the designers and mistakes that arose because of incomplete understanding by the designers. Both the foreman and the engineer are learning about the intentions of the designers. Both individuals learn on-the-job about the kinds of mistakes and slips made by the technicians (and themselves) and utilise this learning in order to help avoid similar problems in future. The engineer is also learning about the amount of time and resources required to achieve the final result.

What this demonstrates is that engineers have devised workable ways to reliably convey their intent and translate their ideas into reality through multiple reinterpretations by other people. However, as mentioned, it is not a simple process. It relies on written documents and successive interpretation of the documents by different people, exploiting differences in understanding in order to pick up oversights and errors introduced earlier in the process. It also relies on the knowledge, skills, and understanding distributed among all the participants who are coordinating their collaborative work and sharing their insights effectively. They also rely on the inherent knowledge and experience of many other people in the form of procedures and standard ways of performing the work.

With a little prior understanding combined with careful observation, you can learn this in the context where you will be working. Notice how some of the work appears to be very tedious, but by picking up problems before they actually occur, costly rework needed to fix mistakes at a later time can be avoided.

It is very important to learn how to do this effectively. There is nothing worse than receiving a telephone call at three o'clock in the morning when you're on holiday with your family because something that you planned has gone seriously wrong in your absence. Being able to learn how to reliably convey your ideas into practical reality and effectively work through the hands of other people is an essential skill that distinguishes an expert engineer.

Practice exercise 13: Learning from experienced engineers

At an appropriate moment, ask a senior engineer to answer this question: 'When you need to be absolutely sure that other people have completely understood the requirements that you specify, with 100% accuracy, with no mistakes, could you please tell me how you manage to achieve this in practice?'

Listen carefully and take notes. As a follow up, ask them to tell you about situations when this was particularly difficult. See if you can classify the collaboration performances that they tell you about.

Practice concept 46: Teams work because of diverse interpretations

Novice engineers often think that ...	Expert engineers know that ...
Teamwork means dividing the work into sub-tasks to be performed individually by team members and assembled at the end. Teamwork requires tedious meetings that waste too much time. Individual effort is not rewarded, and many times, almost everything is done by one person.	Teams outperform the same number of individuals working separately by exploiting the diversity in knowledge, interpretation, and thinking of their members. This enables better ideas to emerge from social interactions and working together.

Many students play on sporting teams and learn that a good team is something quite different from a collection of good players. But what is it that makes a good team?[54]

A cricket team consisting entirely of first-class bowlers, entirely of first-class batsmen, or entirely of first-class fielders, or even slips catchers, would never win a serious competition. Instead, winning cricket teams are composed of a diverse group of high-class players each with different strengths that complement the relative weaknesses of the other players. The team is successful because they trust the other players to excel in their strengths so they can focus on developing their own skills and strengths. Once the batsmen start criticising the bowlers for poor bowling, the team is in trouble.

There is no equal division of effort in a sporting team: a good team works because the people on it put in their best effort when it really counts, and everyone has a backup in case of an error.

We have learnt in the previous sections how we can actually build greater strength from diversity by exploiting the differences in the way each person thinks and acts. This is the essence of teamwork: exploiting diversity.

What it means, though, is that the diversity can only be exploited through the technical and social interactions between the team members, by working together at the same time on the same issues. There are many ways that this could happen.

One of the most effective ways is for one team member to arrange for another to check their work. It is much easier for another person to spot mistakes in your own documents, or your ideas, than for you to find them yourself. This process has been formalised as the 'walk-through' method for detecting programming errors in software engineering. Why does this work so effectively? Another person will think differently from you and use words in different ways. They will spot mistakes that you have not seen, even though you have read and reread the same sentence or calculation step countless times. Engineers with whom we have discussed this in our research tell us that most of the errors they find relate to mistaken or missing assumptions. They find very few calculation errors, possibly because your education has devoted much more time to this aspect of your skills than others.

Earlier, we discussed how relatively few technical issues arrive on an engineer's desk as formal written documents with a complete description. Instead, engineers have to work with other people through conversation and dialogue to discern the nature of the problem and fully comprehend it. Two engineers working together on this can be more than twice as effective as a single individual. First, one can take notes while the other one engages the client and other people in conversation about the issues. The two engineers can even take turns doing this. Each can check the other's notes to compile a more accurate record of the conversations when needed. This is particularly valuable when dealing with clients. Meetings are expensive because of the time cost of bringing everyone involved to the same place at the same time. Keeping accurate records of the meeting is one way to help retain the full value of the discussions at the meeting and make sure that no valuable comments or ideas are lost. Someone has to do this: it is best to take turns, as note taking can be quite tiring.

Another effective way to employ teamwork is in the interpretation of lengthy written documents. Yes, engineers have to frequently work with long documents such as specifications, contracts, scope of work agreements, design documentation, procedures, etc. Interpretation errors are much less likely if two people read the same document and then compare notes. Each person reads and interprets the document differently. By working together and sharing notes, two people can read the same document and compile a much more accurate summary than they could if each one read half of the document on their own.

Rarely in an engineering career can you choose your own team. Even when you're in charge, the people you really want may be unavailable and tied up with other commitments.

Good engineering, therefore, arises from teamwork that depends less on the intrinsic abilities of the individual team members and more on the quality of the social and technical interactions between team members. In other words, a great idea works *despite* the different interpretations of the people who implement it.

As in previous sections, the benefits of teamwork arise through social interactions. Expert engineers intuitively know this, but it takes time for young engineers to overcome the misconceptions that can easily arise through formal education.

Practice exercise 14: Interpretation differences

Ask someone else to check some of your technical work and observe the kinds of issues that they detect. Why did they spot these issues? Were they actually mistakes or were they differences of interpretation? Think about the differences between yourself and the other person and try and understand why they were able to spot these issues. What was different about the way the other person was thinking from your normal way of thinking?

Practice concept 47: Expert engineers spend a lot of time listening and talking with other people

Novice engineers often think that …	Expert engineers know that …
A good engineer works quietly away without talking. Many engineers think that talking is a waste of time.	Engineering teamwork is more effective when team members with complementary technical and commercial expertise are talking and working together to build cooperative relationships and trust. Relationships and trust make collaboration much easier.

The idea that good engineers work quietly without talking to other people is remarkably widespread. The engineering student society at my own university has a Latin motto, 'Non loqui sed facere', which roughly translates as 'the people who don't waste time by talking but get on and do things'.

Many students interpret this as distinguishing the difference between themselves and students in arts, law, and commerce who seem to spend most of their time talking and consequently achieve very little.

Earlier, we discussed how technical coordination takes 25% or more of the working time for an engineer, on average. Influencing other people without any organisational authority, which is the essence of technical coordination, is much easier and quicker if there are trusting relationships already in place. However, most of the time, the only way to build trusting relationships is to spend time working or socialising together.

Many engineers understand this at an intuitive level and take the time to talk with people, often about personal issues, sporting results, political developments, or even scientific issues of general interest. Other engineers see this as an exercise in 'office politics'. Of course, sometimes people use influence irresponsibly. However, engineers can learn to build relationships strategically as well. By thinking and planning ahead, considering the people with whom you will need to collaborate in the future, and building relationships with them over time, you can save yourself a huge amount of time and frustration in the long run.

Practice exercise 15: Relationships revisited

Return to the exercise in Chapter 5 in which you developed a map of technical expertise with relationship strengths. Would you change this map now? Would you add to it? Do you notice the ways in which people strengthen their working relationships, ways that you might not have noticed before you read this chapter?

Practice concept 48: Expert engineers manage their time effectively by the smart use of e-mail and calendars

Novice engineers and many young people often think that ...	Expert engineers know that ...
It is necessary to respond to any electronic messages and e-mails immediately.	A response to a written message requires time to write and can easily be misunderstood. In fact, it is almost certain to be misunderstood. Expert engineers know how and when to respond effectively.

Engineers always work under time pressure: there is never enough time to handle every aspect of every issue that concerns an engineering project. Therefore, engineers need to prioritise and decide which activities are the most important to devote time to. Partly because of this, many engineers report feeling chronically short of time, which is true of many professionals. Leslie Perlow wrote about this in her article *The Time Famine*.[55] The fact that most of the demands on an engineer's time, or interruptions to ongoing work, arise through social interactions with other people can contribute to the perception that this activity is counterproductive and gets in the way of 'real engineering', which is the widespread notion of 'productive work' that engineers associate with technical problem-solving and design. My own research has helped to show how engineers mostly relegate work associated with social interactions to a secondary status.[56]

Even though electronic messaging is less intrusive and potentially more manageable than face-to-face interactions, research evidence suggests otherwise. For engineers, e-mails and other forms of electronic messages usually require between 30 minutes and two hours per day, every day, and often even more time than that. This does not include the time needed to prepare extended written reports, drawings, and other documents. Even though electronic messages do not have to be answered immediately, research demonstrates that most people, either because of innate curiosity or perhaps the desire to deal with something immediately, find it difficult not to respond on the spot.

Although face-to-face contact can seem like a more intrusive interruption than e-mail, research shows that face-to-face discussions resolve technical and many other issues much faster than e-mail correspondence.[57]

Research has also demonstrated that the habit of responding immediately to any incoming message or e-mail can impose a significant extra mental workload because it interrupts other tasks.[58] Most of the tasks that occupy an engineer require some kind of short-term working memory. It might be as simple as remembering the name of a document and where it was stored, or perhaps a particular number representing a property of some material, such as its thermal conductivity. Since most of us do not write down every single item of information that we deal with from moment to moment, we have to rely instead on our short-term memory. Any interruption to the task at the forefront of our mind can disrupt this short-term memory: when the task is resumed, it takes several minutes to return to the state of mind we had at the time of an interruption. That is why skilled technicians take pre-emptive measures to minimise interruptions during critical procedures.[59]

Multitasking, or at least giving the appearance of working on several different issues at the same time, might look impressive. However, the research demonstrates with a high degree of confidence that switching from one task to another damages short-term memory and ultimately requires extra time to complete the original task. In fact, one study produced results that suggest that 70% of office tasks that are interrupted by incoming e-mails are never completed. I expect that, like me, you have had the experience of closing multiple windows on your computer desktop and perhaps clearing papers from your desk at the end of a day, only to find many uncompleted tasks that you had completely forgot about.

The same research shows that setting aside time to answer e-mails and deal with other important tasks can boost your personal productivity by between 10% and 15%. Perlow suggested setting aside time each day when you could be confident that you will not be interrupted.

There are several methods that can help you ease your own personal 'Time Famine'.

One way is to use your calendar or diary: if you need a significant amount of time for 'thinking' or solitary technical work (like I do), then reserve that time in your calendar or diary, perhaps at a regular time of day for several weeks or even months ahead. Make appointments with yourself. Then, when someone asks you to attend a meeting at the same time, you can tell them that you already have a prior engagement, and you will have to see whether it can be changed. It may be possible to shift personal time, but it is likely that at certain times of the day, you will almost certainly be interrupted, so avoid shifting your personal time only to have it interrupted. After some time, maybe a day or so, let the other person know that you have rearranged your schedule; that way, they will respect your time more and will be less likely to be late. In the same way that you would try and get to a meeting on time, make sure you start your personal work at the reserved time, as well.

One of the most difficult traps to avoid, especially for a young engineer who wants to be helpful, well-liked by peers, or 'noticed' by more senior colleagues, is over-commitment, which mainly comes in the form of saying 'yes' to too many people. Learning to say 'no' is one of the most difficult challenges faced by many people, no matter what their occupation happens to be. There is an easier way to avoid that potential conflict, simply say, 'Yes' instead of 'No'. Does that sound confusing? The trick lies in managing to keep a full diary by taking the time to fill in all future commitments, with extra time reserved to cover inaccurate estimations of how long existing commitments will take, as well as allowing for unexpected incursions on your time, such as unavoidable family commitments.

With a diary already filled up with commitments going forward, any request for help can be answered with something like, 'Well, I would love to do that for you. It sounds like something I would really want to be involved with. How about we start next April?' (Or choose another date several weeks or months ahead.) Most people will find someone else to help, leaving you with no extra commitment, but your colleague will remember your positive and enthusiastic response. It works well: I have often used it myself.

Another important lesson, no matter what your profession happens to be, is to learn to recognise the signs of your own fatigue. When you are tired, it is much more difficult to make sound decisions about how to use your time effectively. Take time to relax, exercise regularly, and get to know how to work within your physical and psychological limits. Some companies impose strict working time limits, typically around 44 hours per week. They have learnt through experience that the cost of rectifying mistakes made by tired people working extra hours usually far outweighs the small additional value from those unpaid extra work hours.

Another technique for easing your 'time famine' is to ask a colleague or an intern, such as a student on work experience, to observe your daily activity and prepare a detailed task time record. They write down the exact times when you start and stop every single task during the day, including telephone calls, occasional interruptions, even visiting the bathroom. This exercise helps the colleague or intern learn much more about engineering practice at the same time. You will find that in most organisations, there are predictable patterns. Interruptions tend to be clustered at certain times of the day. You can exploit this valuable data by planning your day such that you work on short tasks when you are likely to be interrupted. Therefore, the chances are that you will be able to complete most tasks before the next interruption. These tasks tend to be ones that require little, if any, short-term memory, so the time required to resume an interrupted task is likely to be minimal. Set aside time at another point of the day for tasks that demand more intense concentration and short-term working memory. When you perform these tasks, make sure that you take breaks. Some organisations have even resorted to software that locks you out of your computer for a few minutes every hour, thereby forcing you to take a break. Granted, this policy can be just as destructive as unplanned interruptions, but these companies understand that fatigue build-up caused by not taking breaks is even more destructive.

Try and avoid sending e-mail or other electronic messages: pick up the telephone or go and visit the recipient instead. Why? As we have already discussed and will explore in more detail as the book progresses, most collaboration requires two-way communication. Every message you send electronically will probably necessitate at least one more message in reply, or at least some type of acknowledgement. With face-to-face or telephone communication, you can be much more confident that the recipient understands what you are trying to convey. These encounters also offer the opportunity to build and reinforce relationships with additional casual conversations. They prompt memories of issues that need to be raised that might otherwise be completely forgotten.

The next method for controlling the use of your time requires that you think about the implications of any message that you send. When does the recipient need to have your response? If you send it too early and the recipient can't be bothered to read your response carefully in order to understand it completely, they may send you another message later that raises questions about your response, questions for which there are answers in your original message. Another possibility is that the recipient will have completely forgotten about your message when they need those answers. Try and put yourself in the mind of the recipient: what do they really need to know and when? If you are not sure, ask by telephone. Then, learn to use the delayed send function that is available in most e-mail services so that your message is delivered at the appropriate time without you having to remember. Alternatively, you can use the drafts folder to prepare a response and then store the message until it needs to be sent.

Try and express your response in three simple sentences:

a)   summarise the present situation,
b)   predict what is likely to happen if the recipient chooses not to respond with corrective action, and,
c)   summarise any corrective actions that the recipient needs to initiate.

Append or attach any further details for reference.

Remember that many people check their electronic messages using smartphones with small screens and often do not scroll down, making it easy to miss anything not covered in the first sentence or two. Take the time to learn how to write something as briefly as possible, expressed in words that are familiar to the recipient. Work out when to follow up on your message and put a reminder in your calendar, diary, or 'to-do' list.

Finally, learn to make effective use of e-mail attachments. Avoid sending a message without a summary of the attachment's contents and what the recipient needs to do with them. It is safe to assume that most people will not open the attachments unless there is a compelling reason to do so that is clearly explained.

Developing effective collaboration skills, particularly the discovery and teaching skills that are discussed in the coming chapter, can also make the use of your time more effective.

Practice concept 49: E-mail and text messages seldom resolve conflicts

Novice engineers and many young people often think that ...	Expert engineers know that ...
E-mail and text messages are quick and convenient and you don't have to look a person in the eye when you need to convey bad news.	When there is any conflict or emotional arousal, there is a much better chance that any message will be interpreted in a way that further inflames the feelings of conflict.

As explained earlier, written communication can be easily misunderstood. Any emotional arousal, whether positive (e.g. affection, happiness, excitement) or negative (e.g. anger, frustration, fear), is likely to increase the chance that the message will be interpreted in a way that reinforces the perception of the emotion. Furthermore, many people who are angry or frustrated want to distribute their response far and wide in the organisation, partly as a means (they think) to garner support and sympathy from others.

Learn to recognise the signs of emotional responses and choose to handle the situation with face-to-face communication, or at least vocal communication by telephone, if there is any sign of strong emotions. From time to time, we all face situations where we have to deliver bad news to someone. These can be very difficult experiences, so be prepared to seek help and support from others. Above all, consider the person receiving the message and make sure that they have adequate support to deal with the message when they receive it.

Practice concept 50: Systems for coordinating engineering activity

Larger engineering enterprises use many formal techniques to coordinate collaboration performances, including written organisational procedures and, increasingly, software such as project management systems (e.g. Primavera, Microsoft Project) and large enterprise resource management systems (e.g. SAP). Maintenance, for example, is often coordinated using a computer maintenance management system and several ERM systems have specialised modules to provide functions, such as CMMS, which companies can use.[60]

Another example of a formal collaboration technique is configuration management: a means to coordinate the production of documents and even physical products. A configuration management system, for example, enables people in the enterprise to find out the exact status of a technical document or artefact. You could use this type of system to find out that an artefact with a particular serial number was built from certain design and manufacturing documents with parts from specific batches, even parts with specific serial numbers, in some instances.

These formal systems often provide tools for large-scale collaboration that enables large enterprises to function. However, there are many limitations and difficulties associated with using them. They are often advocated on the grounds of efficiency and effectiveness, but the benefits of specific systems often lie hidden in contested debates. They can end up being used as little more than a highly complex, automated database, calendar, and scheduling system; gaining the full potential advantages of such systems can be much more difficult than it might originally seem.[61] Often, the focus is on the system itself, rather than on the ultimate purpose, which is to coordinate a large collaborative performance by perhaps tens of thousands of people. However, in reality, to be effective, these systems must work despite the fact that few of the people who use them have any real idea about how the end purpose is achieved. Very few systems actually manage to work well in the presence of this forgivable and understandable ignorance.[62]

There are many such systems used in engineering enterprises. Examples include project management, configuration management, environmental and safety management, computerised maintenance management systems, production management systems, purchasing systems, and inventory management systems.

It is beyond the scope of this book to explore any of these in great detail. Chapter 10 deals with project management, but project management systems are only briefly mentioned. The chapter is mainly about the process of project management, rather than the systems that support it.

Many engineering companies have invested enormous amounts of human effort into developing these systems. These systems represent a great deal of their intellectual capital: their ability to organise and coordinate large numbers of people in engineering activities. To a young engineer, these systems can initially seem daunting, but they are an integral part of your development. Many young engineers spend much of their time interacting with these systems, for example, setting up data that describes activities in a project management system or performing a 'stock take' to ensure that the inventory management system accurately represents the holdings of spare parts and other materials in storage.

All of these systems store human readable information and, as such, represent different means by which people communicate with each other, albeit indirectly. Unlike e-mail communication, a management system will receive information and only make it available when another person requests it. However, these systems are just as important as e-mail or any other kind of communication in terms of organising collaborative activity.

Very large engineering enterprises can only operate with the support of these systems. Furthermore, these systems impose significant limitations on human collaborative performance. It is quite easy to fall into the trap of imagining that these systems will enable an organisation to operate without time-consuming face-to-face, telephone, and e-mail conversations. Ongoing research is helping to reveal many of the shortcomings inherent in using these systems.[63] For example, a failure to allocate sufficient resources for ensuring the accuracy and quality of information held by the management system can severely impair the performance of an engineering enterprise. For instance, computerised maintenance management systems make it possible to organise routine maintenance for very large engineering artefacts, such as large-scale chemical process plants, railways, hospitals, fleets of ships, aircraft, vehicles, and many other systems. In principle, it is possible for these systems to store information provided by maintenance technicians so that engineers can understand the reasons for failures and devise better maintenance strategies. In practice, however, even though maintenance technicians are highly skilled at using the systems, they are often disinclined to provide information about the work they have performed and their observations on why each failure occurred.[64]

COLLABORATION – SUMMARY

Engineering relies on coordinated collaboration between many people. Communication is the way that people collaborate and it encompasses gestures, speech, listening, writing, reading, and watching. We can only understand engineering practice if we can learn how to effectively understand ourselves and others.[65]

One of the greatest impediments that you face as a novice engineer is that your engineering training at university has, most likely, completely overlooked the need for you to learn about yourself and others. This is the greatest challenge that you face on your path to becoming an expert engineer.

Does every expert engineer have to become an expert communicator? Not necessarily. It is a question of degree and technique. First, as we shall see in the coming chapter, highly developed perception skills distinguish experts from others: this aspect of your development cannot be ignored. If you are a shy person and you feel uncomfortable in many social settings, as I was in my student days, you need reassurance that you can develop your social abilities and collaboration skills. Some people learn faster than others – that's an unavoidable reality. One of the most important lessons from the research on expertise is that, so far, there is no measurable human intelligence attribute such as mathematics skills or social ability that distinguishes experts from others. Anyone can become an expert at collaboration and engineering; time is the only variable.

Although it might appear to be a major omission, I have not mentioned most of the conventional teamwork techniques in this chapter. This is an intentional omission. First of all, there are plenty of books and articles that have already been written on the topic. Second, I have restricted the content of this chapter to issues that are not commonly mentioned, particularly the core socio-technical aspects that our research has exposed as being central to engineering practice. In addition, I have covered some philosophical and language issues that emerged from our research as critical elements for understanding our engineering practice observations.

Reading about teamwork techniques will complement the material in the coming chapters.

A final word of caution on teamwork is essential. In our observations of engineers at work, engineers rarely participated in formal teams, except as a member of an over-all enterprise or project team. Most of the social interactions we witnessed or learnt about during interviews were between people who worked together temporarily, perhaps only for a single task, not necessarily people who were designated as members of a particular team. Chapter 9 discusses this in more detail.

How will you know that you are mastering the material in this chapter?

Technical coordination, project management, and negotiations will become easier and less stressful, as well as less time-consuming.

You will find that you spend less time dealing with the consequences of misunderstanding. You will more rapidly understand what other people are trying to explain to you, and others will understand you faster, with less time spent clearing up differences of interpretation. You will have a clearer idea of the different ways that people use identical words, and you will be able to anticipate these differences more accurately.

You will spend less time trying to write clearly, recognising that there is no such thing as a piece of writing that cannot be misunderstood. Instead, you will listen to people to whom you have given instructions or explanations and you will be more skilled at spotting the differences of interpretation. You will be able to act sooner to ensure that these differences do not result in unnecessary delays or the misuse of resources.

Finally, you will have much greater appreciation for the value of organisational processes and procedures that promote collaboration. You may be able to spot ways in which they undermine collaboration and suggest improvements.

NOTES

1. Mitcham (1991).
2. Perry and his colleagues have produced the most recent reports on this worldwide issue, with many others before them. Martin and Maytham were engineering students who made the same observation for themselves (Blom & Saeki, 2011; Dahlgren et al., 2006; Darling & Dannels, 2003; Duderstadt, 2008; Martin et al., 2005; Perry, Danielson, & Kirkpatrick, 2012; Shuman, Besterfield-Sacre, & McGourty, 2005; Spinks, Silburn, & Birchall, 2007).
3. Many surveys ask employers about their perceptions of team skills and engineering graduates about their confidence in their team skills: graduates rate themselves highly, and while employers see room for improvement, their ratings are higher than for several other graduate attributes (e.g. American Society for Engineering Education (ASEE), 2013, p. 36; Robinson, 2013, p. 42).

4. Sheppard, Macatangay, Colby, & Sullivan (2009).
5. Kotta is one of several researchers who has documented this kind of collaboration (2011).
6. Leonardi and Jackson provide a thought-provoking analysis of student behaviour in software engineering team projects (2009).
7. See, for example, Galloway's book, p. 21, where she writes, 'the purpose of communication is to convey information'. She has been the CEO of successful engineering enterprises (2008). At the time of writing, even Wikipedia contains the same misleading understanding of communication.
8. It does not always apply . . . one engineer sarcastically remarked that keeping silent might have improved their marriage!
9. A French word pronounced like 'zhonrer', plural has an 's' but sounds the same. Luzon explains this in the context of technical writing (2005).
10. Orr (1996, Ch. 8).
11. The most senior engineer in a particular discipline is often referred to as the 'lead', for example the 'structural lead' or the 'reliability lead'.
12. Rick Evans suggested using the term 'performance' to describe aspects of engineering practice, especially communication and collaboration (R. Evans & Gabriel, 2007).
13. American Society of Civil Engineers (2008).
14. Engeström (2004) explains the need to recognise collaborative expertise as part of a more general appreciation of expertise based on individual technical performance and judgement, as explained in Chapter 4.
15. Tan (2013) provides an extensive review of this.
16. Hubert and Vinck (2014) present an example in micro-electronics and semiconductor technology.
17. This material is a development from findings presented in Trevelyan (2010b).
18. As explained in Chapter 5.
19. In a wide ranging discussion, Collins and Evans referred to this as interactive expertise, though they seem to reserve the term for STS researchers (Collins & Evans, 2002, p. 252).
20. See also, for example, Darr (2000, p. 208).
21. Sonnentag et al. (2006, p. 380).
22. For example, the use of design elements, in this case appropriate structural design elements, is described in Gainsburg, Rodriguez-Lluesma & Bailey (2010). There are similar reports from other disciplines such as software engineering Sonnentag et al. (2006, p. 377).
23. Often referred to as a project phase gate decision process (Cooper, 1993, p. 109).
24. E.g. Dowling, Carew & Hadgraft (2009, Ch. 3).
25. Younger engineers need up to an hour of explicit help and guidance daily (Bailey & Barley, 2010).
26. Tan (2013).
27. Itabashi-Campbell & Gluesing (2014); Itabashi-Campbell, Perelli, & Gluesing (2011).
28. Robert Bolton has provided extended explanations and suggestions for this in his book, *People Skills* (Bolton, 1986, Ch. 7, Part 3). This exposition conveys the ideas effectively so there is no need to describe this element of collaborative performance.
29. Anderson et al. (2010); Blandin (2012); Domal (2010); B. Williams, Figueiredo, & Trevelyan (2014).
30. Trevelyan (2007).
31. E.g. Aster (2008); Darling & Dannels (2003).
32. Bailey & Barley (2010).
33. Standards Australia (2009).
34. Also referred to as 'organizing alignment' by Suchman (2000).
35. Bailey and Barley (2010).
36. Bransford, Brown, Cocking, Donovan, & Pellegrino (2000).

37. Trevelyan (2010a).
38. Tan has provided an extended discussion on architects and engineers negotiating in both technical and commercial spaces (2013).
39. Meiksins & Smith (1996).
40. ASME – American Society of Mechanical Engineers.
41. Engineering meetings in Brunei, where securing agreement is important, nearly always have food as an important element. Engineers from different Southeast Asian countries can encounter difficulties when moving from one regional culture into another (S. S. Tang, 2012).
42. Trevelyan (2014a).
43. To see an example of this in an engineering context, refer to Petermann, Trevelyan, Felgen, and Lindemann (2007).
44. French post-modernist philosophers can be difficult reading! Some of their material is accessible, however, and others have provided more lucid interpretations (Bennington & Derrida, 1993; Foucault, 1984; Gherardi, 2009; Hoy, 1985; C. Johnson, 1997; Johnston, Lee, & McGregor, 1996).
45. Hamilton (1973).
46. Speech recognition software uses an analogous method without understanding the meaning of words. Instead, software relies on statistical pattern recognition to choose the most likely combination of words to match a sound pattern. Every person has a different way of pronouncing and using words, hence the need for software to be 'trained' to recognise the sounds made by a particular speaker and also the ways that the speaker uses words.
47. Social scientists also use the same term 'lens' to describe how performing observations on people is influenced by the particular theoretical framework being used: there are many such frameworks, such as social capital (associated with Bourdieu, Bandura) and actor network theory (associated with Woolgar, Latour).
48. For a more scholarly explanation of post-modernist ideas, see works such as Bennington and Derrida (1993), Foucault (1984), Hoy (1985), C. Johnson (1997), or Steiner (1975).
49. Experienced engineers can become attuned to the ways in which people choose to interpret information differently to suit their organisational interests (Bella, 2006).
50. Gouws (2013). See earlier publications as well: Gouws and Gouws (2006) and Gouws and Trevelyan (2006).
51. Bechky (2003, p. 319) provided some insightful examples of technical terms with shifting meanings across communities of engineers and technicians.
52. Collins described this idea in more detail (2010, pp. 23–25).
53. Sule Nair (2011) unpublished manuscript.
54. Salas and his colleagues provide an introduction to extensive literature on team performance (2006).
55. Perlow (1999).
56. Trevelyan (2010b).
57. LaToza, Venolia, & DeLine (2006, p. 495).
58. It has been found that e-mail and other interruptions can cause the loss of short-term working memory, typically taking up to 15 minutes to recover if the person even remembers to return to their original task (Gupta, Sharda, & Greve, 2010; F. Smith, 2010).
59. S. Barley & Bechky (1994, pp. 99–105); Jackson, Burgess, & Edwards (2006).
60. Gouws & Trevelyan (2006).
61. Gouws & Gouws (2006).
62. Gouws (2013) provides a comprehensive treatment in the context of maintenance management systems and many of her findings apply equally to comparable systems. We will deal with project management systems in more detail in Chapter 10.

63. Gouws (2013); Gouws & Gouws (2006); Gouws & Trevelyan (2006); Nair & Trevelyan (2008).
64. Sule Nair (2010), in an unpublished manuscript, drew on social theory to explain how the choice of the term 'resource' to describe people, tools, parts, and even the equipment to be maintained can lead to people adopting behaviour patterns associated with inanimate objects.
65. Larsson explains how essential it is to know and understand others, particularly when working in geographically distributed teams. Knowing the depth of their relevant technical knowledge and experience, their tendency to over- or understate the importance of an issue, as well as their trustworthiness on a particular issue provides essential understanding that is needed to interpret the sense of what might be brief comments in a phone call or meeting. Citing a study of medical specialists, he quoted, 'When Jones at the same institution reads it and says, "There's a suspicion of a tumour there," I take it damn seriously because if he thinks it's there, by God it probably is' (Larsson, 2007).

Chapter 8

Informal teaching: More than an interpreter

In Chapter 7, I explained how learning and teaching lie at the core of several collaboration performances. Reflecting on the mass of data collected from our interviews and observations, and studying engineers working in four different countries, it is apparent that difficulties with learning and teaching lie behind most of the factors that frustrate collaboration performances by engineers.

Fortunately, recent intellectual advances in the learning sciences and education psychology have much to offer that can help reduce these frustrations for engineers by providing helpful guidance to improve learning and teaching. Most of the research has been directed at formal learning and education institutions, i.e. classroom learning situations. Collaboration in engineering, however, mostly relies on informal teaching. It should be noted, though, that the fundamental ideas exposed by research on formal teaching and learning are still applicable and can be very helpful.

In this chapter, I will outline some of the results that offer the greatest potential for engineering practice.

Another good reason to learn more about teaching is that only a minority of engineers are comfortable with reading in order to learn, perhaps less than 10%. Given that you have read this far in the book, you could be one of them. For readers and non-readers alike, this book holds a great deal of value: learning and practicing skills makes it easier for everyone involved in an engineering project. By passing on useful aspects of this book to engineers who are not as comfortable with reading, you will not only be able to practice your teaching, but also reinforce the ideas in your own mind and encourage others to read more for themselves.

Most of the research literature on workplace learning has focused on the need to improve the skills and knowledge of people working in an enterprise.[1] Other literature focuses on the development of specific capabilities, for example in communities of practice.[2] Recent research, however, has demonstrated that in many workplaces, people come together to collaborate for a limited period of time in relatively informal organisational structures, most of which are hardly formal enough to be regarded as 'teams'.[3] This can only happen with effective learning and teaching performances. Learning and teaching extend far beyond organisational skills and knowledge development activities and seem to be much more common in the workplace than has been previously acknowledged.

At the same time however, teaching with the aim of skills or knowledge improvement is an important aspect of engineering practice. As humans, we all age and progress from being novices through learning throughout our working lives. Teaching others is a part of being human, whether it is our own children, our peers, novices, or even senior engineers who need to learn something new. Dianne Bailey and her colleagues, in an extensive and detailed study of novice engineers in several different workplaces, found that they were spending up to an hour or so every day, on average, in informal one-to-one teaching situations.[4] More senior engineers were doing the teaching, often spending this amount of time with two or three less experienced engineers, guiding their work and helping them learn more advanced skills. Often, as an engineer, you will be called upon to teach new knowledge to people with much more experience than yourself. Shortly, we will review some of the situations in which these issues present themselves.

First though, a few words about theories of learning and teaching. Developments in the last few decades have provided us with powerful ways to understand what happens when people learn. In learning about these theories however, many engineers may have to think differently about theories in the context of engineering. Theories of human behaviour are not like the familiar theories that have emerged from the traditional physical and engineering sciences. For a start, most are not expressible in mathematical form; instead, they rely on extensive written argumentation. The few theories with quantitative empirical support in sociology and psychology need to be interpreted for their results to be useful.

For most engineers, four significant thinking changes will be needed.

i)   Social theories explain, but generally don't predict, human behaviour. Instead, their value lies in providing concepts and language that help you perceive what is happening. You can then discern patterns of behaviour and learn more quickly from experience.

ii) Instead of a single 'accepted' theory, for example, the theory of thermodynamics, the social sciences offer multiple theories and approaches, all of which may be valid in a given situation. All of them can be informative, although often they conflict with each other and the explanations that emerge from them can be contested. None offer a complete understanding, and the variety of different approaches offers a richer understanding of human behaviour than a single theory.

iii) By and large, most social theorists would now agree on the impossibility of truly independent, objective observations of human behaviour. As we briefly discussed in Chapter 7, everyone sees the same text or human situation through a different 'lens' that is shaped by their prior knowledge, experience, and beliefs. By understanding the ways that other people see a particular situation we can learn to 'place ourselves in their shoes',[5] and begin to appreciate the particular lenses through which we see a situation. An 'objective' description, therefore, is one that considers the many different ways that different people see a given situation from their own viewpoints, as well as an explicit acknowledgement of the writer's own lens.

iv) The terms 'qualitative' and 'quantitative' have different meanings in the context of social science. For engineers, 'qualitative' denotes data that is non-numeric. This would include data from what social scientists would regard as a 'quantitative' survey, in which, for example, participants would respond to multiple-choice questions or rate their agreement with certain propositions on a Likert scale. For social scientists, qualitative data usually refers to open-ended text collected in interviews or observations of human activity. Qualitative data also can include observations about documents, images, drawings, or even natural objects that have been placed in certain positions by people.

Social science theories can provide engineers with a deeper understanding of human behaviour. In particular, a deeper understanding of learning and teaching can make life much easier for an engineer. Our observations of expert engineers underlined the significance of almost entirely self-taught abilities related to teaching. By acquiring some familiarity with social science and education theories, you will be able to get there much faster.

This chapter introduces you to some ideas of teaching and learning that, according to our research, appear to be highly relevant to engineering practice. When you can master them, much of your engineering practice will become significantly easier. The aim is not for you to become an expert teacher, but rather to understand many of the key ideas behind effective teaching. You will learn to identify when teaching performances are needed, who needs to learn, what they need to learn and be able to do, and how you can verify that learning has occurred. If you follow the major references and practice techniques, carefully observing the results, you will find that collaboration performances in your engineering practice become easier and less stressful.

As an engineer with these abilities, people are more likely to do the right thing for you the first time around. They are more likely to trust you and they are less likely to need supervision. You can get more people doing what you need them to do, yet at the same time enrich your understanding by building on the knowledge that other people possess that may exceed your own.

THEORY AND CONCEPTS

Why were you asked to learn so much theory in your engineering course at university?

The answer is simple: theory helps you learn faster, as long as the theory can help explain observations relevant to the task!

In one of the many experiments conducted by learning researchers, psychologists found that students who understood the concept of the refraction of light found it much easier to hit a submerged underwater target using a crossbow while standing by the side of a swimming pool. After one or two trial shots that went wide of the mark, these students realised that you have compensate for light refraction at the surface of the pool. Students who had not learnt about the refraction of light accomplished the same task but took many more attempts before they figured out that you have to aim the crossbow well below the target.[6]

That is a simple example of why it is important for you to begin to understand some of the background theories that can help you learn about teaching techniques.

Some engineering frustrations

Most of the most commonly reported frustrations we found among engineers stem from difficulties in teaching. Engineers often express frustration in these terms:

'I can't get other people to understand simple engineering ideas.'

Here is a senior software engineer who was running a major project at the time.

'We have to make sure that the client's people understand how our systems will help them in their business. It's so easy for their ideas and ours to diverge, for our people to lose track of their needs and you end up with a system that produces little real value for the client. It's a constant struggle.'

Here, you can see two teaching performances. The engineer had to work with the client's people and educate them on how the software could improve their business, while at the same time he also had to learn about the client's business in order to appreciate how the software would help. Then, he had to work with his own software engineers and educate them about the client's requirements, not just once, but continually throughout the project. Without direct contact with the client's people, the software engineers would easily lose track of the requirements, and without continually being monitored and taught, would end up writing software that would not produce a useful outcome for the client. These two performances turn out to be very common in many branches of engineering, particularly in terms of ensuring that the technical intent survives sufficiently intact through multiple reinterpretations by the people who work on implementing it in a way that produces lasting value for end users.

Another example comes from a software engineering R&D team leader who discussed the challenges he had in persuading a group of mining engineers that their software tools needed to be dramatically improved.

'The software engineers can see a bigger picture and advantages that an improved software tool would bring, while our very much overpaid mining engineers and the management group are very satisfied with what they already have and are not eager to change. We prepared a proposal advocating changes and gave examples of other companies that have already adopted better software tools and gained better performance. We became so frustrated trying different strategies and the outcomes. We spent so much energy and time gathering a team of researchers to develop ideas and improve the existing software tools at hand. This proved to me that being technically strong in one dimension can be very good but it's even more important to be an excellent communicator, listener, and a good negotiator. Ability to explain ideas and facts and convincing people to follow is the key of success in any industry.'

DISCOVERY AND TEACHING

As Chapter 7 briefly explains, collaborative learning performances can be classified into two main categories. First, there are situations in which an engineer is learning from other people, but it is reasonably clear that no one has started with a well-defined idea of what needs to be learnt. I have called this performance 'discovery', because learning mostly takes place in the context of working together and involves informal conversations and encounters between people. Oftentimes, the learning emerges almost by accident. Distributed cognition is closely related to this, when new knowledge emerges from social interactions between people.[7]

Teaching, on the other hand, describes performances in which there is a relatively clear idea about the ideas and information that needs to be 'appropriated' by learners through the teaching methods of an engineer.

As we shall see, most effective teaching performances rely on some discovery, which is why I have chosen to include both in this chapter.

So, how do you go about executing an effective teaching performance? How can you convince other engineers that they need to adopt different techniques in order to improve their performance? How can you ensure that other engineers, contractors, technicians, fabricators, and plant operators interpret your ideas faithfully and conscientiously? How can you be sure that your client actually understands what you are proposing to build for them?

Before describing some of the more productive approaches that are useful in discovery and teaching performances, I need to describe some of the important ideas about human learning that have emerged in the last few decades.

Let's start with a common misconception among young engineers. By now, having read Chapter 7, you should no longer have doubts about this.

Misconception 15: A concise and logical explanation is sufficient

Students and novice engineers often think that . . .	Research demonstrates that (particularly for engineers) . . .
A technical idea can be conveyed to another person with a concise and logical explanation.	Bringing other people to a usable understanding of a new idea is difficult, time-consuming, and can be very stressful. Logical explanations are just the start of a lengthy process that involves many steps, takes time to prepare and deliver, and requires careful monitoring to ensure that the new understanding is influencing behaviour in the way that was expected.

Young engineers can be forgiven for this misconception; it reflects the way that they were taught. Engineering students are among the brightest and most eager to learn at any university. They have performed very well at school and are therefore among the easiest young people to teach. Very often, a concise and logical explanation is sufficient for them to grasp an idea and try it out for themselves. Even if this doesn't work, most engineering is taught in this way. The curriculum is dominated by lectures, even in tutorials, during which the scientific principles behind engineering are explained with a few worked examples. The rest is up to the students; those that pass are clearly able to learn a sufficient amount in this way.

There is another important factor: motivation. Students are motivated to learn because they want to graduate and that means gaining as many marks as possible for the least effort. Now, consider the mining engineers mentioned in the quotation on the first page. They are already, apparently, well paid, and improving their performance with new software is unlikely to increase their pay. Also, it will mean that they have to learn something new, so their job will become (at least temporarily) more difficult. Unlike when they were students, there may be little motivation for them to learn.

This is why teaching in the real world is so much more difficult than simply providing concise and logical explanations. Engineers often have to work with people who think they know it all already, and who usually know enough about learning to remember that it is difficult and to be avoided if possible.

Practice concept 51: Planning: the learning objective is a performance

You will need a script.

A teaching performance requires careful thought and planning to avoid frustration. Given that you are working with people, frustration can happen, even with the best thinking and planning beforehand.

Teaching has to start with the idea that the learner or learners are going to have to accomplish a performance with the help of their learning. It may be giving you approval to do something or it may be that they are going to implement some of your ideas in the form of new artefacts or systems. Either way, a teaching performance starts with an assessment of what the learners need to be able to perform and the skills or knowledge that they will need to do that.

We refer to this as the learning objective. A teaching performance starts by defining learning objectives, and also how you, as a teacher, can assess the subsequent performance by the learner or learners to gauge the extent of learning that has occurred.

In other words, what has to be learnt and how can we tell that the learning goal has been achieved?

Practice concept 52: All learning is based on prior knowledge

Students and novice engineers often think that …	However in reality …
People learn by listening or watching and memorising what they hear or see.	People build new knowledge by adding to ideas they already have. People 'discover' new ideas in terms of the ones they already know about. Anything that does not fit in with prior knowledge may be completely missed.

We learn by listening to other people, watching them, reading books, and so on. Yet, in order to discern something that needs to be learnt, we have to understand what other people are trying to convey to us in words or symbols. As explained in Chapter 7, there is no such thing as uniform understanding of words and symbols. We all have different life experiences and memories; in other words, we all have different prior knowledge. This is especially true when you are working with adults in the workplace, rather than in a formal school or university education setting. Therefore, any explanation will be understood in a way that is unique to every learner.

Our language knowledge is what enables us to perceive something that we need to learn.

Again, as explained in Chapter 7, we may all share English as a common language, to a greater or lesser extent, but we all use English in a different way. Engineers have a special vocabulary: some examples are listed in the glossary (online appendix).

However, it is not quite as simple as that. Behind our language, we all have different ideas about reality. We have a different set of mental models and concepts to explain the world around us. When we perceive something new, we try and fit this perception into our existing mental models.

The introductory chapter of the book, 'How People Learn',[8] illustrates this idea using Leo Lionni's children's story *Fish is Fish*.[9] A baby minnow (a tiny fish) in a pond is anxious to explore the world outside. Unable to leave the pond, the fish befriends a tadpole that eventually grows into a frog. The frog leaves the pond for a while, exploring the world outside. Eventually, the frog comes back and describes what he has seen. The book's illustrations show the way the minnow interprets what the frog has told him: cows appear as fat fish with four fins touching the bottom of the pond, with a large pouch (the udder) between the fins of the back-end. Birds appear as fish with extended fins looking like wings. People are imagined to be fish walking on their tailfins. Although it is not in the book, the frog might have had some difficulty explaining rain: what is the equivalent of rain for a fish? We learn from this simple story that each person receives a story about something new in terms of things they are already familiar with. Therefore, if you can tell the story in terms of familiar words and ideas, it is more likely to be understood.

Susan Ambrose and her colleagues in the book, *How Learning Works*,[10] describe how prior knowledge has to be activated, sufficient, appropriate, and accurate for learning to take place. This poses three challenges: first, we need to know about the language used by the learner, and second, we need to know about the ideas already in the mind of the learner upon which we can build. Finally, we need to ensure that the prior knowledge has been activated and brought to the foreground of the learner's mind.

Let's reflect for a moment on how we learn about the places we live and work and how we move around in a rural or urban environment. Everyone has a different way to represent their understanding of their environment, the mental maps they use to find their way around. For me, my map of the city in which I live closely corresponds to a printed street directory or, these days, Google Maps. As I walk or drive through the city, I retain a reasonably clear idea of the landmarks and districts on either side of my route. If I take a wrong turn, I still have a clear enough idea of where I need to go, so I find my way by driving in the appropriate direction even though I may not know the actual streets I need to follow to get there. I may have to ask for directions or check the street maps once or twice.

My wife, on the other hand, remembers a route as a series of streets that take her from home to her destination. She has no clear idea of the direction in which she is travelling or what lies on either side of her route. She just knows that this route will get her to where she wants to go. While I feel comfortable taking an alternate route, partly to explore another district of the city, she can become completely lost as soon as she leaves her regular route.

Many Aboriginal people in Australia who still live in their traditional 'country' learn how to traverse what seems to us to be endless featureless scrub or even desert by learning songs or stories. Each stage of a song or story is related to a particular place, often involving the names of specific types of plants that grow there. With tens of thousands of different native plant species, many that are found in different forms depending on the soil where they grow, combinations of plant names provide the equivalent of a street directory for the Australian landscape. The songs or stories string places together to form a route in a way that can be remembered and passed on orally from one generation to the next.

What we learn from this is that we all have different ways of representing the world around us in our memory. As we get to know the area where we live, we gradually build our own mental representation in a way that is related to our culture and traditions. We learn by incorporating new experiences into our mental representation. A teacher cannot simply feed the map directly into your memory; you have to construct a map yourself in your own particular way. A good teacher, therefore, facilitates that learning or mental construction process. An effective teacher helps the learners accomplish this challenging mental construction task for themselves.

Now, think about the way that you remember the way around your local area. Do you remember the appearances of different places, houses, or shops and use them to recognise where you are? Or do you use the street names to work out where you are? Do you look for major landmarks, hills, or tall buildings? Do you think about the angle of the sun during the day to work out the direction you are heading? We are all different, which means that we each have our own combination of techniques.

Now, imagine that you meet a stranger, someone who has only been in the area for a short time. The stranger has asked you for help to find their way to the local branch of a particular bank. How would you begin to explain? (You may not even know where the local branch of that bank is located if it is not the bank that you use, but let's assume that you do know.) You might begin by asking the stranger how he reached you, and what else he knows about the surroundings. 'Have you seen the tall B&OD building, the one with blue-coloured glass? No? Well, have you seen the huge Shell garage, the one with the McDonald's next door?'

If the stranger speaks broken English with many grammatical mistakes, you will quickly realise that they may not understand English very well. You may have to adapt the way you speak in a way that makes it easier for this stranger to understand, namely by using simple words and speaking slowly. In other words, you are thinking about the stranger's grasp of language, and how to provide directions in a way that the stranger is more likely to be able to understand.

Having confirmed what the stranger already knows about the district, you might then remind them about a particular landmark needed to find the bank. 'Okay, now walking from the Shell garage past McDonald's, you go past three or four shops, and then you will find a jewellery shop on the left-hand side, just on the corner. Have you seen that before?' Having confirmed that the stranger has seen the jewellery shop, you can then direct them along a side street that leads to the bank.

How would you know if the stranger now has an accurate idea of the route to the bank? The stranger might be trying to understand what you are talking about and is not really quite sure about the jewellery shop, but is confident that it can easily be found since it is quite close to the McDonald's outlet. How sure can you be that the jewellery store is still there on the corner? How many shops do you have to go past? Are you sure that it's just three or four shops? Or is it actually several hundred metres along the same street? Maybe the stranger cannot actually remember the large Shell garage next to the McDonald's outlet but does not want to seem completely ignorant about major landmarks in the district. With any of these slips, the stranger will construct new knowledge about the district that is incorrect and misleading. As a result, the stranger may become lost and have to ask for directions again.

Does it really matter if the stranger has an accurate idea? With enough determination the stranger can gradually find the way to the bank by a combination of people pointing in the right direction, as well as simple trial and error.

Returning to the four prior knowledge conditions that have to be satisfied in order for learning to take place.

i)   Appropriate: the stranger knows about landmarks close to the bank branch.
ii) Sufficient: the stranger knows about the Shell garage, the McDonald's outlet, and the jewellery store.
iii) Accurate: the stranger can remember which side of the street the jewellery store is on.
iv) Activated: through conversation, the stranger has reminded himself or herself of these landmarks.

In engineering, teaching performances matter. Facilitating learning by others that is right the first time means that you spend less time and energy having to check later and correct mistakes or misunderstandings by other people.

Finding out about prior knowledge is not as easy as it seems. Merely asking someone if they already know about reverse cycle air conditioners, as an example, is not going to help you. Some people will respond, 'Yes, of course I do'. These people do not want to give the impression that they are ignorant about something as basic as an air conditioner. Then there are others who will say, 'No, I don't'. These people might be concerned that you will then go on and ask them something about how air conditioners work: they don't want to be shown up as being ignorant by being unable to answer your follow-up question. Therefore, being cautious, they deny any knowledge.

Here is an example from an engineer in the oil and gas industry. She found that understanding her client's prior knowledge was crucial before trying to educate them about the engineering work that needed to be done for their projects.

> 'Most of our real clients know a lot about engineering issues and they have their own engineers. However, they often don't know much about construction issues. Often they have a focus on operations and maintenance without understanding how construction affects that. The less experienced clients often are not aware of regulatory issues. For example, if they had not recently built some hydrocarbon process plant they may not be aware that you need to alert the Department of Environmental Protection at least 12 months ahead or their construction work will go nowhere. In subtle ways you need to ask questions to figure out how much they are really aware of. For example, you can casually bring AS1210 (Pressure Vessel Design Code) into the conversation and judge from their responses whether they know about it in detail or not.'

As this example shows, casual conversations are more effective at exposing prior learning. Try starting a conversation about a topic that is likely to be of interest to the learner: it might be about a recent sporting match or a political incident reported in the media. Gently move the conversation to the topic of recent hot weather and how many people are using air conditioners, perhaps by posing a rhetorical question, such as, 'I wonder what kinds of air conditioners are used around here?' A casual conversation like this can help to activate prior knowledge; in other words, the prior knowledge is brought closer to the surface of consciousness. Particularly with adult learning, when you have no real idea of what the other person already knows, accurately understanding prior knowledge is indispensable. You need to be able to assess whether it is sufficient or appropriate. It might be quite inappropriate or it may simply not be enough to comprehend the ideas that you are going to try and teach.

Now think about a situation that you might face as an engineer when you need someone to perform some skilled technical work. For example, imagine that you have received a special purpose hydraulic control valve from an equipment supplier and you need to have it checked out by technicians to make sure that it functions as expected. You will need to provide technicians with instructions on what to do: this is a teaching performance.

At first you seem to have a clear idea of what needs to be done, but when you think about it, there are many detailed aspects that you are not sure about, especially if you are not a mechanical engineer or you are not familiar with hydraulic control systems. For instance, a technician will need to connect the control valve to a hydraulic circuit with the appropriate instrumentation. Is there a suitable testing facility already in the workshop? How will the valve be flushed out to remove any small particles of dirt that could interfere with its operation? Does another hydraulic circuit have to be adapted for the test? Is there instrumentation with the required measurement accuracy? What tests will need to be performed? You may know the answers to some of these and the technician will probably have answers for some of the other questions.

Most likely, you will start by arranging a time to meet with a senior technician and discuss the testing process together. If you have not worked with the technician before, you will spend time figuring out how much the technician actually knows about these kinds of control valves and different testing methods. You may not be very sure about the kinds of tests that need to be performed if you have not worked with similar control valves before.[11] This is typical of an engineering discovery performance, similar to the discovery performance between you and the stranger asking directions to the bank: you are trying to establish some understanding of prior knowledge. At the same time, you may be listening carefully to the technician to pick up some of the language that they are using. You may have no idea of what is meant by a Swagelok fitting for example, so you admit to that ignorance and subsequently learn about a common type of pipe connection used in hydraulic systems.

Once again, think about how you would confirm that the technician's prior knowledge satisfies these requirements.

i)   Appropriate: the technician is familiar with hydraulic testing methods and constructing hydraulic circuits.

ii) Sufficient: the technician knows enough to understand about the testing work you need done.

iii) Accurate: the technician has up-to-date knowledge of the actual components needed.

iv) Activated: through your conversation or recent work on similar hydraulic circuits, the technician can quickly access the required knowledge.

That's why teaching almost always starts with a discovery performance where we engage the learner in a casual conversation about ideas that might be relevant for future learning. However, it is more than casual for the teacher; you will be paying particular attention to the words that the learner uses and the ideas expressed in those words. These observations will form the foundation on which you are going to plan your teaching. You would remember from Chapter 5 about the importance of taking notes and of recording the actual words being spoken. Any message that you want the learner to comprehend has to be expressed using the words that are already familiar in the learner's vocabulary.

In more formal settings, particularly working with several learners at the same time, there may be large variations in prior knowledge between them. As an engineer, you might encounter this when starting up a new project or working with a supporting team of other engineers and support staff. You need to educate them on what's been done so far and what needs to be done for the next phase of the project.

If the aim of the learning is to develop conceptual knowledge, it can be helpful for learners to construct graphical representations of their existing knowledge: sketches, concept maps, plans, or other diagrams that represent their prior knowledge. The advantage of a graphical representation is that you can often quickly appraise the extent of their prior understanding by glancing at a diagram, rather than reading an extended piece of text.

A more useful technique in vocational learning situations is to ask the learners to engage in an activity that will require them to use their knowledge. It is useful to be able to set it up as an entertaining game so that everyone can have a few laughs. However, the game has a serious intent: you get to watch the learners perform. From that, you can judge their prior knowledge. You can even set up a mock judging panel composed of some of the participants and get them to score the others' performances. Obviously, you should rotate participants through the panel so that everyone gets a chance to evaluate others.

It is important to make sure that people are in a comfortable environment for the discovery aspect of your performance. In many situations, sharing a meal together can be a great way to start, because it provides plenty of opportunities for casual conversation. The more listening you do and the more talking your learners do, the more you will be able to discover about what they already know and understand. Equally important, by listening carefully, you may be able to gauge what they *don't* understand. In this book, many student misconceptions about engineering practice have been pointed out. All these have come from listening to students and young engineers talk about engineering practice, writing down what they said, and then thinking about the ideas that might have led them to make those statements.

Any of these techniques could be appropriate for working with the technician in the example above. Think about how best to appreciate the extent of the technician's knowledge, and your own, and the gaps that might have to be made up by learning or experimentation, another discovery performance.

Once you understand what your learners already know, then you can work out what they need to learn. With adult learners, a good way to do this is to engage them in further discussion – see if they can suggest what they need to learn. Later, we will discuss learner motivation. Adult learners who have some say in deciding what they need to learn are much more likely to be motivated. This is particularly appropriate when collaborating with skilled technical workers. Asking them to figure out what needs to be done demonstrates your respect for their knowledge and your desire to learn from them.

COLLABORATIVE DISCOVERY PERFORMANCES

As explained earlier, a discovery performance is one in which the learner engages in conversation with other people and learning happens as a result of that conversation. The conversation may be face-to-face, over the telephone, or even in writing. Neither the learner nor the others involved have a clear idea of what the learner needs to know; those details emerge gradually through the discussions. A discovery performance can last anywhere from a few minutes of a single casual conversation to several months that involve a whole series of meetings and encounters.

Take the example of the engineer-inventor working with the patent attorney. The engineer has a clear idea of the invention but needs to work with the patent attorney who knows how to write in the way that is required for a patent specification. Initially, the patent attorney has only a brief written description of the invention. After sitting with the engineer who can explain it to him, the attorney develops a better idea and then works from the documents to produce the first draft of the patent. Later the attorney would ask the engineer to check the patent specification to make sure that the specification explains all the important elements of the intervention and the reasons why they have been incorporated. Since the intended reader of a patent specification is a similarly skilled engineer or technician, the patent attorney may ask questions on how other engineers might interpret the words of the specification. Gradually, after working with an attorney on several inventions, the engineer develops a clearer idea about how patents are written and may even be able to produce a draft patent specification that requires only a minor amount of further editing. However, the attorney is always more likely to be aware of legal technicalities and the necessity to use certain techniques in drafting, particularly for the critical claims usually listed at the end of the patent.

Here is an excerpt from an interview with an engineer talking about document review, another example of a discovery performance:

'Once an engineer produces a document, then it is checked by another engineer who goes through and redlines[12] the document. He checks every line, and also has a checklist that he has to tick off. Then it would come to an approver – that's me. I add my comments on to it and then it's updated by the originating engineer. Then the documents are circulated for review and approval. Typically I'll get them on the review cycle. I have my opportunity to make changes. But the way it works – because it's kind of multidisciplinary – the subsea people will get it, also drilling, operations and other groups. They see it as well. They put in their comments and we'll take their comments into account if we think they are appropriate. At the end of the day, it's up to me if we take them or not, as an approvalist.

'[When a document gets redlined, what sort of issues are picked up?][13] You pick up all sorts of stuff from just grammar, bad writing, punctuation, all that kind of normal stuff. Plus obvious questions about functionality. We find mistakes in

assumptions – not often with pressures, though. But gas composition, tempera-
tures potentially. You know, that kind of stuff. So you get to check it for technical
accuracy as well. The other issue is where there has not been good communication
between engineering disciplines. You find a missing issue somewhere that you are
aware of but the author of the document has not been made aware of it and has
written the document without taking it into account.'

The person speaking has to check the work of the less experienced engineers writing documents, such as specifications. The document author is learning as a result of seeing the comments written by other engineers and discussing them with his supervisor. The reviewer learns from the author as he reads the document: the author will have investigated technical issues in much more detail than the reviewer has time for. Engineers in other disciplines learn about developments elsewhere in the project team as they read the document, looking for improvements that have to be made with respect to their own discipline area. Finally, as explained in the last sentence, the reviewers also learn about communication gaps in their own organisation, as well as possibly learning about weaknesses on the part of the author.

The final example demonstrates yet another kind of discovery learning performance for engineers.

'We asked the equipment vendors to do some engineering work and that required
a budget, I guess a couple of million dollars each. We had to sell that to the
vendors and our management. It was actually to do some engineering, prior to
them preparing their proposals. In a way, it actually funded their proposals. What
happened was that we wanted to get a better idea of what they thought about
what the equipment should look like and what challenges they foresaw. We paid
them to kind of pick their brains, if nothing else. I think it was money well spent.
I think our management did as well, at the end of the day.'

This engineer is talking about a recent experience in which he and his colleagues had to choose between different vendor firms supplying some highly complex equipment for a special project. This engineer had extensive experience but knew that the engineers in the firms had much more specialised knowledge than he had been able to acquire himself. In some instances, particularly if there is fierce competition between the firms, the suppliers may have been willing to allow the engineers to spend considerable time developing and explaining their proposals. Engineering firms normally see the education of prospective users of their products as a very high priority and part of their business development (or marketing) efforts. In this case however, at the instigation of the engineer being interviewed, the customer firm has recognised the importance of going further than normal. They have paid a substantial amount of money to commission the vendors' firms to perform a more detailed study of their requirements, partly so that their own engineers could learn from the vendors' engineers before starting to evaluate the vendor proposals.

Notice how all of these examples demonstrate how knowledge is distributed among different people in the overall engineering enterprise, as explained in Chapter 5. They also show how discovery learning takes time, which has to be carefully balanced against the likely benefits.

Table 8.1 Some discovery performances in engineering.

Asking another person about the advantages and disadvantages of competing products or services, learning about future developments in competing products or services.

Posing a technical query in order to clarify what seems to be missing information or a mistake.

Seeking to understand more about other people or stakeholders and learning about their opinions, viewpoints, languages, values, and interests.

Ask about progress that has been made on a piece of technical work.

Negotiating shared meaning, a common understanding, taking diverse perspectives into account, aligning stakeholders – all of which have been shown to predict successful collaborative problem solving.

Discussing the need for a particular artefact, solution, or process with people such as end users or sales staff for whom it might provide value to assess their level of interest and their needs.

Working with a patent attorney to prepare a patent specification: a claim for the limited monopoly rights of an inventor for a particular innovation. The engineer understands the invention, but not how to write a patent, whereas the patent attorney understands how to write a patent, but needs to learn about the invention from the engineer.

Arranging for technical work to be reviewed by peers, possibly external to the enterprise.

Effective learning from discovery performances

Table 8.1 is framed around the likelihood that you, the reader, need to acquire much of your learning in discovery performances. How can you use the previous discussion on learning to make this easier for you? Can you identify your prior knowledge in a way that could help you make better use of a discovery learning opportunity? After the performance has finished, you could try and identify where your prior knowledge was insufficient or included misunderstandings.

To make the most effective use of a discovery learning performance opportunity, it is a good idea to prepare with some self-study. Start by reading documents that will be the focus of the discussion or closely examining other relevant artefacts. Practice your drawing skills in order to notice details that might otherwise escape your attention. Above all, identify technical terms that might be unfamiliar or used in different ways. You may need to clarify the meanings during the discussions. Don't overdo preparation either; remember, this is a learning experience. Be prepared to ask many questions, preferably open questions that encourage other people to talk about what they know. For example, if you are exploring how much someone already knows about air conditioning technology, it is not a good idea to ask a question like, 'Is that the liquid trap?'

This is an example of a closed-form question: the answer is simply yes or no. Instead, an equivalent open-form question might be:

> *'This cylinder here, I'm not sure, but it looks like it could be a liquid trap to me, could you tell me about different arrangements that you may have seen and how they compare to this one?'*

This kind of question encourages the other person to talk more, giving you the opportunity to learn from their experience and observations. If, on the other hand, the other person has little technical understanding, the way that you have posed your question, acknowledging your own uncertainty, allows them to be open about their own uncertainties by saying something like this.

'Yes, it might be a liquid trap, but I haven't noticed things like that before.'

Practice your listening skills.

Make sure that you take notes. It's not easy to ask people to remember what they said in these discussions and people don't necessarily repeat themselves when you want them to. Be prepared to share your notes and ask the other participants to check them for technical accuracy later.

Make sure that you remember to learn about the ways in which people use different words. Don't let unfamiliar words, or phrases used in an unexpected way, slip by without understanding how the speaker intends the meaning to be received. It may be different from your understanding of what the word or phrase means. It might seem awkward to be repeatedly asking people to explain what they mean by a particular word or phrase in a meeting. Try to remember that everyone else has been where you are and probably remember their own feelings of ignorance, their reluctance to ask questions, and their regret at not having asked questions when they had the opportunity. Also, more experienced engineers may be even more reluctant than novices to expose their own uncertainties, so they may be very grateful that you asked for clarification.

Ensure that you write notes as people provide explanations, recording the actual words used as accurately as possible. Although this takes time, you will gain the respect of others. The time and trouble that you take to prepare notes demonstrates your respect for their time and trouble in providing you with explanations.

TEACHING PERFORMANCES

Table 8.2 lists several teaching performances in engineering. In reality, however, it is sometimes hard to distinguish individual teaching performances and separate them from discovery learning performances. Real life is inevitably more complex, as different people learn from each other. Sometimes, the teacher learns just as much as the learner, if not more.

Since most literature on teaching and learning focuses on formal classroom learning in formal educational institutions, it is helpful to examine a real case study drawn from engineering practice. The engineer speaking to us in this case study is both a learner and a teacher. She is responsible for a critical component in a process facility using high-pressure, high-temperature fluids that pose a significant safety hazard if any leaks or equipment failures occur. The actual words of the engineer have been edited to make the case study easier to read and to protect the identity of the people involved.

'This special high-pressure high-temperature reactor vessel, I was the initiator of the concept. It was very frustrating initially because the company's ears were closed to that idea, based on prejudice, I think, from other projects, I don't know.

Table 8.2 Some teaching performances in engineering – refer to text for explanations about the columns listing attributes of the performances.

Performance description	Learner	Intent	Timing	Mode	Meeting?	Follow-up
Convey, specify the details of the intent behind the design of an artefact or process so that others can manufacture it with the predicted attributes or operate the process successfully.	performer	implicit	later, synchronous	writing/f2f	y	inspect, monitor
Provide a technical explanation: a verbal performance with the aim of building a level of technical understanding in the mind of another person or audience.	performer	implicit	synchronous	f2f	y	monitor
Advocate the particular advantages of an artefact, solution, or process with the intention of convincing another firm or engineer to purchase it or recommend it for procurement.	potential client	implicit	synchronous	f2f	y	follow-up
Write a report to be placed on record: a succinct statement summarising a technical issue to record what was known and by whom at a particular point in time.	engineer, inquirer	intentional	historical	writing	n	none, self-evaluation
Write a request for quotation: a detailed description of an artefact, information package, service, or material to be purchased, written to enable suppliers to prepare quotations.	potential partner	intentional	later	writing	n	responses
Write a quotation: a firm estimate of the cost to provide specified information, an artefact, a defined quantity of material, or to follow a defined process.	potential client	intentional	later	writing	n	follow-up
Write an expression of interest: a response to an enquiry from an enterprise seeking to procure an artefact, information package, service, or material, indicating that your firm is interested in being approached.	potential client	intentional	later	writing	n	follow-up

Perform a demonstration: a performance in which an artefact or other equipment or materials are demonstrated with the aim of reassuring the audience about the capacity to achieve a given level of performance in service.	decider, peers	intentional	later, synchronous	f2f/video	y	responses, follow-up
Inform: enabling or empowering other people to make use of information.	info user	intentional	later, synchronous	writing/f2f	y	monitor
Help another person to learn how to perform a task or how to do something to help achieve desired objectives.	performer	intentional	synchronous	f2f	y	monitor, follow-up
Provide guidance and reassurance to help a less experienced person develop their technical capability.	performer	intentional	synchronous	f2f	y	monitor, follow-up
Reassure a client that the project will be completed on time and within budget with the required technical performance.	client	unintentional	later, synchronous	writing/f2f	y	responses
Warn: alert someone to a technical issue and the need for pre-emptive action to avoid undesirable consequences.	decider, peers	unintentional	later, synchronous	writing/f2f	y	monitor, follow-up
Stake out or assert your territory: helping people understand your competence and expertise and ability to take responsibility for a given aspect of a technical undertaking.	peers	unintentional	later, synchronous	writing/f2f	y	monitor, responses
Advocate a particular technical approach.	decider, peers	unintentional	synchronous	writing/f2f	y	monitor, follow-up
Influence perceptions: conversing with another person in a way that encourages the person to change their perception about other people or of a technical issue.	decider, peers	unintentional	synchronous	f2f	y	monitor, follow-up

I have been focusing on justifying the decision to use a reduced wall thickness than is normally required,[14] justifying that the risk is acceptable to the operator, I guess. There are still some doubters within the company that it is going to work, ultimately.

'*I am supervising some reliability modelling being performed in London by specialists. I had to develop the detailed methodologies for that, discussing it with the experts in London and also the standards authority. It is a step-by-step process – we develop the concept and we have that subjected to third-party review by the standards authority. I had to visit Norway and London, just bringing the parties together to reach consensus. Now they are doing the modelling[15] and the next step will be to review the results and present those to the company again, and then have those subjected to an independent external review yet again.*

'*I have to make sure that we don't lose sight of what we are trying to achieve. Technical integrity[16] is essential to keep the standards authority people in Oslo on side. Then we have to bring along many other stakeholders[17] within the company particularly the operations people – they are very nervous about the whole thing. It has always been quite a challenge.*

'*I don't have all the technical competency I need, myself. I can't understand all of the nuts and bolts around structural failure mechanisms and stress modelling – it is a very specialist area of statistics and engineering that you just hope that the guys that are doing it know what they're doing.*

'*What we are doing is we're really extending these methodologies – we are not changing the fundamentals. So in that respect, I guess, that's given me comfort that I am not completely out on a limb in terms of proposing this solution.*

'*[You have daily telephone conversations on this?] Yes – what we have now are a slight misalignment is in the modelling results between our own work by the specialists and other work that the standards authority did in the past. So at the moment we are just trying to define the process, step-by-step, by which we can hopefully rationalise the differences.*

'*I have to manage the politics of that because we have two sets of professional people who probably both think they are right. I need to attempt to find some middle ground that we can both agree on. Perhaps somewhere along the line we've missed a piece of detail in our reviews of the methodologies and that might explain the misalignment. When we do the numbers, they have come up different. Why the difference? We are starting to understand that – it wasn't expected.*

'*Ultimately it is the probability of failure which you can then hopefully believe and then you can apply your margins to that number, inherently applying conservatisms in methodology. We are just demonstrating, we believe robustly, that we have a huge safety margin so we don't need those extra millimetres of thickness. That leads to a significant change in the project economics.*

'*[And you're working on this more or less by yourself?] That's right though I have another engineer, one of our own designers who acts as a sounding board for me. He is a very competent engineer so I use him as a reviewer as well.*'

This engineer had realised early on that reducing the thickness of the metal reactor vessel walls by a few millimetres would significantly reduce the cost. In this particular case, with an exotic metal required for the reactor vessel, the cost saving was very large. There were further cost savings than just the material cost: with reduced thickness, the welding process would be more reliable. With a thicker wall, many of the joints would have to be bolted, requiring even more material and high-quality machining operations.

In proposing this design, however, this engineer had stretched the interpretation of the Norwegian standard, which caused concern among her colleagues and senior management. At the centre of this lies a teaching performance; the engineer has to convince her senior management and colleagues, particularly those colleagues responsible for the plant operation, that the design is safe and the probability of failure is acceptably small. The particular standard in question is based on statistical modelling because of the large number of welded joints needed for this complex containment vessel, as well as the fact that many vessels are required.

In order to provide a more extensive justification for his proposal, the engineer had already convinced her senior management to commission an independent expert in London to perform complex mathematical modelling. The engineer had to educate these experts on the complex design of the reactor vessel and the need to be able to perform calculations in order to demonstrate that, even with a reduction in wall thickness, the strength of the reactor vessel would be much greater than required for this application. However, after all the work of setting up computational models, these experts have produced results that differed from earlier studies conducted by the standards authority in Norway.

Notice how this engineer used the term 'politics' in the sense of dealing with disagreements between highly expert engineers in different organisations. Engineers often use this term to refer to difficult situations when opinions differ for reasons that they don't necessarily understand. It is not that they have performed the same calculation and obtained different results. Both have used slightly different mathematical and computational approaches, with different assumptions, and each group of engineers has different prior experience. Their interpretation of the technical issues is therefore different, as you would now understand given the earlier discussion in this chapter, as well as Chapter 7.

The way to resolve this is to engage in a discovery learning performance: an extended series of telephone calls, meetings, and e-mail correspondence, in which all of the actors, including our engineer, gradually learn about their different viewpoints and learn to see the same situation from a different perspective.

Rachel Itabashi-Campbell, in her study of problem solving by engineers in the American automotive industry,[18] revealed similar teaching performances that she referred to as 'aligning stakeholder perceptions of the problem'. She showed how engineers have to spend a lot of effort on this performance. Success in aligning stakeholder understanding, however, was the main factor that distinguished successful problem-solving efforts from unsuccessful ones.

Practice exercise 16: Identifying teaching and learning performances

Take a moment to read the case study quotation above once again. See how many discovery, learning, and teaching performances you can identify in the text. (A self-evaluation guide is provided in the online appendix.)

Variations in teaching performances

Table 8.2 lists a selection of teaching performances commonly encountered in engineering practice. Apart from the description, the other columns list the attributes of each performance.

The 'learner' column describes who is going to be the learner: it is taken for granted that the engineer is the teacher.

The 'intent' column indicates whether the learning will be implicit, intentional, or unintentional. Intentional learning occurs when the learner sets out to engage with a person or document in order to learn something. Implicit learning occurs when the learner engages with the teacher or document produced by the teacher, even though the learner did not necessarily have the intention to learn or the expectation of learning. The act of engaging with the document creates the implicit expectation that learning will occur, even though the learner may not see it as a learning exercise. In several performances, the learning is unintentional. It happens without a desire to learn on the part of the learner. In other performances, the learning is intentional. The learner engages with the teacher or documents in order to improve their knowledge or understanding.

The 'timing' column indicates whether the learning will be synchronous with the teaching as it is in face-to-face encounters or will occur later, possibly without any interaction with the teacher. This book is an example of the latter case; although it might be possible for you as a reader to interact with me by e-mail, I hope that this book is sufficiently self-explanatory in the context of your engineering practice that we don't need to correspond in order to me to clarify ideas in the book. In the case of a technical report, it may be prepared in case it is needed at some stage in the future recording details of engineering work or testing performed for some purpose. The report may never be read by anybody other than the author and those who checked it. The report may even be read as a historical document many years or decades later.

The 'mode' column indicates whether the interaction between the teacher and the learner will be face-to-face (f2f) or through written documents, possibly with e-mail correspondence.

The 'meeting' column indicates whether there will be at least some interaction during the performance in the context of a formal meeting.

The 'follow-up' column indicates how the teacher can confirm that learning has taken place.

Several of performances could take place in modes other than those shown in the table: it is not the intention that the table should include every possible engineering teaching performance.

The first entry in the table might be the most familiar to a novice engineer, who has been educated with the idea that the solution for a problem or a design needs to be communicated to a client who will then use the results of the analysis. As we have seen earlier in this book, engineers rarely, if ever, directly participate in the production of artefacts or delivery of services resulting from their work. Therefore, they can only influence what happens by conveying the technical intent emerging from their work so that other people can perform the work with that knowledge to guide them.

Remember this quotation from an engineer with long experience in maintenance organisation?

> '*No amount of reliability modelling calculations serves any useful purpose until a fitter uses his tools in a different way.*'

Therefore, the challenge for the reliability engineer is to find an effective way to influence the fitter's actions. To do that, the engineer needs to convey technical intent to the fitter and others involved in maintenance.

Most young engineers would imagine all that is needed is an e-mail with appropriate instructions and perhaps a drawing or two. Just because the documents exist, of course, does not mean that people actually read them, let alone act on the contents. That's what teaching performances are all about.

Learning is not easy; it takes considerable physical energy on the part of the teacher and the learner, as well as a significant amount of time. Furthermore, it is not a one-way process as is so often portrayed by the simplistic notion that many young engineers hold, that teaching is simply a transferral of information from the teacher to the learner. Instead, as we have seen earlier, learners add new knowledge by building on their existing knowledge, just like the way that we extend the mental map of our surroundings as we visit places for the first time. If we extend our mental map on the basis of missing or incorrect information or associations, we get lost and need to ask for help. In most learning situations, it is rare for the learner to construct new knowledge without needing some help and guidance to resolve differences of interpretation. The teacher has to observe the learner perform in a way that reveals both correct and incorrect understandings. By reinforcing the correct understandings, the learner gradually develops workable knowledge that will enable an appropriately good performance.

Given the effort needed for any kind of learning and teaching, it is not surprising that motivation plays a big part in securing the required engagement and interest of the learner. Looking at Table 8.2, we can see that many teaching performances involve unintentional learning. These pose a significant challenge, because the learner does not expect to be learning and is not necessarily willing to devote the required effort. As such, understanding how to motivate and engage learners is another prerequisite for effective teaching performances.

Practice concept 53: Value, interests, expectancy, and environment: motivation for learning

In addition to figuring out learning objectives, it is also important to work out why your learners need to learn. Adult learners, in particular, are very strategic about their learning. They only learn when they know they need to. Adults can also be faster learners than young people because they don't bother to learn what they don't need. Young people, on the other hand, often don't have a clear idea of what they will need in practice, so they attempt to learn everything that they can. Unfortunately, they don't always learn it very well.

With formal learning in an educational institution, assessment provides the most powerful motivation for learning. Every student needs sufficiently high marks to pass. Learners soon adapt their learning behaviour to achieve those marks in order to reach their specific objective – to pass, to do well, or to excel. Informal learning has no assessment mechanism, however. Therefore, other motivational factors become much more significant for influencing learning performance.

Learning consumes considerable cognitive energy. Physiological studies suggest that the brain consumes around one third of our body's energy production. It certainly consumes more when it has to learn, and studies suggest that intense emotions, physical exercise, and recovering from infection or injury all compete for our body's supply of energy. That's why motivation is so important: motivation helps direct our energy towards learning.

Learning science has pointed out two critical factors that contribute to the motivation to learn: *subjective value* and expectancy.

Subjective value and interests

Researchers have identified three sources of *subjective value* that drive learning.

- *Attainment value*: the internal and personal satisfaction you gain from mastery and accomplishment of a goal or task, even if no one else notices.
- *Intrinsic value*: the satisfaction gained from simply doing the task itself. Personally, I always used to find building models much more satisfying than completing them; once they were built, I put them away in a cupboard and often never looked at them again. For me, model building had great intrinsic value. Corporate values come into play here. Learning that aligns with corporate values is more likely to provide intrinsic satisfaction.
- *Instrumental value*: the degree to which achieving a goal will help you progress towards achieving later more important goals or achieving extrinsic rewards, such as praise from others, public recognition, money, material goods, an interesting career, or a high-status job. This is where interests become important; demonstrating that the learning has a positive impact on the interests of the learner helps to establish instrumental value, and hence motivation.

One of the most useful ideas here is to identify the principal interests of the learners. Interests are measures of outcomes, either qualitative or quantitative, that provide greater benefits (or reduced penalties) for a person or a group. Many people have an interest in earning more money. Some people are more interested in earning a predictable amount of money, rather than earning more, but having it come with more income security. Some people may have an interest in building relationships, others in experiencing something new and interesting. Many engineers have an interest in learning about new technologies and see this as something that might make one job opportunity more attractive than another.

If the teacher can demonstrate to the learner that their interests will be advanced as a result of learning, the learner is much more likely to be motivated.

One of the best ways to identify the immediate interests of the learner is to start with a simple question: 'What is it about this situation that could be keeping you awake at night?' The ensuing conversation is likely to provide you with a rich understanding on the most immediate interests, or other motivational factors, that will drive learning. It may well tell you that the learner is preoccupied with issues that are completely unrelated to the planned learning, in which case it might be better to wait for another opportunity.

As a teacher, you need to clearly link the effort that learners need to devote to these three subjective values. Demonstrating your own *intrinsic value* is a good way to start; 'Hey everyone, I think this is really cool!' Of course, not everyone is going to find everything intrinsically satisfying, particularly if it turns out to be difficult. Helping people see the *instrumental value* always helps; 'This will be really helpful when you have to diagnose faults with electronic sensing equipment.' Notice how this book is structured around the idea that you can become an expert engineer. By explaining why the ability to teach is valuable for expert engineers, I have provided you with *instrumental value* to help motivate you to learn this material.

Practice exercise 17: Identify your own interests

See if you can identify your own interests with respect to your current employment situation. If you are not working in an engineering job at the moment, what are the interests that would help determine your choice to work with one company or another. Is the location important? Would the prospect of travel be important to you? How much do you expect to get paid? Do you expect to be paid if you have to work more than the normal number of hours in a day? Is the opportunity to work with advanced technologies important to you? Are you seeking stable employment with job security? How much do you value training opportunities? Do you prefer technically challenging work, or are you more inclined to prepare yourself for a management career? Are you interested in developing a reputation for technical excellence, making it difficult for you to compromise when a quick and clumsy solution is adequate?

Expectancy

Expectancy is the second element of motivation. Students need to think that they can achieve the learning goals. In this regard, it is particularly powerful for a student to say: 'Hey people, if I can even get this, then we all can.' Note the use of the 'we' that identifies the speaker as one of the group, rather than a separate teacher who might say 'you'.

One of the best illustrations comes from my own experience. When I started engineering, the dean of the faculty told us in the first lecture to look at the person sitting to our right and our left sides. 'Only one out of the three of you will make it through the course,' he said. And that's what happened. The course completion rate in the engineering faculty at that time was about 33%. This was a self-fulfilling prophecy; any student who was doubtful about their ability to complete the course had their doubts resolved at that moment: they would fail. And they did. Psychology research has now demonstrated that everyone has the potential to achieve any learning goal they set themselves. The only major difference between people is the rate at which they can learn. Some people simply learn faster than others.

It is important to keep reinforcing a simple message: 'We can all do this, together.' Keep repeating that!

The 'together' word is important. Learning takes place within a social context and the people around us help create the motivation we need to learn by giving us encouragement.

Environment

The last factor that contributes to motivation is a supportive environment. This has nothing to do with the strength and durability of the building. However, it has everything to do with the encouragement and emotional support that we provide to others.

Table 8.3 Learning motivation responses.

	Unsupportive environment		Supportive environment	
	Little subjective value	*High subjective value*	*Little subjective value*	*High subjective value*
Expectancy (efficacy) is HIGH	Evading	Defiant	Evading	MOTIVATED
Expectancy is LOW	Rejecting	Hopeless	Rejecting	Fragile

Here's an example of an unsupportive teacher, or even a fellow learner: 'Why can't you get it? Any idiot can understand that!' Here is a much more supportive remark: 'I can see that it's been a struggle to get your mind around this, and I really appreciate you devoting time and energy to understand all this.'

Think about ways in which you can establish the right combination of motivation – high expectancy, high subjective value, and a supportive environment.

It is very important to pay attention to the actions and attitudes of learners. Listen to what they are saying to each other. If you misinterpret what is needed or how they are progressing, you will end up with learners who reject your guidance, evade your teaching, and deliberately do not listen to you. Take a look at Table 8.3.

A major part of environmental support can come from organisational and community values. Here I am using the term 'values' in a different way: it is not the same as the plural of 'value' mentioned earlier.

In this context, values are either stated or unstated behavioural attributes derived from actual behaviour that people respect as being virtuous, which shape the way individuals perceive their environment, relate to other people, and make choices within a community, an enterprise, or an organisation. The relative importance of different values for an individual may contrast sharply with those of another, however.

Here are some examples of values:

Forgiving – recognising mistakes and offences as almost always unintentional
Kind, generous – giving to others without expectation of favours in return
Caring, honest, responsible, trustworthy, fair, firm – but not unyielding
Disciplined – not easily distracted, perhaps difficult to manipulate
Flexible – responsive to needs of others, possibly easily distracted
Status conscious – aspiring to be one who is admired by others and whose opinions one respects
Stable, predictable – feeling of security, often in the form of material possessions or money, avoiding uncertainty

Recent research in economics has helped to demonstrate how powerful values can be in motivating collaborative behaviour within an organisation – far more powerful than financial inducements. Most people devote more effort to their work with an organisation with which they have shared common values. Leading companies have recognised this and devote considerable resources to create statements of their corporate values that most of their workforce can identify with.[19]

IBM, for example, recently decided on a set of corporate values for the first time since the company was founded. Most major companies display their values on a web page that can easily be found using search terms 'corporate values' or 'company values'. Here are some examples:

Delighting our customers
Acting as a single team
Acting with integrity and respect for others
Being safe and environmentally responsible
Delivering value and profit

Companies tend to choose values that other people positively identify with, for example, helping other people, courage, determination, and honesty. Values are powerful determinants of action when faced with choices. For example, when faced with the choice of completing a task as quickly as possible in order to leave work by a given time, or staying behind and completing the task to a high degree of quality, care, and attention, what choice would you make? Some people value conscientiousness and work more highly than family time, and others place more value in spending time with their families. However, by reminding everybody about the company's values, which might include doing everything possible to keep customers happy, people are a bit more inclined to devote extra care and time to their work, knowing that the result will make somebody happier.

Of course, we all have different values and the extent to which we have shared values with the company or organisation will be different from one person to the next. However, by describing the need for learning in terms of the company values, it is more likely that people will understand the need in positive terms and thus devote a little more effort than might have otherwise been the case.

It is important to realise that the espoused values, the values that appear on the company website, may not be well aligned with the enacted values, which are the values that influence the real decisions that people make in the workplace. For example, most engineering companies strongly emphasise health and safety in their corporate values. Unfortunately, the strength with which health and safety appears to be emphasised in corporate values does not always reflect the reality in the workplace. For example, in a televised interview after the catastrophic explosion of the BP Horizon rig in the Gulf of Mexico in 2010, the chief operating officer of BP, Bob Dudley, stated that the company strived for the right balance between safety, business, and performance.[20] In other words, he was saying that the business performance requirement for economic production of oil and gas required compromises on safety. He was reflecting, no doubt, the enacted company value on safety, overlooking the reality (obvious in hindsight) that without safety, there can be no production from a rig that has blown up and sunk to the seabed. In spite of its immense size, he acknowledged that BP came extremely close to complete financial failure as a result of the explosion and the subsequent leakage of oil that caused extensive pollution in the Gulf of Mexico. The company had lost the confidence of investors and suppliers that it could meet its short-term financial obligations; the severity of the global community's mistrust of the company following the disaster was intense.

There are some values that emerged from our research as ones that many engineers subscribe to and identify with. When working with engineers, you may be able to use these to gain an extra degree of engagement. Notice that some may conflict with organisational interests (if not values).

Technical satisfaction – getting the technical solution right for the job and having enough time to confirm it as such rather than having to move on because of budget constraints.

Emergence – seeing plans and ideas take shape and go into effect, artefacts emerge from countless discussions, drawings, designs, and documents.

Elegance – a solution that is simple and aesthetically pleasing from a technical standpoint.

Impact – a perception that one's work has made a positive difference to the outcome that can be pointed out to one's family: 'I designed that' or 'You see that small yellow bit, that resulted from my contributions, if they hadn't had it. . . .'

Innovation – novel or inventive in a significant way, or in advance of what peers would consider 'state-of-the-art'.

Challenge – work which tests one's abilities, technical and/or organisational, which nevertheless is just within one's level of confidence and competence, albeit requiring an advance in learning or discovery relative to one's abilities or knowledge at the start to be satisfactorily completed.

Humanitarian – work that has evident social benefit, particularly for disadvantaged people in poor communities.

One way that you can build on these values to gain extra engagement from engineers is to offer ways for engineers to have more experience with work that they can identify as having these attributes. Again, not all engineers subscribe to these values; there are large individual differences.

Practice exercise 18: Find your company values

Find out the values upon which your company is founded. If you are not working, choose any company that you associate with engineering activities and search for the value statement on their website. Compare the espoused values with the enacted values that you observe in the workplace, or, if you are not employed, research the company's environmental and safety performance from news reports and compare this with the values that they claim.

If you can, carefully consider each of these values and work out the extent to which your own values align with them.

Practice concept 54: Delivering the message – more than being an interpreter

Now, after reviewing the preparatory discovery performances to learn about the languages, prior knowledge, conceptual understandings, and motivations of the learners, we can begin to construct the message. The ideas that we would like the learners to appropriate[21] may have emerged in the context of an entirely different social community, using different language, concepts, and ideas from those familiar to the learner.

Therefore, your challenge in a teaching performance is, in at least one sense, to be an interpreter. You may be trying to explain to a mechanical designer why electrical cables in a robot manipulator need to be constrained to keep them from bending with less than a minimum allowable radius of curvature or you may be trying to cross a much wider gap in understanding. You may be trying to explain to an accountant why it is wiser to purchase the most expensive seals for a hydraulic cylinder, rather than the least expensive ones with a shorter operating life.[22] You may need to provide directions for a cement truck driver who has minimal proficiency in English. Wet concrete cannot be stored and it can be a huge embarrassment if the truck driver delivers it to the wrong location on a construction site. You may be working on the physical layout of high-frequency electronic circuit components and need to explain to a mechanical injection-moulding designer where certain components need to be placed to minimise electromagnetic interference or emission susceptibility.

Remember that most of the engineers you will work with will not have read a book like this. Many, but not all, focus their interests and work in a relatively narrow specialised domain of interests and knowledge. As an aspiring expert engineer, your challenge is to build on their knowledge, skills, and perhaps long experience. However, you can only do this if you translate and reconstruct your needs and requirements into a frame of reference that is familiar to them. You don't need to match their technical understanding of their particular engineering sub-discipline. However, you will need to understand enough of their prior knowledge to be able to describe what needs to be done in terminology they understand and through the use of their concepts.

By explaining how learning helps them with their immediate interests, you are more likely to keep the learners sufficiently engaged to do the hard work that is required to master a new concept.

Why is this more than a normal interpretation performance, translating the message into language that the learner will understand?

First, it is not just a matter of words. The concepts and ideas may be quite foreign to the learner. Therefore, you need to return to the language and ideas of the learner that you discovered (and noted down accurately) earlier. Instead of simply translating the message into the learner's language, you will find yourself reconstructing the message to build on ideas that are familiar to the learner. Instead of asking the cement truck driver to position his discharge chute to pour cement onto the lift shaft footings (part of a building under construction, familiar to the engineers who have an understanding of every detail of the drawings), you might get into the cab with him and show him where to back up the truck for the first concrete delivery. To the truck driver, a construction site ready for a concrete pour is just a mass of holes in the ground with steel reinforcing bars sticking up all over the place.

Thus, the essence of the actual teaching performance is to construct a meaningful message for the learner by using the words and ideas that are familiar to the learner in a way that addresses immediate interests. You can only do this if you have taken the time to listen to the learner and identify the best approach for teaching.

There are many helpful books that can help you deliver a message effectively once you have crafted the content with sufficient understanding about the prior knowledge and ideas in the minds of your learners. However, many of these books overlook some of the barriers that you need to take into account when delivering your message.

Your learners may not be able to listen accurately in the ways that you have learnt earlier in this book. If they take notes, they might retain between 5% and 50% of what you say, and then only if you have their complete attention. Go back and review the chapter on listening skills and remember the cues you can watch for that will tell you if your learners are listening. If not, stop, reassess, and avoid wasting their time and yours. Find a way to regain their attention and focus: 20–30 seconds of silence is not a bad way to try and do this.

Delivering the message in writing is more challenging because you will not be present to answer questions or help with clarification when you see your learners having difficulties.

Engineering students are less accustomed to reading than students in other academic disciplines. From data collected during my teaching career, I have learnt that only 10% of my engineering students read the most important written material that I provide in my courses: the paragraphs on assessment requirements that specify how their marks will be calculated. The rest still know what was in those paragraphs; they come to know which students will have read them and ask them to tell them about the assessment requirements. Only about 2% read material that does not seem immediately relevant.

Do not assume that engineers presented with your written material will have changed much since they were students. Assume that most of them will only read if there is no other way to find what they need. Assume that they will miss a significant proportion even when they do read the material, so remember to repeat the most important messages more than once.

Put yourself in the place of an engineer who is a reluctant reader and needs to find the really important material in a hurry, material that is written in a way that makes sense to an engineer. Many engineers feel more comfortable with diagrams, even if they only serve as convenient place markers showing where relevant writing is located.

Variation method

In its essence, the variation method for constructing a message is so simple that it seems obvious. To explain an idea, think of its intrinsic variations, and then draw the attention of the learner to the part that remains constant through all the variations. It sounds easy, but it is not nearly as easy as it looks.

Here's a simple example. To teach a child the colour purple, we show the child several objects that are all coloured purple, maybe with slightly different shades of purple. We might say 'Henry, this is a purple car, this is a purple tube, this is a purple bottle top, this is a purple plate, etc.'. However, it is all too easy not to notice other similarities. Just by coincidence, we might have chosen four or five plastic objects to teach 'purple' to Henry. However, they might also all have been more or less round (e.g. a plastic toy car that is round, with round wheels). Then, without us realising, Henry thinks that anything that is made of similar plastic material is purple. We didn't notice this because we already had 'purple' in our minds and simply grouped a bunch of purple objects that happened to be nearby.

In their paper, 'On Some Necessary Conditions of Learning', which is challenging yet highly rewarding reading, Marton and Pang demonstrate not only the extraordinary effectiveness of teaching using this basic technique, but also how easy it is, even for highly prepared and experienced teachers, to get it wrong without realising.[23] It takes great care, experience, and insight to achieve an effective teaching strategy.

Zone of proximal development

Let us return once again and remember the brief discussion on teaching the stranger directions to a particular bank branch. Imagine for a moment that the way to get there is rather complicated and involves 25 stages, with the route marked by 30 or more landmarks. Unless the stranger had developed the capacity for total recall of verbal instructions, it is unlikely they would remember more than the first two or three steps of the route. That might be sufficient; the stranger would then stop and ask someone else for directions.

This thinking exercise illustrates the idea of a zone of proximal development.[24] This loosely refers to the extent of learning that the student is capable of at one time, with the help of other learners and the teacher. No matter how hard we try, an individual can only learn a certain amount in one sitting. Everyone is different, of course, and it is not easy to figure out how much anyone can learn at a particular time. However, the principle behind this is important. Gradually, when teaching, we develop a sense of what is possible to learn at one time.

In practice, what it means is that you have to think about the limitations of the learners. If you are going to try and explain something that involves too many conceptual steps and new ideas, it is not going to happen in one session. Therefore, you need to think about time constraints. You may need to build a 'learning bridge', a way of reaching a useful learning outcome without having to learn all the intermediate details that might seem important to you. This means thinking carefully about what you want to learn to be able to do at the end of the process. It may be necessary to simplify the background knowledge required, so that it is just enough for them to perform adequately. Later, they can learn more of the background details that might provide a more in-depth appreciation. Think about your own experience through school and university. You may have had to study similar material many times, maybe twice in primary school and again twice at secondary school and then again at university. Each time, you learnt the subject in more detail and emerged with more insight than before. However, for many purposes, the primary school learning might have been quite sufficient. It got you to where you needed to be quickly.

I faced a similar challenge in writing this book. Our research interviews and field observations yielded around 5,000 pages of text. Putting all the relevant research literature together (books, journal articles, conference papers, and reports) contributes perhaps 15,000 pages of text. Even though this book will run to about 600 pages when it is finished, my aim has been to provide you with as much learning as is reasonably possible from those 20,000 pages of text without losing too much valuable information. I have tried to highlight the important concepts and also misconceptions that can impede learning. In doing so, I have had to construct many conceptual bridges and borrow many more from literature to provide you with something from which you can learn. Also, I hope that you will remember that most of your learning will take place when you learn to interpret the world around you in terms of the ideas in this book.

Collaborative learning

The attraction of cooperative learning in an engineering setting lies in the similarity with distributed cognition processes that occur naturally in the workplace, as discussed in Chapter 5 page 140. Engineers adopt the same kinds of interactions as have been observed in cooperative learning classrooms. Cooperative learning has been demonstrated to be more effective than most other techniques in hundreds of carefully controlled trials.[25]

Cooperative learning[26] describes a large family of methods in which learners help each other learn. There are both formal and informal learning methods. Hundreds of carefully controlled trials have demonstrated large learning improvements compared with conventional teaching methods. (Note that the term 'co-op education' is used in the USA to describe internship or workplace learning. This is simply a clash of terminology; there is no direct connection with cooperative learning.)

Smith et al. (2005) provide a concise description: 'Formal cooperative learning groups are more structured than informal cooperative learning groups, are given more complex tasks, and typically stay together longer.' Well-structured formal cooperative learning groups are differentiated from poorly structured ones on the basis of the characteristics presented in the table. From these characteristics we can distil

Table 8.4 Effects of structure in cooperative learning.

Less structured (traditional)	More structured (cooperative)
Low interdependence. Members take responsibility only for themselves. Focus is typically on a single product (report or presentation).	High positive interdependence. Members are responsible for theirs and each other's learning. Focus is on joint performance.
Individual accountability only, usually through exams and quizzes.	Both group and individual accountability. Members hold self and others accountable for high-quality work.
Little or no attention to group formation (students often select members). Groups are typically large (4–8 members).	Deliberately formed groups (random, distribute knowledge/experience or interests). Groups are small (2–4 members).
Assignments are discussed with little interest in each other's learning.	Members promote each other's success. They do real work together, helping and supporting each other's efforts to learn.
Teamwork skills are ignored. Leader is appointed to direct member participation.	Teamwork skills are emphasised. Members are taught and expected to use collaborative skills. Leadership roles are shared (by rotation, for example) among all members.
No group processing of the quality of its work. Individual accomplishments are rewarded.	Group processes (discusses) quality of work and (evaluates) how effectively members are working together. Continuous improvement is emphasised.

five essential elements in successful implementation of formal cooperative learning groups: positive interdependence, face-to-face promotive interaction, individual accountability/personal responsibility, teamwork skills, and group processing.

MASTERY

Learners don't develop mastery simply by listening to an explanation.

At the early stage of learning, every task can be difficult, even listening to the introduction. The student is consciously trying to understand new ideas in a new context, which requires significant energy. Even elementary tasks require the brain to think in ways that it has not tried before. The brain literally needs to develop new neural connections; this takes time and energy.

With practice, the new connections strengthen and become persistent in the brain. We reinforce the right connections through feedback. The *pleasure from achievement* literally provides chemical signals that reinforce the most recently used neural connections.

Practice makes perfect but only if the learner receives informed and reliable feedback. Without reliable feedback, practice can result in inappropriate skills and connections becoming reinforced and later they can be difficult to undo.[27]

These ideas can be summarised with the learning pyramid, figure 8.1. The diagram is attempting to show how the essence of good teaching is to allow the learners to discover new knowledge for themselves. This is particularly true in the company of other people who provide both a *supportive environment* and also the *extrinsic value* that comes when others recognise the learner's achievement.

Most people only learn effectively when they use new knowledge in the company of supportive fellow learners. Some people can learn very effectively by themselves. They have less of a need for recognition from their peers and teacher. They gain more subjective value from the *intrinsic pleasure* of doing it for themselves and the *attainment satisfaction* that comes from completing learning tasks successfully. When they do attend classes, these individuals can be very helpful, because they often can provide additional support for students who thrive in a classroom environment.

Checklist for effective teaching performances

As explained earlier, you will need a script. By working through this checklist, you can develop your own script to your discovery learning and informal teaching performances.

Identify teaching and discovery performances that will be needed

Listing all the separate teaching and discovery performances will help you appreciate the tasks that lie ahead of you.

Identify the learners and outline the learning objectives. For each learning objective, work out how you will be able to confirm that the required learning has taken place.

Figure 8.1 Learning pyramid – each of the comments reflect increasing degrees of learning from the top to the bottom. The bottom statement emphasises the power of cooperative learning in which each participant has as much responsibility for teaching the others as learning for themselves.

Discover the learner's language and prior knowledge

Confirm that prior knowledge:

i)   is appropriate,
ii)   is sufficient,
iii) is accurate, and
iv)   has been recently activated.

Ensure that you listen carefully to learners to develop an understanding of their language and terminology. Engage learners in casual conversation and try out some of the learning messages, carefully listening to the responses.

Ensure that you can construct the learning messages using terminology and language that is both familiar and comfortable for the learners. Ask peers who are fluent in the language of the learners to evaluate your messages ahead of time in case adjustments are required.

Characterising teaching performances

Assess the degree of learning intention by the learners: will they participate in the performance knowing that they are there to learn something? Or will the learning be incidental, without any intention to learn?

Assess the timing: will the learning take place in a face-to-face environment in your presence? Will some or all of the learning take place later, without you being present? Will the learners be able to interact with you and ask questions?

What will be the medium for instruction? Face-to-face, the use of artefacts, electronic mail, websites, and documents can all be used singly or in combination.

Will the learning take place in the context of a formal meeting, or at least be supported with formal meetings?

Will the learning be monitored and evaluated by the inspection of documents, or by monitoring specific performances by the learners?

Motivation: values and interests

Identify specific interests of the individual learners and develop an appropriate script that builds on these interests, particularly to highlight the instrumental value of learning.

Ensure that the script identifies the attainment value, intrinsic value, and instrumental value of learning.

Expectancy and environment

Identify ways to build a supportive learning environment: one that creates the expectancy among the participants of sufficient learning to achieve their objectives.

Ensure that the script contains messages to promote positive expectancy, using 'we will ...' rather than 'they will ...' so that the teacher is placed in the community of learners and is not separated or placed on an authoritative pedestal.

Ensure that your script contains hints and suggestions on creating a supportive learning environment.

Identify corporate values so that your learners may share and develop an appropriate script that links motivation with these values.

Presenting the message: more than interpretation

What are the ideas and words that are familiar to the learners with which you can construct the message they need to appropriate?

Do they need to see objects to understand? Did they work with sketches or images to explain their ideas to you, hand signs? Did they take you to the actual places to show you something or could they describe it in words effectively?

Construct the message using familiar words, places, ideas, and images. Do you need to introduce them to a new idea? What might already be familiar to the learners that you could use to explain the idea?

Can you use the variation method, illustrating what is constant through all possible variations in terms of context and background?

Zone of proximal development

Carefully consider the learning objectives and, with an understanding of prior knowledge among the learners, decide whether you need to construct 'conceptual bridges' so that useful learning that leads to satisfactory performances by the learners is achieved in the required time.

Attention span

Construct a learning schedule that recognises limitations on the attention span of the learners. For most people, it is wise to keep the attention requirement under 15 minutes for a single session. Ensure that your script enables you to summarise the learning objectives in 30 seconds.

Rehearsal and recording

Like any other performance, a rehearsal is likely to greatly improve the quality, particularly if the rehearsal takes place in front of a friendly audience. Even before the rehearsal, practice your teaching performance many times in your own room, imagining that there is an audience sitting in front of you. The more often you do this, the more confident your presentation will become.

The more often you can practice your explanations, even to the extent of boring your friends and acquaintances, the more likely you will be able to deliver a polished performance when it is really needed.

Record your presentation, especially for yourself, so that you can criticise it and improve it later, but also so that you know exactly what you said. You can also make the recording available to others if they need it later.

Monitoring learning

Watch for learners nodding their heads or saying, 'Yes, I understand that', which indicates that the learners think they have learnt something. However, what they have learnt might be quite different to what you expect.

Only by carefully listening to their spoken responses and subsequent actions can you determine whether any actual, useful learning has taken place after the session is over. This requires the teacher to follow-up with discovery performances similar to the ones required to assess prior knowledge. This will enable a teacher to find out whether the learners have changed the way they think about what they have learnt, or whether their perceptions have appropriately changed. If the aim was a skilled performance by the learners, it will be necessary to observe the performance to see whether the learning has had a legitimate effect.

As we shall see in the next chapter, follow-up and monitoring requires significant amounts of time from engineers.

Mastery

How can your learners practice to further develop learning? Remembering that effective learning through practice requires a means of evaluating performance, work out how learners may be able to evaluate their own performances. If this is not possible, further teaching or mentoring may be essential.

SUMMARY

Most young engineers assume that other people learn the way they did at school: the teacher presents a logical explanation, perhaps using a set of PowerPoint slides, and that is enough to solidify the concept.

By now, you probably realise that teaching – being the interpreter – is anything but simple.

However, more than anything else, effective teaching lies at the core of expert engineering practice. We shall see in the following chapter how critical teaching is for the most common collaboration performance for engineers – technical coordination.

Thanks to recent research, mostly since the start of the 1990s, we now have some powerful techniques that can help you learn how to teach effectively. There is not space to introduce them all here, but the references will lead to you other effective techniques.

Once you have read this chapter, find some friends who are schoolteachers and discuss the ideas explained in this chapter with them. You may be amazed how much you can learn from them that will positively impact your own approaches to sharing knowledge and teaching others.

How will you be able to tell whether your discovery learning and teaching performances are improving? If you have improved your discovery performances, you will have more comprehensive and accurate notes to refer to, with keywords and ideas highlighted so you can use them for teaching. When you present your message to your learners, they will quickly show you that they have understood, and may even thank you for explaining something in language they can easily understand. They will echo your message accurately. While this does not show they can remember it accurately, at least you know you have mastered the first part of a teaching performance.

When you have improved your teaching performances, you will notice how much less time you spend correcting mistakes made by learners and, yet again, showing them how to do it right. You will spend less time fixing up the consequences of mistakes and incorrect actions that happened because people did not act in the way you expected them to.

Don't expect your skills to show definite improvement every time. I teach students year after year and every cohort is different; they react in unexpected ways and the explanations that seemed to work every time before somehow lead to confusion. It might be something small that you have done differently, or it might simply be that the learners are different from the previous group.

The important thing is to keep trying. With the material in this chapter, you should be able to outperform most other engineers, until they catch up with the help of the lessons in this book.

NOTES

1. Ashforth, Sluss, & Saks (2007); Eraut (2004, 2007); Fuller & Unwin (2004); Hager (2004); Katz (1993); Korte (2007); Marsick & Watkins (1990); McGregor, Marks, & Johnston (2000); Wenger (2005)
2. Wenger (2005).
3. Engeström (2004); Rooney et al. (2014); Trevelyan (2007).
4. Bailey (2010) Personal communication, Also see further details reported in Bailey and Barley (2010).
5. Martha Nussbaum(2009) has refocused attention on the capacity to understand a situation from the point of view of someone else, and this has become a significant issue in recent discussions on design. See for example Oosterlaken and Hoven (2012). Lucena and

Leydens have also drawn attention to this capability in their notion of contextual listening (Lucena & Leydens, 2009).

6. Bransford et al. (2000).
7. Itabashi-Campbell et al. (2011); Lam (1996); Nonaka (1994).
8. The introductory chapter of the book *How People Learn* describes the view of learning in which new knowledge is constructed on foundations provided by previously learnt knowledge (Bransford et al., 2000).
9. Lionni (1970).
10. The book *How Learning Works* provides references with links to contemporary research in education psychology and neurological science (Ambrose et al., 2010).
11. Chapter 5, Figure 5.10, shows a map of specialised technical knowledge.
12. Each company adopts its own convention for marking up changes and corrections, such as using red lines to indicate errors in assumptions or logic, green for improvements to English expression, etc.
13. Square brackets denote a question by the interviewer.
14. Thickness of the metal walls of the pressure vessel.
15. Modelling to predict the likely behaviour of the pressure vessel when subjected to varying pressures and temperatures causing additional stress due to thermal expansion and contraction, coupled with the possibility of stress-induced corrosion affecting the strength of the welds joining the metal parts.
16. Integrity here means the process by which engineers assure themselves (and external reviewers) that an installation will meet its design requirements through construction and its anticipated service life in operation.
17. Stakeholders are people or groups of people who can influence a project or whose interests could be influenced by the project. In this context, they would include the engineer's management team, senior managers in the company, the likely vendors supplying the reactor vessel, the London experts and their management, the operations engineers and their management, and other members of the project team.
18. Itabashi-Campbell & Gluesing (2014); Itabashi-Campbell et al. (2011).
19. Akerlof & Kranton (2005).
20. ABC Four Corners, March 11, 2011. For an extended discussion, refer to the PhD thesis written by the chief investigator for Piper Alpha, Exxon Valdez, and the BP Horizon blowout disasters (Bea, 2000).
21. Appropriate is a verb used in education psychology that refers to the multiple mental processes that lead to the development of new knowledge, understanding, skills, and capabilities in the learner.
22. Seals, relative to the overall cost of machinery, are relatively inexpensive items. However, if they fail prematurely, they can be very expensive to replace because of the time required to disassemble and reassemble complex hydraulic machinery. More expensive, longer life seals are likely to be much more cost effective than cheaper ones.
23. Marton & Pang (2006).
24. The concept originated in the 1920s from the work of the Russian child development psychologist, Lev Vygotsky (Bransford et al., 2000; Brown et al., 1993).
25. K. A. Smith, Sheppard, Johnson, & Johnson (2005).
26. Felder & Brent (2007); D. W. Johnson & Johnson (2009); D. W. Johnson & Johnson (1999).
27. Draper (2009a, 2009b).

Chapter 9

Technical coordination: Informal leadership

In the previous chapters, we discussed communication skills such as listening, speaking, seeing, reading, and writing, as well as collaboration performances, including seeking approval, teaching, discovery learning, and assertion. In this chapter, we will build on these ideas to develop the next of the four major combined performances that are used extensively in engineering practice: technical coordination. Two other combined performances will come later: project management and negotiation.

We are not going to simply learn about a more complex collaboration performance; there is another critical difference in this chapter.

From now on, the aim is not just to communicate, nor even just to collaborate. From this point forward, the overriding objective is *action*. Engineers can only deliver on their promises by getting other people to perform *appropriate* actions, at *appropriate times*, and with *appropriate care and skill*.

Let's start from the beginning: what is technical coordination?

Technical coordination is a performance by which an engineer gains the willing and conscientious collaboration of other people to contribute skilled performances according to a mutually agreed upon schedule, mostly without involving any form of organisational authority.

Technical coordination seems to be a major part of the day-to-day practice for practically all engineers, and it is an intrinsic requirement in any engineering enterprise that requires coordinated performances by different people.

Technical coordination is all about getting things done: informal leadership is an important part of that.

When employers talk about the characteristics they value most in engineers, they often cite the ability to get things done, above all else, as the quality that defines a good engineer. That is often what they mean by 'teamwork skills' and 'leadership skills'.

However, and this might come as a surprise to you, not one of the engineers interviewed for our research mentioned technical coordination as being part of their work.

The term 'technical coordination' might be new for many engineers and readers. Yet, when I discussed the idea with our engineer participants, they invariably acknowledged technical coordination as being highly significant in their daily work. Once the concept was explained, they could see technical coordination as a performance that embraces much of what they had previously dismissed as 'not real engineering work', while at the same time being essential for 'getting things done'. Several engineers expressed great relief: they had imagined that somehow, other engineers did not have to get involved in chasing people to do things. They were relieved to find that it is an element of daily work for all engineers.

Because technical coordination is a new concept for most engineers, I have provided many notes with references to research literature to help establish the theoretical basis for this idea. If you are interested in a deeper understanding about how to influence the perceptions and actions of other people, then the literature references will give you a starting point for your own exploration. Otherwise, all you need to know is that there are many ideas in the research literature that can help you understand more, but you can safely postpone exploring them for the time being.

Employers often look for good 'team working' skills. In our research, we found engineers working in teams, but most of their collaboration performances were informal and extended beyond a particular team. Engineers were often working on several projects at the same time, frequently as a member of those respective projects' teams and often in diverse supporting roles.

There are many excellent books on teamwork skills.[1] Most, however, sidestep the technical issues that so commonly arise in engineering work. For that reason, this chapter describes how these technical factors influence collaboration and coordination. For example, one of the crucial differences between the literature related to teams and actual engineering practice is found in the subject of team membership. In engineering, where technical knowledge is essential, team membership (even a role within the team) is primarily determined by a person's availability and the team's need for specialist technical and business knowledge and experience. On the other hand, while organisation behaviour literature advocates the use of personality measures, such as in the work of Belbin and Myers Briggs, engineering relies on distributed knowledge, and most engineering team membership choices reflect this priority. Team leadership, on the other hand, requires someone with the appropriate personality, skills, and experience: this chapter will help you develop those essential skills.

Why, then, do employers place so much importance on team skills? Most likely, it is simply a short way to express the requirement for collaboration skills, such as technical coordination.

Although there have only been a few limited research studies in Australia, India, Pakistan, France, Portugal, and the USA, technical coordination seems to dominate engineering practice, accounting for between 25% and 35% of working time.[2] Unlike other aspects of engineering practice, the amount of time spent on technical coordination seems to be relatively consistent between different engineers, practice settings, and engineering disciplines – even from one week to the next for a given individual engineer. Many engineers see it as non-technical 'admin stuff' or even 'chaos' that takes time away from their 'real engineering work', even though the language and most of the interactions are highly technical in nature.[3]

As explained earlier, research studies did not confirm the widespread subjective perception among engineers that their work involves interaction that is more personal and less technical as they progress in their careers. Instead, we found that novice engineers spend roughly the same amount of time interacting with other people as more senior engineers: for all engineers, social interactions dominate your practice, at least in terms of time.

Gradually, like others, I came to the conclusion that there is something intrinsic to engineering practice that explains these observations, particularly the large proportion of time spent on collaboration performances. At the moment, the best way to explain the notion of distributed knowledge, as explained in Chapter 5, is that engineering relies on knowledge and expertise that is unevenly distributed among the participants, both engineers and non-engineers. Engineers gain access to this distributed knowledge by mobilising skilled performances by people with the required knowledge.[4] Time constraints impose limitations on what individuals can learn from each other during any one engineering project. Therefore, it is usually more effective for people that already have the knowledge to make skilled contributions than it is for engineers to learn the knowledge from others and then apply it.

Mobilising these skilled performances, including identifying the knowledge and skills required and then arranging for people to contribute their knowledge and skill, is a major part of technical coordination.

It is also important to understand from the beginning that technical coordination is different from project management. Project management is a formal process that requires significant documentation and formalised relationships that we will discuss in the next chapter. Unlike project management, technical coordination is an informal process that relies largely on undocumented social interactions. However, technical coordination can also take place using e-mail communication; like its face-to-face equivalent, e-mail messages are relatively informal.

It is also possible to confuse technical coordination with teamwork. While teamwork requires technical coordination, some of the technical coordination performed by any given engineer (if not most of it) involves people who are not part of a designated team. They may even be people outside the work group, or possibly even the firm. However, following our definition of an engineering enterprise in Chapter 5, these people are considered as a part of the enterprise. Engeström calls this 'knotworking, transient networks of collaboration'. He referred to this as a new (meaning newly recognised) form of expertise, 'not based on supreme and supposedly stable individual knowledge and ability but on the capacity of working communities to cross boundaries, negotiate, and improvise "knots" of collaboration in meeting constantly changing challenges and reshaping their own activities.'[5]

Technical coordination is not the same as management, even though many engineers associate this kind of work with managing people. Management is based on formal recognition of organisational authority, and a manager takes on long-term responsibilities for subordinates.[6] Technical coordination, on the other hand, like knotworking, mostly involves short-term informal relationships in which people collaborate without organisational authority.[7]

Technical coordination is a type of informal leadership.[8] Informal leaders need to gain the respect of those whom they lead in an organisation. We shall see later in this chapter how technical coordination relies on influencing people without resorting to organisational authority.[9] However, patterns of influence depend on the local culture in which engineering practice is situated. In the USA, for example, the research literature suggests that informal social relationships required for influence are strongly based on transactions, such as performing a favour for someone with the expectation that it will be returned. In Australia, at least among the engineers interviewed for this research, there was less emphasis on this exchange of favours. Instead, influence seemed to be associated with respect, also known as referent power.[10]

Once you gain recognition as someone who can get things done, you will almost inevitably be sought out for higher levels of responsibility. Many engineers have expressed frustration at being passed over for others who receive organisational responsibilities such as project management positions, often people with non-engineering qualifications.[11] Learning and practicing coordination skills is one way to avoid this kind of frustrating situation.

I will now explain technical coordination in terms of a real engineering situation described in the extended quote in Chapter 8 pages 261–264 in which we found numerous discovery learning and teaching performances. I have summarised this situation in Figure 9.1, which shows the major and extraneous relationships through which technical coordination takes place.

Anne, our engineer who described her coordination activities in Chapter 8, sits at the centre of the diagram. Remember that she has developed an innovative design for a high-temperature, high-pressure reactor vessel that is to become part of a new process plant. She is working as part of a project team and reports to the senior engineer, Dave, who I have assumed to be acting as the project manager. It is quite possible that another engineer would take on the role of the project manager and Anne would report to Dave as the lead engineer responsible for the technical aspects of the project.

Senior management (such as Yevgeni and Charlie Ho) are concerned with the risks inherent in this design. Anne is coordinating work with a group of London specialists. She is mainly working with Henry, whom she has known for some time: he acts as

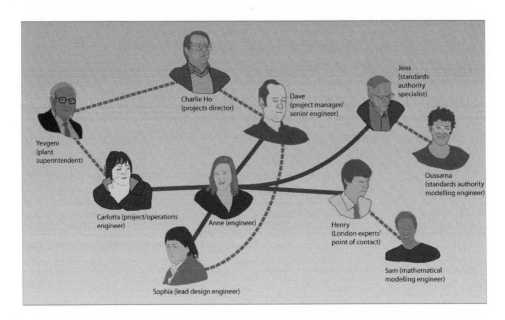

Figure 9.1 Simplified illustration of coordination relationships described in the quote from Chapter 8. Solid lines represent coordination relationships, while dashed lines represent a few of the other relationships that influence people in the diagram.[12]

the point of contact, handling the contract from Anne's firm to perform mathematical reliability modelling calculations and analysis. This is one of several contracts for which Henry is responsible. Henry doesn't actually do much of the modelling himself: on this contract, Sam, another engineer, does most of it. It will be necessary to reassure the project financiers and insurers by demonstrating that the high-temperature, high-pressure reactor vessel complies with well-recognised international standards, otherwise the cost of raising capital may be too high. Anne is confident that she has correctly interpreted the current international standard, in this case a standard that has evolved from the Norwegian standards authority, DNV. However, to be sure, she wants Jens, the senior technical specialist with the standards authority, to check that the mathematical reliability modelling calculations being performed in London produce similar results to calculations performed independently by the standards authority. Oussama is the standards authority engineer doing this. As we saw in the quotation, the standards authority obtained different results from the London experts, although the differences were not large. Anne now finds that she has to coordinate Henry and Sam to share their results with Jens and Oussama so that they can agree on a common interpretation to explain the differences. Even though Henry is working under a contract between Anne's firm and the London expert consultancy, Anne has no authority to tell Henry what to do: the contractual authority usually does not extend to individuals. At the same time, Anne has to work with Carlotta, an operations engineer working in the existing process plant. Carlotta reviews the new design and provides feedback on day-to-day operational issues, such as maintenance, instrumentation layout, accessibility, and installation. Carlotta reports to Yevgeni, the production superintendent in charge of the existing plant. Finally, Anne also coordinates with Sophia, who reviews her designs and provides additional technical feedback. Sophia doesn't actually report to Anne: they both report to Dave, the project manager.

Anne is coordinating the work of all these people in order to build up a sound case to support the use of the new high-temperature, high-pressure reactor vessel. Her direct coordination relationships are with Dave, Sophia, Carlotta, Henry, and Jens. She is coordinating Dave, even though he is her immediate superior in the organisation, because Dave can more easily gain the essential support of Charlie Ho and other senior management and, perhaps more importantly, the funding to support the mathematical modelling calculations being performed by the London experts working with Henry. She is coordinating Carlotta, even though Carlotta is in a different operating division of the firm. Jens is a senior technical specialist in the Norwegian standards authority; even though he has known Anne for some time, Anne cannot tell him what to do. The actual modelling work is being performed by Sam and Oussama, who report to Henry and Jens, respectively.

The intricate network we see here is typical of technical coordination: Anne has no formal authority, so she cannot simply tell someone else what needs to be done. Instead, she has to gain their willing and conscientious collaboration. Not only does she not have authority, she also doesn't have enough knowledge herself to give clear and detailed instructions for the work that needs doing. She doesn't know enough about the firm's power relationships to tell Dave how to secure support from Charlie Ho, nor does she know enough about reliability modelling calculations to tell Henry what needs to be done. She doesn't even know enough to figure out why Sam and Oussama get different results. Yet, unless she can coordinate all these people to provide the substantiation she needs for her high-pressure, high-temperature reactor vessel design, with its exotic metal construction, the firm will proceed with an older, more expensive, but better-understood reactor vessel. Anne does not want her ideas and all her work to go to waste. She has support for her efforts, of course. Her design will make it easier for Dave to deliver a new plant with superior performance.

In Chapter 8, I asked you to identify teaching performances in this particular situation described by Anne. These teaching performances are actually part of technical coordination performances that we will talk about in this chapter.

In our research, it was interesting that even highly experienced engineers had never noticed how much time they spend working outside formal lines of authority, as this quote from a male engineer illustrates:[13]

> *'[Could we come back to the issue of supervision? Most of these interactions seem to involve supervision without direct authority, don't they?] Okay, initially an engineer will be coordinating the work of other people. Later, he will be given supervision authority over non-engineers … Now, of course, officially, there is a line of authority. The engineer can take work to his head engineer who can pass the work on to the head drafter who can direct the drafters what to do. But you don't want to rely on that because it involves too many people and it's too slow. It's quite unwieldy and that's why horizontal interactions are essential. That's why engineers spend most of their time using informal coordination methods, working with people over whom, yes, I see what you mean now, people who they don't have any control over working along the traditional lines of authority … This kind of situation illustrates why an engineer spends a lot of his time managing up, managing sideways, and managing down all at the same time. It's a bit like dealing with a powerful woman: you need lots of subtle negotiation. Resorting to authority is a total waste of time, as it only creates resistance, and the lines of authority may not even exist.'*

'Managing up' describes the coordination relationship between Anne and Dave: Anne needs Dave to provide high-level organisational support when she needs it. To do this, Anne spends time with Dave, talking through the issues and seeking his advice. In doing so, she is building up a very necessary collaborative relationship and, at the same time, is helping Dave gain more trust and confidence in her work. This is important for Anne, particularly when securing assistance from other people in the company. The degree of support she has from Dave enables her to gain the help of others who are busy with other day-to-day issues in the existing process plant, while also helping secure funding to support the mathematical modelling computation work in London.

'Managing sideways' describes Anne's relationship with her 'peers', Carlotta, Sophia, Jens, and Henry. She needs their help and collaboration. She could secure Carlotta's help by asking Dave to arrange with Yevgeni to instruct Carlotta to provide her with the advice and assistance she needs. However, as explained in the quote above, it is much faster and more effective to 'interact horizontally' by approaching Carlotta directly and later ensuring that Dave talks to Yevgeni to secure tacit approval for this. In the same way, Anne could ask Dave, in his formal capacity as project manager, to ask Charlie Ho, in his formal capacity as project director, to instruct the London consultancy firm to pass on directions to Henry, and consequently to Sam. This would be tedious and slow because all the people in between are extremely busy and would not place a high priority on passing on a message that could be better handled at the working level. Once again, horizontal interactions with peers are essential: Anne has to coordinate with Henry and Sam and gain their willing collaboration.

Not only does Anne need collaboration from peers in order for them to perform the tasks that she needs done, she also needs their collaboration to get the tasks done in the required time. Jens, Henry, Sam, and Oussama are all busy on other projects at the same time. They have many urgent requests associated with those projects and have to fit in Anne's work between their other priority tasks. It is easy for them to forget about Anne, far away on the other side of the world in Perth, Australia.

Just as Anne is busy coordinating Jens, Henry, Carlotta, Sophia and Dave, they are all busy coordinating others and are likely to be coordinating Anne at the same time. For example, Henry's firm of consultants in London would like to expand their business with Anne's firm: Henry is building on his relationship with Anne to seek out other opportunities for mathematical modelling work. He needs to develop other contacts in the firm through Anne and has asked her to see if she can locate some public documents dating back a few years.

In our research, we found many different kinds of coordination performances, listed in Table 9.1. Some of them appear in the example above, but there are many others.

WILLING AND CONSCIENTIOUS COLLABORATION

In Table 9.1 and the illustrated example discussed before, we can see that coordination extends to several situations in which there is organisational, or at least contractual, authority between the coordinator and the peer. In this chapter, I will use the term 'peer' to denote the person whose work is being coordinated.

In Chapter 7, I noted the common misconceptions among students that a boss tells people what to do. This misconception persists among young engineers who think that the existence of a contract or the fact that someone is being paid is sufficient to make that person perform the tasks that are part of their job. There is a belief that engineers can simply tell technicians what needs to be done and technicians will do it. As one engineer responded, *'That's their job, isn't it?'*

This is where we need to pay special attention to the quality of collaboration or, in the case of a relationship based on organisational or contract authority, the quality of compliance.

Consider, for example, a maintenance crew who is required to repair several leaking bolted joints in a gas pipeline. As an engineer, you may be required to supervise their work. Before the repair work takes place, the pipeline is isolated by closing valves at each end and removing high-pressure gas, perhaps even flushing the pipe with an inert gas to reduce the risk of an accidental explosion or fire. The technicians then dismantle the leaking joints by loosening and removing the bolts so that the internal seals can be replaced. Then, the technicians replace the bolts and tighten them using a torque wrench to ensure that the bolts are tightened correctly.

As the supervising engineer, you are responsible for the repairs being performed correctly. How do you ensure that this happens?

Table 9.1 Coordination performances.

Performance type	Description
Coordinate insiders, mentoring	Coordinate the work of peers, subordinates, and superiors. Perform technical checks on work and watch for emerging technical issues and roadblocks, while possibly providing advice and feedback, reviewing technical competence, and assessing training needs, as well as providing informal training, when appropriate.
Supervise staff	Supervise and support staff for which the engineer has line management responsibility.
Coordinate outsiders	Coordinate with outside organisations, such as other contractors working on the same project, community organisations, etc.
Coordinate with client	Liaise with the client, discover the client's needs and requirements, expedite solution review and acceptance, coordinate installation, commissioning, and monitor acceptance testing.
Advocacy, negotiate shared meaning†	Advocacy for a particular technical or commercial view, negotiating a shared understanding or meaning.
Site engineering	Coordinate and supervise work on-site: perform inspections, watch tests, ensure that work is performed according to drawings and specifications, plan site work, and coordinate with site supervisors.
Supervise contractors and suppliers	Coordinate and supervise work performed by contractors: perform inspections, watch tests, and ensure that work is performed according to specifications and requirements.
Assist contractors or suppliers	Recognise the need and provide technical assistance in order to help a contractor or supplier when needed.
Reverse mentoring	Providing mentoring, guidance, coordination, training, and supervision to more senior or experienced personnel.
Cross-cultural coordination	Coordination and supervision of people from different cultural backgrounds.
Report progress	Report to supervisor, team leaders, and peers on project progress, solutions, and financial and resource consumption in verbal or written form, or in meetings. Represent interest area.
Delegate technical work	Allocate responsibility for technical work: balance technical expertise and experience against cost and availability and decide whether to employ additional staff or contractors, etc. Select appropriate working methods and tools.
Delegate supervision*	Allocate appropriate technical supervision capacity for a given activity to ensure that required performance and quality standards are measured, maintained, and recorded.
Build and lead team	Build and lead a project team. Create shared vision/objective, monitor team members, and support team members.
Networking	Networking: develop and maintain a network of contacts to help with performance of the job.
Organise socials	Organise recreational and social activities within the organisation.

†A teaching performance discussed in Chapter 8.
* Normally performed by experienced engineers.

One option is for you to supervise each technician involved, telling them not to proceed to the next stage of the repair until you have checked each step. This option would be extremely time-consuming. Realistically, it would probably be impractical for you to supervise more than two technicians at one time who were both working in the same location.

Instead, you would more likely allow the technicians to perform their job, knowing that they probably have far more experience performing this kind of work than you do.

How, then, can you tell that the work has been done correctly? A bolt that has been tightened correctly looks no different from one that is too loose or too tight. Perhaps you could use your own torque wrench to check the tightness of one or two of the bolts. However, you would not be able to tell whether the bolts had been overtightened, and then loosened, and then tightened to the correct setting. Would this matter even if it had occurred? The answer is that it probably would matter. Particularly in a high-pressure pipeline, ensuring that the bolts are correctly tightened and not overtightened is essential to avoid fatigue damage to the bolts, as they are subjected to very high stresses in normal operation. Overtightening can permanently damage a bolt, causing invisible elongation, damage to the threads, and even a slight, but significant, enlargement of microscopic cracks in the metal.

Therefore, even though you are responsible for the repairs being performed correctly, you cannot actually observe every technician performing their work. Instead, you have to rely on the willing and conscientious collaboration of the technicians. You are relying on their willingness to perform the work correctly and conscientiously, taking care to ensure that their work is done correctly.

Returning to the example from earlier in the chapter, Anne and Henry both rely on Sam to perform the modelling calculations correctly. Henry has sufficient knowledge of the kinds of things that can go wrong with computer calculations to remind Sam to perform the necessary tests to confirm the accuracy of the calculations. However, everything relies on Sam, ultimately, to perform those tests diligently and conscientiously. Henry could spend a great deal of his time watching Sam as he does this, but that would add to the cost of the work.

What we can see here is that the technical quality of the work performance affects the technical quality of the result in a way that cannot necessarily be observed by a supervisor, even one watching at the time. As we have seen before, most engineering is performed under strict time and budget constraints: the work has to be completed according to an agreed upon schedule and cost. Therefore, a great deal of technical work relies on both the conscientious performance of the work by individuals taking care to make sure that there are no mistakes, and the levels of delegation and checking to minimise the likelihood of mistakes.[14]

The classic book, *Working: People Talk About What They Do All Day and How They Feel About What They Do*, illustrates how the *feelings* of people towards their work influence what they do. Here's a quote from a steelworker:[15]

> *'It isn't that the average working guy is dumb. He is tired, that's all. I picked up a book on chess one time. That thing laid in the drawer for two or three weeks, you are too tired. During the weekends you want to take your kids out. You don't want to sit there and the kid comes up: "Daddy, can I go to the park?" You got your nose in a book? Forget it.*

'Yes. I want my signature on 'em too. Sometimes, out of pure meanness, when I make something, I put a little dent in it. I like to do something to make it really unique. Hit it with a hammer. I deliberately fuck it up to see if it'll get by, just so I can say I did it. It could be anything. Let me put it this way: I think God invented the dodo bird so when we get up there we could tell Him, "Don't you ever make mistakes?"

'And He'd say, "Sure, look." (Laughs.) I'd like to make my imprint. My dodo bird. A mistake, mind. Let's say the whole building is nothing but red bricks. I'd like to have just the black one or the white one or the purple one. Deliberately fuck up.'

Ethical behaviour builds respect

Terkel's insight leads us to realise that the respect that others accord you can be critical here. Your reputation for ethical behaviour and fairness will significantly influence the *quality* of conscientious collaboration. The respect that a person holds for an organisation, or a community, also has a similar influence. There is instrumental value here in maintaining high ethical standards: a means to achieve higher performance. High ethical standards help to build respect that others hold for you, and this can *reduce* the monitoring work you will need to perform. You can be more confident about the quality of work that others will do out of respect for you.

This is a very tangible benefit that accrues from ethical behaviour, a theme that we will return to in the concluding chapter of the book. You might think that this only applies when people get to know you ... and you would be quite wrong, as first impressions are often critical and take a lot of time to change. There is a second instrumental value in ethical behaviour.

While there are millions of engineers across the world, engineering is an occupation that consists of thousands of small, specialised communities of engineers who have a global reach. With engineering projects now increasingly conducted in several different locations, instances of unethical behaviour tend to become remarkably well known rather quickly. As one engineer put it in an interview, '*Whatever happens these days, when you really stuff something up, they will get to know about it on the other side of the world in a few hours.*'

Sooner than you think, your reputation will precede you, thanks to global communication and networks among engineers that have worked together across the world.

Practice concept 55: Technical coordination process

Technical coordination can be seen as a three-part process. However, as a prior relationship between the coordinator and the peer makes coordination more effective, I have shown this as an additional preparatory step in Figure 9.2. Many engineers are proactive in developing workplace relationships, even outside their own firm or organisation, partly in anticipation of future collaboration. They spend time socialising or in casual conversations, often around the coffee machine.

The first coordination phase consists of several discovery performances and organisational steps. The second phase consists of monitoring while the task is being

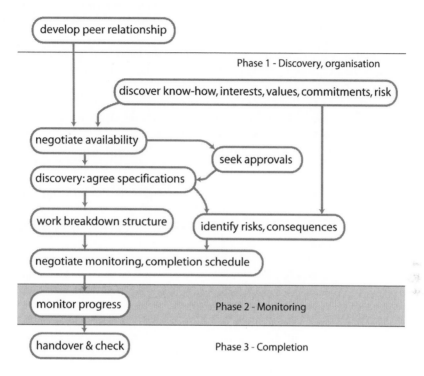

Figure 9.2 Technical coordination process.

performed, and the third phase occurs when the task is completed and the results are checked. As an informal process, there may be no clear separation between the different phases.

Since technical coordination is based on willing and conscientious collaboration, it works best on the basis of what Raven and French, working in the 1960s, identified as *referent* power, one of several different ways in which people influence others:[16]

- Coercion is power derived through the threat of punishment for non-compliance.
- Reward is power derived through the promise of reward for compliance. Personal forms of reward lead to influence derived from the threat of rejection (coercion) or personal approval (reward).
- Legitimacy is power derived through an organisational structure that defines a power relationship (formal legitimacy) through obligation (legitimising reciprocity, 'You should do that because I did that for you.'), through equity (legitimising equity, 'I suffered, so I have a right to ask you to do something to make up for that.'), or by legitimising dependence (powerlessness, there is obligation to help those who cannot help themselves). Legitimacy also has strong cultural links, for example, 'On' in Japan, which describes obligations towards one's parents, country, and teachers.

- Expert power is derived from the acknowledgement of expertise. This can have the opposite (negative) effect when the target of influence thinks that the expert has personal self-interest and is abusing expert power.
- Referent power, which is derived from respect or the need to identify with someone, can equally be negative in the case of someone whom the target holds in great disrespect or disregard and does not want to identify with.
- Information power derives from the power of persuasion based on independently understanding the need for action, which can be more effective if the information is presented indirectly in the form of a story: 'Of course I'm not an expert but I distinctly remember, if my memory is correct, that this worked rather well in case X.'

Coercion and reward depend on the continuing social presence of the source of power: the source of reward or punishment needs to be continuously visible in direct or symbolic form, and can lead to worker resistance, expressed eloquently by Terkel in the quote earlier in the chapter.

Other forms of social power help with coordination because the source of power does not have to be physically present all the time.

As we have seen in the examples discussed earlier in this chapter, technical coordination relies on willing and conscientious collaboration and therefore relies mainly on expert and referent power. Even if there is a contractual or employment relationship that requires performance of the task by the peer (i.e. the possibility of using legitimate power), willing and conscientious collaboration is almost always needed, and therefore details of the task will be coordinated at the working level without resorting to contractual authority. It is rare for people in an engineering enterprise to have just one task to which they can devote all their time and energy. Therefore, part of the coordinator's job is to keep the peer focused on the agreed task to the extent necessary to ensure it is completed satisfactorily within the agreed schedule.

Phase 1: Planning, requirements, and organisation – discovery performances

The absence of authority requires that the task be first agreed on between the coordinator and the peer. Even before this takes place, the coordinator needs to discuss current commitments with the peer and also consider the peer's know-how, values, and interests. This requires a discovery performance, preferably a face-to-face discussion, so that the coordinator can confirm that the peer is the right person to perform the task and that they will have enough time to complete it on schedule.

It may be necessary to negotiate the peer's availability at certain times. Often, the peer would perform this by discussing task commitments with other peers and working out how to arrange an informal schedule to create space for the new task.

It may also be necessary for the peer or coordinator to seek approval from the coordinator's supervisor, also possibly the coordinator's project manager, if the peer is a member of a project team.

Having cleared the way for the peer to devote time to the task, the next step is agreement on the details of the task between the peer and the coordinator. Once again, this is another discovery performance, because it is rare for the coordinator to be able to completely specify the task without discussing it with the peer. The peer usually has special resources, such as technical knowledge, skills, facilities, tools, or access to equipment: this is the reason why the coordinator needs the peer to perform the task. The coordinator needs to learn from the peer at the same time as the peer learns about the requirements from the coordinator. Together, they 'firm up' the requirements as they improve their respective understanding of one another, gradually eliminating mutual ambiguity, negotiating meaning.

Usually, the requirements will include a budget for resources, especially in terms of hours of work performed by the peer. Sometimes, the peer's support will already have been approved, as in the case of Henry and Sam in the example above. However, Henry was still accountable for the time he and Sam spent on Anne's work, and coming to a clear understanding on this would help Anne as well. Coordination and monitoring, as we shall see later, can be time-consuming, and this resource allocation needs to be allowed for *in addition* to the technical work that will be performed.

Like any teaching performance, however, an expert coordinator takes many precautions, carefully constructing messages that make sense to the peer by using concepts and terminology that the peer is comfortable using. An expert coordinator takes enough time for the peer to learn about the requirements and to ensure that the peer has sufficient understanding to commence the performance of the task.

If the task is complex, the coordinator and the peer may develop a written work breakdown structure. In the case of a contractual arrangement, this kind of documentation is probably essential and may need external approval.

The last step in the organisation stage is to agree on a schedule, not only for the task performance on completion, but also for monitoring. The coordinator will monitor the task more or less frequently as the work progresses.

At the same time, at least privately, the coordinator needs to anticipate issues that may arise during the task performance and assess risks that could affect the task performance schedule or quality. The coordinator needs to foresee events that are usually intrinsically unpredictable and have some idea of how to handle these events if they occur, namely, a series of backup plans. Of course, if there are health and safety issues, then it is essential to discuss these with the peer as well, and possibly others.

About 40 different technical coordination performance attributes emerged during our research and they are listed in Table 9.2. Most are relevant during the initial phase of coordination and several are relevant during the monitoring phase. Each one is described in more detail later in the chapter.

Even though this sounds complex, most technical coordination is arranged informally, without extensive documentation. Even if there are e-mails involved in the coordination, they tend to be informal in nature since the whole basis of coordination is willing collaboration.

Phase 2: Monitoring – a discovery performance

The monitoring phase of technical coordination is usually the most time-consuming. Figure 9.3 illustrates this as a cyclic process that starts once task performance commences. Deciding on the repetition frequency is critically important.

Table 9.2 Checklist of coordination performance attributes, factors that influence collaboration quality (see last part of chapter).

	Beforehand	During
Quality of Willing and Conscientious Collaboration		
Ethical behaviour builds respect	*	
Socio-technical environment attributes		
Material and tool quality	*	
Environment, maintenance standards	*	
Supporting documentation	*	*
Quality culture	*	*
Peer attributes		
Health, fatigue	*	*
Prior training	*	
Task attributes		
Ergonomics, human factors	*	
Technical challenge	*	
Subjective value, expectancy	*	*
Work that allows free thinking	*	*
Clean work	*	*
Coordinator and work preparation attributes		
Expertise	*	
Resources, supervision	*	
Location	*	
Quality of supervision and informal support	*	*
Scope, documentation of task, accurately conveying intent	*	
Prior organisation	*	
Understanding of constraints	*	
Performance quality assessment	*	*
Work organisation attributes		
Interruptions	*	*
Shift work	*	*
Safety	*	*
Clothing	*	*
Self-organisation	*	*
Handling problems	*	*
Territory	*	*
Timing	*	
Presence	*	
Exposure to consequences	*	
History	*	
Affective factors		
Opportunity for recognition and identification	*	*
Relationship, trust	*	*
Prestige	*	*
Learning	*	*
Developing others	*	*
Social benefits	*	
Reward	*	
Contractual terms	*	
Ongoing work opportunities	*	*
Payment already received	*	

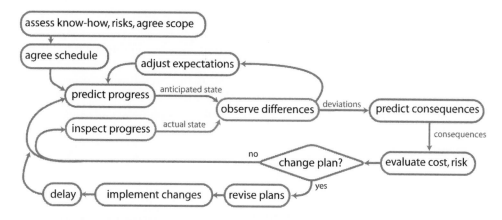

Figure 9.3 Monitoring process for technical coordination. Once the peer signals that the task has been completed, the third phase, completion and handover, will begin.

Monitoring starts with a prediction: the coordinator attempts to predict the current task status and progress, usually quite informally. Technical knowledge helps a coordinator make predictions that are more accurate and know what to look for during inspections.

The coordinator would arrange to meet the peer or at least discuss the task by telephone. Monitoring by e-mail is a last resort and is unlikely to be an effective way to rectify performance issues. The 'inspect progress' block in the diagram represents another discovery performance: a discussion with the peer about the task that results in the coordinator learning how the task is progressing. In doing so, both may also discover misunderstandings about the requirements and adjust their expectations.

Having assessed the actual state of the task, the coordinator then reflects on the difference between the predicted and actual state and, in doing so, adjusts expectations for future progress prediction. Technical expertise is essential for this.

As explained earlier, one of the aspects of technical coordination is that it is a temporary relationship: the coordinator and the peer may never have worked with each other before and may never do so again. If they have a previous working relationship, then it is more likely that the prediction will be more accurate, but if there is a significant difference, it may be due to significant changes in the work environment or unexpected demands from other commitments.

Next, the coordinator thinks ahead and evaluates the consequences of issues that have arisen as a result of meeting with the peer. For example, there may have been a misunderstanding about the requirement for quality: the peer may be taking much more time than was originally foreseen, ensuring that the work is performed to the highest standards through extensive checking. The coordinator may not have clarified the expected quality standard for the work. Also, engineers prefer to take time to seek the best possible results and check their work carefully. In the early phases of the project, particularly if the coordinator needs only a rough estimate from the peer, there may have the expectation for the work to be done quickly.

The coordinator may have to think about different kinds of consequences and risks. If the task is being performed on a budget, he may have to renegotiate the budget or make allowances elsewhere. If the completion time is important, the coordinator may have to change the scope of the work or even find someone else to help move the process along.

If there has to be a change in plans, it takes time to discuss this with the peer, revise the current plans, and then implement changes. Bringing someone else in to help might sound like a good idea, but it almost invariably requires a significant amount of time: the two peers also may take more time to establish a productive working arrangement on the task. There is almost always some delay as a result of changing plans.

Mostly this is all an informal process: monitoring involves hardly any documentation or notes.

The most important decision for the coordinator is to decide the frequency of monitoring. Too much monitoring can undermine trust: the essence of coordination is willing collaboration and the coordinator has to display that confidence to the peer.

If the monitoring is too infrequent, however, then issues can arise that may seem unimportant to the peer but that may be very important for the coordinator.

Another kind of issue is a technical difficulty that the peer feels confident in solving and continues to work through, relying on that confidence. They may try several different solutions before asking for help or alerting the coordinator. The peer may need assistance from someone else, but that person fails to respond as expected. The result is an unexpected delay. In the following example, we hear from a senior technical expert about this issue. He is a senior control system software engineer acting as a technical consultant to several different project teams. He does not have any direct management responsibilities. Instead, his job is to work with application engineers producing control system software to help them understand the requirements, how to produce work that complies with the firm's technical standards, and also how to avoid wasting time and causing cost overruns. This happens when an engineer tries to solve a technical issue without the required breadth of understanding. The application engineer may spend week after week without making significant progress – a situation referred to as 'churn'.

> 'I enable things to be done which otherwise could not be done and I am a prime driver of consistency and quality, meaning that things are done in a neat, logical way, and the software is consistent and well structured. In terms of costs, I am a major element in terms of avoiding waste, meaning that people do not churn in terms of doing jobs. If an engineer spends a lot of time struggling to work out how to do something, that's what we regard as waste and I make sure they are equipped and the roadblocks are removed before they get into the job. I have seen it before. If you have a 150-hour job, an engineer can easily get up to 400 or 450 hours before they realise that they have a major problem. I know what people are capable of now and the rates at which they can work, the sorts of things they can turn out in a certain time, if they are equipped with the skills that they need and the major bottlenecks are removed, they have a clear design path to follow, and a clear architecture to work within. My contribution is to avoid cost overruns on between a third and a half of all the jobs that we do by a factor of two or three times.'

Elsewhere in the interview, this engineer talked about the value of being able to foresee technical issues that would cause difficulties for the application engineers and devise ways to work around the issues should the need arise. Of course, if the application engineers managed to do this for themselves, then there was no need for his intervention.

In another example, an experienced contracts manager talks about coordinating contractors. In his early career, he had qualified in several trades, including welding.

> '[In essence, your technical knowledge allows you to save money because you can spot the mistakes that the contractor has made.] Yes, it not only saves money, but you can actually go and talk to the contractor while you walk through that paddock and ask, "Are you really serious about what you are trying to do here? What are you doing? This is wrong. I think you are doing it the wrong way round. Maybe you should try this." That's important for the process.'

Here, he was talking about welding work being performed by the contractor. Earlier in the same interview, he had shown how revealing aspects of his own knowledge helped to build informal respect and authority, thereby reducing the need to resort to formal lines of authority.

> 'I think that one of the things in any contracting relationship is that too often we talk about an "us and them" scenario, too often it is confrontationalist. One of the important issues is to get the respect of both parties in the discussion. A contractor that succeeds means that me, as a client, has succeeded. He gets a good result and I get a good result. If he fails, I have generally failed. Rarely does a project work where everybody has not succeeded, that you have actually got a successful project outcome. You are either over budget or behind time.'

In the last part of this quote, he is emphasising the need for all stakeholders in a project to feel they have succeeded, that they end up respecting every other stakeholder's need to succeed as well. In the next passage, he finally refers to costs. The 'threshold of pain' indicates the minimum cost at which the contractor would perform the work adequately. 'Living with the outcome' implies being able to predict the technical consequences caused by accepting compromises on the work performed by the contractor.

> 'So the issue is, if you can both respect each other and you can both talk at the same level technically and then commercially you can help each other get a result, then you hit that threshold of pain where it is not (quite) enough for him and it is (a little) more expensive for you, but you can both live with the outcome. That's a good result.'

In the next interview, we hear about monitoring frequency from an expert construction engineer with three decades of experience.

> 'The other factor is response time. If there is a mistake, something going wrong, the foreman should spot it within 30 minutes and the problem should be corrected

within an hour. The site engineer will go round the site twice a day. If there is a problem that has not been corrected by the foreman he can pick it up and correct it within a single working day. The project manager, on a similar basis, will correct the problem within one to two days. Now the construction manager is operating on a minimum of a weekly cycle. It will take him at least one week to detect the presence of the problem. He will then have to go into extra investigations, calling subcontractors, the client, sniffing around, which means it will take him probably another two weeks to get the problem identified and then fixed. The construction manager is supervised by the business manager operating on a monthly cycle, so he can take at least three months to fix a problem. Worst of all, the poor managing director will be operating on an annual cycle and if he doesn't get it right to begin with, it will take longer than his tenure to fix the problem!'

At each level, the people involved want to try and fix the problem for themselves and have confidence that they can. They tend not to want to elevate a problem and reveal their own inability to deal with it. However, it can be easy to underestimate the number of attempts that will be needed. This is where technical insight can be a critically important advantage, including awareness of the limits of one's personal knowledge and when to seek specialised help from others. The cost of seeking help is usually less than the cost of yet another fruitless attempt to fix the problem without help.[17]

As a rough guide, therefore, monitoring frequency depends on the likelihood of an unexpected issue occurring that the peer cannot solve by themselves, and the 'float time', which is the allowance in the schedule for unexpected delays. You must also take into account the probability that if the peer cannot solve the problem, then it may be even more difficult for you to solve it: you may have to seek help from someone else.

Use the checklist in Table 9.2 as a guide to consider factors that would influence your decisions on how often and what aspects of the task to monitor.

Contriving casual encounters

Naturally, most people are uncomfortable with excessive monitoring. If it is too frequent, monitoring can undermine trust and confidence. However, frequent monitoring is often the only way to keep a task on track, especially when the peer has numerous other distracting priorities to attend to. During our research, we uncovered some intriguing strategies that allow for more frequent monitoring without it seeming overtly obvious.

Often, the monitoring only serves to keep you, the coordinator, in the mind of the peer so they don't forget about the task they are performing for you. It is possible to do this in other ways, by contriving casual encounters.

One engineer told us: 'When I need to coordinate some extremely busy people, I try and position myself with a view of the corridor that leads to the bathroom or the coffee machine, preferably both. Then, when I spot someone who is likely to run behind on a job, I'll arrange to show up at the coffee machine or bathroom at the same time, seemingly by coincidence. Often, I will make sure I talk to someone else: all that is usually needed is for the other person to notice me and that makes them remember they're running late.'

Another technique is to find a good reason to walk past the peer's cubicle, within clear view, and give a casual, friendly glance in the peer's direction when they are looking up.

E-mail and SMS (Short Message Service)

Although some degree of monitoring is possible by e-mail and SMS, it is relatively ineffective and it is easy for a peer to provide reassuring messages just to keep you from visiting or calling by phone. Expert engineers rely on direct, face-to-face or voice contact to a much greater extent: the cost is negligible compared with the cost of schedule delays.

Social culture

Naturally, any effort in exercising social power and influence will be affected by local social norms. We will return to this in more detail in a later chapter. So far, the extent of our research has not been sufficient to provide more than tentative suggestions on the factors that local culture can influence.

On the Indian subcontinent, for example, it is well known that technical work requires more continuous presence by supervisors. Engineers have told us about spending almost their entire days directly supervising just a few production line workers: work quickly stops whenever the supervisor is absent. Language and social divides may explain this.[18] Other observations that suggest a greater need for monitoring include reports by engineers that their work requires two or three hundred phone calls on most days.

Phase 3: completion and handover

The last phase in technical coordination involves careful checking by both the peer and the coordinator to ensure that everything has been completed to their satisfaction. It may take time: they may both agree on testing with the expectation that any remaining performance issues identified as a result of testing will be rectified later.

It is a good idea to avoid a premature declaration that the task is complete. Remembering that technical coordination relies on willing collaboration, it can be embarrassing to go back to a peer and confess that you did not check the work thoroughly enough when you initially accepted it. However, this can sometimes happen, even with the best intentions.

As with monitoring, your technical knowledge is essential to predict the consequences of accepting any remaining deviations, including the ultimate risks of doing so.

The most important aspect of the completion phase is to remember that task completion is not the only valuable outcome. As the coordinator, you and the peer will have strengthened your relationship. Relationships endure much longer than temporary coordination arrangements. Therefore, it is important to maintain the relationship. Otherwise, you will risk giving that person the impression that the relationship was only important in order to get a particular task completed.

This does not mean that you need to spend an excessive amount of time with the peer. However, it also does not mean that you can completely ignore the other person

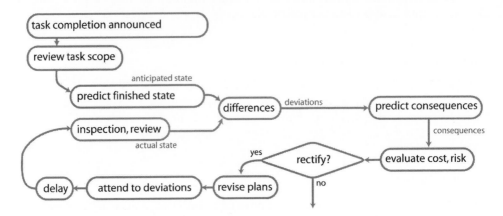

Figure 9.4 Completion and handover phase. The coordinator has to evaluate the consequences of any deviations from the agreed upon task scope, as well as the cost and risks. If necessary, the coordinator will insist on deviations being rectified, accepting that there will be a delay and another inspection.

indefinitely, or until the next time you require their collaboration. Take the time to visit the peer occasionally and view this as an important aspect of your work: building and maintaining relationships.

The chapters at the heart of this book, those focusing on neglected communication skills, collaboration, teaching, technical coordination, and project management, are all about developing your ability to collaborate with others. As we have seen in the early chapters, engineering is a collaborative enterprise and technical coordination lies at the heart of this, taking up 25–35% of your time. That's why personal relationships are so important, as we heard earlier from one of the expert engineers interviewed for this research.

In the next chapter, we will move on to project management. In principle, project management is very similar to technical coordination. However, project management enables a coordinated collaboration performance on a much larger scale.

TECHNICAL COORDINATION PERFORMANCE ATTRIBUTES

We can now look at other factors that influence the quality and timing of work performed by other people – most often, your peers. By understanding the influence of the attributes, you can improve the quality and efficacy of your coordination performances, and also improve the quality of work performed by other people.

The results of your coordination efforts will play a large part in determining how others judge your performance as an engineer.

Some of these are attributes of the working environment and the organisation, while some are attributes of the peer, who is the person performing the work. Others are attributes of the task to be performed and the way this task is seen by the peer. Finally, there are attributes that you, as an engineer coordinating the work, can readily influence.

You may be surprised by the number of attributes, many of which you can directly or indirectly influence. Of course, there are many others that you cannot influence easily. Understanding the effect of these influences can help you appreciate the difficulty of predicting work performance. This is why monitoring the work in progress is an essential component of technical coordination. We will return to this theme later.

Socio-technical environment attributes

This set of factors is often predetermined, because it is almost impossible to shift an organisation's culture in the time needed for a typical technical coordination performance, which is usually completed within a few weeks, at most. By being aware of these factors, you can plan for them and adapt your monitoring frequency and focus accordingly. You may even be able to exert some influence, for example, by obtaining an agreement for the use of certain tools and materials in advance.

Material and tool quality

Material, tool, and equipment quality (the result of previous selection processes and purchasing decisions) affects the quality of technical work and its timing. High-quality tools and materials, for example, enable better quality results, because operators don't have to compensate as much for tool or material deficiencies. Better-quality software helps engineers work more productively with fewer mistakes and less rework. Working with high-quality tools, materials, and equipment is attractive: a peer is more likely to choose to do this work first because of the intrinsic enjoyment and satisfaction gained by performing it.

Environment, maintenance standards

The working environment also influences task performance. If the temperature, humidity, noise, vibration, wind, dust, smoke, fumes, or other environmental factors move outside the normal, accepted limits in the workplace, it is likely that performance will decrease – both quality and time may be affected.

In a similar way, maintenance standards influence technical quality. High-quality tools and equipment require routine maintenance and will not be satisfying to work with if they are in poor condition. The condition of materials, tools, and equipment is the result of earlier handling, maintenance, storage, and cleaning by peers and other people. This extends to the surrounding building as well. A building that is well maintained, meaning that it provides a clean, tidy, well-lit, and comfortable working environment, is much more conducive to the performance of high-quality technical work.

Supporting documentation

The availability and accuracy of supporting documentation and reference materials also influences the quality of technical work.[19] In many organisations, it is common to find that drawings are incomplete, inaccurate, and out of date, and the technical manuals that describe how to use and maintain equipment are not updated regularly.

Once this is well known, engineers and technicians will spend time looking for updates and checking the drawings before they even commence their technical work performance. This adds to costs and decreases the confidence of engineers and technicians in performing their tasks.

Even if there is documentation available, if there is no place to put it or conveniently view it (as most documentation is now electronic), it may as well not exist. I have seen heavy documentation resting on delicate and easily damaged electrical cables and connectors in an assembly and factory test area simply because there was nowhere else to put it. This problem contributed to defects that later showed up as a premature failure of the equipment that was not only expensive to build, but also extremely expensive to install and recover for repairs.

Quality culture

An organisation with a highly developed quality culture will help individuals take more care with their work. Such an organisation will probably provide recognition for meeting quality expectations, which may vary according to the client's needs and risk assessments. Recognition can be informal (e.g. from the coordinator), formal (from the organisation), or both. Striving for the ultimate level of quality is something that many engineers take for granted, without necessarily understanding that a faster, more economic, and more adaptable response is often needed. An organisation with a well-explained quality culture is also likely to provide peer support, constant reminders, and a variety of techniques for quality checking.

Error culture in organisations can also influence conscientious working habits. Acceptance that deviations (rather than being called 'errors') are a normal part of human behaviour reduces pressure during work execution.[20] A culture in which mistakes are treated as normal events is more likely to provide encouragement and support for a peer to ask someone else to check their work, which that person would be more likely to check carefully, using review methods designed to detect and correct normal human deviations.[21] We also need to recognise that value perception plays a part in this as well.

Peer attributes

Among all the other factors, only two relate directly to the peer performing the task. You may have some control over these, particularly the first one. A peer who is tired or unwell can postpone your task: be prepared to take the time to inquire respectfully about the peer's health and personal life.

Be prepared to provide training and skill development if it is needed, or source it from someone else. You may only discover the need for it as you monitor progress with the task, but it is never too late. As adults, many people are much more prepared to learn new skills when they see a pressing need to do so, and can be energetic and responsive if given appropriate encouragement and support.

Health and fatigue

The peer's state of physical and emotional health has a major influence on the quality and timing of technical work. The use of medications, stimulants, and semi-legal or

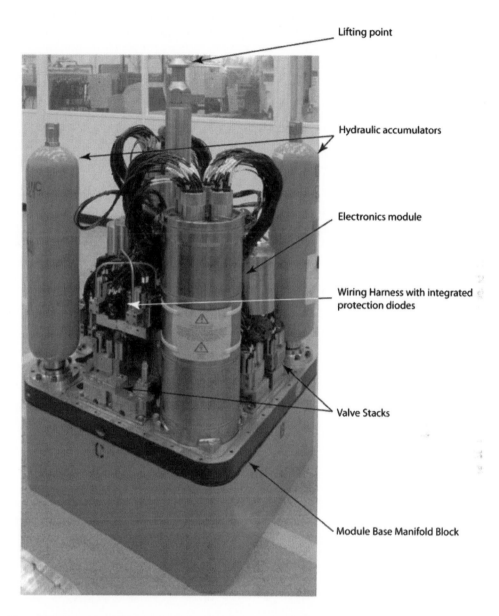

Figure 9.5 Complex and expensive hydraulic control equipment observed in a factory integration test facility without anywhere to conveniently place documentation. Documentation was available, but was kept in a nearby cabinet and rarely accessed (*source*: confidential, permission to use image for education purposes provided by originator).

illegal drugs has been an increasingly significant issue in many workplaces. The only effective way to ensure that work is performed under the best conditions is to secure the willing cooperation of the peer. Although drug testing has been used, it is expensive, invasive, and can significantly decrease the perception of trust in the organisation.

Fatigue or dehydration, often induced by factors other than the task itself, can have just as significant an effect as drugs such as alcohol.[22] This aspect of work performance has been extensively researched. However, it can be difficult for people to monitor their own levels of fatigue.

Prior training

The peer's expertise and skill (including recent training and practice) is one of the most significant influencing factors that affects performance quality and timing.[23] This includes knowledge about where to obtain appropriate tools, equipment, and materials in order to obtain the required level of quality. Understanding the actual level of quality required is also an important performance parameter.

If past performance requires a peer to make on the spot decisions to deal with unexpected results or deviations from the norm, then the peer will need to be able to diagnose and resolve technical issues. The quality of conceptual models available to the peer will determine whether their responses are appropriate or not. We can judge the quality of conceptual models in terms of their applicability, utility, and effectiveness, but not necessarily accuracy or scientific validity. Engineers and technicians have often been observed using inaccurate or even invalid conceptual models, yet they are still able to perform satisfactory corrective actions.[24]

Task attributes

The following factors, while related to the peer, are primarily determined by the nature of the task. It is important to recognise these because they will influence task performance. You have only limited scope for influencing these factors, but you may be able to influence the way they are *perceived* by the peer.

Ergonomics and human factors

Many factors that determine the quality and timing of work performance are derived from the nature of the task itself. Ergonomics (known as 'human factors' in the USA) provide extensive literature on how tasks can be designed to reduce the degree of physical or intellectual effort needed to perform the work relative to the peer's capabilities.[25] Realistically, however, informal technical coordination often involves tasks that are performed only once, or just a few times, and almost certainly not on a long-term, repetitive basis. This makes it harder to justify the time and energy needed to re-engineer the task.

Technical challenge

Technical challenge in the task can provide intrinsic motivation, particularly for engineers who want to demonstrate their technical abilities. If the task involves working near or beyond currently accepted boundaries, the level of challenge, prestige, or motivation is likely to be greater.[26]

Subjective value and expectancy

Intrinsic fun, entertainment, or excitement involved with the task may result in more or less careful work.[27] This is related to the perception of value and the peer's confidence that they can perform the task (expectancy, self-efficacy). We reviewed this as one of the motivating factors for learning in the previous chapter.

Work that allows free thinking

Often, task performance is influenced by the way it is seen by the worker, for example, whether it is considered intrinsically interesting and varied, perhaps requiring skill and care, or is thought to be boring, mundane, and repetitive, perhaps requiring no skill at all. Many people prefer tasks that don't require much thought so that they can be free to use their imagination to get them through the working day.

> 'When I first went in there, I kind of envied foremen. Now I wouldn't have a foreman's job. I wouldn't give them the time of day.'[28]

Clean work

Historically, in industrialised countries, there was an expectation that men who performed dirty work would be supported by women at home washing their clothes. In extreme cases, washing and cleaning was accepted to be an employer responsibility, such as in many coalmines. This may still be the case in many industrialising and low-income countries today. However, especially in low-income countries, few people who perform dirty work like to be seen with the evidence on their hands. While 'getting your hands dirty' is a term used by engineers who value hands-on technical work, it is more a metaphor than an accurate description. It expresses the value of being on-site, in visual contact with the work, rather than physically getting one's hands dirty. In many societies, aspiring to be an engineer means aspiring to the end of washing dirty clothes and bodies at the end of a long sweaty working day. In most societies, dirt is seen as a sign of low social status.[29]

Therefore, finding ways to keep people and their clothes clean can significantly influence task performance and the motivation to perform it conscientiously. Often, the only way to appreciate this is through monitoring task performance. Many engineers will discuss tasks with technicians in clean, air-conditioned environments, with little appreciation of how the task is performed in practice.

> 'I had been observing Dave and Scott, who have been working for about an hour "out there" in the chemical plant. Dave comments that we are lucky, because it is a mild day; it is not one of those really hot days we experience in Western Australia. I can barely hear him. There is a continuous throb from the huge web of machinery within the plant. Scott and I are wearing earplugs; Dave is not. We head towards the area where they will replace the pump; the new one has already been delivered, waiting to be craned up to where it will be fitted. I follow Dave and Scott up the metal steps through the web of pipes and pumps. Dave tells us to wait, that he'll be right back. He is searching for something. He comes back with a thick wooden stick he finds near one of the huge vessels. They use it as a lever to lift the pipes up

while they dismantle the old pipe (so the flange bolts will come out easily). Scott has another of his coughing attacks. He explains to me that he was exposed to two different types of contaminants at the plant on two different occasions during the morning and that he has been coughing frequently since then. A few minutes later, a leak occurs from an overhead pipe, and we are all covered in a black dust. The two men were directly under the pipe and close to the leak. As a result, not only their clothing, but also their faces are covered in black dust, aggravating the cough Scott already had. Despite all these incidents, he completes the task and when he is leaving he says: "And all this for $29 an hour", pointing to his face. Dave says, "We all become blasé after a while".[30]

Coordinator and work preparation attributes

The next set of factors is within your ability to influence as coordinator, but you need to think about these at the commencement of the task. Your ability to influence these may be reduced or even be negligible once the task has started.

Expertise

The peer's perception of the coordinator's level of technical know-how influences task performance. If the peer respects the coordinator's know-how and experience, then the peer is likely to be less inclined to take shortcuts that would be detectable by the coordinator.[31] The peer will also place more value in any appreciation expressed by the coordinator.

Resources and supervision

The level of resources available to do the work influences task performance (time allowance, material availability, and quality), particularly the level of expert assistance and supervision when required, not necessarily from the coordinator. The work is much more likely to be completed on time and at a high quality if there is expert supervision available when necessary.

Location

Travelling time can influence alertness and the capacity to perform a task, particularly if the peer has had to spend a significant amount of time travelling on the same day. Depending on the level of support services available at a remote site, there may be significant influences on the task performance, especially if the level of support is meagre. Access to expert assistance and supervision, as explained above, has a significant influence and may not be available at a remote site. However, new technology, such as smartphones with still and video imaging capabilities, can provide ways to help overcome these limitations. More likely, the belief in the effectiveness of this technology can lead to situations where work is performed at a remote site with inadequate supervision or inadequate communication bandwidth to support the use of these devices.

If the work has to be performed remotely, it significantly helps if the coordinator has been to the remote site, can demonstrate understanding of the local constraints, and has ensured that suitable support is available.

Quality of supervision and informal support

Beyond the peer's skill and expertise, one of the remaining significant factors influencing task performance quality and timing is the skill, expertise, and experience of the immediate supervisor. In many instances, the coordinator does not necessarily provide immediate supervision. For example, you may have requested some drafting design work to be performed by a particular drafter. However, the quality of the work performed by this drafter will not only reflect their skill, but also the quality of supervision provided by the drawing office supervisor or manager. As explained in the introduction and again in later chapters, one of the most important cost factors overlooked in thinking about labour costs is the cost of supervision. Not only is supervision a significant cost, but it is also a significant enabler of high-quality performance. Even if there is no direct supervisor available to provide support, other highly skilled people can provide appropriate levels of superior support. The competence of an individual is strongly influenced by the competence of the people they work with.

Although some people are happy working on their own, others feel uncomfortable without social support from companions, even if those companions are working on entirely unrelated tasks. Other people don't really know whether they work best with people around them or on their own. Some people prefer company for certain kinds of work and solitude for others. Either way, it is important to appreciate the significance of social support in terms of task performance.

Scope, documentation of task, and accurately conveying intent

As explained later in this chapter, technical coordination performance requires a teaching performance for the peer to understand the requirements. The quality and timing of the peer's performance is influenced by the ability of the coordinator to provide a concise description of requirements in a form that the peer can understand, taking into account likely issues, and foreseeing reasonable risk factors. This, in turn, depends on the coordinator's knowledge of the task and relevant know-how. Since technical coordination is often a way in which skilled people contribute their knowledge and expertise, the coordinator may only have limited understanding and knowledge of the task and be learning as much or even more than the peer.

An accurate and concise definition of the scope is also helpful because it reduces the likelihood that the peer will perform unnecessary work, which adds to cost and time. Naturally, if the peer is more knowledgeable or experienced than the coordinator (which is often the case), the coordinator will need help from the peer to define the scope: a joint performance.

As explained in Chapter 8, it is often highly desirable for the coordinator to sit down with the peer and work through all the requirements, step-by-step, to confirm accurate understanding by the peer and to help each other ensure that all predictable issues have been discussed.

While it helps to specify requirements in writing, it is essential to confirm that the peer comprehends them and can accurately perceive the requirements from documents, drawings, and sketches. Obviously, it also helps the coordinator to have a clear and detailed understanding of the work required.

Depending on the task, the peer may have significant latitude regarding interpretation, particularly if the coordinator has only a limited understanding of the task.

Of course, it is essential to establish a good working relationship so that the peer knows that they can contact the coordinator when necessary, in order to discuss variations.

For a task that is well understood, a step-by-step written procedure or checklist can be valuable in helping the peer remember all the required steps and checks that must be performed. However, if the procedure or checklist is not easy to hold, view, or read at the work location, it will probably not be used.[32]

Prior organisation

Task performance will be heavily influenced by the degree of prior organisation, such as whether the peer has planned and prepared the task beforehand. For example, preparing tools and instrumentation (e.g. sharpening and calibrating), as well as ensuring that all tools and materials are available in the workspace, can make a significant difference.[33]

Understanding of constraints

One of the essential aspects of technical coordination is that the work is performed in accordance with an agreed upon schedule. This can be the most difficult aspect to control. It helps for the peer to have an understanding of the constraints within which the coordinator is working (e.g. total cost, the need to coordinate a task schedule with other people).

It is also important for the coordinator to understand the other priorities and tasks faced by the peer, particularly if there are critical deadlines for the other tasks that will arise during the peer's work for the coordinator. A coordinator should ask, 'Are you likely to have other deadlines come up before this is finished?'

Performance quality assessment

Task performance is influenced by the perception that quality and completion in the required time will be assessed. However, experience demonstrates that it is important for peers to be able to use the quality assessment instrument (or procedure) for themselves, rather than having the work passed on to an external assessor later. This allows the peer to adjust their own performance to meet the quality expectations. You will have prior experience of this yourself: as a student, it is much easier to prepare for an examination when you have access to previous examination papers or questions with worked solutions.

You can also reduce the likelihood that the peer will perform the work to a higher quality standard than is needed, thereby adding extra time and cost.

Work organisation attributes

The last set of factors may be within your control, or ability to influence, to a greater or lesser extent. You can draw on these factors if you need to find ways to speed up work without compromising the quality of collaboration: you may even be able to improve it.

Interruptions

Interruptions can have a very large effect on the performance of skilled technical work, particularly work that requires short-term or intermediate-term memory. While it is possible to design tasks to reduce memory loading,[34] there is a limit to which this can be done when tasks are informally organised, as is often the case with technical coordination. Skilled workers will take steps to eliminate or minimise interruptions.[35] Interruptions caused by incoming e-mails can significantly influence office worker productivity.[36]

In a situation where the peer is likely to be interrupted, you may be able to negotiate times when interruptions can be minimised or even eliminated. If the work demands concentration and accuracy, then it may be reasonable to request that the peer switches off their mobile phone and closes messaging applications on their computer screen.

Shift work

While shift work has its own special issues, particularly in managing fatigue, an issue of significant technical importance is the handover between a person on one shift to a person on the next shift. Because technical coordination relies extensively on informal interactions, it can be difficult to avoid this problem except by trying to avoid task performance on a shift basis.[37] The same issue can arise when the work is strenuous and two or more people have to perform a task in rotation so that they get adequate rest and hydration. Manual landmine clearance is a good example of this, which requires physical effort coupled with high mental alertness.[38]

Safety

Workplace safety, particularly the perception of safety by peers, also has a significant influence on task performance.[39] Safe working is strongly emphasised by most major organisations, but small organisations and individual contractors working in large organisations often miss the importance of safety. Many engineers have inherited cultural practices from organisations that see a trade-off between productivity and safety: the more resources that are devoted to safety, the less that are available for production. Many disasters have demonstrated the fallacy of such a trade-off: Piper Alpha and the BP Horizon platform disasters illustrated this basic safety argument. Safe working conditions both enable and promote productivity.

Clothing

Clothing influences task performance in many different ways. Even if protective clothing is needed for safety reasons, it may not be worn, especially if it is uncomfortable or hinders sight, hearing, or movement. Wearing protective clothing may decrease task performance capacity. Make a point of wearing the same protective gear as any peer would. That way, you can appreciate their difficulties and, more importantly, they will appreciate that you understand what it's like to wear it.

In less obvious ways, clothing influences the way people feel about each other and themselves. Wearing clothing that marks an individual as being of higher status can help with task performance as it helps to build self-esteem. Why else would so many business people and lawyers, both male and female, wear expensive business suits, even in extremely hot weather?

Self-organisation

Particularly with highly skilled work, task performance is influenced by the degree of control imposed externally. Although it can be difficult to appreciate the best trade-off without previous experience, task performance is usually improved when the peer can decide when and how to perform the task.

Handling problems

What is the peer expected to do when something unexpected happens in the performance of the task? There are several possibilities. The peer may stop the work and ask the coordinator or supervisor what to do next. The peer may assume an answer and carry on with or without reporting the issue to the coordinator. The peer may wait and do nothing (or work on another task) until the coordinator turns up. This last possibility can cause significant delays, especially if the difficulty has not been foreseen. Unfortunately, this last choice is surprisingly common, especially if the peer lacks confidence or technical support from supervisors or other peers.

Territory

Task performance is significantly influenced by the perception of territory: 'Is this job mine or really someone else's?' Julian Orr demonstrated that the performance of photocopier machines that were regularly serviced by particular technicians, and were seen to be the responsibility of those technicians, were significantly better than other machines.[40] Territory is often defined by technical expertise within engineering practice. It is important that others acknowledge a peer's 'personal territory'. Working on others' territory will reflect the personal relationship with the territory 'owner' and needs to be carefully negotiated.

Timing

Task performance can easily be influenced by the time of day when it is performed. For example, a task that demands short-term memory will take much longer if it is performed at a time when there are likely to be interruptions. It may also take longer if the peer is tired at the end of a long working day; paradoxically, this can also be a time when there are the least number of interruptions. Everyone has their 'best time' of the day: yours might be mid-morning; mine tends to be early morning. If the peer works on the task during their best time, the results can be significantly better.

The other factor that influences task performance is the chosen starting time. In his famous book, *Parkinson's Law*, Northcote Parkinson observed that work always expands to fill the time available.[41] Thus, if the object is to save time, it can be a good idea to start as late as possible. This is a principle that many university students adopt, also supported by the knowledge that the later they leave an assignment, the more likely it is that other students will be able to help them.

From a more serious point of view, however, starting late is more likely to cause problems in completing a task on time. If unexpected issues arise that require consultation between the coordinator, the peer, and perhaps others, there are inevitable delays. Since one can never completely predict these issues, it is always best to ensure that work starts as soon as possible to allow as much time as possible to rectify misunderstandings or deal with unexpected developments.

Presence

The degree of involvement and regular (or continuous) presence of the coordinator can make a big difference, particularly in ensuring that the work will be completed on time. Here is an example from an interview with an engineer in the oil and gas industry:

> 'While I was there, I noticed that projects that had no client present weren't resourced as consistently. That company runs late, they have a reputation for that and all firms run late to some extent – it is always an issue of resources and schedules which don't always coincide in the way that they should optimally. If you're not careful they'll start working on someone else's job. But if they know that the client is on the site, they'll work much more conscientiously on your job.'

It is not necessary for the coordinator or the client representative to be watching what is happening all the time. That can be destructive, because it undermines confidence and trust between the coordinator and the peer. However, if the peer notices that the coordinator is physically present and nearby, it serves as a reminder to keep the coordinator's task at the forefront of decisions related to allocating time and other resources. This is why major firms know that they have to keep a representative on-site whenever major contractors are performing work for them. This is a common industry practice.

In some cultures, such as South Asia, work stops completely in the absence of supervision. This is not the same as the coordinator being present, however. It seems to have more to do with the enormous social gap between workers with dirty hands and their supervisors, a social gap that often includes language, social culture, education, ethnicity and, of course, remuneration.[42]

Obviously, physical presence on-site comes at the cost of reducing the capacity to coordinate other work in person.

Exposure to consequences

Task performance quality is also influenced by the extent to which a peer is influenced by the consequences of neglecting quality. For example, photocopier technicians who know that they will have to return to look after a particular photocopier if it fails again tend to take more care with their work. If they partially destroy the head of the screw, for example, the chances are good that they will be the one affected on a later occasion when they have to extract the damaged screw from the machine. Therefore, they are less likely to damage screws and more likely to replace a damaged screw immediately, rather than leaving it for someone else (or themselves) on a subsequent visit.

At the other end of the spectrum, this factor influences the work of project managers, who are often rewarded with projects that are more prestigious if they manage to complete a project within the planned time schedule and budget. These rewards also increase the temptation to reduce the scope of the project in order to meet the time schedule and budget, as this example demonstrates:

> '... It was a fast tracked project, so the amount of planning that went into the project was minimal to begin with. Midway through design and construct, a quarter of the project was thrown away because it was going to cost an extra 2 million dollars ... Our project manager works on a time and cost basis and gets rewarded for those things – not on the basis of 'does this work?' So it was not difficult for him to make the decision. A new capital works project is now in the wings to fix everything that is wrong (pause) so an 8 million-dollar asset is going to end up costing at least $12 million dollars when you include the additional modification and repair costs now required. However, because the project was delivered on time and budget, our project manager has been rewarded with a much larger and juicier contract.'

This last quote illustrates a major issue for engineers: designing contract targets, payment schedules, and incentives that help achieve beneficial results for all concerned. All too easily, contract frameworks result in perverse incentives that can frustrate the aims of an enterprise. However, this is a topic well beyond the scope of this book.

History

Past practice is a strong determinant of task performance, good or bad. An engineer inherits the consequences of decisions made by many other engineers in the past, particularly when it comes to coordinating the work of peers. When you first appear on the scene, the peer may be remembering previous engineers from your part of the enterprise and thinking, 'here comes yet another ignorant and arrogant smart-arse engineer'. Anyone attempting to change expectations may be derided as an outsider 'trying to change the culture, the way we do things around here'.

Affective factors

As explained in Chapter 8, organisational values can have a strong influence on task performance and motivation, and the influence is almost always more significant over the long-term than financial remuneration. Whether the person doing the work feels valued or not, part of the organisation or not, has high personal regard for the coordinator or not – all these can influence task performance.[43]

Another related influence is the extent to which a peer identifies with the particular task. A task is likely to be performed sooner and more conscientiously if the peer can perceive strong subjective value in the work or identifies with the work. If the peer considers the work to be 'beneath' or 'not belonging' to their social identity framework, then the task is likely to be deferred, completed late, or even not done at all.[44] This explains why document checking and review tends to be relegated to secondary status by most engineers. They don't see it as being 'productive engineering work'.[45]

As explained in Chapter 8, however, as a coordinator it is possible to change the way that the peer views the subjective value of a particular contribution. As will be explained in Chapter 11, many engineers have only a tenuous understanding of commercial value and therefore find it hard to understand why the checking and review of documents should add to the value of a project when checking does not seem to contribute any significant design progress but does cost money and time. It should be easier to understand why checking increases the value of the project with the help of the material found in Chapter 11.

Opportunity for recognition and identification

A peer with less status in the organisation may welcome the opportunity to gain recognition by performing a task for a more senior coordinator, which may provide strong motivation. Success may increase the peer's visibility and/or job security, providing a significant amount of motivation for conscientious work.[46]

Relationships and trust

A prior relationship between the coordinator and the peer can strongly influence task performance, particularly in terms of maintaining conscientious vigilance. It is often difficult for a coordinator to accurately scope and specify a task. Therefore, a peer who has a strong pre-existing relationship is more likely to know where to look for gaps and mistakes in scope, documents, or drawings. They will feel more comfortable drawing this to the coordinator's attention and will have a better appreciation of the constraints under which the coordinator is working. In short, new peers require new relationships that take time to build and require more monitoring until a high level of trust has been established.

This explains why engineers strongly prefer people with whom they already enjoy a strong collaborative relationship.[47] This is not necessarily the same as a social friendship, but having a social friendship can make it easier to develop a professional collaboration. The trust that has been built up through the relationship enables a coordinator to have more confidence in a peer and therefore spend less time on direct supervision and follow-up monitoring.

The main value of maintaining relationships is their enduring character: they make it so much easier to collaborate in the future. Take this example from an engineer:

> '[What are the lasting effects of your work? (the contributions that will remain after you leave)] Consistency. A few bright ideas. Most importantly, some people who know a little bit more and have been inspired a bit, learnt something about life and engineering, and who have taught me a bit about life and engineering, so it's all about personal relationships. That really should go first, it's much the most important.'

Many young engineers do not emerge with good relationship skills from their education: their focus has been on achieving the necessary grades, supported mainly by solitary proficiency in technical analysis skills. For them, the best way to learn to build relationships is to spend time working with other people, even though it can seem frustrating and tedious at first. Later, they can look back on these experiences and laugh at how inept they were. That is nothing to be ashamed about; it is simply a product of the education system as it stands at the moment. The engineer quoted above confessed that he was a self-absorbed nerd when he emerged from university studies. It took him a long time to learn about the importance of relationships in engineering practice. Perhaps because of that, he later acknowledged the importance of relationships.

Prestige

The prestige of a project to which the task contributes can influence task performance. Most engineers strongly value the opportunity to work on a prestigious project, one that will be recognised by other people as being significant. Engineers also value the opportunity to work with people with whom they strongly identify as role models.[48]

If, as a coordinator, you can bring a senior engineer to take a look at ongoing work so the peer feels that you have increased their visibility to people 'higher up the food chain', then you may notice a distinct increase in collaboration and productivity.

Learning

Most engineers also positively identify with tasks that enable them to learn new skills: this can be seen as having instrumental value in the sense that new skills will open up new opportunities in subsequent projects. However, it can also be seen as an intrinsic value simply because many engineers enjoy expanding their knowledge on technical issues, whether or not that knowledge will prove useful later in their career.[49]

Developing others

Many engineers also respond positively if a task provides an opportunity to mentor or help a less experienced person develop new skills and abilities.[50]

Social benefits

Although secondary compared to other influences, the fact that the project will benefit society as a whole does provide a positive motivational influence for some engineers.

> 'I think one of the bigger drives is the importance to society. I [see aerospace] as a field that will probably benefit humanity as a whole.'[51]

Reward

The essence of technical coordination is willing and conscientious collaboration. A financial reward, even a non-financial one, may not be significant in influencing the degree or quality of cooperation. However, it can be more challenging to secure the collaboration of people who are not being directly rewarded, especially if the coordinator is receiving a significant reward for the role.

Contractual terms

Most engineers don't like looking at contracts, as this quote demonstrates:

> *'The best kind of contract is one that stays in the bottom drawer, one that you never have to look at again.'*

This does not mean that this engineer sees little value in a contract. On the contrary, this engineer is explaining that a well-designed contract is one in which all parties are completely clear in terms of performance expectations and what will happen in different 'predictably uncertain' circumstances. The contract never has to be read, once it has been agreed upon, because everyone has clear expectations.

Most contractual and work disputes arise from failures to anticipate foreseeable but uncertain events, as well as failures to negotiate an agreement on how to handle these events before the contract goes into effect. Disputes over contracts and requirements can be extremely destructive in any working environment, undermining trust and compromising the possibility of collaboration in the future.

Engineers with experience regard a 'good contract' as fair when it clearly identifies where risks (unpredictable changes) can occur and how each party responds to changes. In practice, however, many companies enter into contracts that are seen to be inequitable and, as a result, often end up with a dispute.

At my university, unfortunately, young engineers develop a strong dislike of lawyers, nurtured by a long history of friendly, and sometimes unfriendly, competition between the Law faculty and the Engineering faculty. Therefore, it may take longer for our engineering graduates to appreciate the value of a legal mind in drafting contracts that account for every foreseeable but uncertain event. However, experienced engineers also appreciate that contracts are too important to delegate entirely to lawyers: learning to develop 'good' contracts is valuable for any expert engineer. Once again, this is an advanced topic beyond the scope of this book.

Ongoing work opportunities

If the peer perceives the opportunity for ongoing remunerated work as a result of working with a coordinator, it is more likely that the peer's performance will be better than normal. This is not necessarily a good thing. This is the kind of financial reward that can undermine the essence of technical coordination – genuinely willing and conscientious collaboration. This kind of collaboration, motivated for a temporary financial reward, can lead to unrealistic future performance expectations.

Payment already received

One of the most important lessons for any young engineer to learn is not to approve payment for any work performed by a contractor until it has been thoroughly checked and, preferably, independently tested. Once a contractor has received the final progress payment, it can be very difficult to persuade them to return to rectify mistakes, unless there is a maintenance agreement in the contract or the strong possibility of obtaining further work from the firm.

SELF-ASSESSMENT

You will have plenty of opportunities to practice technical coordination and informal leadership, even if you are not working as an engineer. As explained in a later chapter, anyone working in a voluntary organisation has opportunities to learn leadership and coordination skills. No one can force volunteers to do something; instead, gentle sympathetic leadership is needed to build willing and conscientious collaboration.

Technical skills are essential, and coordination may offer many opportunities to deepen your own technical knowledge, as you will often be coordinating people with more knowledge and experience. Learn as much as you can.

So how can you monitor your performances?

The easiest way is through a journal. Keep track of the time you spend informally coordinating the work performances of others, not necessarily every performance, but at least on a selected sample.

If your performance is improving, you should notice a gradual reduction in the time needed for any given performance and improvement in the quality of the results. You will gradually learn more about that most valued attribute of an expert engineer: the ability to get things done.

As your coordination performances improve, and the complexity of tasks being coordinated gradually increases, you will see growth in your ability to define tasks and coordinate technical engineering work performed by other people while still being able to work without undue stress or long hours.

You may also find that other people notice your leadership abilities, and you will steadily gain respect for your ability to get things done. This might not be obvious but can take the form of being asked to take on much more challenging coordination tasks, perhaps those that require working with difficult people.

This chapter will enable you to see structure in your technical coordination work, and by seeing that, you will be able to focus on aspects where you can improve your own performance.

NOTES

1. See for example Johnson & Johnson (2009), now in its 10th edition.
2. Trevelyan (2007); B. Williams et al. (2014).
3. Quotes are from interview data (see also Anderson et al., 2010, p. 161).
4. 'Mobilising' is a term noted by Blandin (2012).
5. Engeström (2004).
6. Management in the context of engineering has been extensively discussed by Badawy (1983, 1995), Samson (1995), and Mintzberg (1973, 1994, 2004). In other cultures, for example, Japan, management patterns differ but still rely on organisational authority (Lam, 1996; Lynn, 2002).
7. Newport & Elms (1997).
8. See Pielstick (2000) also Kraut and Streeter (1995).
9. Cohen & Bradford (1989, 2005); Kendrick (2006); Perlow & Weeks (2002); Shainis & McDermott (1988).

10. Raven (1992) is explained in the coming section on the Practice Concept 55: Technical coordination. Cropanzano and Mitchell (2005) provide a scholarly review on social exchange.
11. A large survey of American engineers revealed this specifically (American Society of Civil Engineers, 2008).
12. All names are fictitious to protect the identities of engineers who contributed to our research.
13. Some readers may object to the apparent sexism in this quote. I would request patience and understanding from them. It is better to accurately quote the speaker rather than edit the quote at the cost of losing authenticity.
14. The effectiveness of a senior engineer depends on their ability to delegate technical work, and also to select aspects to be reviewed and what can be accepted without detailed review. A good senior engineer will soon recognise the capacity to coordinate technical work sufficiently well so that it does not have to be reviewed at a higher level.
15. Terkel (1972, p. 22). Collinson and Akroyd (2005) provide a scholarly review of resistance and dissent in the workplace.
16. Raven (1992).
17. Another instance of 'churn', explained before.
18. Trevelyan (2014a).
19. Orr (1996), also personal communications from Adrian Stephan (2006–2012).
20. Dekker (2006); Kletz (1991); Stewart & Chase (1999).
21. Mehravari (2007).
22. See, for example, a study of accident reduction efforts in the context of landmine clearance (Trevelyan, 2000). See also, for example, Lawson (2005), Lucas, Mackay, Cowell, and Livingston (1997), Reason and Hobbs (2003), and Willers (2005).
23. Dekker (2006); Reason & Hobbs (2003).
24. Orr (1996); Schön (1983); Vincenti (1990).
25. Leveson & Turner (1993); Lucas et al. (1997); Malone et al. (1996); Reason & Hobbs (2003); K. H. Roberts & Rousseau (1989); K. H. Roberts, Stout, & Halpern (1994); Weick & Roberts (1993); Willers (2005).
26. Anderson et al. (2010); Han (2008); Kendrick (2006).
27. Eccles (2005).
28. Terkel (1972, p. 153).
29. Ansari, Kapoor, & Rehana (1984).
30. Sule Nair describing observations of technicians working on a continuous process chemical plant, unpublished manuscript (2010).
31. Social power from expertise (Raven, 1992).
32. Reason & Hobbs (2003).
33. Orr (1996); Reason & Hobbs (2003).
34. Reason & Hobbs (2003).
35. S. Barley & Bechky (1994).
36. Gupta et al. (2010); Iqbal & Bailey (2006, 2008); Jackson, Dawson, & Wilson (2003); LaToza et al. (2006); Perlow (1999); F. Smith (2010).
37. Reason & Hobbs (2003).
38. Trevelyan (2002).
39. Cross et al. (2000); Bea (2000, p. 402).
40. Orr (1996).
41. Parkinson (1957).
42. Roy (1974); Trevelyan (2014a).
43. Ashforth & Mael (1989).
44. Ashforth & Mael (1989).

45. Mehravari (2007); Trevelyan (2010b).
46. Han (2008); Kendrick (2006).
47. Tan & Trevelyan (2013).
48. Kendrick (2006).
49. Anderson et al. (2010); Kendrick (2006).
50. Kendrick (2006).
51. Anderson et al. (2010, p. 168).

Chapter 10

Managing a project

Today, the 'project' is the method of choice for organising most engineering work. As one engineer explained in an interview, 'We only do projects in this organisation, nothing else. Everything is a project.' There are, of course, other ways to organise engineering work. For example, operations and maintenance are typically organised on a continuing service basis. This kind of organisation is needed for running continuous process plants and may also be used in manufacturing. However, even within this kind of organisation, project organisation may still be used for specific time-limited activities, such as plant shutdowns for maintenance.

Your reputation, namely the extent to which people trust you to spend their money, depends on your ability to deliver on your promises. Every expert engineer has to be able to manage projects effectively and to deliver on promises. The challenge you will face is that delivery depends on the interpretations and subsequent actions of many people, working with incomplete information and large degrees of uncertainty in terms of actions and timing. This means that you have to reliably deliver predictable results in the face of ever-present uncertainty. How is that possible?

As explained earlier, there is a common misperception that engineering is a precise scientific activity with no room for uncertainty. It is easy for you to have gained this impression during your undergraduate studies, because in most engineering courses, uncertainty is only discussed in a very limited mathematical sense.

WORKING WITH UNCERTAINTY

Expert engineers deal with uncertainty all the time and have devised elaborate precautions to limit the consequences of that uncertainty as much as possible. It is partly their apparent success at doing so that leads people to think that engineering has become so predictable. At least in the industrialised world, for example, one never pauses to wonder whether the lights will come on when you flip the light switch.

We will discuss many ways to limit the consequences of uncertainty in this book. At this stage, it is important to appreciate some important ideas that can help correct common misconceptions for novice engineers.

One of the techniques for working with uncertainty is risk management.

Most students today now learn about risk management at some stage of their university course. It is worth remembering that the fundamental notion of a risk is associated with a particular unpredictable event that can affect outcomes, both advantageously and disadvantageously. The magnitude of the risk depends on both the probability that the event will occur, and the consequences of its occurrence. The relative importance of a given risk, therefore, depends on a combination of probability and adverse consequences. Opportunities (uncertain events with favourable consequences) can be assessed in the same way. Different risks have different kinds of consequences. Some risks are primarily economic, such as an unexpected and significant rise in the price of a product from an engineering enterprise that dramatically improves profits. Others might be reputational risks, such as a high-profile employee being involved in court action related to allegations of misbehaviour. Still others might be technical risks: the possibility that a particular solution will not provide the performance that has been predicted. Risk management is a formal method in which engineers systematically examine all of the potential risks and opportunities, deciding on appropriate planning to maximise the opportunity to take advantage of 'upside' risks and to limit the undesirable consequences of 'downside' risks.

Now, let's look at some fundamental ideas behind the management of uncertainty in engineering.

Practice concept 56: Reliable performance requires a high level of predictability

Student engineers often think that …	Expert engineers know that …
50% is okay to pass and 70% is a distinction. Getting it 100% right is for nerds.	99.99% predictability is required (or more) for reliable performance.

This almost seems trivial. However, the grading system that governs your behaviour as a student means that you rarely get the chance to learn how to perform engineering technical work to an extremely high level of predictability and reliability. You need to remember that engineers work under similar time pressure as you did when you were a student. There is simply never enough time to do everything. How, then, does an expert engineer manage to achieve such a high level of reliability in technical work performed by people who are no more predictable in their professional working life than they were as students?

As unbelievable as that might sound, you can also learn how to achieve this.

Practice concept 57: Uncertainty in engineering mostly arises from unpredictable human behaviour

Novice engineers often think that …	Expert engineers know that …
Uncertainty is a probability, most often following a statistical normal distribution.	Uncertainty is mostly due to unpredictable differences in human interpretation, subsequent actions, and timing.

To the extent that uncertainty is part of an engineering curriculum, it is almost always considered to be a statistical phenomenon, often receiving very little attention in most subjects that make up the engineering course. It may arise in the form of 'noise', 'signal to noise ratio', or 'the Weibull distribution, which describes the distribution of component failures over time'. Uncertainty is only described in terms of inanimate causes. If people are considered at all, the consequences of human behavioural uncertainty can be minimised by rules, protective clothing, fences, gates, warning signs, regulations, and laws.

At least until the first half of the 20th century, most of the uncertainty faced by engineers was due to natural causes or limitations in our knowledge of engineering science. Today, much of that has changed. Engineering science now allows us to make extraordinarily accurate predictions ... most of the time. Obviously, engineered structures can still be overwhelmed by natural disasters, such as storms, floods, volcanic eruptions, earthquakes, tsunamis, and wildfires. However, these events are very rare in statistical terms. Much more commonly, today at least, uncertainty and limitations on predictability arise from human behaviour and interpretation. The 2011 Japanese tsunami was predictable to natural scientists. However, people had decided that the risk was so small that it was not worth the expenditure to create sufficiently strong defences.

Humans are intrinsically unpredictable, though we all have mostly predictable aspects, as well. Engineers and all the other people who work in engineering enterprises are no more predictable than other people.[1]

Therefore, expert engineers have a large number of special techniques that take human characteristics into account to ensure that the results of engineering are reliably predictable.

Risk management practice helps to confirm the overwhelming influence of human behaviour in contributing uncertainty to contemporary engineering projects. A typical risk assessment for a medium-sized project might list 200 or more sources of uncertainty, of which 5–15 will normally originate from natural variability, and perhaps two or three will be completely unpredictable. For example, weather is a natural source of uncertainty for any outdoor engineering activity, but modern forecasting almost always enables bad weather to be predicted days in advance, long enough to avoid any risks. The other 190 or so uncertainties will involve human unpredictability, such as failure to perform maintenance, obtaining the wrong fuel or materials, not following appropriate procedures, tired operators who are more likely to make mistakes, running out of spare parts, collisions involving vessels, aircraft or vehicles controlled by people, and many others.

As we shall see below, expert engineers know that understanding how human behaviour contributes to uncertainty is a critical aspect of practice. Every day, engineers have to confront uncertainty, most of it originating from human behaviour. However, some of the best ways to reduce the consequences of uncertainty lie in making use of the same human diversity that leads to uncertainty in the first place.

How does human behaviour cause uncertainty, you may ask?

Remember that technical work is largely autonomous: it relies extensively on human interpretation, human cognition, and human action, as well as social interactions. It is not uncommon to find people in an engineering enterprise with 60 or more simultaneous issues to deal with: requests for assistance, coordination of other people, decisions that have to be made, collecting information to help with decisions, e-mail correspondence, and administrative issues. What people do each day, the order in which they do it, when they choose to start, what they remember to do and what they forget to do, when things get interrupted by the actions of others, when people forget to return to actions that were interrupted: all these contribute to the unpredictability in engineering work performed every day. Beyond those common examples, there are human errors or mistakes. These can be unconscious slips made without any awareness, deliberate incorrect actions taken because of incorrect perception, deliberate incorrect actions taken because of interpretation differences, deliberate incorrect actions taken because of habit, failure to acknowledge incorrect actions and take corrective action in time, and incorrect actions taken with the intent to deceive and keep them hidden.[2]

Engineering techniques to limit the consequences of human uncertainty include education (to understand the purpose of the enterprise or project), skill development and training, organisational procedures and checklists, risk management, project planning, individual skill, experience and knowledge, such as how to prevent interruptions that lead to slips, practice to perfect skills, technical standards to guide action, monitoring, the use of appropriate tools, appropriate organisational culture and attitudes towards quality, and documentation.

Engineering enterprises are also vulnerable to unpredictable human behaviour from external causes, such as changes in the economic or political environment, changed financial circumstances affecting investors, colleagues, contractors, and suppliers, industrial action by trade unions, and the consequences of various actions that spill over and affect other people.

Practice concept 58: Human diversity can help reduce behavioural uncertainties

One of the most effective ways to limit the consequences of unpredictable human behaviour is to make use of the diversity in perception, interpretation, thinking, and action that is intrinsically human. Different people often see the same thing differently: that is why it is best for documents to be checked by someone other than the authors.

The mere expectation that work will be systematically checked by senior engineers can improve performance by raising expectations on the expected standard of work.

Novice engineers and students seem to be very reluctant to engage in this. In our research, we have observed how many experienced engineers, judging from their actions and interview responses to carefully designed questions, think that systematic checking is a non-productive chore for which, they say, they don't have enough time.[3] Without understanding how reducing uncertainty contributes to value perceptions, it may not be easy for engineers to understand how checking produces any real value. In practice, the earlier a mistake is detected, the cheaper it is to correct. Something that might take four hours in a design office might require hundreds of hours to fix if it remains undetected until construction is half-completed. One of the factors that most often contributes to unexpected project delays is discovering mistakes that should have been detected earlier.[4]

Novice engineers often think that ...	Expert engineers know that ...
Checking is just a chore that takes away valuable, productive time. The more you check, the more mistakes you will find.	As an example, systematic checking is essential and best done by another person, as they will see things in a different way. This helps expose alternative interpretations and reduces the chances of undetected mistakes, thereby increasing value by reducing uncertainty.

The same principle can be applied in risk management. Involving non-engineers and external people in the assessment of environmental and social risks, in particular, is now part of established practice.[5] Non-engineers and outsiders can often spot mistakes and omissions made by engineers much easier than engineers can and can point out some of the unexpected social implications of what otherwise might seem like optimal technical solutions. They are not necessarily any smarter than engineers: they just think in different ways and have different ways of perceiving the world. In other words, they see things differently. They can think of obvious things that can go wrong that engineers easily overlook.

Later, we will look again at review processes and checking in engineering practice.

Practice concept 59: Project management enables predictable results in the presence of uncertainty

Novice engineers often think that ...	Expert engineers know that ...
Project management consists of planning, preparing a Gantt chart, and is largely an administrative non-technical function, something that anyone who can't handle 'technical stuff' is able to do.	Project management means leading and guiding people to faithfully translate technical ideas into an engineering reality in line with project objectives.

Project management means coordinating an enterprise to reliably deliver predicted results, on time, safely, and within the planned resource budget. As an expert engineer, if you can reliably deliver predicted results in the face of uncertainty, then people will trust you with a lot of their money. It is not easy to do this without knowing how. In practice, as we shall see, it is often impossible to achieve every project requirement; instead, a project manager may have to choose between multiple conflicting requirements.

Many engineers who prefer to focus on building their technical capabilities dismiss project management as something that anyone can do, particularly anyone who is not a technical expert. 'They're just project managers; they can't hack the technical stuff, so they get other people to do it for them.'

Sooner or later, however, every expert engineer who focuses on technical subtlety or complexity comes up against the challenge of bringing ideas to reality at a scale that is beyond the capacity of one or two people, sometimes on a gigantic scale with a budget in the billions of dollars.

That's when project management becomes vitally important: it is the main process by which engineers deliver results of lasting value, the critical delivery thread of engineering practice, which was the large bottom half of Figure 3.7.

You will find project managers working in every organisation at every level. As the discipline has evolved, some people have emerged as project management specialists: project managers who find themselves able to manage any kind of project. Many engineers have encountered frustrations in working with professional project managers in charge of the projects in which they find themselves. This has arisen partly because engineers, as a professional group, have largely forgotten that delivering results is all about coordinating large numbers of people. This doesn't happen simply by writing documents and issuing them. It takes persuasion and gentle influence, also known as leadership, to guide people so that they all take part in a well-coordinated performance.

Some may argue that it is better for people with good relationship skills to take on project management responsibilities and leave engineers to do what they are better at, such as technical design, calculations, and modelling. However, there are three weaknesses with this argument. First, engineers need to develop 'influencing skills' in order for their ideas to be accepted and implemented faithfully, as we have already seen and discussed. Second, as explained in Chapter 4, anyone with the proper determination can become an expert: even engineers can become expert influencers and project managers, provided they are willing to engage in deliberate practice and seek the help of experts who can evaluate their performance. Third, technical understanding is helpful, if not essential, in delivering complex technical engineering work on time and budget.[6]

Almost all formal courses on project management focus primarily (often exclusively) on the production of documents, plans, and schedules. Unquestionably, that is an important element. Effective project management is a formal process and documentation is a part of it. However, that is probably the easiest part of project management. The more difficult part is not always included: dealing effectively with uncertainty through a formal risk management process. A good risk-management plan will provide a means to deal with almost all foreseeable, but unpredictable, events that can have a significant effect on the project outcome.

The truly difficult part of project management is making sure that the work runs according to the plans: it frequently does not, so the plans have to be modified to ensure that the objectives are still achieved. Therefore, an effective project manager knows how to draw up plans that can be changed, and how to monitor the progress of work and detect unwanted delays early enough to do something about them. This can be difficult, as many people do not want to even admit to themselves that they are running late. Understanding the different ways that people behave is essential for effective monitoring.

Finally, a good project manager knows how to set up systems to ensure that everything is checked, which ensures that when the project is completed, there is nothing that has been forgotten.

Obviously, financial controls are just as important as controlling the progress of technical work. As we shall see later in this chapter, monitoring finance is a critical part of monitoring the progress of the project as a whole. There are many technical and other techniques to learn: they will all come in a later chapter. Understanding why people behave the way they do helps you make the most of these techniques.

There have been many debates on the extent to which project managers need to have an understanding of technical issues. However, that argument is irrelevant for this book. Expert engineers need to be able to understand both technical issues and project management issues. Understanding how they overlap and how you can take advantage of that dual knowledge is something that you will develop with time and patience.

An engineering project starts with negotiations to clarify the project objectives and the client's requirements, particularly to understand the client's tolerance for uncertainty and financial risk. The technical aspects follow the typical sequence shown in the stack of blocks up the middle of Figure 3.10. These blocks describe activity by people which, as we have seen, involve countless unpredictable elements, people who are often tired, bored, anxious to get home, or simply distracted by other priorities.

Project management is a systematic combined performance that enables engineers to obtain predictable results in the presence of all this human and natural uncertainty. Project management happens in the big cylindrical tube surrounding the stack of blocks in the centre of the diagram: part of the tube has been cut away so that we can see the stack. Informal coordination is portrayed on the left side of the tube, while formal project management is portrayed on the right side (along with many other engineering management systems, of which project management is just one).

In essence, project management provides the guidance that keeps the technical engineering work on track and properly coordinated: the thin threads joining the tube to the stack represent this guidance, influence, collaboration, and coordination.

Figure 3.10 separates the core technical activities from the coordination and management that is taking place outside: this is purely to help understand the notion of a combined performance. Part of the work involves coordinating the performance. In reality, however, it is often the case that the same people perform both the technical and project management tasks.

The previous chapter described technical coordination: an informal process used by engineers all the time, usually in one-to-one collaboration relationships with other people. Technical coordination is usually undocumented, except for occasional diary entries and e-mail correspondence, in certain cases.

Project management, on the other hand, is a more formally documented process invoked when the scale of activity in cost, complexity, space, and time goes beyond what a single person can effectively coordinate in an informal manner. Project management heavily relies on written documents to supplement human memory.

Practice concept 60: Written knowledge needs to be learnt

While project management relies on documents, it is essential to understand how documents are used. Many of these documents provide the scripts for the overall combined performance. Most people in engineering enterprises, including most engineers, are reluctant readers, at best. Providing information in documents, no matter how well they are written, is often as good as keeping the information secret. The information can only influence actions when it is in people's heads, and that requires learning.

Therefore, in the context of project management, documents mostly serve only as a record of discussions, which are often discovery performances by people (see Chapter 8) to build up knowledge and understanding about the project in their heads. Documents can extend human capability by storing information that is difficult to remember, but this is only useful if a person knows where the information can be found when it is needed. Documents also provide evidence to other people about the issues considered during the planning process. This can be important for seeking approval from government regulatory agencies later in the planning phase.

There are many different ways to approach project management: there are countless styles of documentation, planning techniques, and software systems for storing information. Most engineering enterprises have their own preferred ways of doing it. I am not going to describe any of these preferred ways, as you will learn them in due course. Instead, I'm going to take you through a long sequence of questions to make you think about how we can achieve a predictable result from a very large number of individual human performances, each of which has a certain degree of unpredictability. All of these questions are relevant when managing an engineering project.

You have to do most of the work: you will need to find answers for these questions and learn how to implement the answers within the project management system used by your enterprise.

Most likely, you are probably reading this chapter because you are facing some kind of project management responsibility. Your job is to seek the answers to these questions that are pertinent to your situation, here and now, by discussing them with your colleagues. Writing down these responses will provide a record of those discussions, evidence that they took place, and proof that your colleagues have worked with these ideas. The knowledge is therefore with them too. By the time you finish, you will have a workable plan, along with having learnt a lot about project management and how it fits into engineering practice.

Practice concept 61: Project management performances can be improved

Engineers around the world have an unenviable record in engineering project delivery, particularly in the energy, manufacturing, and chemical process industries that are increasingly critical for environmental sustainability. Performance is particularly troublesome in many developing countries, and larger projects are more likely to expose project delivery weaknesses

Ultimately, these failures are failures in engineering project management: a failure to deliver the promised results on time, on budget, and with predicted performance, safety, and reliability. It is possible that the many fundamental misconceptions related to engineering practice described in this book contribute to these failures. We have seen many instances of the misconceptions in this book directly contributing to project failures. In Chapter 11, we will closely examine the failure of a major engineering project to expose more of these issues.

While every guide to project management recognises the importance of communication, most restrict the understanding of communication to simple information transfer, extending from the project manager to the participants. Remember Practice Concept 40 in Chapter 7? If only everyone would simply read the plan and follow it!

Fortunately, by reading the last five chapters, you should now understand that effective collaboration between people relies on much more than that. It is not easy for information in written documents (or e-mails) to be learnt by people so that their actions are influenced appropriately. Building effective working relationships lies at the heart of collaboration, and building relationships requires much more attention to the idea that we are working with people, all of whom have different ideas and different interests. Influencing perceptions is crucial for influencing the actions of other people. In other words, we need a much richer understanding of communication.

Working with other people, especially working with lots of other people, can be a highly stressful and draining experience if you don't know how to do it properly. Like anything else, however, as you develop more expertise and you understand more, it can become one of the most enjoyable, rewarding, and simultaneously intellectually challenging aspects of engineering. Once you start to do it well, the improvement in your abilities will be impossible to miss.

Practice concept 62: The language of project management shapes behaviour

Project managers have evolved their own language and project management has become a discipline in its own right. There are some very interesting terms in contemporary project management literature.

The first is *resources*.[7] Project managers include other people, along with equipment, machinery, space, and consumable materials as 'resources'. For decades, 'resource allocation', 'resource levelling', or 'resource optimisation' have been a part of the vernacular of project management. That being said, recent research has started to draw attention to one of the drawbacks of referring to people as 'resources'.[8] People referred to in this way start to behave like inanimate resources. They don't move without being pushed. They resist efforts to push them and prefer to stay in one location. This resistance is silent: people don't complain and they don't talk about it, but they also don't respond.

That is why in this chapter, I do not refer to people as 'resources'. Instead I use the term 'human effort': work that has to be performed by people. When planning a project, we try and estimate the amount of time that will be needed, which is ostensibly the number of hours of human effort. However, in the last chapter, we saw how the number of hours of human effort is often completely unrelated to the nature of the task. People find effective ways to finish work quickly when they want to and other ways to make the work last much longer in different circumstances. The extent to which they have a positive relationship with the person coordinating and leading the work has a large influence on this factor, as described in Chapter 9.

Another interesting term is *subject matter expert*, a term used to refer to technical specialists such as engineers in project management literature. Subject matter experts are consulted and provide information, but seldom take any active role in decision-making. You should learn about project management so you don't get sidelined as a subject matter expert. A 2004 survey of several thousand civil engineers in the USA revealed extensive dissatisfaction when engineers found themselves working under 'unqualified' project managers. The difficulty is that many engineers do not learn the appropriate skills to deliver practical results from their work and can therefore easily fall into this trap.

PLANNING, ORGANISING AND APPROVAL

Managing a project, like technical coordination, can be described as a three stage process:

1. planning and organisation,
2. monitoring progress, and
3. completing the project.

Most of the chapter is devoted to project planning, emphasising that a managing a project requires a detailed plan.

Practice concept 63: A project plan is a living document

The next part of the chapter presents a series of questions (italics headings). Several have sample answers (in italics) that will help you supply your own answers. The aim is to help you develop your project-planning skills by following a project in which you may already be involved. By seeking answers from your workplace colleagues in a series of discussions, which will likely be 'discovery learning performances', you can learn how to prepare your own project plan.

As you answer the questions, you are going to start to construct a project plan as the main practice exercise for this chapter.

However, even the best project managers don't get everything right the first time. Instead, every document that you create is a *living document* that evolves and is progressively elaborated during the course of the project as you, your colleagues, and often the client and other stakeholders get a clearer idea of the requirements. Some documents are still being elaborated even at the very end of a project.

A well-run project team will regularly review the project scoping document, specifications, and project plans throughout the project, every few days, weeks, or months, depending on the project duration, especially when the scope, specifications, or plans need to be changed.

Particularly when you are planning a project for the first time, be prepared to revisit all of these questions many times over. Each time you come back to these questions, you will find that you need to elaborate each of the documents that were created along the way. You will think of more details to add, just as in writing a chapter of a book, I think of more details to add to the text every time I look at it. However, engineering is pragmatic: there is always an endpoint to a project. There is always a time to move on and accept that we have done the best we can, but it is never complete.

At the same time, always look for ways to simplify the requirements without losing detail: keep documents as short as possible.

Defining the project scope and creating a plan: the project definition phase

What is a project?

This is one of the few questions with an answer! Be warned.

A project is a temporary endeavour undertaken to create a unique product, service, or result.[9] A project is temporary because every project has a definite beginning and a definite end. The end is reached when the project objectives have been achieved, it becomes clear that the project objectives will not or cannot be met, or the need for the project no longer exists and the project is terminated.

A project creates unique *deliverables*: products or physical assets, information, services, or results.

A project can create:

- A product (artefact) that has measurable, quantifiable attributes and can be either an end item in itself or a component of another item,
- The capability to perform a service, such as business functions supporting production or distribution, or, the performance of a defined service.
- A result such as information, outcomes, or documents. For example, a research project develops knowledge that can be used to determine whether or not a trend is present or if a new process will benefit society.

Uniqueness is an important characteristic of project deliverables. For example, thousands of office buildings have been developed, but each individual facility is

Figure 10.1   Project definition and planning phase. While there is a clear sequence, in reality, the process is at least partly iterative. As significant technical issues and uncertainties become clearer and as more investigation by the project team provides insight, it may be necessary to renegotiate the scope and risk sharing. It may even be necessary to revisit some of the planning once the project gets under way, as shown in Figure 10.3. Most of the major blocks in the diagram appear as subject headings in the text that follows.

unique – different owner, different design, different location, different contractors, and so on. The presence of repetitive elements does not change the fundamental uniqueness of the project work.

Do you have access to the planning documents for a previous, similar successful project?

Never start planning from scratch if you can adapt aspects of planning documents produced for earlier successful projects. Learn from others who have done that type of work before.

Negotiate and define the scope

This is the single most important job for a project manager: negotiating and reaching an agreement on defining the scope with the client and other stakeholders (see below). Some projects do not involve a paying client: they may be sponsored by someone with appropriate authority in your own firm, or may be a part of your education, in which case the sponsor is your lecturer. The sponsor instigates the project and can therefore be treated as equivalent to the client.

Mostly, especially in engineering projects, the sponsor or client has only a sketchy idea of the requirements. In the early phase of the project, therefore, your main role will be to find a way to meet the client's requirements and specify the necessary details.

Project scope is defined in one or more documents that provide a succinct definition of everything that is to be accomplished by a project.

While everything else can be progressively elaborated, be extremely cautious about elaborating anything in the project scope. This is called 'scope creep', a series of small extensions to the scope, each one being apparently insignificant. We will come to this issue again. Adding anything to the project scope almost invariably means increasing the cost and time to complete the project. Finalising the project scope definition right from the start is one of the most crucial skills of a project manager.

What is the project going to achieve? What are the project objectives?

Examples:

'*Improved access to our factory to reduce the time and cost of unloading incoming parts, materials, and supplies, while also reducing the cost of loading finished products for delivery to customers.*'
'*Remove hazardous waste material from one or more designated locations and safely dispose of the material using appropriate methods.*'
'*Design, construct, and evaluate the performance of, and customer attitudes to, a novel prototype machine.*'
'*Evaluate the feasibility of implementing a new automated classification algorithm in a computer vision inspection machine.*'

Answering this question, like many others in this chapter, requires a discovery learning performance. One or more discussions with the client or sponsor will help both parties learn about the objectives. Why is this needed for the client or sponsor who, presumably, must have thought about this earlier? Often, they have not actually clearly stated the project objectives in terms that make sense in an engineering context. Most of the time, they will not be engineers and will therefore need your help to do this.

How will you know that you have achieved these objectives? What are the acceptance criteria?

Write a carefully worded statement to explain how you will be able to confirm that all the project objectives have been completely achieved. Whose opinion will be most important if there is any disagreement?

Once again, this may be part of the discovery performance with the client or sponsor.

What will be delivered to the client, or what task will be performed for the client (tasks and deliverables)?

Write a list of every item that needs to be delivered to the client, and when it is needed. Several items will probably have to be delivered well before the end of the project, especially documents, such as progress reports.

What are the important attributes of project performance (the project drivers)?

Schedule, quality, cost? Which of these three is the main *driver* for this project? The project plan and budget will be strongly influenced by the need to:

a) keep to a particular time schedule,
b) maintain certain quality standards, or
c) minimise the initial capital expense.

You need to understand which of these drivers is most important for the client. These are the three basic drivers; project performance attributes and the cost implications of each one will be explained in more detail in Chapter 11. Another driver is the degree to which the client has defined the project objectives. If the objectives are only vaguely defined and can't be tied down until later, it can be very difficult to forecast the actual budget without a large allowance to account for cost estimate increases once the requirements become clearer at a later date.

What are all the main activities that need to be completed to achieve the project outcomes?

Later, we will create a work breakdown structure (WBS) which is a complete list of work activities to be performed by everyone concerned with the project. At this stage, it is only necessary to list the major activities that will be part of this. Remember to include documents, procedures (work needed to secure relevant government and regulatory approvals), and document checking by the client.

What will the client pay for?

It is easy to assume that the client will pay for all activities, but this may not actually be the case. The client may have assumed that certain activities will be paid for by other stakeholders, such as government agencies, or even the contractor.

What are the main activities required for the project that the client is not paying for and who is going to pay for them?

This is crucial: a project manager must know who is paying for all the activities needed to complete the project.

Are there any main activities that no one is going to pay for?

If there are, find a way to remove them completely from the project scope.

Note that the project scope does not have to specify every single activity in the project. The project scope is always written with the client in mind: put yourself in the position of the client and ask yourself the same questions again. Review your answers and, if necessary, reword them as the client would write them.

What is the time schedule for the deliverables or services to be performed?

- When does the client expect the outcomes of the project to be delivered?
- Where are the *deliverables* to be provided?
- In what form are they to be provided?
- If they are documents, will they be in electronic or printed form?
- If electronic, which formats are to be provided?
- Who is responsible for the indefinite storage of a reference copy of each of the documents? (This is particularly important for electronic documents.)

What is the project budget?

As I will explain in Chapter 11, there is nothing simple about estimating project costs and setting an appropriate budget. Cost is affected by other aspects of the requirements: the need to keep to a particular schedule and the need to maintain certain quality standards. There is also the definition of project cost: is the client talking about the total life-cycle cost, including costs incurred long after the project is completed, such as maintenance and operating expenses? Or is the client simply referring to project cost, and wants to minimise the expense during the project itself, regardless of costs incurred later as a result of doing that?

Expect that project management activities, such as defining the project, developing the project plan, and monitoring the project after the work has started will cost 2–4% of the total project budget.

Also, take into account that it may not be possible to accurately forecast the cost of the project until most of the plan is in place, and at least part of the detailed engineering design has been completed. Sometimes, when a project is repeating something very similar that has been done before, accurate estimates of the total cost may be possible based entirely on size parameters of the objectives, such as the length of a freeway given the number of traffic lanes or the number of drawings that need to be completed for a structural design project.

Apart from financial resources, what is the client providing for the project?

- Are there people who work for the client who have to be involved directly in the project?
- Is the client going to provide land, space, materials, equipment, information, or software?
- What are the expectations for this?
- When are these resources going to be provided?

Is there a possibility that these resources may not be available when promised?

- If so, how will equivalent resources be provided for the project when needed?

What is the contractual arrangement between the project team and the client?

- If there is no existing contractual arrangement, how will one be prepared and put in place?

How much is the client going to pay and when?

- What proportion of the total cost is to be paid in advance of any work being completed and accepted as complete?
- What precautions can be taken to minimise the risk that work that has been paid for in advance cannot be completed by the contractor(s)?
- If the client is going to make payments before the project is completed, what conditions have to be satisfied before the client makes these payments?

Practice concept 64: Specifications

The scope is not complete without specifying details of the artefacts, information, or particular services that will be provided for the client. Often, the specifications are separate documents referred to by the scope document. The specifications may be written by the project team and negotiated with the client early in the project but gradually refined throughout the course of the project. Alternatively, they may be provided at the start of the project as non-negotiable documents.

Often, the specifications are part of a contract defining the project, binding the enterprise delivering the project – remembering that we are using the term 'enterprise' to include the client, end users, and the organisations that actually perform the work.

The specification usually does not define every detail; instead, it provides sufficient technical information so that the deliverables will be fit for the intended purpose(s). Like the scope, it is often up to the project team to write the specifications and then reach an agreement with the client or sponsor.

The specifications will also refer to drawings, or, more likely, digital representations of the object to be created, encompassing most, if not all, aspects of the design.[10] These representations of the complete object also provide the details needed to construct the *bill of materials* and *work breakdown structure*, both of which will be described later.

So, the next logical question is: how is a specification structured as a written document?

For this, we need to understand the purpose of a specification: it defines what is acceptable and how we can determine whether the particular deliverable is acceptable or not.

Remember Practice Exercise 9, when you attempted to write an outline specification for a beam? A real specification has to include a definition that will allow an engineer to decide if the artefact, information, or service meets the requirements.

Although the client does not need to understand every detail contained in the specification documents, the people who are responsible for delivering the artefact, material, service, or information will need to understand. As an engineer, one of your primary responsibilities is to decide the level of detail required for the specification document in order for the client's requirements to be met at the lowest possible cost. Chapter 11 includes a case study that shows how a minor error in the level of detail included in a specification resulted in additional costs that amounted to millions of dollars for a single item of instrumentation.

There are two basic types of specification: test and method.

Test specification

The first type of specification is known as *test specification*: we describe the testing and inspection that will confirm that an objective has been achieved (in other words, the result complies with the specification) or has not been achieved (because the result does not comply with the specification).

For example, we might be required to construct a beam. The specified acceptance test might consist of a defined loading on the beam and measurement of its deflected shape. Hence, the specification might define:

a)   how the beam is to be positioned and supported,
b)   the applied load(s) and how the loads are to be applied (e.g. a testing machine, weights, attached steel ropes, etc.),
c)   allowable deflection limits in certain directions and rotational movements at given positions on the beam,
d)   the length of time for which the load is applied,*
e)   maximum and minimum temperature and relative humidity limits for testing,*
f)   storage conditions for the beam prior to test,* and
g)   the maximum allowable wind (if the test is done outside).

Items marked with an asterisk (*) are important for non-metallic structures, particularly fibre-reinforced composites, plastic materials, or natural materials, such as wood.

If possible, it is preferable to use standard tests so the results can be compared with previous tests. For example, the American Society for Testing and Materials (ASTM) is an international standards organisation that develops and publishes voluntary consensus technical standards that focus on testing methods; they publish a large catalogue of testing standards.

The weakness of a test specification is that the testing is only performed once, when the artefact, information, or service is provided. However, we might need an artefact that performs reliably for 30 years with only annual inspections and minor maintenance, such as repainting. We might require information that will be valid for at least five years without requiring an update. We might need a service that can be provided in the same way 30 years later with no reduction in the quality of the service received by the end user. How can we be confident that an artefact that passes an acceptance test on delivery will still pass when tested 20 or 30 years later?

One option is to incorporate what is known as *accelerated aging* into the test specification. An artefact intended for outdoor application in a coastal environment could be subjected to alternating high and low temperatures, high and low humidity, intense ultraviolet radiation, salt spray, and vibration, perhaps simultaneously. Degradation that might take a decade or more in the actual location might be reproduced with two weeks of testing by using accelerated aging. Obviously, the method used to accelerate degradation depends on the application. Electronic circuits might be aged with a combination of heat, humidity, dust, electromagnetic radiation, and vibration. However, building a test facility to provide the required ageing in a reproducible manner is not cheap. Civil, hydraulics, mining, and aeronautical engineers who work with large structures often have to rely on scale model testing to keep the cost reasonable.

This kind of testing, therefore, has limitations. Another kind of specification, method specification, provides an alternative approach that can help overcome this weakness.

Method specification

The second kind of specification is known as *method specification*: we describe the method, process, or procedure by which an objective is to be achieved. A method specification will often include detailed specification of the tools, materials, and monitoring of the production process.

A method specification gives us confidence that an artefact will perform as intended by ensuring that it is constructed in a known manner by suitably trained and experienced people, it is made from materials of known quality, and it is inspected during the manufacturing processes using standardised techniques.

Sometimes, a specification requires a combination of these two approaches. For example, a method specification may require certain intermediate acceptance tests to be performed during construction, manufacturing, or assembly processes.

One of the most difficult aspects of writing a specification is to consider how you can be confident that the product or process will perform as expected throughout its service lifespan. Acceptance tests and inspections can be performed only once, when the product or service is initially delivered, before it commences operations.

There is a common weakness in acceptance testing procedures defined in specification documents that requires careful thinking and writing to overcome. Along with specifying methods and tests, the specification must also define what is to happen if any of the inspections or tests reveal non-compliance. For example, the hydraulic controls unit shown in Figure 9.5 was subjected to a long sequence of acceptance tests before leaving the manufacturer's factory. When the tests were performed on the first units that came off the production line, some of the tests revealed failure conditions, often associated with the bundles of black cables visible at the top of the photograph. The test engineers requested that the cables be replaced before repeating the test that had produced a failed result. However, they did not return to the start of the test sequence: they simply repeated the failed test. If the unit passed, then they continued with the remaining parts of the prescribed test sequence. Many of these units subsequently failed within a few weeks of being installed at the end user's facilities, even though they were designed to have a service life of at least 30 years. The procurement engineers working for the project management organisation forgot to include a requirement to repeat all the tests after any repairs to rectify failures detected during the factory acceptance testing. The cause of the failures was unreliable cable connections. When the cables were replaced, new faults were introduced that were not detectable unless the entire testing sequence was started from the beginning.

Here are a few questions to elicit some further aspects of specifications:

- What is the predicted technical performance, safety, and environmental impact of the artefact to be created?

... or ... for information or services to be provided:

- What are the technical attributes, safety, and environmental impact of the service or information to be delivered as a result of the project (including those as a result of its application by end users)?
- How can these be described in a meaningful way for the client?
- How can these be described in a meaningful way for the people who will create them?

For example, a client may want electrical cabling to be removed and replaced in order to upgrade the power handling capacity of the cabling.

To do this effectively, it will be necessary to specify many details that the client is not particularly interested in. Even if the client has not specified a particular standard for the cabling, the project team will almost certainly specify a national (or international) standard. There are many minor technical details that can be very time-consuming and costly to specify in detail. For example, we could specify every joint, bend, and junction in the conduit (tubing) that will enclose the cables after they have been installed. Alternatively, we could elect to specify AS3000, the Australian wiring standard. A contractor responsible for installing the cabling will then be able to work out all the details required for the conduit, because this particular standard provides all the necessary guidance, and you will be selecting a contractor who is already familiar with it.

What needs to be specified for this project?

This may require even more discovery performances in the form of discussions with the client or project sponsor, as well as other stakeholders, such as the engineers or contractors who will participate in the project. It is possible for a project to be over-specified, which will entail unnecessary, extra costs.

Which codes and standards will be invoked to guide the performance of technical work during the project?

While the requirement to follow particular codes and standards would normally be stated in a method specification, most test specifications also call for compliance with similar codes and standards. These can be a combination of local, national, company, industry, and international standards.

Testing and inspections

Who will be responsible for performing any tests or inspections?

Tests and inspections will confirm that the project objectives have been met and the deliverables comply with the client's expectations. However, the supplier has an interest in passing the specified tests and inspections, so if these tests and inspections are performed by staff from the supplying organisation, there will be a tendency for them to portray the results in the most positive way possible. In the same way, engineers working for the procurement or project management organisation want to avoid, at all costs, a situation in which they accept equipment or services only to find that later, there are defects that they overlooked. Therefore, they may tend to interpret any test or inspection result in the most negative way possible, adding to the costs sustained by the supplier and possibly leading to contract disputes. It is also possible (and this has happened regularly) that the engineers working for the procurement or project management organisation have received inducements from the supplier organisation to adopt a lenient or generous opinion when interpreting tests and inspection results.

For these reasons, it is not unusual for acceptance tests and inspections to be performed by a neutral third party, one without an interest in portraying the results in any particular way.

- Is the person performing the acceptance tests or inspections a genuinely independent third party, someone who is employed neither by the client nor by the project team? In other words, is this a person who cannot readily be influenced by either the client or the project team?
- Has the involvement of this person been agreed on by the project team and the client, such that both will accept that person's judgement and evaluation?
- How can we describe the level of quality that the client expects in a way that is meaningful to the client?
- How do we describe this level of quality in a way that is meaningful for the people who will be performing the work?
- How can we measure or evaluate quality during the project, as well as at the conclusion?

Intellectual property (IP)

- Discuss any new intellectual property that may be or will be generated as a result of the project.
- To whom will this new intellectual property belong when the project is completed?
- How will any new IP that is generated be protected, and who will pay for this protection?

Keeping the client informed

What does the client expect to be told during the project?

- How are we going to keep the client informed about progress related to the project objectives?
- What happens if there are unexpected setbacks or unexpected opportunities?
- How and when do we inform the client about these?

There is always the possibility that a project may not succeed.

- How do we describe the client's expectations about the risk that the project might not succeed?

What documentation does the client expect to receive?

- When does the client expect each document to be provided?

Checking and reviewing the scope

Once you have answers to all these questions, you will have nearly completed your scoping document.

Provide your document to the client and ask the client to review the document carefully and provide feedback and suggested changes. This is yet another discovery performance: expect the subsequent discussion to provide plenty of learning opportunities.

Do not expect the client to send you detailed comments and suggestions for changes. This might happen, but often, it will not occur. Instead, be prepared to meet face-to-face with the client's representative and go through the scoping document line-by-line and sentence-by-sentence to ensure approval that the client is comfortable with everything written in it.

Write a list of all the changes that you have agreed upon with the client. Incorporate the changes and send the revised document to the client with a further request to check it carefully.

At this stage in many projects, there may be a renegotiation on the budget for the project, and also on the acceptance criteria: how you and the client will know that the project objectives have been achieved.

Close alignment usually requires extensive discussions to clarify differences of interpretation and to gain insight into the other participants' understanding of the objectives. A well-written scoping document usually only emerges after these types of extensive discussions.[11]

Stakeholders

Write a list of all the individuals, organisations, or other groups of people that are:

- actively contributing to the project,
- whose interests may be affected as a result of project execution or completion, or

- who may also exert influence over the project objectives and outcomes, as well as the ability of the team to achieve them, should they choose to do so.

Each of these people or groups of people is considered a *stakeholder*.
Examples of stakeholders:

- *Yourself – project manager or engineer in charge*
- *Client or project sponsor*
- *Project team members*
- *Competing engineering firms*
- *Contractors and unions*
- *Local community*
- *Media*
- *Regulators (in government departments)*
- *Suppliers: companies supplying materials, components, and hiring equipment*
- *Local government authority*
- *Social and environmental interest groups, NGOs (non-governmental organisations)*
- *Customer/user – note the possible multiple layers of customers, for example, a pharmaceutical product is prescribed by doctors and taken by patients who may be compensated by insurers. The customer may be the person paying for the project and the users may be separate people who use one or more products.*

Which of these 'stakeholders' is most affected by, or could exert the greatest influence over, the success of the project? (Key stakeholders) Identify the interests, values, requirements, and expectations of each stakeholder

In Chapter 8, we discussed values and interests in the context of motivation for teaching performances. Use the same concepts and examples here. Expectations are outcomes that are anticipated by a stakeholder, but which may not have been stated explicitly. For example, engineers assigned to an advanced technology project might expect to be involved with technology development. However, if they instead find themselves performing routine administration work in which the advanced technology plays no part, and the engineers require no understanding of the technology to perform their work, then they will perceive that their expectations have not been met. As another example, most clients expect to be informed immediately about any developments that significantly affect the likely project outcomes. They expect not to encounter nasty surprises in project review meetings.

How can you identify the interests, values, requirements, and expectations of each stakeholder? Express these in the language used by the particular stakeholder

Answering these questions requires discussions with each of the stakeholders. These discussions are further examples of discovery learning performances. It is crucial that you prepare accurate notes so that the statements of values and interests are constructed in the specific language of each stakeholder. It can be handy to include actual quotations so that other stakeholders can better appreciate the context of these words.

Examples of interests, values, requirements, and expectations:

- *Companies supplying materials, components, and hiring equipment*

    - *Interests: attributes of project outcomes that provide measurable benefits or penalties, e.g. i) business cash flow, ii) profit, iii) opportunities for publicity . . . 'Our product is being used for project X', and iv) opportunity to learn about new industry techniques.*
    - *Values: attributes of the business that owners wish to emphasise or identify themselves with, e.g. i) profitable business, ii) good customer service experience, iii) long-term business relationships, iv) ethical practice, v) development of local community, and vi) social justice or help for less fortunate people.*
    - *Requirements: i) payment within 28 days or the application of interest on outstanding accounts and ii) customers demonstrate an ethical record of doing business in the past.*
    - *Expectations: i) reasonable notice of delivery deadlines, ii) customer applies, stores, and uses products within the guidelines provided by a supplier, and iii) customer ensures products are only used by staff with appropriate training, protective equipment, and who maintain appropriate safety precautions.*

- *Regulators (government departments)*

    - *Interests: i) orderly conduct of business operations complying with both the intent and details of regulations, providing maximum community benefits and minimum harm to community interests, ii) good relationships with operating companies, iii) avoiding unnecessary disputes on regulations and time-consuming investigations, and iv) opportunities to learn about new industries and processes.*
    - *Values: i) maximising overall community benefits, ii) supportive environment for businesses, iii) long-term business relationships, and iv) ethical practices.*
    - *Requirements: i) businesses apply for permits and approvals well before they commence operations that require such approvals, and ii) applications for approvals comply with the state requirements and procedures.*
    - *Expectations: i) businesses demonstrate a commitment to not only meet, but also exceed regulatory requirements, ii) businesses maintain adequate records of operations to demonstrate both practical compliance and the intent to comply with operating requirements, and iii) business are open and provide immediate access to information and their operations following reasonable notice and requests.*

Explain the 'value proposition', defined as how the project outcomes add value or provide measurable benefits to the client and other stakeholders, as well as any costs or penalties

This question is crucial because the responses from discussions with stakeholders provide the essential understanding necessary to handle risks and opportunities: unpredictable events that will influence the project outcome.

How can the risks and opportunities (time schedule, financial, regulatory, community, environmental, health and safety, political) be shared amongst stakeholders?

Uncertainty cannot be eliminated, but it can certainly be managed, as we shall see in the later section on risk management. We can plan the project in a way that reduces the likelihood and consequences of risks that can negatively affect stakeholder interests. At the same time, we can also plan in a way that is likely to maximise benefits and opportunities that arise from unpredictable events.

Few engineering firms, and certainly not individual engineers, have the necessary financial capacity to be able to pay for the consequences when a project overruns its budget or runs late, thereby incurring financial penalties. Therefore, it is critical to negotiate the sharing of these risks so that when unfortunate events happen, stakeholders who are accepting the financial and other consequences are well-prepared. Naturally, these same stakeholders would normally be the ones receiving the greatest benefits when the project succeeds.

Understanding the ability of the project sponsor or client, in particular, to accept financial risks makes a big difference in the way that the project is planned.

For example, consider a project to develop a highly profitable product that can easily be copied by competitors. The project sponsor hopes to capture the market quickly and sell enough to make a reasonable profit. However, if the project runs late, there is a much greater risk that competitors will enter the market first. Therefore, the project sponsor can probably tolerate a significant budget overrun, provided that the product is ready on time. Therefore, the most important priority in the project will be to maintain the planned project schedule.

Alternatively, consider a project to refurbish a shopping centre in a mature property market. The shopping centre will have a long life with predictably secure financial returns from shop owners who will pay rent on long-term lease agreements. However, being a relatively low-risk investment, the financial return will also be relatively low, so it will be very important to restrict any budget increases as much as possible. On the other hand, maintaining the project schedule would be significantly less important.

In the former case, the stakeholders will be more prepared to accept a budget increase, whereas in the latter case, the stakeholders are more likely to accept late completion in order to avoid a budget overrun.

The most important aspect here is to discuss the risks with the stakeholders who are most affected to make sure that you have their support and confidence.

Chapter 11 provides a much more extensive discussion on engineering investment decisions.

What are the best locations for stakeholder discussions?

Stakeholders are more likely to feel respected if discussions take place at or near their 'home territory'.

How do we best maintain a relationship and communicate effectively with each stakeholder?

- What is the most effective communication channel: face-to-face, telephone, e-mail, documents, news media, or through community groups?

What can you and the project team do to influence the perceptions of each of the stakeholders?

Creating a positive perception would mean that each stakeholder sees the project outcome as providing benefits, and therefore contributes to the success of the project.[12]

Are there stakeholders who cannot be adequately represented at discussions about the project?

This might be the case because they are too far away, cannot communicate effectively, or are not yet present, as in the case of future generations.

The environment can be included as a stakeholder without a voice, although there will often be vocal groups in the community who will claim to represent environmental interests.

Ethical issues also present themselves in this context. With the knowledge and expertise we have as engineers, we are often in a privileged position to be able to anticipate situations where the benefits of a project may be shared unfairly. Professional ethics demand that we safeguard the interests of people, and the environment, that will be affected by any project but cannot be adequately represented to present their point of view when project decisions are made.

What will be the effects of the project work on the people who are involved in the project?

The effects occur either because they are employed to work on the project or because they live nearby, meaning that they will be close enough to experience the effects of the project.

What kind of meetings or negotiations will be needed to align the project objectives with the interests and values of all the stakeholders?

This is the sustainability agenda; we will learn more about this in Chapter 12.

What is the best way for each of the stakeholders to have some means of influencing project decisions during the course of the project, should they feel that to be necessary?

What special knowledge could stakeholders contribute to help make the project be more successful?

The work breakdown structure

Do you have access to the work breakdown structure for a previous, similar project?

Never start building a work breakdown structure from scratch if you can adapt aspects of documents produced for earlier, similar projects. Learn from others who have done that type of work before.

Write a comprehensive list of all the physical assets to be created and the work activities that need to be performed

The reference point for the work breakdown structure is, most likely, a digital CAD model of the objects to be created or services to be provided.[13] This data provides a detailed list of components that enable the definition of all activities and procurement that have to be completed in order to achieve the project objectives, as defined in the project scoping document. Include activities to be performed by all contributors, not just the project team. Include the review of documents by the client and processing of approvals by regulators. Use the responses to the questions in the scoping document as a starting point.

Each activity should normally take at least one day to complete. Activities that last more than 20 days (four working weeks) should be subdivided into shorter activities. Otherwise, it can be difficult to monitor progress on the project. We will return to this issue later.

Specify the location for each activity and physical asset. Almost always, work conducted at two or more separate locations should be included as separate activities, with at least one activity in each of the respective locations.

When calculating costs, distinguish between the direct cost of physical assets and the indirect cost of activities needed to get them in place.

Devise a way to group activities to minimise complex interactions

Do your best to arrange these activity items so that the list is organised and the activities are arranged in groups that make sense in the context of the project.

For example, activities might be arranged in groups according to:

- obtaining approvals and permits,
- the technical expertise necessary to perform them,
- the contractor who will perform them,
- the location at which they will be performed, or
- the phase of the project: early planning, project definition and agreement, preliminary design, detailed design, procurement, manufacture, commissioning, operating, or decommissioning.

Normally, a good project manager will group the activities to minimise the complexity of interaction between the work that needs to be performed in each group. Thus, as far as possible, difficulties experienced in one part of the project defined by a group of work activities will have the least possible effect on other parts of the project.

When you are evaluating the project, identify every activity in the WBS that was added to the list after the initial planning stage. The fewer activities added later, the better the quality of the initial planning and the WBS.

How long will each activity take to complete, and what will it take?

For each activity, estimate the duration and what will be needed to complete it in terms of human effort (hours of work needed), financial expenditure, and material resources. Also, identify other preliminary activities that need to be completed (either fully or partially) before each activity can be started (activity constraints).

If these estimates are uncertain, estimate the least likely value and the greatest likely value as well.

Prepare a bill of materials listing every item needed for the project and how it is to be obtained, transported, and stored

- Obtain Material Safety Data Sheets (MSDS) for every material that might be considered hazardous in any circumstance.
- Create a management system to ensure that every person handling hazardous materials has received appropriate training and wears appropriate protective equipment when handling the materials.

Arrange procurement, suppliers, and contractors

For every item that needs to be purchased or obtained from another organisation, either purchased or leased, identify possible suppliers or equipment hire companies. With their help, determine delivery lead times: how far in advance do the necessary purchase or procurement contracts have to be completed in order for the items to be available when they are needed?

Similarly, for each contract for supplying services to the project, identify a reasonable number of possible contractors who will be able to supply the services at the time they are needed. For each contractor, estimate the cost of the services and any support requirements that have to be provided by the project organisation (such as office space, desks, parking, special tools and equipment, and training).

Define the process by which the project organisation will choose which suppliers will provide the procured items and which contractors will provide the services. Make sure that the project plan includes activities for the project team to define this process, as well as to perform the necessary selection work and any negotiations that might be required. Also, consider prequalification: suppliers need to demonstrate that they can achieve certain standards before even being considered for selection.

Who will be responsible for monitoring each activity?

Wherever possible, assign a project team member who will be personally responsible for monitoring progress and making sure that every activity is successfully completed.

How would someone know that the activity is completed (are there deliverables or intermediate milestones)?

How long will the project take to complete and what is the likely cost?

Based only on the constraints of each activity, determine the most likely schedule for the activities required to complete the project. There are standard techniques for this, which are described in detail in project management texts and are available in software packages, such as Primavera, Microsoft Project, Libreplan, and many others.[14] Even Excel can be used with appropriate macro programming.

Using information about procurement items and services to be supplied by contractors, estimate the cash flow and payments required for each month (or a shorter period) of the project.

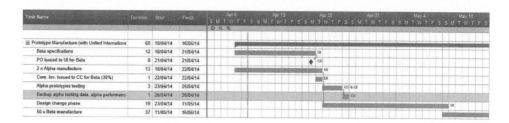

Figure 10.2 Small part of a project Gantt chart.

There are standard ways to represent the activity schedule in diagrams. The most common (and in many ways, the most limited in usefulness) is the Gantt chart, which displays each activity as a horizontal bar with a time scale along the horizontal axis. Wikipedia has a wealth of reference material on these techniques.

Many project management texts focus considerable attention on calculating the likely time schedule, human effort requirements, and cash flow. However, this aspect of project management is almost routine and can be easily arranged with one of the many software packages that are now available. The difficult part is answering all the questions that we have discussed in this chapter: the responses lead to the data that you need for the software packages.

Project locations

Where is the project work going to be performed?

- How much space is required?
- Have members of the project team visited the site and surveyed the available space?
- What access is available to transport supplies and equipment to the location? How is this access likely to be affected by weather conditions, especially heavy rain or flooding?
- What is the maximum vehicle length that can access the site?
- What is the maximum weight of a vehicle that can access the site? What is the load capacity of the access route? What is the clearance, the maximum load size that can be accommodated by wheeled transport? Are there power lines that have to be lifted when large loads need to enter or leave the site?

> The Swedish government donated a 70-tonne demining machine to the Cambodian government in the 1990s. At the time, Cambodia had a major catastrophe with landmines left over from the conflict following the Pol Pot regime. Unfortunately, the Swedish authorities discovered that the bridge linking the port where the machine was unloaded with the rest of the country could only support a load of 50 tonnes. The machine had to be shipped back to Sweden until the bridge could be strengthened.

- Who owns the land, not only where the work is to be performed, but also the land immediately adjacent to the site?

- Is the land, or land in the immediate area (that may need to be accessed for transport to and from the location), subject to a contested claim or perhaps is it of special significance to indigenous or tribal people?
- Are there any significant resource constraints on the project, for example, limitations on the amount of water or land available for the project?
- If some activities are located in a foreign country, how will logistics, customs clearance, import permits, and dangerous goods be handled?
- How long will it take to obtain each of the approvals and permits for the project?

What kind of approvals will be needed for different kinds of operations at the site, as well as for transport movements to and from the site?

- What restrictions are in place that could affect operations at the site, including transport to and from the site?
- What are the weather conditions at different times of the year at this location?
- Where will the water required for the project (and for ongoing use) come from? What is the quality of the water? Is a separate drinking water (potable water) supply needed?

What is the sustainability of water and energy supplies in the area?

- How will used water and rainwater be disposed of? Will any of the water be contaminated, thereby requiring special holding ponds, treatment, or special disposal methods?
- How will sewage be handled at the site? How many people will be living on-site and do the existing sewage disposal facilities have enough capacity?
- If the location is subject to a wet season or occasional downpours, how will unusually heavy rain be handled? Is there a possibility of rain entering the site because of flooding, leading to the site becoming contaminated and the water needing to be treated or prevented from leaving the site? This is particularly relevant for mining operations in the northern part of Australia.
- What is the status of electric power supplies at the site? How much power will be needed for lighting, heating, cooling, site operations, and construction? Remember to include the power required for offices and living accommodation at the site.
- If additional power is required at the site, how will it be supplied? Is solar power an option? Can the site's power consumption be varied to suit a renewable source, such as wind or solar power? If a diesel backup generator is needed, how will the fuel be transported and stored? How will the fuel storage be protected to ensure that there is no contamination from leaking fuel?
- What kind of security fencing and access control measures are needed at the site?
- How will you communicate with people at the site? Does the existing communication infrastructure have sufficient bandwidth and how would you know? Will you need additional satellite communication capacity? Can an optical fibre be laid at the site, and if so, from where and at what cost? Are there any restrictions on electromagnetic radiation at the site (especially if it is in an area where there are radio telescopes or other sensitive instrumentation)?
- How will rubbish be collected and removed from the site?

- What kind of accommodation will be needed for people at the site? Will they need to work shifts? Will the workers fly in and fly out? Where is the nearest airfield to enable that to happen? What will be the transit time for workers travelling to and from the site? What kind of advance reservations will be needed to make sure that the necessary people can travel to and from the site when needed, possibly on short notice?
- How will work at the site be monitored and inspected when required? What will be the cost of monitoring the work and performing the necessary inspections?
- What kinds of excavation or drilling are required at the site?
- List all the approvals and permissions needed for operations at the site.
- What kinds of safety precautions will be needed at the site?
- Who will have access to the site and under what conditions? Will visitors to the site require special safety inductions, and will they need to be escorted?

How will medical services be provided at the site?

- Where is the closest hospital that can handle serious medical issues?
- How will people be transported to the hospital in case serious medical treatment is needed?
- What kind of training will be needed for staff at the site to make sure that adequate first aid can be provided?
- What kinds of health issues exist at the site for people or animals? What kinds of disease could people or animals be exposed to? What kinds of immunisation will be needed?
- What storage facilities will be required at the site?
- How will security at the site be maintained, both for materials and people?
- What kinds of recreation facilities are available close to the site?

Practice concept 65: Risk assessment and management: working with uncertainty

Identify risks and opportunities

As explained at the start, engineering work is intrinsically unpredictable to a certain extent because so much of it depends on interpretations and subsequent actions and performances by individual people. Each person involved in engineering projects may be involved in up to 60 or more simultaneous, overlapping activities, many of them not even on the same project. The attention needed for each of these activities can interfere with the work on all the others. Many, if not most, people in engineering organisations have a great deal of autonomy in deciding what they do each day and when they choose to start work on any particular activity for a given project. People can easily forget activities to which they are supposed to contribute, partly because they have so many disparate responsibilities.

There are other sources of uncertainty that can influence any project. Broadly, we can classify these as 'opportunities', which are unpredictable events that bring potentially beneficial consequences for a project, and 'risks' or 'hazards', which generally connote negative or undesirable consequences for a project.

Technically, 'risk' can be associated with a potentially advantageous event just as much as a potentially undesirable one. However, most people instinctively associate 'risk' with undesirable consequences. Therefore, it is probably more appropriate to use the term 'opportunities' to denote unpredictable events with potentially beneficial consequences.

Particularly in an engineering context, it is essential to create a broad understanding of the risks involved in any venture. It is particularly important for the engineering contractor to ensure that the client understands the risks and that appropriate allowances are made for them. This provides a way to share the risk, rather than expecting either side to carry the full liability.

Risk management has been executed successfully if there are no surprises along the way.

What are the unpredictable events that could affect project outcomes?

For each group of activities in the WBS, we need to identify intrinsically unpredictable events with consequences that could influence the outcomes, the time, or the resources needed to complete the activity group.[15]

Each event with potentially undesirable consequences can be listed as a risk. Each event with potentially desirable consequences can be listed as an opportunity.

Try to avoid grouping distinctly different events with different consequences or impacts on the outcomes. Note that this is not easy to do in practice.

Who can identify risks?

Very often, as engineers, we are not the best people to identify risks and opportunities. Perhaps it is because we tend to think logically and are often better at convergent, linear thinking rather than lateral thinking. Other people, even people without any technical background, can be better at predicting the unpredictable. For example, local community members can often be better at predicting events with the potential to harm the environment. The best strategy is to seek a diverse range of people to participate in risk identification.

Until around the year 2000, expert risk management consultancies were called upon for their specialist knowledge to identify the key unpredictable events that could affect project outcomes. However, research and public experience suggested that expert risk assessment was often wrong. For example, consider the UK-France Channel Tunnel opened in 1995. From the earliest stages of design, the possibility of a fire in the tunnel was taken very seriously. Designers went to enormous trouble to reduce the risk of fire and experts were happy that the risk had been reduced almost to zero by the time the tunnel was opened. However, within 12 months of opening, a serious fire broke out in the tunnel, causing several fatalities and closing the tunnel for several weeks. The subsequent loss of business was catastrophic for the tunnel consortium; the British and French governments had to rescue the project.[16]

There is increasing research evidence that organisations are less capable of assessing risks to their operations than we might expect. They are insufficiently able to learn from their own experience and find learning from the experience of others even more difficult. We do not understand the reasons for this, but the evidence is impressive.

Ultimately, the general public carry the cost of engineering failures. The Piper Alpha platform disaster cost Occidental (the platform operator) about 1/10 of the total loss sustained by the British Government and the wider public. Therefore, the recommended practice is to involve diverse stakeholders, particularly the local community, in wide public consultation on the assessment of risks. Researchers now recognise that non-experts can provide information that matches or exceeds the validity of that provided by internal experts. This is counter-intuitive, given the wide variation between subjective and objective perceptions of risk in ordinary people and experts. One would expect that objective risk analysis would always provide a better result than subjective assessment. Unfortunately, experience indicates that this is not the case; a careful combination of the two provides better results.

Needless to say, effective listening skills are essential for this collaborative discussion. Listening is all the more difficult for engineers who have devoted so much effort towards their favoured solution, especially if there is a risk that the solution might have to change to account for unforeseen risk factors.

Risk analysis

There are several techniques available in the literature.[17] Most project risk assessment uses qualitative descriptors, rather than quantitative probability analysis. Experience has shown that this approach works better for nearly all projects because there is seldom a sufficient amount of trustworthy data to perform a quantitative analysis.

Most organisations adopt a 'risk control' approach: risks (or opportunities) from unpredictable events are carefully and comprehensively defined, categorised, prioritised, and where possible, influenced by introducing control measures to improve likely outcomes. This approach is very useful in the context of engineering management, particularly for project management of a major contract. Before this approach was adopted, unpredictable factors were accounted for by adding a fixed contingency sum to the price of a contract, for example, 10% of the total costs. The percentage was decided on the basis of experience. Given that many engineering contracts are highly competitive and are priced with very small profit margins, mistakes in costing or judging the contingency allowance can make the difference between profit and loss. Replacing a fixed percentage contingency with a risk-based allowance can provide benefits for both the contractor and the client. Often, the client and contractor will agree to share different proportions of the risk for a given contract. For example, the client may agree to cover risks arising from unknowable factors, such as the nature of the ground deep under a construction site, whereas the contractor would normally agree to cover risks over which they can influence some control, such as late delivery of materials or equipment.

The engineering approach to risk management has been further refined in the field of reliability and maintenance engineering. In recent years, computers have enabled the use of quantitative methods for predicting the likelihood of failure. Complex computer models have been used on inventories of plant equipment listing, for example, every pipe valve and fitting in a process plant, together with the inventory of chemicals that they carry. The computer can readily identify hazardous combinations of materials and chemicals and predict inspection times to detect corrosion or wear that will be early enough to prevent a serious failure. Like other computer methods, this approach relies on comprehensive data and accurate modelling. There is still considerable debate on the effectiveness of these methods: whether the savings in maintenance actually outweigh the costs of establishing the database, particularly in the case of an older plant for which construction data is not necessarily available. However, these methods are valuable for designing a complex plant to improve the overall reliability.

What is the likelihood of each of the unpredictable events?

We classify the probability of each event using a qualitative scale:

A: Highly likely within the project timescale, almost certain to occur
B: Moderate probability, probably will occur once or twice within the project timescale
C: Low probability, has been known to occur, but not on most projects
D: Remote possibility, has occurred on rare occasions

What are the consequences of each of the unpredictable events?

What are the classes of risk and opportunity consequences?

We classify the severity of the consequences (+ or −) into a qualitative scale:

1:   Minor: relatively inconsequential
2:   Medium: significant, but a small part of project value
3:   Serious: significant proportion of project value
4:   Major: several times greater than the project value
5:   Catastrophic: very large compared with project value

Any unpredictable event can have different classes of consequences:

Reputation: the consequences can affect the reputation of the project team, as well as the clients and maybe other stakeholders as well. Reputation affects an organisation's ability to acquire financing for projects.
Schedule: the consequences will affect the time needed to complete the project.
Health and safety: the consequences can cause harmful effects to people working on the project, as well as end users, outsiders, bystanders, spectators, and community members. The consequences can be immediate or long-term; the latter is particularly relevant in the case of nuclear radiation or the release of contaminants into the environment.
Social, regulatory, and governance: the consequences can affect perceptions by members of the community that can consequently influence the actions of governments and agencies, such as regulators.
Financial: the consequences can affect the cost of the project or the level of benefits obtained from the project. The consequences can impact stakeholders unequally: some may be affected much more than others.

Tables 10.1 and 10.2 show examples of the criteria used to evaluate the consequences of undesirable risks for different companies involved in different industries.

Table 10.1 Sample risk consequences table for information systems in a health facility.

	Health	Operations	Community reputation	Financial impact	Compliance
Catastrophic (5)	Data error causing incorrect medication dose, patient has permanent health consequences	Incorrect code causes a system failure or loss of data affecting patient records	National current affairs investigation by newspaper or TV network	Corrective action causes operating loss for the year, permanent effect on financial return	Incident reported, restrictions on operating licence, temporary or permanent closure
Major (4)	Data error causing incorrect medication dose, health consequences, but patient recovers	Incorrect code causes system failure for two hours or less	Adverse newspaper article in national newspaper or on the front page of the local newspaper or TV news bulletin	Corrective action costs less than $1 million, ongoing financial impact significantly affects profit	Incident reported, health regulator conducts investigation, no ongoing effect on operating licence
Serious (3)	Data error causing incorrect medication dose, no health consequences	Incorrect code prevents system from being operated for 30 minutes or less	Single adverse article in local newspaper, not on front page, or multiple letters of complaint on a similar issue	Corrective action costs less than $100,000, ongoing impact less than 0.1% turnover	Incident reported with consequent follow-up by health regulator, no adverse consequences
Medium (2)	Systematic data error that medical staff can recognise and correct	Incorrect algorithm causes systematic errors that do not impact patient health	Patient or relative writes formal letter of complaint with a copy to government organisation	Corrective action costs less than $10,000, no ongoing financial impact	Incident needs to be reported to regulator, no follow-up required
Minor (1)	Isolated data error that medical staff can recognise and correct	Incorrect data causes incorrect billing for one or two patients	Single telephone complaint from patient or relative	Corrective action costs less than $1000, no ongoing financial impact	Incident does not need to be reported

Table 10.2 Sample risk consequences table for a chemical handling facility (financial not included).

	Health	Safety	Environment	Community reputation	Compliance
Catastrophic (5)	Debilitating lung condition to multiple people	Multiple fatalities or major third-degree burns to multiple workers	Contamination of local waterways of environmental significance	National current affairs TV program about chemical handling issues	Prosecution with the threat of losing operating licence
Major (4)	Permanent lung damage impacting standard of life	Totality or third-degree facial burns	Groundwater contamination	Repeated letters of complaint in regional newspapers affecting credibility	Licence breach with prosecution from government agency, significant fine, operating restrictions
Serious (3)	Bronchitis or breathing difficulties with hospitalisation	Chemical burns with hospitalisation resulting in lost time injury	Release into catchment pond	Group of neighbours request a meeting to discuss spill and response	Improvement notice received from environment agency
Medium (2)	Nausea and vomiting with local treatment	Minor chemical burns requiring medical treatment	Contained chemical spill immediately recovered	Neighbour writes formal letter complaining about spill	Incorrect incident reporting leads to late response
Minor (1)	Inhalation causing coughing	Skin and eye irritation	Chemical spill does not require specific recovery actions	Single phone call by neighbour to complain about chemical spill	Spill contained within licenced area

A single event can result in consequences with different ratings in the different classes. For example, failure to perform maintenance on a pump might lead to an unscheduled plant shutdown necessitated by a hazardous chemical leak arising from the failure of the pump seals.

The amount of leakage may be very small, perhaps detectable only by monitoring instruments within the plant itself.

The safety consequences are either none at all or minor, with no threats to the safety or health of any people.

The reputation consequences are negligible because the incident does not get reported to anyone outside the plant.

The social, regulatory, and governance consequences are also negligible because the incident does not get reported to anyone outside the plant, with the possible exception of environmental regulators, to whom the leakage may have to be reported.

However, an unscheduled shutdown of the entire plant just to replace leaking seals on a pump may have major financial consequences, particularly if the product from the plant has to be delivered on time to an important customer. Thus, the economic consequences of the seal breakdown are significant.

In evaluating the risk, therefore, we need to consider the greatest level of consequence in the different classes in order to decide the relative priority of dealing with this risk.

Given the likelihood and consequences, what is the risk (or opportunity)?

The normal approach is to use a lookup table to classify the risks according to the probability and consequence ratings for each unpredictable event.

L: Low
M: Medium
H: High
E: Extreme

Table 10.3 provides an example of such a look-up table, usually known as a 'risk matrix'. A project team will choose an appropriate risk matrix for their analysis. Most organisations adopt a standard risk matrix.

Table 10.3 Example of a risk matrix.

Consequence/ Likelihood	1: Minor	2: Small	3: Medium	4: Large	5: Extreme
A: Highly likely	L	M	H	E	E
B: Moderate	L	L	M	H	E
C: Unlikely	L	L	L	M	H
D: Remote possibility	L	L	L	L	M

What are the most important risks and opportunities to focus on?

The lookup table provides a simple measuring tool for the importance of each risk or opportunity. Usually, a project team will evaluate between 100 and 300 unpredictable events, but often far more on a larger project. The risk register, a spreadsheet containing details of all the unpredictable events, including their likelihood and consequences, is a critically important document for the project.

The simple classification of risks into four categories allows the project team to focus on the most important ones first. Usually, low and medium risks are disregarded or are handled with a single contingency fund.

How can we treat the important risks?

Finally, the risk treatment options are examined for every risk in decreasing order of severity. In any real project, the number of risks becomes very large, so this is a tedious process requiring attention to detail. Obviously, risks categorised as L, and possibly M, may not be worth considering for treatment depending on the number of high or extreme risks that need to be considered. The options are listed below. Sometimes, a combination of options is chosen for a particular risk. The costs for each treatment need to be clearly specified.

Option 1: Do nothing

We accept the risk, although we also possibly make a financial provision for it. This is often called a 'contingency'. In a contract, there might be a 'contingency amount' set aside to cover unexpected events, such as delays caused by bad weather, delays in the arrival of materials outside the control of the supplier, etc. The contingency amount might be 5% or 10% of the total contract value. Often, an external person (arbitrator or commercial facilitator) will be called in to decide whether the contingency amount can be used to cover the consequences of an event if the stakeholders cannot agree. A contract agreement might also share the benefits of achieving a lower than expected completion cost between the contractors and the client.

Option 2: Reduce the probability of the risk

We can sometimes do this through implementing technical precautions, using different materials, providing safety barriers, and utilising instrumentation to provide early warning, as well as sensors to detect when people are likely to take unnecessary risks.

We can also reduce the possibility of unpredictable events by obtaining more data, carrying out more detailed surveys, and performing more analysis. However, this comes at a significant cost.

There are many other options for reducing the likelihood of an event that arises from human behaviour. Most of these provide both productivity improvements and risk reduction.

- Planning and organisation
- Designing simple, effective procedures

- Checklists to reduce the risk of forgetting items
- Adopting standards, following standard procedures, and complying with relevant codes
- Training and practising with staff in using procedures
- Supervision and frequent reminders to reduce 'risk-taking behaviour'
- Checking and review by other people
- Configuration management – accurate documentation
- Coordination and monitoring
- Teamwork, backup plans, and redundant equipment
- Regular testing and inspection, with frequency in proportion to the risk (risk-based inspection)
- More frequent and diverse data backups to reduce the probability of data loss
- Increased frequency of progress payments by client to reduce the chances that engineering work will be performed when the client cannot pay

Option 3: Reduce the consequences

We can use personal protective equipment (PPE), such as helmets and eye and ear protection.

We can include protection or safety measures in the design, or use energy-absorbing materials, etc.

Other examples of undesirable consequence reduction measures include:

- Creating a low dam wall (or bund) and sealing the ground around a storage tank to reduce the consequences of a spill from a tank: the fluid is retained in the area immediately around the tank and does not leak into the ground.
- Providing automatic fire extinguishers to reduce intensity or extinguish an accidental fire completely.
- Arranging for payment in advance before undertaking work of equivalent value: if the client cannot pay, then work will not be started.

There is a possibility that a particular manufacturing process will result in a significant proportion of defective parts.

One way to control this risk is to anticipate that, for example, up to 10% of the parts will be defective; therefore, order 10% more parts than are needed, so at least the required number are likely to be available.

Another complimentary way to control this risk is to provide for defective parts to be repaired once they have been identified by quality control inspection. The latter control method requires an alteration to the project activity sequence: additional activities need to be inserted.

Of course, in the case of opportunities, we want to increase the consequential benefits, such as a potentially valuable discovery or innovation:

- Ensure that innovations that prove to be effective are rapidly and effectively shared with other divisions of the company. This is almost always best achieved by the people who implemented the innovation successfully (not necessarily the inventors). Plan on transferring these people to other company divisions (temporarily) to help others use the idea. Also, consider patents, registered designs, and other forms of intellectual property protection.

Option 4: Avoid the risk

We can avoid a risk by choosing a different strategy or solution to a problem (applies only to negative risks).

Option 5: Transfer (or share) the risk

We can pursue this option by taking out insurance (for negative risks), by agreeing with another stakeholder for that stakeholder to accept the risk, or by agreeing with other stakeholders to share risks and limit the extent to which each will cover the risk.

- Take out an insurance policy to cover the risk: this transfers the risk to an insurance company, but you have to pay a fee (called an insurance premium) to persuade them to do this. The fee will be determined by their assessment of the probability of the risk, plus their assessment of the cost of administration and the likely cost of investigating any claims you might make. This means that insurance is often an expensive option.
- Share the risks between several stakeholders. This is now almost a universal practice on larger engineering contracts and much of the negotiation is not so much on price as on the ways that risk will be shared. The two extreme cases are:
  - Cost-plus contract: the client pays the costs incurred by the contractor, plus a small agreed upon profit margin. The client carries all the risk, so the profit provided to the contractor will be small. It may be possible to negotiate a decreasing profit margin for work extending beyond an agreed completion date to ensure that the contractor has an incentive to finish the project on time.
  - Fixed-price contract: the client pays a fixed price, regardless of the outcome. The contractor makes a big profit if things turn out well, but may take a loss otherwise. This works well on routine engineering work where the costs are predictable and controllable.

Usually, the risks are shared according to an agreed formula between the ultimate client, the main contractor, and the sub-contractors. Clearly, the amount of risk being shared depends ultimately on the financial strength of each party, because it is not in anyone's interest for one of the parties to go bankrupt as a result of a project failure. There are many ways to structure projects to mitigate commercial risks; however, these techniques are beyond the scope of this introductory chapter.

It is all too tempting to transfer the risks to an organisation that is prepared to accept the risk, but actually may not have the capacity to handle it if anything does go wrong. In the end, the result of that situation is that the transfer is ineffective because the task of handling the risk will fall back on the party that transferred the risk in the first place. For example, if transferring a risk to a contractor results in the contractor going bankrupt, then the work will have to be completed by the project organisation, often at a much greater cost. This is a common project pitfall.

How do we need to modify the project plan to build in the risk control measures?

Once we have designed control measures for all of the major risks, we need to go back to the stakeholders, particularly the client, and negotiate changes to the project plans and possibly the project budget.

Here, it is important to understand the value proposition that we mentioned earlier. How does the client, in particular, as well as other stakeholders, benefit from the risk control measures that we have planned? While it is tempting to reduce everything to a financial measurement of risk, this is usually not possible. The best guide for these negotiations is the homework that we did earlier when we identified interests, values, requirements, and expectations for each of the stakeholders.

Practice concept 66: Approvals

Engineering has become critical in supporting the social structure of our civilisation, with a large influence on the lives of every human being. Engineering projects can cause significant, unintended consequences for the community. Because of past failures, engineers now find themselves under detailed scrutiny by government regulatory agencies, which mostly employ engineers. Today, engineers find themselves devoting considerable effort to writing applications seeking approval for their projects from government agencies.

Even a small project can require 20 or 30 separate approvals, both within the project organisation for the plan and the budget, and from government regulatory agencies. A large project in an industrialised country may require as many as 150 or even 200 separate approvals from several different local, regional, and national government agencies.

The kind of planning we have discussed in this chapter provides the groundwork for these applications. Most agencies use some form of 'due diligence' criteria in assessing applications. In other words, has the project team considered all the relevant issues at a level of detail that would be appropriate for a well-run engineering project? Has the team conducted sufficient consultations with stakeholders, particularly all the people in the community who will be affected by the project?

If the planning has been done well and comprehensively, if the stakeholders have been involved in setting the project scope and constraints, including identifying, managing, and sharing the risks, and if these discussions have been appropriately recorded and summarised in the form of documents, then all the supporting evidence for most approval applications will already exist.

Once again, the most appropriate strategy is to consider the regulators as stakeholders in the project. Rather than preparing an application for approval towards the end of the planning phase, a good project leader will involve the regulators as stakeholders in discussions from the start of the planning phase. Developing constructive and helpful personal relationships with individuals in the regulatory agencies early in the project can avoid tedious objections and what seems like petty obstructive behaviour when an application for approval is rejected.

This is another kind of collaborative performance that draws on the ideas presented in Chapters 7 and 8.

What does the regulator expect in an application for approval?

Like other stakeholders, regulatory agencies have their own interests, values, requirements, and expectations. By listening early on in the project, it is possible to understand how they see the project. This understanding will help you write an application for an approval or operating licence that takes their expectations into account, written in a language, an English sub-dialect, which is meaningful for the regulators.

A good place to start is to find a previously successful application that was approved with only minor, if any, amendments.

Final Investment Decision (FID) approval

The most critical approval is the one that authorises real work to commence on the project, beyond the initial project definition and planning that has been described up to this point in the chapter.

Experience shows that projects with detailed plans prepared are much more likely to achieve their objectives. The concept of '*front-end loading*', developed by Merrow, reflects this. Front-end loading represents the degree of effort devoted to planning and preliminary design for a project in order to maximise the chance of achieving the stated objectives.[18] It is reasonable to expect that 10% of the project budget will be spent *before* the final approval to proceed is given. If the amount is substantially less, and the project is not one for which detailed plans could be adapted with minor changes from an earlier project, then you can expect to encounter trouble ahead.

Once final approval is given to proceed, work on the project activities listed in the work breakdown structure can commence and the monitoring stage of project management also commences at that point.

MONITORING PROGRESS

Once the project work commences, project management shifts from planning to monitoring.

As explained before, the essence of project management work is coordinating collaborative work performed by many different people, possibly in many different locations. In order to do this, it is essential to monitor the progress of every activity, which is not as easy as it looks.

Practice concept 67: Monitoring is a cyclical process

Monitoring requires a cycle of activity. The cycle is repeated regularly, but the time interval depends on how the project is being managed and by whom.

Many describe the monitoring process as a simple, endless, repeating cycle: plan, do, check, act, plan ... One starts with a *plan*, and then people follow (*do*) the plan, someone *checks* the results against expectations, one takes any necessary corrective *action*, and then adjusts the *plan*, if necessary, repeating the cycle until the project is completed.

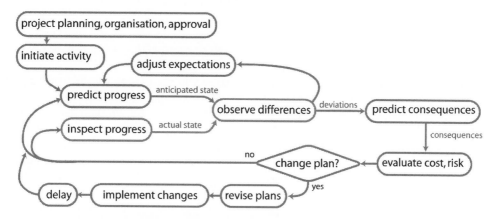

Figure 10.3 Project monitoring is a cyclical activity, identical in principle to the monitoring phase discussed in Chapter 9 on technical coordination.

Figure 10.3 illustrates the same idea: you will recognise this diagram as one from the previous chapter. In principle, there is no difference between the formal monitoring needed for managing a project and the informal monitoring needed in technical coordination. However, since a project involves many distinct activities that happen simultaneously, we need a different way of organising the monitoring.

Any decision on monitoring frequency will depend on the criticality of a given activity. Activities on the critical path, where any completion delay will delay the entire project, are obviously monitored more carefully than ones with several weeks of float time. An expert project manager will assess each activity and allow for the time needed to fix the consequences of foreseeable, but unpredictable, events that could affect schedules.

Monitoring occurs simultaneously with different levels of detail, scope, and frequency.

At the site where the activity is taking place, a foreman or supervisor will check on progress at least once a day, but usually two or three times. If there is a site engineer, they will review progress with the foreman or supervisor every day. Note that the site engineer is often younger and inexperienced: part of the responsibility of the foreman or site supervisor is to help the site engineer learn all about managing a site where there is practical engineering activity taking place. This can be a design office with several CAD workstations operated by drafting experts, a construction site, part of a factory, a software production house with programmers, a maintenance workshop, an electrical substation, or a power plant: there are countless different locales where engineers learn how to deliver real results.

A project engineer or project manager who is not located at the site will typically check on progress every week. They will visit the site and spend the first hour or two with the site supervisor or site engineer, working through every activity currently in progress and reviewing records of progress. Then, together, they will walk around the site looking at locations where there are hold-ups, unexpected issues, or mistakes that need to be rectified, as well as simply conversing with sub-contractors, craftspeople, tradespeople, and technicians.

Perhaps you remember from the brief description about C. Y. Connor in Chapter 4, how he spent days at a time, up to 12 hours each day, inspecting work first-hand on the goldfields pipeline and storage dam at Mundaring.

Before each inspection visit, a project manager will predict the results from each currently running activity: what they expect to see. By comparing actual progress with expected progress, one can 'calibrate' one's expectations so that predictions become better and better over time.

Critical to all of this, of course, is establishing some way of assessing progress on each activity. Anything that results in tangible production is relatively easy to monitor. The difficulty with a lot of engineering work is that it is intellectual and cannot be directly observed. Intermediate results often give little indication of actual progress. Design work ultimately emerges in the form of drawings and documentation, but these appear relatively late in the process. In the early stages, the best one can expect to see is often just a series of sketches backed up by conversation with the designer. Here, one can begin to appreciate how important it is to think in advance about the questions posed earlier in relation to each activity.

How can we assess progress, quality, and other attributes of the work in question well before it is actually finished?

How can we tell if it is going to be completed on time, at an acceptable quality, on budget, and sufficiently long before it is actually needed, so corrective action can be taken, if required?

And finally,

How can we tell that it actually has been completed with sufficient quality, accuracy, and within the allocated budget?

One of the most difficult aspects of monitoring progress is a tendency for many engineers (and others) to engage in self-deception, particularly with engineering technical work that is highly intellectual and depends on abstract thinking, such as design and planning. William James foresaw this in his essay, 'Sentiment of Rationality'.[19] He distinguished between abstract rationality and concrete rationality. The former is more an emotional sense of satisfaction, a sense of inner peace, or the feeling that one has reached a resolution and that it all fits together rationally. The latter is 'reaching what is right', and requires that all the details be resolved in a way that others can comprehend sufficiently. Often, we only discover how much more there is to resolve when we attempt to convert abstract rationality into concrete rationality in the form of drawings and documents.

Many design engineers tend to underestimate the proportion of the design and planning work that has actually been completed. They create a reality in their minds in which everything has been resolved, except for the final stage needed to produce concrete reality: the drawings and documents. I have often experienced this myself. We will assure others that the work is essentially complete, maybe 80% done, without the slightest doubt in our minds. Only someone with an intimate knowledge and deep experience can ask us questions to expose the gaps that we convince ourselves have been covered and resolved.

Often, a project manager does not have the depth of technical expertise in every aspect of the project to be able to ask these questions, which means that they go unanswered. Yet, as time drags on, we gradually realise that the 80% completion estimate was more like 30%. As weeks turn to months (and sometimes even longer), and completion times are pushed out further each week, the project gradually unravels in front of us.

The difficulty is that we were not lying when we said that the work was 80% done! We completely believed that at the time. We could have sworn an affidavit in front of a judge without the slightest twinge of doubt.

As a project manager, how can one detect these misleading situations? Particularly if one does not have the same technical expertise as the people one is supervising?

The first step, of course, is to ensure that there are appropriate supervisors with the technical knowledge to ask the questions that need to be asked.

Another technique is to carefully monitor the actual expenditure of time and resources on each activity. Most engineers are comfortable completing timesheets each week to account for their hours: they report the approximate number of hours spent on different aspects of one or more projects on which they are currently working. Even though engineers are notorious for misreporting their hours, carefully working out in advance what they are expected to report and then providing information that aligns with expectations, rather than necessarily with reality, this data can still reveal a lot of useful information about progress.

If the actual expenditure in money and hours, compared with the budget, reflects 30% progress, while the people performing the work are simultaneously reporting 70% progress, then there is a clear discrepancy. Very few engineering tasks can be completed that far under budget. That is a major warning sign.

Another way is to carefully monitor scope creep: an issue that we mentioned earlier. Scope creep appears in the form of a few additional documents that appear on the list of deliverables, or even additional activities that were somehow overlooked at the planning stage. Here, a good project manager learns to distinguish between activities that were accidentally forgotten at the planning stage and activities that have been introduced into the plan because of technical or other complications that were not fully anticipated or appreciated earlier in the project. The former can be covered by an overall contingency: an established allowance because some aspects of the project will always be forgotten. The latter, however, can be warning signs of a project that is getting out of control. It takes some time for the information about additional documents or activities to be included in project plans. By the time they are noticed, there could be even more requests for additions, more scope creep that is already in the pipeline. What initially seems like a few overlooked details becomes an avalanche: it is revealed that whole areas of technical complication were overlooked because of insufficient time to explore the details or insufficient experience with the technical aspects of the project.

Also, as I have explained, those closest to the technical work itself can often be the least able to appreciate the reality of the situation.

One progress monitoring technique that has emerged in the last decade is known as 'earned value' monitoring. Each activity forms part of the completion value of the project in proportion to its total cost. A certain proportion of the activity value is said to have been 'earned' when the activity passes certain milestones, such as 50%, 80%, and 100% completion. For example, the project team may agree that 40% of the activity value has been earned when it is assessed as 50% complete. By carefully comparing actual spending on the project with the total 'earned value' from all the activities that have passed these milestones, a project manager can foresee future delays and budget 'blow-out'. If the earned value is significantly below the total spending, then a delay or cost overrun is likely.

What is the current predicted project completion plan (dates, expenditure)?

Once each round of inspections is completed, the project manager needs to update all the future predictions: when each activity is likely to be performed and when the project is going to be completed, as well as the latest estimates of expenditure. It is also necessary to review and update the project risk assessment. Each of these tasks can be tedious, but they are essential pieces of documentation to monitor progress. This is where computer tools can be really useful on larger projects, by updating these predictions automatically.

If there are delays (or opportunities), is it worth changing the plan?

At this stage, the project manager can evaluate the consequences if activities are running late or over budget. If the situation is likely to seriously compromise the interests of influential stakeholders, especially if there are changes in the likelihood or consequences of significant risk factors, then it may be necessary to modify the project plans.

Changing the plans inevitably introduces even more of a delay. First, there is the time required for everyone who is affected by the change to provide feedback and comments on the implications before any final decision is made. Particularly when equipment is hired and transported to a remote site, or when work is held up so that contracting teams have to stay longer at a remote site, the consequential costs can be very significant.

Once the plan has been changed, it takes time for everyone to become familiar with the revised plans and adapt their activity schedules accordingly.

So, as we can see from this brief discussion, the monitoring phase of a project, with its repeated prediction, inspection, evaluation, and review steps, is time-consuming. A project manager's job is not easy. Also, it can be difficult for others to appreciate the amount of time and effort involved. To those people performing the technical work itself, project managers often seem like expensive sources of overhead. Spending money on a project manager is often seen as a waste: 'It would be better to spend the money with us; that way we could achieve much more!'

Practice concept 68: Contract variations

Significant changes in the project plan that require new work, or rework that was not accounted for in the original estimates, will often lead to negotiations on how the additional cost is to be covered. The client will often argue that the work should have been foreseen by the engineers and covered in the project contingency, which is the item in the project budget that covers uncertainties. These issues are referred to as *variations*.

In most projects, any addition to the scope required by the client becomes a 'variation'; it is not uncommon for the costs of variations to be estimated at a much higher level than the original work. Mostly, the additional costs are fully justifiable. The original work will have been carefully planned so that the specialised tasks requiring skilled labour have been grouped together. Bringing people with special skills and equipment back onto the site, perhaps for a job taking just a few minutes, can impose a large cost, particularly for a remote site. The situation is even worse if other people are in the way and the skilled people have to wait around doing nothing, simply because they can't access their equipment to perform the work.

In some industries, project delivery firms will negotiate a low price, perhaps even below the actual cost of the project, knowing that the client will request variations on which they can subsequently charge a much higher price so they can make their profit.

Variations almost always result in unpleasant cost increases for the client. By spending more time with the client during the planning phase, you can ensure that all the requirements are included in the project scope documents and specifications before the project commences. During the project, the client needs to be repeatedly reminded not to make any further changes – clients can be their own worst enemies in this respect when they make late changes to the scope of the project. If the client resists this temptation, then there is less chance that variations will be needed and a greater chance that the client will be happy with the outcome. Happy clients lead to repeat business and an enhanced reputation that brings work from new clients.

One common trap for young engineers is to try and be 'nice' with a client by accepting what seem to be minor scope changes without first checking the implications with more experienced engineers. Without painful experiences in the past to learn from, it is easy for a young engineer to overlook some of the consequences that can result in painful additional expenses that the client may not pay for.

COMPLETING THE PROJECT

Practice concept 69: Project completion is also a cyclical process

A project is completed when all the agreed upon objectives have been achieved and accepted by the client. Completion is relatively straightforward if the project definition phase included all the necessary details on how the completion of every task was to be certified with appropriate acceptance tests and inspections. Inspection and testing activities would normally be included as an activity in the work breakdown structure. Therefore, completion of the project corresponds with the completion of all the activities in the work breakdown structure.

Figure 10.4 illustrates the kind of repetitive process needed for the completion of each activity.

Once the person responsible for the activity tasks announces that they have been completed, the project manager would normally review the scope in order to anticipate what should be evident at task completion (the anticipated state). The project manager

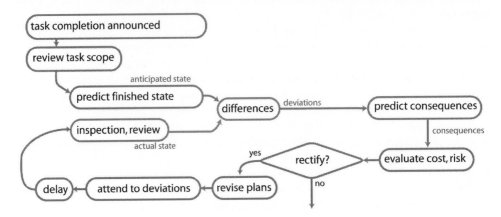

Figure 10.4 Activity task completion.

will then perform an inspection to determine the actual state, will have tests conducted, or will inspect independent tests and inspection reports to verify task completion. Depending on the contract arrangement, the client may also be responsible for verifying task completion.

Any differences between the anticipated state and the actual state are noted as deviations. All of the deviations noted for a particular task will be entered into a *punch list*, which serves as a record of outstanding deviations. The project manager, collaborating with the client for important issues, will then review the punch list and the consequences of leaving deviations as they are. Depending on the consequences, items on the list may have to be rectified, in which case the project plan is updated to indicate that the activity tasks are not yet completed. Additional time has to be scheduled for the deviations to be rectified.

Once they are reported as having been rectified, some or all of the tests and inspections will be repeated before certifying that the activity has been completed.

In reality, most projects end up with substantial lists of items that need to be inspected or rectified. The benefits of detailed documentation at the start of the project now become apparent: without details, a project manager can easily end up in protracted negotiation with the client on how to complete each activity. Naturally, the project manager wants to keep the overall expenditure within the budget or as close to the budget as possible. The client, on the other hand, has a vested interest in minimising payments to the contractor (the project organisation).

Normally, the project manager is under intense pressure to deliver the agreed objectives within the original project budget. Unless project documentation is sufficiently detailed, this tension presents the project manager with the tempting option to remove non-essential items from the scope, or to simply ignore them, particularly if they will cause the project budget or completion timescale to be exceeded. It is not easy for a project manager to negotiate an extension to the budget or time schedule as the project draws to a close unless there are well-established reasons that were documented earlier in the project.

This is perhaps the most fundamental weakness of the conventional method for organising a project. Project managers are rewarded for completing projects on time and budget, which can conflict with the interests of the organisation that inherits the assets and other deliverables created by the project. The next chapter describes a case study that illustrates this problem. A large process plant was constructed at a remote site by a team that did not fully understand the technology with which they were working. As the client's engineering team became more aware of the operating requirements for the plant, engineers in the client's organisation insisted on many changes late in the construction. For example, they insisted that adequate access be provided for maintaining equipment, such as the pumps. While this might seem to be a normal and reasonable requirement, these details had not been included in the original project documentation. Eventually, the construction project team ignored further change requests from the client, effectively locking them out of the site until all the construction was completed. When the client's operating team eventually took over the plant, there was a long series of major and minor plant upgrades to address all the issues that had been raised by the operating team. These upgrades were still not completed by the time the plant closed a few years later. With the plant already constructed, it was much more expensive to carry out these upgrades, but the operating budget only allowed for limited engineering work to be performed.

In our research interviews, engineers reported on several occasions that project managers, who had apparently completed projects on time and budget, had subsequently been rewarded with responsibility for even larger projects. However, they only managed to do this by deleting major items that would otherwise have caused a delay or a budget overrun. In doing so, they had been able to exploit weaknesses in the initial documentation of the project: gaps and inconsistencies in the project scoping documents and specifications allowed the responsibility for these items to be transferred from the project team to the operating organisation that would inherit them after the project completion. For example, access platforms and lifting devices needed for maintaining pumps and other equipment can easily be removed from the construction phase of a process plant without affecting its ability to operate for the first year or so. In doing so, the project team passes responsibility for installing the access platforms to the operating organisation: the platforms will be needed eventually.

Several improvements to project contracting arrangements have been proposed to prevent such transfers from happening. For example, a client can provide incentives that reward a project delivery team if the operating cost expectations are met for the first twelve months but these provisions can be very difficult to negotiate and evaluate. Here's an example from the offshore petroleum industry. The interests of the client organisation, typically specialising in operations and selling petroleum products, are:[20]

- Maximised shareholder returns
- Contractual obligations to customers (maximising profit and minimising penalties/damages for non-performance of obligations)
- Plant/facility efficiency with minimal downtime
- Minimum OPEX (operating expenditure)
- Minimum CAPEX (capital expenditure)
- Safety

On the other hand, the interests of the project delivery team are:

- Maximised shareholder returns
- Contractual obligations to clients (maximising returns to shareholders and likelihood of repeat business)
- Lowest possible expenses to achieve objectives
- Completion in minimum time
- Minimal risk (reducing possibility of claims)
- Safety

Even though there are clear conflicts between the interests of the client organisation and the project delivery team, there are also common interests, including safety, completion in the minimum time, lowest possible expense, and the likelihood of repeat business in future. The last item, in particular, is much more likely if the client and the project delivery team reach a happy compromise. The best way to achieve this is to spend more time on the project definition phase – maximising front-end loading.

NOTES

1. Some non-engineers might contest this, citing the common stereotype that many engineers are far too predictably boring, without necessarily understanding what engineers do and how we think.
2. Human error in organisations has been extensively studied and engineers can learn from this literature. For a discussion of operator errors, see Kletz (1991). Dekker (2006) provides a more modern discussion from an organisational standpoint. Other informative references: Bea (2000), Bigley and Roberts (2001), Bloch (2009), Busby (2006), Hobbs (2000), Klein, Bigley, and Roberts (1995), Leveson and Turner (1993), Malone et al. (1996), Reason and Hobbs (2003), K. H. Roberts (1990), Stewart and Chase (1999), S. Tang and Trevelyan (2010), and Weick (1987).
3. Mehravari (2007) and other unpublished studies by the author.
4. An experienced engineer reviewing this chapter commented that, in his experience alone, this factor had accounted for numerous multimillion dollar claims against his engineering firm.
5. Consultation with all stakeholders, including the local community, is the recommended standard practice reflected in ISO31000 and national standards (e.g. Standards Australia, 2009, pp. 14–16).
6. There are differing views on the need for technical capabilities amongst project and organisational managers (e.g. Grant, Baumgardner, & Shan, 1997; Hysong, 2008).
7. Project management terminology in this chapter has been set in bold type to help you find explanations for the terms later.
8. Sule Nair (2011), Unpublished manuscript.
9. Definition based on the *Project Management Body of Knowledge* (Project Management Institute, 2004, p. 5), see also (Hartley, 2009).
10. (Whyte & Lobo, 2010).
11. (Merrow, 2011).
12. (Project Management Institute, 2004, p. 26).
13. Whyte and her colleagues have provided insightful descriptions of what this entails in construction projects. Their accounts detail aesthetic issues, project coordination around fully detailed digital models, and also the limitations of these models when physical prototyping

is necessary (Ewenstein & Whyte, 2007; Whyte, 2013; Whyte, Ewenstein, Hales, & Tidd, 2008; Whyte & Lobo, 2010).

14. Hartley (2009, Ch. 6) provides some further details.
15. Standards Australia (1999, 2009, 2013).
16. Waring & Glendon (2000).
17. E.g. Standards Australia (2013).
18. Merrow (2011).
19. James (1905).
20. David Roberts (2008), Personal communication.

Chapter 11

Understanding investment decisions

Like earlier chapters, this one does not tell you everything that an engineer needs to know about money, finance, accounting, and investment decisions. Instead, it is here only to raise your awareness on some important issues for your career. The real learning will come when you talk with other people in your organisation and start to see how these issues influence your own engineering work. Read this chapter and then start to ask questions on related issues that affect your organisation. If you find you need more detailed knowledge, there are many books on engineering management that cover these issues in much more detail.[1]

In the introduction to this book, I explained that engineering is a wonderful career. You get to spend lots of money, which is much more fun when you are spending other people's money rather than your own.

However, before you encourage people to place their trust in you to spend their money, it's a good idea to learn about the basics of project management and finance, and also to understand why people invest their money in engineering and what they expect. By doing so, you will find it easier to satisfy their expectations and avoid disappointment, frustration, and possibly a nasty and premature end to your career. Bypassing an understanding of finance and investment could be what many engineers call a CLM – a career-limiting move.

Our research provides strong evidence that an appreciation of commercial value is not well developed among engineers.[2] According to recent research, engineering students think that money and finance is a side issue of only remote interest for engineers.[3] It is something for accountants and economists to be concerned about, maybe managers, but certainly not something that engineers need to be concerned with.

This is one of the more serious misconceptions that is so easy to develop in higher education.

FINANCE

In engineering, interesting things don't tend to happen without lots of money being involved. Money is what makes engineering fun; the sooner you understand the basics, the happier you will be as an engineer. You don't need to be a financial wizard or an accountant; once again, a reasonable understanding of people provides you with most of the important ideas that you will need.

As explained in Chapter 1, engineering cannot happen without investment. Engineers spend money provided by investors to create new artefacts and information, either as products in themselves or as a service. It is only when the products or services reach the end user that the ultimate value of the investment is obtained. In the case of public investment, the products or services may be provided free of cost and the value is obtained in the form of public and social benefits. In the case of private investment, money can be collected from the end users. In both cases, the benefits flow back to the investors long after the money was originally provided.

So what persuades an investor to provide money under these seemingly less-than-ideal conditions?

For most investors, the main motivation to provide the money is the expectation of future profits. They expect that the total amount of money collected from end users will be substantially greater than the initial investment. Later, we will learn just how much greater it has to be.

However, not all investors are motivated purely by profit. There are investors, particularly governments, which are seeking public benefits or environmental improvement. For example, many people invest in solar panels on their roofs because they want to do something to improve the environment, or to be able to use electricity without thinking about the consequences of burning fossil fuels. For others, it may be a demonstration of their commitment to improving the environment.

Either way, investors need to be sufficiently confident that the ultimate results can be achieved before they commit their money. This is where engineering comes in; part of the job of the engineer is to provide investors with the best possible advice before they come to the decision to spend the money. Also, a good reputation in the community reassures an investor, which increases the amount that they are prepared to risk spending with you.

As an engineer, one of the most important enabling capabilities you can possibly acquire is the ability to persuade people to invest money in an engineering enterprise. What this requires is the ability to reassure investors and help them feel confident that their money will achieve the results that you, as an engineer, have promised.

It follows that the next important enabling capability is to be able to devise an achievable solution for the client and prepare a realistic budget, and then set up a team that can deliver the promised results for investors on time, on budget, with predicted performance, safely, and within acceptable limits for environmental and social consequences. That is why the preceding chapters on technical coordination and project management are so important for your long-term development.

In this chapter, I will cover some basic ideas about finance and investment; skip any sections where you feel confident that you can perform the analysis by yourself, but make sure that you test your knowledge with the quiz before moving on.

Practice concept 70: Engineering needs money from investors

Novice engineers often think that ...	Expert engineers know that ...
Finance is a business or management issue, not an engineering one; engineers only have to think about the budget they have been given.	Finance enables engineering. Engineering relies on investment from people. The cost of finance and amount available depends on the risk perceptions of investors. Engineers prepare the estimates from which investors make their decisions.

Describe a recent experience with financial decision-making. If you have not been exposed to this in your work, describe the most recent major financial decision you had to make in your private life and how you made it. (Prepare to copy and paste into the online learning system.)

Most people who make investment decisions have neither the time nor the technical insight to fully evaluate the benefits and risks before making an investment. No one can forecast the future with any great certainty, particularly with respect to economic conditions. Even the best economic modelling cannot predict, for example, when an unexpected political development on the other side of the world might cause panic buying and speculator-driven price rises in gold or oil, for example. In the past few years we have seen large and rapid changes in energy costs. The international price of oil, for example, has varied between around US $40 per barrel and US $140. This makes it extremely difficult for anyone to have great confidence in forecasted costs and benefits from an engineering investment that may take several years in planning and construction before any income can be earned at all.

The main factors that govern investor behaviour are based on risk perceptions: what is the relative certainty that the investment will provide the expected benefits and financial returns? What is the level of confidence that the solution will work as expected? How great is the confidence in capital and operating cost estimates? What is the confidence level in the estimates for earning income from the venture? All of these perceptions are influenced by data that is provided by engineers, except for sales estimates, which usually come from marketing studies. For infrastructural projects, even the income data and traffic forecasts come from engineers.

We need to appreciate the importance of confidence on a qualitative scale. Lending money to a government is one of the least risky investments, particularly the government of a wealthy country that has been prudently governed in the past. At the other extreme, investing in a mining company that has staked a claim to an, as yet, unmapped mineral deposit in a country facing political upheavals and conflict would be one of the more risky investments that you could make. Somewhere between the two extremes lie many other investment alternatives, such as investing in the shares of large corporations; usually, this involves a low to moderate risk.

High-risk investments have to offer investors a reasonable chance of making a large financial profit relatively quickly to convince them to provide the money. On the other hand, with lower-risk investments, investors will provide money for a long time with relatively modest profits because they know that there is a high certainty that the benefits will reliably come over the long-term.

Investors are people, and we all have different preferences that change with circumstances. Older people in many industrialised countries are less comfortable with risky investments because they need to have confidence that they will be able to have money to support their retirement. Younger investors often like to invest in more risky ventures; they are attracted by the excitement of making a large and quick profit. Most investors like to have a modest amount invested in high risk ventures; there will always be some that pay off and they also tend to be more exciting investments to follow. Most investors prefer to allocate much more of their capital to low- and medium-risk ventures than high-risk ones to safeguard their family's long-term financial security and provide reasonable peace of mind.

Practice concept 71: Value is subjective

The concept of value is fundamental when thinking about money, which leads us into the world of finance and economics. However, there are some difficult concepts that lie here, right on the doorstep as you enter.

For engineers, the word 'value' is especially problematic. You have been educated systematically for many years with the idea that 'value' is a precisely defined number, the 'value' of a particular variable at a particular time. 'Value' is therefore associated with 'precision', something that is precisely defined to a given number of decimal places. There are other meanings as well. For example, you have your own personal 'values', such as honesty, integrity, and studying hard early in life in order to bring greater benefits later. If you did not have these values, then you probably wouldn't have graduated from a university engineering course.

Largely because of your education as an engineer, you are particularly susceptible to the belief that 'value' is something precise, meaning that it is absolute and incontestable.

In the context of money and economics, the word 'value' takes on quite a different meaning, as explained earlier in Chapter 1. It can be confusing, however. There are different ways to use the term, even in the context of economics.

For example, what is the value of money? Take a $5 bill, a 100 rupee note, or a €1 coin. Surely, the value is the number printed or embossed on the coin, no? And surely that's precise and unambiguous? Relatively invariant? A $5 bill cannot represent any other value than $5, surely. This is what we refer to as the 'face value' of currency – currency being dollars, euros, pounds, rupees, rials, renminbi, or dozens of other currency forms, depending on the country or region you happen to be in at the time. By now, especially if you have travelled to another country, you are aware that the *relative value* of different currencies changes over time. Even while writing this chapter, our Australian dollar varied between US $1.09 and US $0.89. Only a few years ago, it was worth US $0.60.

Another way to describe 'value', however, is as an amount of money that we would exchange for 'goods' that we desire; for example, the 'price' we would pay for a bottle of water. In the context of normal life in an industrialised country, even that can seem invariant and fixed. You walk into a supermarket and choose a bottle of water from countless options on the drinks shelves. The prices are fixed; they are clearly marked on price tags adjacent to each brand and type of bottled water on sale. When you go to the checkout, laser scanners detect the product barcode on the bottle label, and the price automatically appears on the cash register display. Everything, it seems, is predetermined.

However, in a different context, the concept of the 'value' or 'price of goods' can change.

For a start, it depends on how much we desire the goods. Imagine that we are in a dry mountainous desert and have been walking for hours without any water. We feel that we are dying of thirst. We stumble over yet another parched rocky ridge and below us, under a scrubby tree, is a man sitting at a small table with a few bottles of water for sale. With no other alternative source of drinking water, we would be prepared to pay him far more for a bottle of water than we normally would at the supermarket. Perhaps we might even pay him our last $100 for a few bottles.

What has changed here?

First, our desire for water is much greater; in economics, this is called 'demand'. Normally, in an Australian city or town, safe drinking water comes from any convenient tap, even in a garden, for free. Turn on the tap, cup your hands, and drink for free. Why is bottled water sold in bottles in a supermarket at all, you may ask, when a drink can be freely obtained almost anywhere? And bottled water is often more expensive than petrol. Of course, advertisers have created the subjective impression that having a bottle of 'Brand X's Fresh Spring Water' in your bag is associated with wealth and status. Or maybe buying it from an automatic drink vending machine saves you time and is more convenient than trying to find a water tap nearby. Or, if you're like me, you drink the spring water and then refill the bottle when you happen to pass a tap somewhere; the bottle is a useful water container when you have forgotten to bring one with you.

What we can see here is that 'demand', or the desirability of 'goods', which is a bottle of water in this case, depends on what we are thinking at the time: it is dependent on a range of subjective perceptions.

To reinforce this idea, imagine that just over the next parched rocky desert ridge from the scrubby tree with the man selling water, as I mentioned just a few moments ago, there is a large supermarket with shelves full of water bottles on sale at the normal Australian supermarket price, around $1 per litre for example. If we had known that, we would not have paid the man our last $100 for a couple of bottles. Instead, we would have climbed the next ridge, gone to the supermarket, and bought many other 'goods' as well.

'Value', therefore, depends on subjective perceptions of 'demand' as well as both perceptions and the reality of 'supply' or availability.

We need to remember that when we exchange money for goods that we desire, then the price we pay represents the relative value relationship between money and the desired object or service. However, because money is almost universally exchanged for many different kinds of goods, we tend to see money as having a fixed value and the price of goods changing from time to time, especially with changes in the perceived demand for, and supply of goods.

However, money itself can change in relative value when a large proportion of the goods we buy changes in price more or less together, at the same time. This can happen, particularly in times of social unrest and uncertainty.

Obviously, any major change in the relative value of money and the objects or services that we purchase in exchange for money can cause great difficulties for people. It can be very discouraging if you save money to buy food, only to find that the amount of food that you can buy with that money has significantly decreased. This does happen in times of high inflation. In extreme cases people have had to exchange bags full of banknotes just for a loaf of bread; for example, in the period of hyperinflation that occurred in Germany in 1919. Inflation, decreasing the purchasing power of money, can be caused by social unrest, and can make it even worse.

This is particularly an issue for engineers who are expected to predict future costs accurately, and deliver results within their allocated budget.

Figure 11.1 Australian retail price inflation (Australian Bureau of Statistics, Reserve Bank of Australia data).

For this reason, governments and central banks exercise firm control over a nation's economy so that there is stability in the value of money. However, this stability can never be absolute; value remains a subjective human perception.

Practice concept 72: Time changes the value of money

The next important concept that we need to understand is the idea that the value of money gradually changes over time.

The main reason for this is 'inflation'. Each year there is usually a slight rise in prices and incomes. While some commodities may fall in price, most will slightly increase every year. Economists measure inflation by the average percentage price increase over one year. There are many different ways to measure inflation. The consumer price index (CPI) is one way to do this. Economists monitor the price of a representative collection of goods that an average family would need to purchase, including food, accommodation, energy, utilities, transport, education, clothing, communications and entertainment. Another measure of inflation is the labour price index (LPI), which is calculated from a representative set of employment costs including salaries and indirect costs such as renting the space occupied by employees. Usually, there is a gradual improvement in labour productivity each year, often as a result of productivity improvements from using new technology. This means that labour price increases can be slightly greater than consumer price increases. This means that, over a period of time, everyone can buy a little more than they could the year before.

Economists advise governments to carefully control the amounts that they collect in taxes and spend each year because an imbalance can cause inflation, especially if a government increases its spending without increasing taxes. In many countries today, central banks exert independent control over financial interest rates, influencing that rate of price inflation and reducing the potential for unwise political decisions to cause inflation or undue economic distress.

Ideally, the inflation rate should be between 1% and 3%. The reason for a slightly positive rate of inflation is to avoid the risk of deflation, which is reflected by a reduction in prices. Deflation is normally considered undesirable because consumers will delay purchases if they think that the prices will decrease in the future, causing instability and stagnation in the economy. Other external influences can also affect inflation, such as changes in major commodity prices, particularly the price of oil and other forms of energy. Therefore, governments only have limited control over the rate of inflation.

In spite of the best intentions of governments, however, there have been periods of high inflation from time to time in the recent past. In Australia, for example, the inflation rate reached almost 18% in the 1970s, partly due to government policies that encouraged workers to seek higher wages and partly triggered by restrictions on oil supplies from major oil-exporting countries.

Therefore, under normal conditions of low inflation, I can expect that $100 worth of goods purchased today will cost around $102 this time next year. In other words, the value of my $100 will be slightly decreased over the coming 12 months. We describe this by saying that $100 earned 12 months in the future is equivalent to earning $98 today. In other words we apply a *discount rate* to calculate the value of future income or expenses in terms of today's money, and in this case the discount rate due to inflation is 2%.

However, in an engineering project, we usually have to work with a much higher discount rate. To understand why, we need to understand a little about 'interest rates', or the cost of finance, involved in borrowing money.[4] When we borrow money from a bank in the form of a 'bank loan', we have to pay sufficient interest to cover the bank's costs. There are four components that influence the interest rate:

a)   The interest cost for the bank to borrow money in order to be able to lend it to us.
b)   An allowance to cover the expected future inflation rate, if this is higher than a) above.
c)   The administration cost: employing bank staff to borrow money and also to assess our proposal to borrow money (the loan application), the cost of processing regular interest payments and repayment of the capital amount.
d)   The possible cost to the bank if the borrower defaults (i.e. is unable to pay interest or even repay the loan).

In most countries, there is a central bank (or reserve bank) associated with the government. Normally, the central bank enjoys a considerable amount of independence from political pressures. In most countries, the main commercial banks are strictly regulated by the central bank to make sure that they can survive the most severe financial crises. In 2008, in the USA, Britain, and several other advanced economies, several large commercial banks nearly failed, and a few actually did fail. Governments had to borrow large amounts of money to enable banks to meet their trading debts and to cover loans that could not be repaid (bad loans) as a result of the global financial crisis at that time.

The interest cost, the first component of the interest rate the bank charges for providing finance, is directly affected by the central bank cash rate – this is the interest rate that the central bank pays the commercial banks to keep their cash reserves. The commercial banks have to pay a slightly higher rate in order to attract savings deposits from ordinary people, even companies looking for the greatest level of security for their deposits. If the expected inflation rate is significantly higher than the central bank cash rate, then a commercial bank has to include this additional factor in order to attract deposits from savers. This, in turn, is the money they lend to borrowers.

In Australia today, the central bank cash rate is about 3%. In many other Western countries that are experiencing severe unemployment, particularly where governments

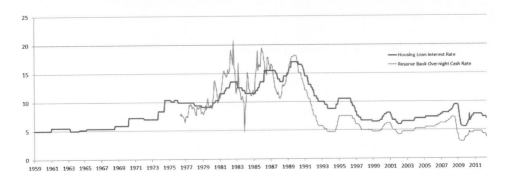

Figure 11.2 Interest rates in Australia (Data from Reserve Bank of Australia). Note that policy changes have altered the relationship between the overnight cash rate and the housing interest rate over time.

have borrowed a lot of money, the central bank cash rate might be as low as 0.5% or even zero.

The most secure form of lending for a bank is a housing loan; if the borrower defaults, the bank can sell the house to recover the outstanding loan and interest payments. The house serves as 'collateral', a physical asset that can be sold in case of default. Usually, the value of the house has increased since the commencement of the loan and so it is rare for a bank not to be able to cover the cost of default. Of course, the defaulting borrower loses their home. Therefore, the interest rate for a housing loan is only slightly higher than the interest rate that the bank pays for savings deposits, typically between 1% and 2%, to cover the cost of administration, the second component of the bank interest rate for the borrower. Figure 11.2 shows how the central bank cash rate and the housing loan interest rate have varied in Australia since 1959. In the early 1980s, the Australian government made major changes to the banking system, gradually relinquishing direct political control over the banking system, and deregulating many aspects of it. Comparing Figure 11.1 and Figure 11.2, you can see that there has been no obvious predictable relationship between inflation and interest rates.

I could invest my $100 in a savings bank and it would pay me an interest rate that is slightly higher than the central bank cash rate. However, I could also invest my $100 in shares or some other kind of more attractive investment. These investments provide a higher rate of return but, at the same time, there is a risk that the value of my investments might go down instead. In order to attract me to purchase this kind of investment, the company has to offer me a higher rate of return to compensate me for the risk that I might lose part or all of my money. The difference between this higher rate of return and the inflation rate represents a 'risk premium', an incentive that is sufficient to persuade me to invest.

When a bank lends money to a company, there is often a much greater risk. The company may not have enough physical assets or land to provide adequate collateral for the bank in case of default. The company may go bankrupt and be unable to keep up with repayments. Therefore, the interest rate for commercial lending is much higher than for housing because the risk that the bank would lose money if the company cannot keep up its repayments is higher (this is called a *loan default*). The extra component of interest is the risk premium and serves the same purpose as explained in the previous paragraph. The bank manager makes a judgement on the risk of lending money to the company, taking any collateral into account, and sets the risk premium accordingly. What this means is that the cost of finance for a typical engineering project is much higher than the housing interest rate, possibly up to 20% higher. Banks know that many engineering projects fail, often for reasons that are unpredictable, and other times because they are poorly managed. Economic conditions can change with little warning, despite even the best economic forecasting.

Because of these factors, when a company is considering a new project, it may be difficult to estimate what the bank may charge for interest on any finance that may be needed. The finance may be supplied by another division of the company, or a group of associated companies. The company may even supply the finance from revenue if the project is small enough. The finance may be raised by asking shareholders to invest money in the company, or a combination of all of these.

While we can guess what the interest rate might be, we cannot know for sure. Therefore, companies have evolved another method to evaluate the financial outcome of a particular project. Companies estimate their own 'average cost of capital' depending on their circumstances. For larger companies, this is typically around 10% higher than the current commercial interest rate for housing loans. For small companies, the interest rate may be up to 20% higher than housing loan rates.

As I will explain later, a new project is rarely considered in isolation. Most companies have many, perhaps hundreds or even a few thousand different project ideas at any one time, and not all can be supported – indeed, not all should be. In order to predict which will be the most attractive option, companies calculate the expected value of each project, which is called the 'net present value', in terms that reflect the relatively high cost of financing these projects.

Instead of allowing for the expected interest rate directly, it is assumed that finance for the project will have a high-risk premium. However, this is treated in a way that is analogous to a high inflation rate: future income and expenses are discounted as if there really were inflation at that rate.

This seems artificial and unrealistic. However, it reflects the fact that most engineering projects involve a large up-front investment, often for machinery and tools. The company incurs a debt to its financiers. Money is recovered as sales provide income over the succeeding years. The sooner income is received, the sooner the debt can be repaid, which will reduce the interest payments on the debt.

Practice concept 73: Net present value

The examples that follow represent very simple idealised projects to illustrate the basic idea.

In the first example, the investor spends $1000 on a machine to make plastic tubing in the first year, and the tubing manufacturing provides a net income of $300 every year after that, after taking into account money received from sales of the tubing and all the consumable, maintenance, insurance, rent, and labour costs incurred in running the machine.

Table 11.1 Calculated income and expenses with discounted income and expenses for a simple project.

Example 1	Sum	Year 0	Year 1	Year 2	Year 3	Year 4	Year 5	Year 6
Income	1800		300	300	300	300	300	300
Expenses	1000	1000	0	0	0	0	0	0
Difference (yield)	800	−1000	300	300	300	300	300	300
Discounted income	997.65	0	250	208	174	145	121	100
Discounted expenses	1000.00	1000	0	0	0	0	0	0
NPV, future yields	−2.35	−1000	250	208	174	145	121	100

Table 11.1 shows the annual income and expenditure for our very simple project. The rows of the table show:

Income: money received each year, zero in the first year, and $300 every year after that.
Expenses: $1000 to purchase the machine in the first year and zero for later years.
Difference: the difference between income and capital expenses each year, or **yield**.

The column headed 'sum' shows the total amount in each row for all seven years of the project.

Discounted future income and expenses have been calculated by: amount/$(1 + d)^n$ where n is the year number, and d is the annual discount rate, 20% in this example. Notice how the discounted income gets smaller and smaller when we look into the future.

The sum of the discounted yields over the life of the project is called the *Net Present Value* (NPV); in other words, the calculated value of the whole project over its full life expressed in today's dollars. The last row shows the future discounted yield for each year, the difference between the discounted income and discounted expenses.

Figure 11.3 illustrates this graphically. The dark bars show the forecast income and expenses, and the pale bars show the discounted income and expenses in terms of today's dollars. The graph clearly shows how the future value of money decreases in terms of today's values.

Clearly, the NPV depends on the value we choose for the discount rate. If you have access to the electronic version of the spreadsheet used to calculate these tables and graphs, then you can experiment for yourself and discover the effect of changing the discount rate.

Another way to show the effect of changing the discount rate is shown in Figure 11.4. This graph shows the NPV for a range of discount rates between zero and 100%. Most projects show a similar characteristic: NPV decreases with an increasing discount rate.

Using the spreadsheet, we can calculate the 'internal rate of return' – this is normally defined as the discount rate at which the NPV is zero. This is often used as a measure of the viability of the project; the higher the internal rate of return, the better the value of the project. However, it is not necessarily the most reliable indicator, as we shall see.

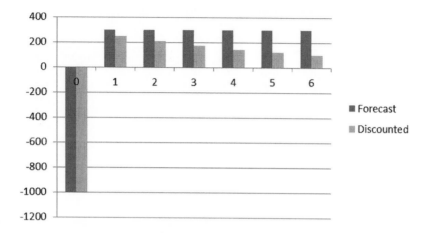

Figure 11.3 Forecast difference between income and major expenses over 7 years of a project (dark), and discounted difference between income and expenses (lighter). There is no discount applied for year zero.

NPV for different discount rates

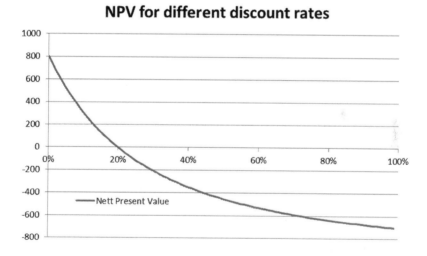

Figure 11.4 Effect on NPV of changing the discount rate on project shown in Table 11.1 and Figure 11.3.

Table 11.2 and Figure 11.5 show a different kind of project. In this project, the machine is less costly but results in significant site contamination. Expenses are incurred in both Year Zero and Year Six. This represents a situation in which there is a significant decommissioning cost; we have to spend money at the conclusion of the project in order, for example, to remove the original installation, remove contamination, and restore the site to its original condition.

Table 11.3 shows a project with an even higher decommissioning cost; this might represent the kind of situation faced by the owner of a nuclear power reactor (although

Table 11.2 Project with decommissioning costs in the final year (discount rate 20%).

Example 2	Sum	Year 0	Year 1	Year 2	Year 3	Year 4	Year 5	Year 6
Income	1800		300	300	300	300	300	300
Expenses	1200	600	0	0	0	0	0	600
Difference (yield)	600	−600	300	300	300	300	300	−300
Discounted income	997.65	0	250	208	174	145	121	100
Discounted expenses	800.94	600	0	0	0	0	0	201
NPV, future yields	196.71	−600	250	208	174	145	121	−100

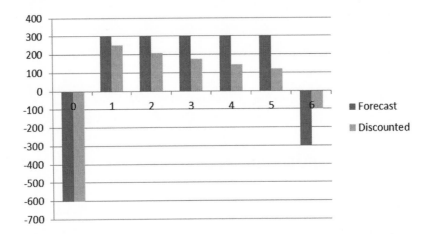

Figure 11.5 Project with decommissioning costs in the final year (discount rate 20%).

Table 11.3 Project with high decommissioning costs in the final year (discount rate 30%).

Example 3	Sum	Year 0	Year 1	Year 2	Year 3	Year 4	Year 5	Year 6
Income	1800		300	300	300	300	300	300
Expenses	1800	500	0	0	0	0	0	1300
Difference (yield)	0	−500	300	300	300	300	300	−1000
Discounted income	792.82	0	231	178	137	105	81	62
Discounted expenses	769.33	500	0	0	0	0	0	269
NPV, future yields	23.49	−500	231	178	137	105	81	−207

it is only a tiny fraction of the cost of a nuclear plant). Figure 11.6 shows the influence on NPV of changing the discount rate. Notice how the relationship between NPV and the discount rate follows quite a different trend from the more conventional project shown in Figure 11.4. In this case, the concept of internal rate of return is meaningless because there are two discount rates at which the NPV of the project is zero.

However, this is not the only reason for being wary about performing NPV calculations.

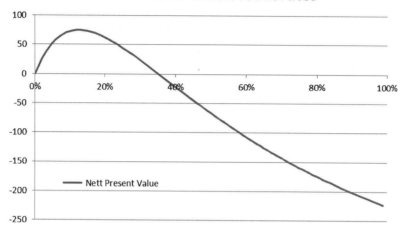

Figure 11.6 Project with very high decommissioning costs – effect of changing the discount rate on NPV.

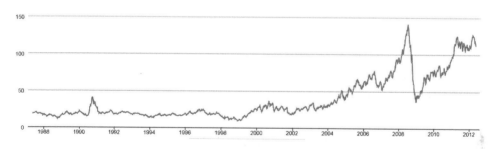

Figure 11.7 Series data showing Brent (North Sea) crude oil price since 1987 in US$ (source http://www.eia.gov/, accessed May 23, 2012).

Notice how we have assumed that the discount rate remains constant for every year of the project, reflecting a constant interest rate, risk premium, inflation rate, etc. We have also assumed we can predict income and expenses accurately for every year of the project.

This is extremely unlikely to happen in practice. Economic changes can come without warning and can be dramatic. Between 2007 and 2011 the price of oil, a proxy measure for the cost of energy, changed between about US $40 and US $140 per barrel.

This does not mean that calculating NPV has less validity.

Instead of calculating a single NPV for a project, engineers perform a sensitivity analysis. In other words, they calculate how NPV is influenced by changing important project parameters such as the choice of equipment, the operating life of the equipment, its reliability, the location, future changes in expenses and income, and many other external factors that could influence the financial results of the project. If they are clever, they will find ways to reduce the influence of these external economic changes on NPV.

It is vital to remember that the NPV calculation *is only used as a comparison technique*. The method provides a way to work out which is the best way to arrange a given project, to compare different technical options, and to compare the many different projects that the company could pursue. The aim is to select only the most attractive projects; to do that, the company needs a systematic and reasonably simple technique to compare them. Calculating NPV serves that purpose and has been so successful that there are very few organisations today that do not use it.

COSTS AND EXPENSES

In an engineering enterprise, it is engineers who provide most, if not all, of the estimates on what a particular project will cost. This data and the degree of confidence in the estimates are a vital input in financial decision-making.

Cost estimation is essential for NPV calculations that are used for comparing different projects or different ways to conduct a particular project. It is also a critical part of planning and managing a project. While we covered this briefly in Chapter 10, we will now examine this in more detail.

As engineers, we need to think in terms of the benefits and costs of a project *as a whole*, in the context of the land and society that will support it, including environmental and social costs and benefits. We will explore these latter issues in a later chapter on sustainability. Here we will focus on the economic costs: what has to be paid for during a project.

Practice concept 74: Capital and operating expenses

Estimating costs for an engineering enterprise relies on foresight: the ability to anticipate how costs are incurred and then the ability to forecast those costs.[5] Chapter 10 showed how engineers negotiate the scope of a project and then produce a detailed work breakdown structure; this is the first step. We forecast costs in terms of:

i) the cost of components, materials, and energy required to produce the finished product(s);
ii) the cost of human effort and machinery needed to transform the components, materials, and energy into the finished product;
iii) the cost of human effort needed to conceive, design, plan, organise, coordinate, manage, and supervise that transformation; and
iv) the contingency for unpredictable costs.

Item (i) is the most straightforward. Engineers work from a bill of materials and calculate the cost of every item using standard industry rates. Every engineering company maintains its own data on standard rates for estimating. The other items are more difficult to estimate.

Project life cycle costing is a relatively standard approach for estimating costs. There is an Australian standard (AS4536) that provides many helpful details and worked examples.

There are many different expenses, even in a relatively simple project. However, we can broadly classify these into three categories:

- Capital expenses (CAPEX), or 'acquisition cost': Money that has to be spent on planning, construction and all other preparations needed to ensure that an engineering enterprise is ready to commence its intended operations.
- Operating expenses (OPEX): Money that has to be spent during normal operations to produce the products and/or services that the enterprise was created to provide.
- Decommissioning expenses: We need to account for the cost of decommissioning at the end of a project, including the possibility that we may have to clean up contamination. Environmental regulations are steadily tightening up the requirements for engineering activities. Contamination that might have been considered acceptable at the start of a project might require costly site remediation 30 years later.

The last item, the contingency, covers the cost of unpredictable factors or risks. Chapter 10 explained risk analysis. The contingency allowance covers risks that have to be accepted, that cannot be avoided or transferred to others. Risk analysis deals with 'known unknowns', which are foreseeable unpredictable events. However, the contingency also has to cover 'unknown unknowns', unpredictable events that have not been foreseen or considered in the risk analysis.

In this chapter, I can only introduce you to some of the expense categories of which, as a young engineer, I found that I had little appreciation. You can learn about all the others if you make the effort to become involved in the planning and administration of a major project. This may not sound like engineering work at first. However, remember that reasonable familiarity with all aspects of finance is one of the distinguishing aspects of an expert engineer, particularly if you want to influence decision-making.

Capital expenses

Capital equipment usually includes large items of equipment such as vehicles and machines that need to be purchased for the project. It is also possible to arrange to rent or hire equipment, just like you would rent a car. We will cover this later. However, it is usually more expensive to rent equipment for the full life of the project. Equipment renting is much more attractive if the equipment is only needed for a relatively short time. An example might be an unusually large crane to help with construction for a few months.

The first expense associated with capital equipment is the purchase cost. Usually it would be necessary to pay a deposit when the order is placed so that the equipment supplier has enough money to arrange for the transportation and installation of the equipment, and its return in the event the purchaser decides to cancel the order. This might be between 10% and 20% of the purchase cost. The remaining amount would be paid once the equipment has been delivered, or has been installed and set up ready to operate, depending on the arrangement with the suppliers.

In addition to the purchase cost, there are usually additional costs to transport equipment to your site and then install it. There may also be warranty costs, which are optional contracts with suppliers to provide the required maintenance and spare parts in the first few months or first year of operation. Oftentimes, equipment has to be installed by specialised technicians with skills and tools that are only available to the equipment suppliers.

The purchaser may also have to pay for insurance to cover the possibility of loss or damage to the equipment in transit.

If the equipment is imported, it may be necessary to pay taxes or import duties and almost inevitably an additional fee for an agent to arrange for the clearance of the consignment through customs.

Finally, most equipment will require specialised spare parts to replace worn or damaged components through the service life of the equipment. While it is not absolutely necessary to purchase these at the time that you order the equipment, and it can be expensive to store these parts, it is always possible that the original manufacturer may not be able to supply spare parts when required. Reputable manufacturers provide spare parts for many years after selling machines and equipment. Sometimes spare parts can be purchased from third party companies after the original manufacturer stops supplying them. However, against this, one has to consider the possible cost of production interruptions if spare parts are not available. Even if spares are available from the original manufacturer, it can take many weeks, or even months, to obtain them when needed. There have been many occasions when extremely expensive machinery has been out of action because of the lack of spare parts costing only a few dollars each.

Opportunity cost

The next cost associated with capital expenses is the income that could have been earned by investing the purchase amount in another way with very low risk. It is almost invariably expressed as a percentage of the capital cost, usually as an equivalent annual interest rate.

Note that this cost is not included in net present value calculations because the discount calculation takes the cost of capital into account.

Depreciation

Equipment does not last indefinitely; every item of equipment has an expected lifespan and will eventually need to be replaced. One can consider depreciation as a form of enforced saving; setting money aside every year so that when the equipment finally needs to be replaced, there will be enough saved to purchase the new equipment.

The calculation is easy, in principle. If equipment that costs $350 million is expected to last 10 years, the depreciation is $35 million every year.

Once again, depreciation is not taken into account in an NPV analysis. However, once a project goes into operation, depreciation is accounted for.

It may not be so easy to forecast what the replacement cost would be. With inflation, one might be tempted to estimate that the replacement cost will be somewhat higher than the purchase cost. However, it is not unusual for the price of equipment to decrease over time relative to inflation. This may not reduce the replacement cost, however. It is more likely that the performance of a replacement machine will be greater than the original one, even though the price is about the same. This introduces a further complication. It may not be possible to purchase identical replacement equipment as manufacturers often update their designs, just as car models change every year or so. New equipment may not fit and a series of modifications may be needed to install the replacements. It may need more or less space, different connections, and the spare parts stored for the old equipment may no longer be compatible with the new equipment.

Estimating the life of the equipment can be just as difficult as estimating the replacement cost.

First, tax policies often encourage companies to depreciate capital equipment at an artificially high rate. From time to time, it is possible to obtain an additional tax deduction by claiming an accelerated depreciation for equipment purchased by companies. For example, a computer may be depreciated by one-third of its purchase price every year, even though it might last for five or six years.

Second, a decision to replace equipment may depend on many unforeseen factors. Equipment might have been used much more or much less than expected. It might wear out sooner than expected. A decision to save money on maintenance by reducing the frequency of maintenance inspections might lead to premature failure of the equipment. It is also possible to spend too much money on maintenance, replacing parts well before they have worn out, simply because it can be desirable to follow the recommendations of the equipment supplier. It is often possible to substitute longer-lasting components for the original ones supplied by the manufacturer, thereby greatly extending the life of the equipment.

For these reasons, the decision on the lifetime of the equipment and the replacement cost to calculate depreciation can be quite subjective. It certainly helps to have experience and to seek the advice of other people.

Taxation and accounting policies may determine the depreciation rates for different equipment – ask your accountants.

Of course, many companies continue to operate their capital equipment long after it has been fully depreciated, after the end of its 'accounting' lifetime. Once the depreciation allowance has been made, the annual cost of using the equipment is less and profits can increase as a result.

In Australia today, consumers are facing steep increases in electricity supply costs, but not because of increases in the raw energy cost needed to generate the electricity. The distribution infrastructure – power lines, transformers, and switching equipment – needs to be replaced and extended. This has happened long after the original intended lifetime. For many years, customers have enjoyed cheaper electric power because the old network infrastructure had been fully depreciated. The utilities have to install expensive new capital equipment over the next few years, and the depreciation cost of the new equipment will need to be paid by electricity users for many years to come. One would expect that most utilities would be gradually replacing and extending their infrastructure every year. What happened in Australia was that electric power utilities had very conservative policies until the 1980s. The distribution networks had considerably more capacity than was actually needed because the utility engineers planned for more future growth in demand than was actually needed. Smarter ways to distribute electric power and more commercial policies meant that the utilities were able to reduce their investment in new network capacity for several decades. Unfortunately, history has caught up with them and the efficiency gains have been used up. The old infrastructure has reached the real end of its lifetime and will fall apart if it is not replaced.

There are many other cost factors that I have not mentioned here; a list appears at the end of the chapter.

Operating and sustainment costs

Keeping the equipment going requires money. Most equipment needs operators or at least a proportion of the time of an operator. Operators need training, holidays, and supervision, not to mention unpredictable absence due to sickness or injury.

Most equipment also requires energy and consumable materials; for example, electric power, fuel, and lubricants. In a manufacturing operation, the cost of consumables can be very large, perhaps the largest operating cost.

From time to time, it will be necessary to inspect the equipment to see whether maintenance is needed. Then, if maintenance is required, it needs to be carefully planned so that it can be performed while the equipment is not needed. Some equipment can be maintained without being taken out of service, but that is relatively unusual.

The next question is whether the maintenance is going to be performed by people already employed by the enterprise or by contractors who come in when needed. It is often possible to negotiate a maintenance contract, sometimes with the supplier of the equipment, and other times with an outside company. The advantage for you is that you know exactly how much you're going to spend on maintenance. However, you are transferring the risk of unanticipated maintenance to an outside enterprise, so they will have to charge you to cover the cost of that risk. On the other hand, if you use people already employed by the enterprise to perform the maintenance, you have to pay them whether they are needed or not.

Either way, people with experience can approximate what proportion of the purchase cost should be to set aside to allow for maintenance. One would normally allow between 1% and 2% for well designed and constructed buildings. For machinery and vehicles one would allow between 5% and 15% (the higher figure applies to equipment requiring extensive maintenance, such as mining equipment) to cover the cost of maintenance.

The cost of maintenance depends on how it is planned. Unplanned maintenance – fixing things when they break – is inevitably much more expensive than planned maintenance. When maintenance is planned, equipment can be taken out of service during production shutdowns. Therefore, there will not be a loss in production. When equipment breaks down unexpectedly, however, costs escalate rapidly. Not only may production have to be sourced from somewhere else, even from a competitor at a greatly increased cost, but the entire workforce may also have to be paid during the enforced shutdown of the plant.

The combination of operational support and maintenance is often referred to as 'sustainment'. This is a term that originated with defence forces that need to carefully consider the cost of expensive assets, such as aircraft and ships. Sustainment refers to a systematically planned and managed process to ensure the highest likelihood that the equipment will be ready to operate when it is needed.

It can be difficult to manage maintenance well and many companies, even well managed ones, occasionally incur large losses due to maintenance and operating mistakes, as well as from inadequate training, supervision, or investment in appropriate supporting systems. These losses can amount to between 30% and 50% of turnover. However, as a great deal of that cost only appears as lost production opportunities, it often does not show in company accounting systems unless the systems were designed to capture this information.[6]

Equipment hire costs

As explained before, if equipment is only going to be used for a relatively short time, it can often be more cost-effective to rent it. The company providing the equipment, of course, has to make enough money from renting it to cover the cost of purchasing the equipment and having it available when you need it. They cannot expect to be hiring out the equipment all the time. Therefore, the rental company takes on the risk that the equipment may only be used for short time periods, and therefore passes on the cost of this in the form of the hourly cost to hire the equipment.

There is another alternative: leasing equipment. However, this is simply an arrangement in which the capital acquisition cost is spread out over time. In all other respects, leasing is the same as buying the equipment, except that in some leasing arrangements, the leasing company takes back the equipment at the end of the lease (or repurchases at a discount) and may also offer some maintenance arrangements.

Accommodation and land

Every engineering enterprise needs space. Therefore, part of the cost is securing access to the space.

Of course, the first step is to secure the necessary approvals for using the desired space. For example, one can't simply start digging up a mineral deposit, refining it, and then transporting the product to a nearby port.

In Australia, a mining or industrial project can require as many as 180 different approvals from various governmental agencies, quite apart from the owners of the land and possibly other stakeholders such as the local community. While there are fees for many of these approvals, the fee usually reflects the cost of evaluating and deciding on the particular approval application. The cost of preparing the necessary documentation is much greater than the cost of the licence fee.

Once the necessary approvals have been secured, it will usually be necessary to pay some kind of monthly rent for the space or land that is being occupied. Sometimes, additional fees are needed to gain access to the necessary sites. For example, if trucks and vehicles need to cross private land in order to access a site, the landowner may be entitled to compensation in the form of access fees.

Labour costs

One of the most frequently misunderstood factors in estimating the costs of engineering project is the cost of labour.

It is easy to focus on the direct salary or wage cost of the labour: what a person gets paid every week, month, or year.

As explained in Chapter 1, it actually costs much more to employ a person than what that person receives as a wage or salary.

The risk of misunderstanding is compounded by the same terms being used in different ways, such as 'direct' and 'indirect' costs. Always ask what these terms refer to in any situation.

Firstly, there are standard 'on-costs', which are usually represented by a percentage of the wage or salary, with the following common components:

- Annual leave and public holidays: most workers are paid when they take their leave, so there is an additional cost in proportion to the amount of time taken as leave to ensure that someone is present in the workplace. For example, if workers receive four weeks' annual paid leave, then a company must allow an additional 8.33% to cover the cost of their pay while they are on leave. In some countries, workers may receive extra pay while on leave to cover the cost of travelling to and from their family homes. Allow 8–11% for leave.
- Sickness leave: it is normal for people to be absent due to sickness for a few days every year. Different enterprises allow for this in different ways. Allow 2–3% to cover sickness absences for the worker or for them to care for their dependents.
- Employer liability insurance: depending on government legislation, employers are required to contribute to an insurance fund that compensates workers for industrial accidents. In many countries, an employer will also be required to contribute (or it will be customary for the employer to contribute) to health insurance for workers and possibly their immediate family. Normally, allow 3%, but in some cases, the cost of this can be up to 10%.
- Superannuation or pension fund contribution: again, depending on government legislation, an employer will be expected to contribute towards a pension entitlement for each worker. Workers are also usually encouraged to pay part of their salary into a pension fund. However, this is separate from the contribution paid by the employer. Allow 3–10%.
- Payroll Tax: again, depending on local government legislation, a company may be required to pay a tax proportional to the salary of every worker employed. Allow 3–5%.

Total on-costs are typically 25–35% of the wage or salary that is paid to a worker, but can be higher if extra hours are worked regularly, if a regular bonus is paid, or if workers receive a special payment upon the termination of their employment.

Next, there are several indirect costs associated with employing labour. These depend on the situation, particularly if the location of the work is remote.

These include the cost of administration, accounting, and complying with all the necessary regulatory requirements associated with employing people. Larger companies employ specialist payroll accounting and human resources staff to perform these tasks. The administrative and managerial cost can be 10–15% on top of every worker's salary.

Although it might have been covered above in the cost of accommodation and land, one should not forget the cost of office accommodation or workspace; no one can do productive work for long without sufficient space and a comfortable working environment.

One of the most expensive components of support costs is supervision. This can be the most expensive factor in a developing country where the cost of employing unskilled or semiskilled labour may appear to be very low. Good supervisors are in demand everywhere, particularly in a developing country, where most workers will stop working if there is no one to supervise them and tell them what to do. The job of a supervisor can be very complex and requires a great deal of experience, particularly in an engineering production environment.[7] The cost can be 30% of the worker's annual salary in an industrialised country environment. In a developing country environment, good supervisors can cost considerably more, as shown in Table 11.1.

Many workers will require special tools, machinery, clothing, and protective equipment. Companies are often expected to provide uniforms for their workers, which can also contribute to morale and prestige, particularly if the enterprise has a good reputation in the local community.

There are still further costs to take into account.

If the work is being performed at a remote site, it will be necessary to pay for transport and accommodation for people working at the site. They will need reasonable security. If the work is being performed in an environment where there is social or political instability, the cost of providing security can be one of the largest components of the cost of labour. If the need to provide security interferes with the work itself, then the lost time and production is an additional cost.

Part of the time spent at work will be needed for safety briefings, transport between work locations on-site, rest breaks, eating, refreshment, and toilet breaks. This is often labelled as 'non-productive time' (as opposed to 'time on tools'). It is easy to see socialising during breaks and around a coffee machine as non-productive time. However, as we have seen and discussed, engineering enterprises rely on distributed knowledge and collaboration; the knowledge required for the enterprise is carried and transformed by the minds of the people who work for it. Social contact is critical for this. Therefore, time spent socialising is a necessary component of working. Obviously, there has to be some agreement on what is reasonable and what is unreasonable. Different people will require a different proportion of their time for a satisfactory level of socialisation.

Another portion of the time spent at work will be needed for training. Supervisors will need to spend face-to-face time with production workers; part of that time will be used to develop their skills and abilities.

A normal 38-hour working week provides almost 2000 working hours per year. However, it would be unrealistic to assume that a normal worker would achieve more than 1500 productive working hours in a year (125 hours per month, about 29 hours per week).

Next, the labour estimate is adjusted by multipliers to cover project-specific 'multiplier' factors (greater than 1) that influence individual worker productivity. For example, a project that requires two or more shifts[8] will normally require some workers, particularly supervisors, to be present for handover briefings for people on the following shift. Time-critical tasks may require additional labour to be present to ensure that they are done on time. Some work may also require time spent on cleaning up the workplace.

Further factors are applied to account for interruptions caused by, for example, wet weather on a construction site.

Table 11.4 Example of hourly labour cost calculations.

	Expatriate skilled trades (industrialised country)	Local skilled trades	Local labour
Direct cost			
Hourly rate with on-costs	80.00	7.00	6.00
Indirect cost			
Recruitment	4.00	0.30	0.20
Supervision	25.00	25.00	25.00
Training	2.50	1.20	1.40
Non-productive time	8.00	0.70	0.70
On-site shelter	2.00	1.00	1.00
PPE[9]	0.80	0.80	0.80
Workshop equipment	4.00	1.00	1.00
Small tools & consumables	6.00	2.00	2.00
Light vehicles	9.00	5.00	5.00
Site office overheads (light, power, first aid, security)	2.00	2.00	2.00
Accommodation	8.00	–	–
Total indirect costs	71.30	39.00	39.10
Administration (10% of direct and indirect costs)	15.13	4.60	4.51
Total cost (USD)	166.43	50.60	49.61

Hourly labour costs for an engineering project in a low-income country (in US $) that are adjusted for costs in 2014. These calculations are based on research notes, not including profit margin. Note that supervision accounts for 50% of local labour cost, while the direct rate of pay with on-cost is only 12% of the total local labour cost.

Finally, we need to remember some important lessons from Chapter 9: the amount of work performed by a person in a given time is highly variable and depends on many factors, including training, motivation, supervision, appropriate tools, equipment and materials, and interactions with other workers. In a low-income country, local workers are usually much less productive than skilled workers brought in from an industrialised country because they have had less training. Their hourly cost may be multiplied by a factor of 1.5–2.5 to account for this.

It is much more important to think about productivity than direct labour costs.

The cost of labour, including indirect components, is usually only a small component of an engineering enterprise. The labour cost proportion ranges from around 8% in a low-income country to 15% or 20% in an industrialised country. Capital costs such as equipment, depreciation, land, materials, energy, transport, and financing costs make up the rest. Labour productivity, therefore, rests on the ability to make effective use of these other resources.

Our research shows that most engineers, and many other people, only perceive labour costs in terms of the most obvious direct wages or salary component. As we can see from this explanation, this can be a smaller component of the total labour cost. This misperception can have serious and often unfortunate consequences, as we shall see in a later chapter.

For example, a large earthmoving machine can cost $150–200 an hour. Machine productivity depends much more on the skill of the driver than on what the driver gets paid. Part of the driver's skill is the ability to drive the machine carefully to minimise non-productive time for machine maintenance: the fastest driver is not necessarily the most productive.

As with the capital expenses, there are many other expense categories not mentioned here (see the life cycle costing guide in the online appendix for a list of other categories). Experience plays an important part as well; experienced engineers develop an understanding of how much labour is required to achieve a given result in a given situation. Estimating is a special skill and expert estimators are vital to any engineering enterprise.

Practice concept 75: The cost of an engineering venture depends on policy drivers

Novice engineers often think that . . .	Expert engineers know that . . .
Estimating the cost and time to complete a project is a relatively simple aspect of engineering. It is simple because faculty staff members don't see the need to teach it. You only need to be able to add up a column of numbers.	Estimating the cost and time to complete a project is anything but simple. Engineering projects inevitably trade-off four main policy factors: quality, scope, schedule, and cost. The appropriate trade-off needs to be chosen to match the objective and client needs.

What could be simpler? Once you know the cost of all the components required to make something work, all you have to do is add up the costs of each of the components to work out the cost of the finished artefact.

This idea, perhaps exaggerated slightly, describes how most students think about costs in engineering. Reducing the cost means finding cheaper components or improving the efficiency to reduce consumption.

Working out the real economic cost of an engineering venture is much more complex.

At this stage, all I want you to learn is that the cost of an engineering venture depends on many policy factors, often known as 'policy drivers' or 'policy settings', and most are related to human behaviour issues. Policy settings emerge from the most fundamental constraints and needs that must be met by an engineering project.

Most projects are influenced by four different policy factors:

Scope: what the project intends to achieve. For example, how long is the road to be constructed? How wide is it, how many lanes of traffic will it carry, and how wide is each lane? What kinds of vehicles will use the roads, and how fast do they need to travel? What kind of soils will provide the foundation for the road? What are the constraints imposed by the desired route for the road – are tunnels or bridges needed? How many bridges will be needed, how long do the bridges have to be, and how high will the supporting piers be?

Timescale or project schedule: When is the road needed? Do we have access for construction 24 hours a day, or do we have to do the construction at night with minimal interference with existing traffic on the roads where the bridges will be constructed? Will entry and exit points be added? Is the land available now? Is there appropriate access for machinery and materials to be carried to the road construction site when needed?

Finance and budget, trade-off between CAPEX and OPEX: Do we have to reduce the capital cost of the road to an absolute minimum? Can we afford round-the-clock construction to open the road as soon as possible? Or, can we afford to spend more on road construction now in order to reduce maintenance costs in the future? What is the timing for capital investment in the road? Does the money have to be spent in the current financial year? What kind of monitoring and reporting is required to satisfy the requirements for financial auditing?

Quality: What is the quality requirement for the road? What will be the maximum vehicle speed, and hence the surface quality? What guarantee do we have to provide that the road will not deteriorate under given traffic conditions or even under extreme traffic with overloaded vehicles? What is the requirement to minimise road noise for surrounding residential or recreation areas like national parks? What are the requirements for aesthetic appearance when the road is completed? What landscaping and graffiti-prevention treatments will be needed?

Within any one project there will be a mixture of constraints. For example, there may be pressure to reduce the cost in some parts of the project in order to be able to spend more in others. There may be a tight timescale for some aspects of the project that need to be completed early, while a more relaxed schedule can be applied to other less urgent aspects. Quality may be an issue for some critical components that, if they fail, have very severe consequences. Many projects have failed because cost-cutting techniques were applied to some critical components that subsequently failed because they were procured from low-quality suppliers to save a few cents. Saving 20 cents on a seal can cause premature failure of a pump worth millions of dollars; it is clearly preferable to impose high quality requirements on these critical components.

There are many other factors that influence the cost as well, some of which were mentioned under the four main categories above. Minimising the risk of environmental and social impact can also influence the cost; expensive precautions may be necessary to avoid the possibility of harmful effects on the environment or disruptions to the social life of communities nearby.

One of the most critical tasks for an expert engineer is to align the written statement of project objectives and requirements with the needs of the client, even though the client may not be able to clearly explain what they really need. The statement must clearly explain the client's needs. One of the most important attributes that needs to be described is the client's tolerance for uncertainty and risk. This is related to the cost of capital for the engineering work, which determines the discount rate used in NPV calculations.

For example, high risk investors in a mineral processing plant in a remote and politically unstable part of the world are not really interested in the long-term operating reliability of the equipment. They want to be able to extract the mineral products as quickly as possible to take advantage of high prices and local tax concessions, and thereby provide the investors with a fast financial return. With a high cost of capital

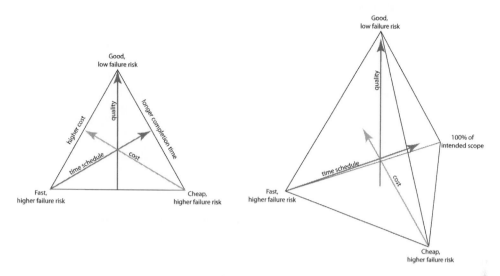

Figure 11.8 Trade-offs in an engineering project. A given project will be represented by a point located somewhere inside the triangle or tetrahedron. A project with the highest quality result *may* take longer and cost more. A project performed with minimum cost (initially) *may* take more time than a project completed as fast as possible. The left-hand diagram is a simple triangle showing possible variations in quality, timescale, and cost. The right-hand diagram portrays a three-dimensional tetrahedron in order to include scope as well; non-critical scope items are often sacrificed to improve cost, timescale, or quality. These are only illustrations to demonstrate that the main requirements are all interlinked.

and a consequently discounted rate, deferring expenses and starting to earn income as early as possible helps achieve a positive NPV more easily.

On the other hand, long-term reliability is probably much more important for a government investing in expensive defence equipment. A fighter aircraft may have to stand idle for much of its life with only minimal flying for training purposes. Fighter aircraft are very expensive to fly and consume a lot of fuel. However, in the event of a conflict, the security of the state and its entire population may depend on being able to operate the fighter aircraft in extreme conditions unanticipated by its designers, years after it first went into service. Therefore, long-term reliability is extremely important.

Ideally, an offensive fighter aircraft should never be used as its designers intended, to destroy other aircraft and kill people. One of the best ways to avoid this outcome is to demonstrate the real possibility that they might be used, and that they would be reliably capable of rapidly destroying any other aircraft threatening a state's security. This also means that the reliability and performance of a fighter aircraft needs to be periodically demonstrated, without harming people, to ensure that a potential aggressor remembers that launching an attack would be a very big mistake.

Effects of policy settings on cost

Understanding what happens when projects are constrained by each of the four factors will help you see how each of these affects the cost in a different way.

Budget: least capital cost

Some projects have strict constraints on the capital expenditure budget, usually because the cost of capital is higher and therefore a larger discount rate applies in the NPV analysis. Therefore, to save on capital expenditure, an engineer will arrange to procure equipment that is cheaper even though it will not last as long, may not be as reliable, may come from a less reputable supplier, and may even have inferior technical performance. Once production has started, and the venture produces a positive cash flow, it would be possible to replace this equipment with more reliable and technically superior equivalents. At least, that might represent the initial intentions. The likely difficulty is that cheaper equipment may break down without warning, interrupting production, possibly even necessitating the purchase of an equivalent product from a competitor in order to fulfil customer orders. In addition, it may not be that easy to substitute more reliable equipment once the plant is in operation. It is quite conceivable that equipment from a more reputable manufacturer with better performance and reliability may not be available any more in the same size as it was when the plant was initially constructed. Simple substitution may no longer be feasible. Typically, in-service plant modifications cost 3 to 5 times as much as they would have if they were planned and part of the original plant construction.

Therefore, saving capital expenditure by purchasing cheaper equipment does not just defer the cost of more reliable equipment until later in the project when it can be afforded. The cost of substituting more reliable equipment can be far higher. Nevertheless, many projects go ahead on this basis; these can be tough decisions that often frustrate engineers who cannot appreciate the influence of the high cost of capital, particularly for smaller companies undertaking risky projects.

Many of the issues arising from the requirement for least capital cost can be avoided. With sufficient effort on detailed planning, searching for the most cost-effective design option, systematic checking, and careful negotiation with suppliers, it is possible to significantly reduce capital cost without necessarily compromising on quality, reliability, or even technical performance. Of course, this means spending more during the early engineering and design phase of the project before anything has been built, with the objective of long-term savings. This is also known as a life cycle costing approach, the idea being to minimise the total cost of ownership throughout the life of the plant. In practice, however, this can be difficult. The difficulty for an engineer arises in trying to help a client make the best decisions with an appropriate policy that takes into account the cost of capital and long-term planning.

One of the strongest arguments in support of accepting a higher capital cost lies in the inherent uncertainties of predicting operating costs. Investing more at the start in order to reduce later operating costs reduces potentially large uncertainties because operating costs can be so hard to predict, simply because the forecasting timescale is much longer. Over a period of several years, interest rates, energy costs, and labour costs can vary in completely unpredictable ways, as we have seen earlier in the chapter. Skilled labour can quickly become scarce if many similar projects begin at the same time. While the cost per hour may not change much, lower skills quickly impose high costs because of down time and schedule interruptions. Later, we will see how reducing uncertainty adds value, so the higher initial cost can be offset by a higher commercial value if the project works as expected.

The reluctance, particularly by small companies, to accept a higher capital cost can also be explained by the desire to retain control of a project. Accepting a higher capital cost may be possible if the company can borrow more money, and a strong argument based on reducing uncertainty may be helpful in securing the loan. However, if much more investment funding is needed, the original investors may lose their influence.

Cost implications of keeping to a strict schedule

There are many occasions when the project schedule is more important than the capital cost. As we have seen over the last few years, commodity and energy prices can fluctuate wildly when markets are unstable. Quite often, there are large profits to be made just by being in operation a few months earlier than anyone else.

Some government projects may also require a strict schedule to ensure, for example, that a project is completed before a coming election.

When the schedule is rushed like this, it is usually not possible to complete many aspects of the design before plant construction commences. A product may not have been completely designed and tested before the production line has to be installed. Inevitably, in these circumstances, there are many design changes, and often some of the work has to be scrapped and redone. For example, there might have been insufficient time for a full set of site measurements and soil samples before constructing a major installation. Once the construction work starts, site engineers will report difficulties with obstacles and unexpected soil conditions that require changes to the original design.

By the time construction has started, there may be even more design changes.

Usually, these changes are much more expensive than the cost of the materials and labour might initially suggest. Contractors often impose high additional charges for variations to the design, because they know that making changes to an existing structure is much more difficult. Also, most design changes disrupt a construction schedule and subcontractors may have to be compensated for schedule changes. An important aspect in any construction project is the need to have subcontractors committed to the project at certain times when their work will be needed. It is only possible to do this by arranging a detailed schedule in advance. Once the schedule is changed, some subcontractors may have to decline other projects in order to be available at a different time to the one originally scheduled, and will therefore need to be compensated for the loss of income. Special equipment like cranes, which are hired and transported to the site, may have to be returned and then brought back again later, which incurs a large additional cost.

Cost of maintaining original design scope

If the project scope cannot be negotiated, if it is not possible to make design changes, the cost of the project can be much higher than originally expected. Often, this occurs because of unanticipated design issues.

Here is an example from an interview with an engineer recalling a recent offshore gas project.

'Our engineers wrote a performance specification that required equipment to operate between −27.1°C and +73.1°C. This performance specification was *reproduced by the equipment manufacturer when ordering components from other contractors. As a result, some of the contractors and the equipment manufacturer decided to include the cost of specialised environmental testing equipment when they submitted their cost estimates for approval. These were hurriedly approved without detailed checking. Later it became apparent that the cost of the equipment was far higher than originally expected. The cost increase was caused by the necessity to demonstrate satisfactory performance with a temperature tolerance of 0.1°C since the original documents specified temperature with equivalent precision. It would have been just as easy (and so much less expensive) to specify an operating temperature range from −30°C to +75°C. The contractors and the equipment manufacturer would then have been able to use conventional environmental testing equipment. As it was, three specialised environmental testing facilities were constructed when only one would have been sufficient: one for the equipment supplier and two others for contractors. By the time this was discovered it was too late to make any changes. Engineers working for the equipment supplier did not question the temperature specification. Instead they said "they must have known what they were doing if they defined the temperature range so precisely." Also, the equipment supplier was able to make a significant additional profit because of the increased equipment cost.'*

This example demonstrates how small details in the scope of an undertaking can have a large impact on cost. As an engineer, it is important to be able to ask contractors to comment on the specification in the early stages of the project and point out which aspects are going to have a large effect on cost. Documents that call for significantly more precision than is required (perhaps because of inadequate checking or oversight) inevitably add significant cost.

Some contractors will offer to negotiate the scope of the specification in order to achieve significant cost savings. They will suggest seemingly small adjustments to performance criteria and offer a large cost saving as a result, often to secure a competitive advantage and win the contract.

Cost implications of quality assurance

Quality requirements can also have a similar effect and increase capital cost, particularly in defence acquisitions.

One important aspect of tight quality assurance is the requirement for traceability and documentation, particularly in the aircraft industry. This means, for example, recording the detailed origin of each component, even down to the batch of material from which it was manufactured.

Subsequently, if there is a component failure on an aircraft, all the components manufactured from the same batch of material can be tracked down and replaced before they cause similar failures. This requires that manufacturers maintain a vast database of information on how, when, and where components were manufactured and stored and in which aircraft they have been installed, even the details of subsequent repairs or modifications. These measures enable modern aircraft to achieve extraordinary reliability: major failures are so rare that they are reported in the international news when they happen.

Not only is the documentation and supervision required for this level of quality assurance very expensive, but the requirement for certification also limits the number of contractors who can be employed to handle this kind of work.

Obviously, contractors who are regularly supplying components or systems to the aerospace industry could be expected to maintain high levels of quality in their products. However, there are many other companies that have gained certification for compliance with certain quality assurance standards, such as the ISO 9000 series. Unfortunately, this does not mean that they also can consistently produce high-quality products. Certification only requires a company to be aware of the quality level it maintains in its products by maintaining appropriate documentation and inspection systems.

Company culture is critical in delivering high-quality performance. Everyone involved has to adopt an extremely cautious and conscientious approach in order to maintain high levels of product quality. There is little room for shortcuts and cost saving ideas that could have long-term implications. Therefore, it is difficult for a single organisation to be able to respond to clients with different levels of quality expectation.

The quality issue demonstrates that project choices (and therefore costs) are not determined by economic factors alone. Defence applications are just one example in which cost considerations may be secondary to performance and effectiveness as a deterrent. Public safety, particularly the perception of safety, minimising any perceived risks to the environment, and safeguarding the reputation of the firm are also important factors in choosing an appropriate strategy for a particular project. All these factors influence the cost of undertaking an engineering project.

Practice concept 76: The investment decision

Having reached this stage where we have forecast all the costs and income as accurately as possible, with clearly understood uncertainties, it would be easy to think that we now have everything we need to make an informed investment decision. Assuming that the net present value analysis using the required discount rate demonstrates that the project is the best on offer, and after taking all the foreseeable project costs into account, the investor will surely decide to proceed with the project. . . .

Unfortunately, life doesn't work like that. In my research, I have found many engineers who have reported feeling disillusioned and frustrated with investment decisions made by engineering enterprises. Oftentimes, engineers think that the firms are run by lawyers and accountants who possess very little understanding of engineering matters and therefore make inappropriate decisions. By understanding how financial investment decisions are made, engineers can develop a clearer understanding that will help them influence decisions more effectively.

In the last phase of my work with the robot sheep-shearing project in the late 1980s, I used to wonder why the wool industry found it so difficult to move towards making the long-term investment required to commercialise the technology they had spent so much money developing over a decade or more.[10] Even the highest estimates of the cost to commercialise robot sheep shearing in Australia were in the range of $100 million. At the time, this amount represented just the cost of the bottom three or four floors in any of the dozens of office towers that were currently under construction in major cities around Australia. There was obviously plenty of money available to invest

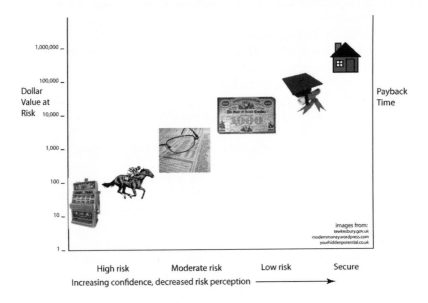

Figure 11.9 Family investment choices.

in city office buildings, without any firm assurances that the office space would actually be needed. Why was there so much reluctance to invest in technology to harvest wool that would reliably grow every year on sheep, particularly when the compensation cost of injuries and long-term partial disablement for shearers doing the job manually were taken into account? This puzzled me at the time and I could not explain this. It puzzled me even more when, in 1990 and 1991, there was a sudden collapse in property values and many of the people who had invested money in expensive office towers lost significant portions of their investment.

Since then, I have learnt much more about investment in engineering projects.

Engineering investment is no different in principle from the investment that your parents and families, and even yourself, have made in your education and upbringing, in economic terms, as well as in personal devotion, energy, effort, and time. By the time you start working productively as an engineer, perhaps three years after graduation and the start of your first job, perhaps at age 25, your parents and the community could have invested up to AU $2 million in you, which includes your education expenses, foregone earnings, community services, and societal infrastructure. Your employer will have invested perhaps two years of your salary in developing your skills and knowledge. This estimate is based on research in Australia and USA; the magnitude of the investment varies in different regions. It will take a long time for you to create benefits that would make that investment look like a wise decision. Of course, parents and the community do this for non-economic reasons as well.

Where, when, and how much people choose to invest simply cannot be predicted in advance; everyone has different criteria, and these change from one moment to the next. However, it is possible to observe overall patterns of behaviour, and Figure 11.9

attempts to capture some of these patterns, at least from the point of view of a private investor.

Most people see houses and land as the most secure form of investment; for most people, this is where they make their largest investments. It is, after all, a place to live, so it serves two purposes – a long-term investment and a personal residence.

At the other end of the scale, most people like to gamble every now and again, betting on a horse race or throwing a few coins into a poker machine, more for entertainment than for serious moneymaking. The entertainment value is important because, as mathematicians and statisticians will tell you, you have to be very lucky to make money when the dice are so heavily loaded against you. Institutionalised gambling is a profitable business, and those profits come from the gamblers themselves. In the long run, it is the owners who make money and governments that collect taxes. Gamblers pay both.

There are two obvious differences between these different forms of investment.

First, investors perceive houses and land as very secure forms of investment. It would be foolish to pretend that gambling on horses is a secure investment. In other words, there is a large difference in risk perception by most people.

Second, most people spend far more investing in houses and land than they do in a gambling casino or at the racecourse. In other words, most people spend relatively large amounts on secure investments and relatively small amounts on risky investments.

Third, few people would gamble at a racecourse or a casino if they had to wait 12 months for their winnings. The entertainment value depends on being paid almost immediately. At the same time, few people would expect to make a short-term profit from a house and land investment. They are prepared to wait many years or even decades.

The diagram attempts to capture this. On the left-hand side is a logarithmic scale that represents the money that a person would invest. The horizontal axis represents a risk perception with investments that are perceived to be more certain on the right-hand side. The right-hand vertical axis represents the repayment timescale. There are no numbers on the horizontal and right hand axes because the diagram attempts to represent qualitative ideas. The line of images represents an intuitive risk boundary: anything largely above the line is essentially too risky.

It is also important to appreciate that the financial return expected from a given investment is related to the perceived risk. Remember the concept of 'risk premium' from the earlier discussion on interest rates. If the perceived risk is negligible, such as lending money to a stable government with guarantees, the financial return expected would normally be equivalent to the central bank overnight cash rate. On the other hand, when investing in a company, investors factor in the possibility that they might miss out on a profit or even lose their investment completely. They expect a significantly higher interest rate, or return, on their investment. If the rate is around 10% higher than the central bank interest rate, we could guess that investors are thinking that there might be a 10% chance of losing all their money. The bond certificate image in Figure 11.9 is positioned lower on the confidence scale than a home, as lending money to a company is perceived to be somewhat more risky. Shares are usually seen to be more risky than bonds; if a company goes into receivership (i.e. runs out of money and defaults on payments), creditors, including workers that are owed wages and entitlements, and bond holders may be paid before shareholders get any money back. Of course, shares have to provide a higher return to investors to compensate for this greater perception of risk.

Wise investors spread their risks. They invest part of their money in different forms of investment at the same time: a family home, education, bonds, shares, and even small amounts on gambling.

What I am trying to show in Figure 11.9 is the general relationship between perceived risk and the amount invested in each form of investment. On the whole, people will invest the greatest part of their money in more secure, less risky investments, and less money in more risky, speculative investments. While everyone likes the thrill of a speculative investment that suddenly returns a large windfall profit, most people place the greatest part of their financial assets in 'blue-chip' investments. These are companies that most people view as being well-managed companies that are likely to provide relatively secure dividend payments and reliable capital appreciation over the long-term. Depending on market conditions, financial advisers will suggest gradual movements between property, government bonds, and cash investments, and blue-chip share investments. At the same time, they will often maintain a modest holding of more exciting speculative investments. In other words, they follow the same characteristic as we see in the diagram – more investment funding is directed at investments that are expected to produce a more reliable return on investment.

Education is also an expensive investment, just like houses and land. Even if you are not paying fees yourself, either your family or your government or both have been paying large amounts of money for you to sit in classrooms for many years. Depending on the state of the economy, you could have been earning lots of money working instead of sitting in those classrooms. Both of those expenses (actual and potential) have to be taken into account as the imputed cost of education. In Australia today, education and child rearing can be seen as a combined private and state investment that amounts to approximately $2 million by the time you are working productively with a university qualification. It is widely perceived to be a secure form of investment and fits the graph rather nicely. There is a long payback time; it takes decades, well into or even after your working career, for the benefits of this investment to be realised.

Sometimes, this general behaviour pattern changes across an entire society, even across many countries at the same time. For one reason or another, investors change their ideas about risk, and may gradually become more and more confident about their investments. This often happens when share prices seem to be rising more or less all the time. Rising confidence seems to affect large firms as much as individuals; firms start investing more of their funds in much more risky ventures. It also happens because fewer and fewer people with first-hand memories of the previous financial crash retain sufficient influence on investment decisions. It is often referred to as 'herd mentality', manifested as a decreased sensitivity to risk, partly because of the idea that 'everyone else is doing it; they can't all be wrong'. They can, of course, as we saw most recently in the GFC, the massive global financial crisis of 2008. With the benefit of hindsight, we can see how banks managed to drop their usual reluctance to lend money to people who would have difficulty repaying it. I was in the USA in 2006 and I watched TV in utter amazement: in prime time, one after another, large banks with well-known names were offering to lend Americans 120% of the value of their home, with no interest payable for the first 12 months, with minimal security. I thought to myself, 'This has to end in tears'. It did, approximately two years and four months later. At the time,

I expressed my concerns to one of my friends, who worked as a senior banker. His response?

'Oh, our investment people have told me it's quite okay, nothing to worry about.'

The same general principles apply when an enterprise decides to invest in a project, even an engineering project. Naturally, the amount of money that directors are prepared to risk may be quite different from the household investment situation. A multinational oil company or bank may be able to direct several billion dollars at a single investment once they have assessed the risks and the likely return. Why, then, will that same company not approve a relatively modest investment, perhaps a few hundreds of thousands of dollars, to develop new engineering technology? This is a question that frustrates many engineers. Engineers are able to see possibilities for technical improvements that would greatly increase the profitability of the enterprise. Often, an engineer will prepare detailed proposals and a full NPV and analysis demonstrating the required rate of return on investment. Somehow, however, their ideas often seem to be ignored.

Practice concept 77: Engineering finance depends on value creation

Students and novice engineers . . .	Expert engineers know that . . .
Mostly do not know how to think about the value that engineering creates.	Engineering creates value by reducing the need for human effort and resource consumption, and also from improved predictability, reduced uncertainty, and often from technical ingenuity, as well.

What does this mean for you as an engineer?

The first implication is that most investors have very little idea about the engineering work that is required for a given project. Therefore, their decision to invest depends on how they *perceive* your ability to make the venture work in a way that provides value for them. Value, as explained earlier in this chapter, almost always has an economic dimension, but also involves reputation, prestige, safety, and a positive social contribution as well. Understanding how a particular client perceives how their project will create value (sometimes called a 'value proposition') is immensely helpful. Often the only way to find out is to engage the client in conversation, talking about each of these aspects in turn. In any organisation, there are many different people, so it is a good idea to discuss value with several people at different levels of the client organisation; you may be surprised at how different the results can be. Remember to utilise your accurate listening skills and take notes!

Achieving a result with the least requirement of human effort and resource consumption, which are usually measured in economic terms, is essential, of course. However, uncertainty plays a big part in the perception of economic value, not only external uncontrollable uncertainties, but also on how reliably investors *think* that you can make your promises come true. Therefore, the more you can reduce the apparent risk in an engineering venture, the more likely it is to receive financial support and the more money is likely to be invested.

Most investors like to know that they are investing in something similar to what has been done before, which greatly increases the likelihood of it working again.

Remember that reducing uncertainty usually helps to reduce the direct cost of a project as well as influencing the perception of investors. As an example, consider a simple bridge consisting of a beam supported at each end. Being able to accurately estimate the maximum loading on the bridge enables a designer to avoid using excess material to allow for uncertainty in loading estimates. It may also be possible to use a lower factor of safety, the margin by which the maximum design load is increased to cover unpredictable factors. This means that less material is required for the bridge beam, the supporting structure does not require as much strength, transporting and installing the beam will be less expensive, and time may also be saved.

Therefore, the challenge for you as an engineer is to find ways to ease the concern of investors, to make them feel more confident, and to change *their perceptions* about the risk of investing in your capacity as an engineer.

Naturally, this also means that you have to be very good at working with human uncertainty and perceptions, reducing the real risks that something will go wrong, and taking advantage of upside opportunities when they present themselves.

Practice concept 78: Staged project decision-making

In practice, when it comes to investment decisions, this principle leads to a series of step-by-step decisions, which is the contemporary staged project decision-making process adopted by most large corporations.

A project has to pass through several stages of decision-making before a company will commit the full amount of money required to complete the project. Figure 11.10 illustrates the process. First comes the idea for a project. The first 'decision gate' answers the questions, 'Is this idea worth investigating?' and 'Is it the kind of project that we do in this company?' If the answers are 'Yes' the company will commit a modest amount of effort (and money) to investigate whether the project could be worthwhile for the company. The next decision gate answers the question, 'Do we need this project?' Again, if the answer is 'Yes' the company will invest a more substantial amount of money and effort to conduct a financial feasibility study and prepare a business case for the next decision gate. At each subsequent stage, the company has to decide if the amount of money needed to take the project to the next stage is worth the risks. In other words, is the ultimate earning potential of the project enough to justify spending more money on it? One of the fundamental principles behind this approach is that the company only moves on to the next stage once a clear decision has been made to proceed.

Most major companies have their own variation with a different name for what is, in essence, exactly the same approach.[11] At the time of writing, Rio Tinto called theirs a 'capital project framework' with six stages: conceptual, order of magnitude evaluation, prefeasibility, feasibility, implementation, and completion. Woodside Petroleum labelled their process OPREP; it has seven equivalent stages: project initiation, a feasibility study, concept selection, basis for design and project specification, design-construct-install, commission-start-up-handover, and operate. Alcoa had fewer

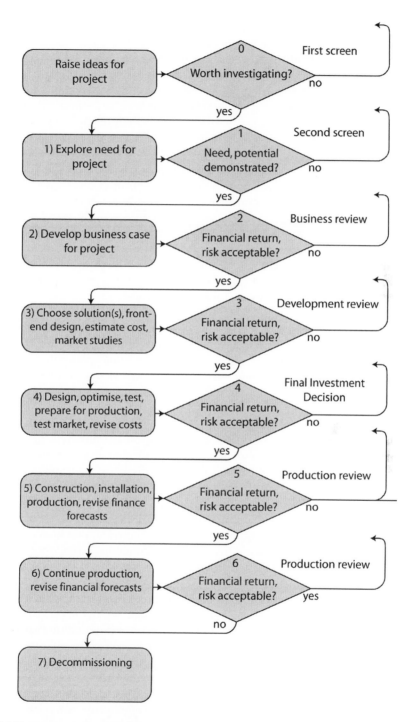

Figure 11.10 Staged project decision-making process. Each project stage is followed by a 'stage gate' decision based on the expected financial return, risks, and opportunities.

stages: a business case, a blitz process to define boundary conditions, the 'contract book' when the project is fully defined, the 'organise and implement' stage, and a review. Chevron had their five-stage project definition and execution process (CPDEP, pronounced 'chipdip'): identify and assess opportunities, generate and select alternative(s), develop preferred alternative(s), execute, and operate-evaluate. Each of these is an instance of what is more widely known as a staged project decision-making process.

Each decision gate has two or three possible outcomes: 'Yes' – proceed to the next stage, 'No' – terminate the project, or 'No' – return to the previous stage and gather more information. A 'Yes' decision represents a commitment to spend a considerably greater amount of money on the next stage of the project than the previous stage. Even by the time a project reaches the fifth decision gate shown in Figure 11.10, the 'final investment decision', only about 10% of the entire project budget will have been spent. This particular decision, the most important in the whole process, involves a commitment of the remaining project budget. This is a decision to spend around ten or more times as much as has been spent on the entire project so far.

The five stages before this are usually called the project 'front-end' during which all the feasibility analysis and planning for the project takes place. The final investment decision is also called 'the point of no return'. Once a project reaches the execution stage, most of the money will be spent, for better or worse.

If we look at the decision gates, each seems almost identical to the last. The decision makers have to assess the expected value that will be created by the project and the investment required to realise that value. The major part of the decision lies in evaluating the uncertainty; both the uncertainty in the value created by the project (opportunities) and uncertainty in the cost required to realise that value (risks). Naturally, as the amount of money to be committed increases for a particular project, the decision has to be taken at a more senior level of the company. The final investment decision on most major projects will only be made by the Board of Directors.

Here we need to realise that most large companies generate hundreds or thousands of ideas for new projects every year. What we are describing here is a screening process. A company has to decide on the best of all the hundreds or thousands of possible projects to develop and invest in. It follows that company staff will be working on many different project proposals at the early stages of investigation. Most will be discarded. However, by adopting a systematic decision-making process, a company has a better chance of selecting projects that it can pursue with good financial results.

The major flaw in this process lies in the growing attachment developed by people in the company to a particular venture. Since the key issues at each decision stage concern uncertainties and confidence in forecasts, it is possible, and all too often rather easy, for individuals to discount information that does not confirm their current assessment of the situation. Any commercial venture requires a healthy degree of optimism to keep going after experiencing setbacks, but this optimism can result in decisions that later can be seen as mistakes – sometimes catastrophic ones. Expert engineers do their best to seek opinions from outsiders, also known as 'third party reviews', whenever possible to gain independent views on a difficult decision. The case study at the end of this chapter shows what can happen, even when third party external reviews play a part. There have been many recent similar examples where projects have gone ahead and have then failed despite well-argued cautionary advice that subsequently turned out to be accurate.

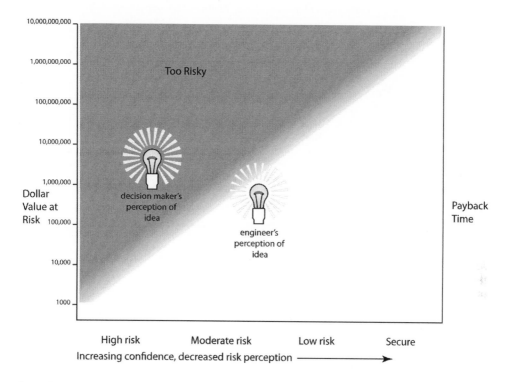

Figure 11.11   Decision-maker's risk perceptions of an engineer's proposal. The decision-maker expects it to be riskier than the engineer describes, as well as more expensive.

At each decision stage, engineers will need to explain how they have developed their confidence that the outcome of the project will work with the required degree of technical performance and reliability. Relying on past experience is always preferred, but innovations will require testing and technical analysis to build the necessary confidence. While capital costs can usually be predicted with a high degree of confidence, operating costs are much harder to predict, especially future operating costs. Even capital costs can be hard to predict over a longer time period. For example, skilled labour may become difficult to find, especially if many similar projects go ahead at the same time.

Now we can return to deal with the engineer's feelings of frustration and address how to improve the chances that a project will succeed. The ideas expressed in Figure 11.9 still apply but the amount of money that a large firm is prepared to invest will likely be much greater. Just like ordinary people, companies are likely to invest more money in projects for which the perceived risk is low. The payback time axis has been deliberately left blank; this diagram illustrates a qualitative relationship. Just like a family investment decision, a company prefers to minimise the payback time when the risk is greater.

It is important to remember that the horizontal axis on this diagram attempts to capture the *perceived* risk. It is the risk perception that is most important in an investment decision.

Figure 11.11 attempts to show what happens when an engineer pitches a new project to his colleagues or managers. The engineer will generally have a high degree

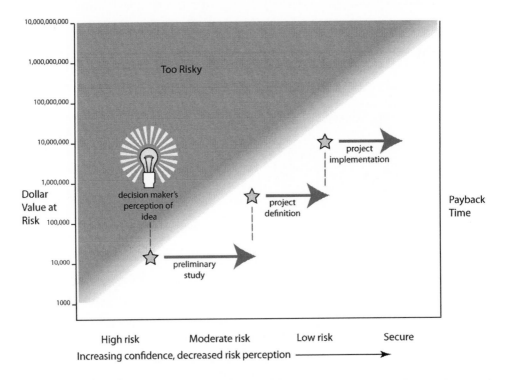

Figure 11.12 Staging a project with a preliminary feasibility study, a project definition phase, and final implementation. Each stage reduces perceived risks and enables more funds to be invested, while still staying within the acceptable region for investment at a given level of risk.

of confidence in their own idea. Other people, however, will almost inevitably have less confidence, particularly at the start. Many may also think that the engineer has underestimated the costs, meaning that they expect an engineer to present an overly optimistic case for their proposal. The light bulbs in the diagram show how the engineer's perception of the risk makes the project acceptable. However, because his colleagues or managers think that the risk is higher, and possibly the cost as well, their view of the idea falls in the upper left shaded area, outside the acceptable range of investment for a given risk perception.

Figure 11.12 shows how an engineer can respond effectively. Rather than trying to persuade managers that the risk is lower than they think, a better strategy is to propose a modest feasibility study to gather more information. On the diagram, this is represented by the star located almost three orders of magnitude lower on the 'value at risk' axis on the left of the graph. Figuratively, this means proposing a feasibility study costing perhaps 0.1% of the expected project budget. The engineer proposing the project (the proponent) may even have to devote considerable unpaid time at this stage, simply out of enthusiasm and dedication to 'due diligence'. This will make it much easier for a company to accept.

The feasibility study, largely composed of information gathering, provides evidence that helps everybody appreciate the value of the project and the risks more accurately. Confidence has increased and the perceived risk has been reduced. Because of this reduced uncertainty, a further, more extensive feasibility study can now be contemplated, involving considerably more resources than the initial investigation. Each study starts with a greater budget than the last, with each step helping people in the firm to become familiar with the knowledge that will be needed if the project goes ahead. By following a series of similar steps on the graph, each of which decreases risk perception, a project can reach the final investment approval stage.

Notice how I have deliberately shown that the final project implementation investment, the amount of money 'at risk', is higher than the light bulb location that represented the managers' initial assessment of the project. This is not uncommon: projects often turn out to be more expensive than anybody initially imagined. However, the principle is simple. By reducing uncertainties and particularly the perception of risk, the amount of money that the firm is prepared to invest will be much greater. Much of the risk reduction may take the form of finding solutions that use 'commercial, off-the-shelf' (COTS) components. Using tried and tested components helps to reduce both the perceived and actual risks.

This explains why so many engineers spend much of their time reducing uncertainty by carrying out investigations that do not seem to directly contribute much to the ultimate design or development of a project. Many investigations contribute data that results in a decision to discontinue a project at the screening stage. These provide more accurate information to enable investment decisions to be made, either to proceed with a project or to discard it.

Practice concept 79: Risk reduction adds to project value

Many engineers think that …	Expert engineers know that …
Time spent on checking, reviews, and inspections adds little (if any) value. It is essentially unproductive. Productive engineering has to involve innovation and cost reduction to add value.	Reducing perceived risks increases the project value and also the likelihood that a project will proceed. Careful and systematic checking also helps to keep projects on schedule with less stress. An expert engineer will 'do it once, do it right'.

This model of commercial engineering decision-making demonstrates that it is often easier and more effective for engineers to work on reducing uncertainty than it is to reduce the cost of a project. The simplified graphical model helps to explain why a firm will invest even more money than was originally considered, provided the risk perception is sufficiently reduced. Reducing the risk means avoiding untried technologies, even if the cost is greater. It means pursuing ideas that the company knows it has experience with, rather than untried ideas from somewhere else, even if these might be technically superior. These decisions run against the grain for many engineers who value innovation and enjoy taking on more ambitious technical challenges than anyone else has done before.

Reducing perceived risk, therefore, can be just as valuable in the context of an engineering enterprise as creative design and developing new technology, often seen by engineers as the pinnacle of engineering achievement. The model I have presented here explains how risk reduction adds value. By moving the *perceived risks* of a project further to the right, the amount that a company is prepared to put at risk by investing in the project increases. Returning to the opening sections of this chapter, you will remember that the value of goods corresponds to the price that someone is prepared to pay for them. By decreasing risk perceptions and increasing the amount that a company is prepared to invest in a project, *an engineer has added to the value of a project.*

Many engineers see checking, inspection, and reviews, which is the detailed technical work needed to reduce uncertainty, as unproductive. They will often label these investigations as 'tick the box' exercises: 'We're just told to do it. But it's not advancing the state of the art, is it?' If we were to look at the work of an individual engineer in isolation, it can be difficult to appreciate the significance or value arising from that activity. Take, for example, an engineer who checks specifications and incoming components to ensure that they comply with safety standards. Taken in isolation, the work seems to have contributed no real value. The engineer has not contributed any design or performance improvement or cost reduction. However, in the context of an enterprise, the engineer has reduced the perceived risk of failure just because the checks have been made, even if no non-compliance was found. In doing so, he has increased the economic value of the enterprise.

Preventative maintenance is often seen as a routine, mundane, and low-status aspect of any engineering project, even to the extent that engineers responsible for maintenance are reluctant to take on a title that might associate them with maintenance. A clearer understanding of the influence of risk perception and value could help maintenance engineers properly explain the inherent commercial value of their work.

A senior consulting engineer related his experience reviewing an extensive design study for a mine in a pristine tropical location:

> 'I had to review the work done in the previous year. They had spent tens of millions on engineering, but they had added no value because they had not dealt with any of the ten showstoppers: risks that could cause a major release into the environment, mostly from flood events.'

This engineer has directly connected the elimination of risks with increasing the project value for the client who was still seeking finance to implement the project. However, the engineers working on the project seem not to have understood the link between the need to eliminate major risk factors and the availability of financing to develop the project.

The explanations in this chapter should help you to understand these connections.

The model explains why the wool industry was so reluctant to invest in the robot sheep-shearing technology that we created so many years ago. Office towers had been built for decades and real estate agents eventually found tenants to rent the office space. It had been done before. Robot sheep shearing had never been done before and, regardless of the best engineering and economic forecasting, investors would always see it as being much more risky. They would want a much greater return on a shorter timescale and would have been reluctant to commit to a final investment decision without several intermediate steps.[12]

Why projects fail – a lesson from recent history

Failures provide some of the best learning opportunities. In this case study, it will be helpful for you to identify the factors that influenced decision-making in this project and how they contributed to the project failure.

This case study describes how Minprom,[13] a multinational mining company, built a mineral processing plant in Central Asia that ended up as a major business failure; in addition, some workers died in accidents and more were injured. A few years after the plant was closed, demolition teams used special machines to cut thousands of tonnes of steel used to build the plant into scrap. The ultimate cost for Minprom was several billion dollars.

The project originated with political pressure on foreign mining companies to increase the value of exported minerals by increasing local processing. This 'pressure' came in the form of large tax incentives. The regional Knutzkoy government also wanted to create local industries based on newly discovered natural gas reserves. Minprom saw an opportunity to convert less valuable nickel ores into an enriched, semi-refined product that could be readily sold. At the time, most sales were of higher grades of ore mined by the company. Low-grade ore had been stockpiled, as it was not considered to be worth exporting until market conditions improved.

Minprom decided to use a relatively new metallurgical process invented in Canada for treating crushed low-grade nickel ore in an atmosphere of hot, high pressure hydrocarbon gases, mainly methane. The resulting enriched nickel powder could be compressed and sintered into easily transported short rods which could be directly fed into electric arc furnaces producing steel alloys. The powder was supposed to flow like a liquid through reactor vessels when aerated with gas flowing from beneath.[14] However, some of the minerals in the ore formed a sticky coating on ore particles at high temperatures, which increased the tendency to form clumps that could clog the reactor vessels.

Minprom sent a consignment of low-grade ore from Knutzkoy to Canada, where it was processed satisfactorily. The results were so successful that Minprom decided to proceed with the urgent construction of a full-scale plant at Knutzkoy so they did not miss out on the largest possible tax incentives. In these circumstances, it would have been prudent to first construct a pilot plant operating at about 5% of the capacity of a full-scale plant, large enough to be confident that issues affecting a full-scale plant would become apparent and to characterise the interactions between process variables.[15] The estimated cost of the pilot plant was $50 million. However, Minprom were concerned that it would take a year to construct a pilot plant and a further year of testing before they could be sufficiently confident to proceed with the construction of a full-scale plant. Therefore, they decided to proceed not with one full-scale processing unit, but with three full-scale processing units on the same site, all being constructed simultaneously.

Safety was a major concern, however, and the Knutzkoy government imposed tough licencing requirements. A leading US firm specialising in chemical plant safety engineering prepared the safety case for operating the plant. This was a detailed independent engineering investigation to ensure that the plant could be operated safely at all times. The US firm was acknowledged internationally as having some of the best expertise available at the time.

The contracting arrangement to build the plant soon ran into difficulties. To properly understand the intricacies of this case, you need to be aware that there are three main kinds of construction contract arrangements.

With a fixed price contract, the construction company undertakes to construct the plant for an agreed price, and the means to handle unknowns and any risk factors are carefully worked out in advance and agreed before the contract commences. Since this was only the second major plant of its kind in the world, and was being designed at the same time as it was being constructed, the risks were simply too high for any construction company to offer a fixed price.[16]

An alternative arrangement, a 'cost-plus' contract, is one where the construction company spends money on behalf of the client and is permitted to make an agreed profit percentage (typically 5%) on all money spent. Because the construction company has an incentive to spend more money than is actually needed in order to make a higher profit, this kind of arrangement requires careful auditing and monitoring by the client. Minprom were not experienced with this kind of arrangement for a chemical process plant.

There were no local engineering companies with sufficient experience for the project, so a local firm worked with a major international engineering contractor in an alliance contract. In this type of arrangement, both companies work as partners and contribute staff to a new organisation specially created for the project. Both companies share information, even commercially sensitive information. Before they start, they negotiate how to estimate the cost and how any profits will be shared. 'No-loss' clauses protect partners from large project risks.[17] One of the advantages of a good alliance arrangement is that money can be spent as it is needed without time-consuming extensive authorisation processes because of the high degree of trust developed between the partners.

Unfortunately, Minprom had little experience in alliance contracting, which was new at the time. The main construction contractor had little experience with chemical process plants and did not know how to set up an effective quality assurance and monitoring process. Since the alliance partners had no interest in operating the plant, they were both anxious to complete the project as quickly as possible. Issues that would affect plant operation later were of little concern to them.

With inexperienced teams, there were inadequate financial controls in the alliance organisation.

In order to accelerate construction, Minprom decided to design the plant at the same time as it was being constructed. The foundations were designed first, while the instrumentation was left for last. Once the foundations were designed, construction started even though what was to be built on them was still unclear. Design errors discovered after the drawings went to the construction site had to be ignored, or were left to be fixed after the construction was finished. In this way, the construction crew had only just enough information to work with and new drawings kept arriving as the plant was being constructed.

However, the greatest problem was that design changes kept arriving long after the final deadline set by the alliance. These changes resulted from reviews by the Minprom plant operations engineers who would eventually run the plant and maintain it. They spotted many shortcomings and deficiencies that would prevent them from operating the plant effectively.

In the end, the construction alliance 'locked the gates' and prevented Minprom engineers from 'meddling' with the project. The plant commissioning phase following construction was intended to be an opportunity to test the plant thoroughly before going into production. Instead, most of the time was spent fixing obvious faults, such as trying to join pipes that were meant to line up with each other but did not align properly once the plant was installed. These mistakes reflected the rush to complete the design in time, with insufficient checking to provide the contractors with drawings that did not contain significant errors.

The relationship between Minprom and the construction alliance steadily deteriorated. When the operating engineers and technicians were finally allowed access to the site, they discovered that all the special-purpose lifting equipment that they would need to access the towering reaction vessels up to 100 meters above ground level had been removed by the construction alliance contractors. The contract specified that this equipment belonged to the plant owner, but it was gone by the time the operating workforce arrived on site.

Most aspects of the technology were new, both for Minprom and for the local engineering enterprises. Special components intended to withstand the high reaction temperatures arrived on-site, but failed in operation because the designers had not been fully aware of the plant's operating conditions. Replacements had to be built by inexperienced local contractors. The instrumentation was inadequate to monitor the processes taking place inside high-temperature, high-pressure reactor vessels.

Canadian and other consultants who were brought in to review the project during construction and commissioning had expressed strong concerns about the viability of the project and the lack of a performance margin, which is an allowance in case plant performance efficiency is below expectations. However, Minprom dismissed their concerns and continued funding the project.

Two and a half years after construction started, the first process plant went into operation. However, Minprom could not obtain the same results they had achieved with the ore that had been sent to Canada four years previously. Within just a few weeks, there was a serious fire resulting from a failure in isolation valves that were supposed to protect the plant in case the operating temperature exceeded safe limits. Glowing fluidised ore powder poured out of one of the damaged reactor vessels, bursting into flame in the outside air, causing millions of dollars of damage.

The plant was shut down and even more Canadian experts were hired to decide what to do. In the end, they decided to incorporate expensive modifications. The total cost increased by another billion dollars or so.

Ultimately, one of the main difficulties facing the project was the inability of Minprom to attract engineers with sufficient experience and background knowledge to work on the plant in Knutzkoy. Minprom was mainly a mining company with little experience beyond handling rocks at room temperature and some 1960s steel-making technology. The Knutzkoy plant was an intrinsically dangerous, complex, high-temperature, high-pressure, continuous petro-chemical processing plant. This was a field of engineering in which Minprom, as a company, had no previous experience.

As explained in Chapter 4, it takes at least 10–15 years for an engineer to become an expert in a particular field. The lead engineers[18] needed for this project tended to be married men with teenage children. Their wives would arrive in Knutzkoy a few weeks after they had started working there. The summers were hot and the winters were bitterly cold, with temperatures plunging to −40°C. Dust clouds that were blown off stockpiles and conveyors at the nearby mines often swept through the town. The dust was harmless, but extremely annoying; a fine reddish powder penetrated every piece of clothing, even underwear. Some of the wives would not even leave the airport and instead demanded to be put on the next flight out of town with their children. Their husbands left soon after. Few engineers stayed longer than 15 months on the project, and most required expensive training in Canada with the company that invented the process.

Most of the systems needed for the plant were designed and built from scratch by engineers with little experience of this kind of process. Many were not ready on time, especially the safety management system. These difficulties were compounded by a worldwide shortage of experienced commissioning engineers because many other projects were being commissioned around the world at the same time. Those on-site were trying to work 70 hours every week to make up for the lack of trained people. Fatigue, sickness, and turnover were the inevitable results, especially with summer temperatures ranging up to a maximum of 48°C in the shade. They even had to improvise thermal protection for electric motors and temperature sensitive instrumentation electronics using shade cloth and domestic fans.

Because the operations staff had been excluded from the construction site, there were numerous plant modifications that were needed. For example, they found that pumps had been installed, some on elevated platforms, without sufficient space around them to permit people to access them for disassembly and repairs. The plant was difficult to maintain when it was handed over.

When the plant was put into operation, there was pressure from the company to ramp up production as quickly as possible. Normally, the rate at which a new chemical process plant increases its production rate is constrained by the novelty of the process. A new process ramps up much more slowly than one which is well understood, as shown in Figure 11.13. Because the desired financial performance of the plant depended on achieving full production within a few months, the production team came under severe pressure as delays mounted, and plant modifications had to be scheduled on short notice.

A complex chemical plant requires a good social environment in which senior engineers, technical staff, plant operators, and maintainers mix together socially. Without a collaborative social environment, the knowledge that is distributed among them does not reach the people who need it, as explained in Chapter 5. A complex plant demands a team-based culture in which everyone is prepared to be flexible to help each other when needed. The plant cannot operate safely and effectively without the free interchange of distributed knowledge and skills.

Unfortunately, Minprom's company culture at the time contrasted with this ideal. Hierarchical management engaged in frequent confrontations with unions still adapting to the new oligarchic capitalism that had displaced communist regimes in power for most of the 20th century. Unions argued with each other over demarcation disputes, partly fostered by some managers who feared a more collaborative culture. Company operating divisions had few interactions; the operations division was separate from maintenance and both were separate from the division administering financial controls. The construction alliance organisation had a similar culture.

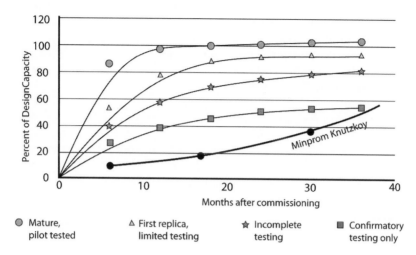

Figure 11.13 Process plant ramp-up curves, showing approximate Minprom nickel plant performance relative to expectations.[19]

As a result, the working environment at the plant reflected Minprom's habitual confrontational silo-based workplace relations culture, far from ideal for a plant of that kind. There were four major groups operating the major sections of the plant responsible for gas generation, the high-pressure reactors, material handling, and sintering. There were supposed to be cross-functional teams to handle health and safety issues, instrumentation and control systems; however, the divisions and culture gaps between the major groups made this difficult.

The risk management process was focused on large numbers of relatively inconsequential risks that were likely to be encountered. Most engineers were more comfortable with the kinds of hazards encountered in a mining enterprise and many had very little practical experience. Inexperience deterred these engineers from dealing with major process risks that could have had disastrous consequences. As a result, these risks were not planned for adequately and the consequences when they did occur severely affected the project: physical consequences, loss of staff morale, and loss of reputation.

Perhaps in an attempt to erase the early history of the project, Minprom renamed the entire enterprise, but the name change could not make up for the earlier mistakes in the project.

RAMS – the combined analysis of plant Reliability, Availability, Maintainability, and Safety – was adopted by the project team, but only after the plant went into operation. That was far too late to gain any advantage from this technique; reliability has to be designed into a plant at the early stages. Once a plant has been built, the reliability issues are predetermined and most improvements are too expensive to justify.

In the end, the critical factor that affected the whole plant was the slim financial operating margin. Minprom had only planned on a 3% return on its original investment. As the total spending slipped past $3 billion, the original financial targets were unachievable. The operating budget was insufficient, partly because there was much more equipment, particularly instrumentation, than the original estimates had allowed for. As a result, no one could ever be confident that the plant would continue operating, and morale among the operating workforce worsened. When the second major accident occurred four years into production, the plant was shut down pending a major safety review and never went back into operation.

Minprom commissioned a technical enquiry by expert US and Canadian metallurgists who concluded that significant scientific discoveries related to nickel ore processing had been made as a result of the company's experience. This was seen by the plant engineers as an attempt to divert attention from all the other issues that contributed to the project failure. The company has never publicly discussed the project.

Ironically, by the time the plant was finally closed down, nickel-refining technology had advanced around the world and low-grade ore was being traded at nearly the same price as higher grades (per tonne of nickel content). Much of the justification for the plant's existence had been bypassed by technology developments elsewhere.

Practice exercise 19: Learning from failures

Most of the engineers involved with this project found it to be a traumatic and depressing experience, though many learnt a great deal that would be useful in their subsequent careers. They insisted that I write down this story in a way that would help a new generation of engineers learn from their experiences. You can learn from this and avoid involvement with projects that are obviously going to fail. You can learn that these failures can be anticipated; you now have the knowledge to see this in any of your future projects.

Write a summary of the key factors that you think defined this failure and compare your responses with the evaluation guide in the online appendix.

You might be thinking that this was a unique experience that would not happen in your own country. You would be quite wrong; this project represents just the tip of a much larger iceberg representing failed engineering projects in all major countries around the world. Few have been well documented. Even after this failure, Minprom went on to yet another disaster of near equal magnitude – organisations tend to be slow learners.

For example, according to recent estimates by independent assessments of capital expansion projects worth USD$60 billion annually, Australia's performance seems significantly worse than other industrialised countries with equivalent economies.[20] However, this is only because most of the very large engineering projects (over USD$1 billion) have been in Australia in recent years. In the case of major capital expansion projects that the mining, gas, and petroleum industries depend on, between 10% and 20% of projects are written off by their owners, costing many billions of dollars annually worldwide. A recent Australian example is the Ravensthorpe nickel project by BHP, which was sold for less than 10% of the cost. Two-thirds of projects over USD$1 billion fail in the sense that they will provide less than half of their planned return on the money provided by investors. Seventy percent of these projects fail to reach their design capacity. Civilian infrastructure projects are among the better performers; the worst performances seem to be in the manufacturing, mineral, and resource processing fields where technical complexity is traditionally greater.

These issues don't just affect private companies; government outsourcing of engineering is affected in the same ways.[21]

Unfortunately, most of these project failures occur in the private sector and are well hidden; those responsible don't want their failures to become known, especially to shareholders. As an example, in one of my recent research interviews, an engineer stated that, 'We probably went down around 1.5 billion on that one [referring to a recent project failure], but because commodity prices rose so spectacularly that year, the shareholders never noticed and we capitalised the loss.'

Understandably, these issues are discomforting for any engineering community. News media and engineering associations point to 'management failures'.[22] There is little doubt that these difficulties lie with project delivery, the practical aspects of converting engineers' ideas into reality so they can provide real value for investors. Often projects are rushed through approval processes in companies to take advantage of temporary commodity price rises, and project documents go out to engineering firms with much of the detailed engineering design still to be completed. Engineering firms hired to construct expensive plants and equipment often don't get to see the final project results; their concern is just to deliver part of the overall project on time and on budget. Therefore, many engineers find it hard to accept that there really is a problem on the scale reported.

These failures affect governments and firms with household names, firms that pension funds rely on. As a community, we share the ultimate cost of these failed investments, even though we never get to know the details. Usually, we only get to hear about large government project failures. As a result, many think the problem lies with the government, rather than with private firms.[23] However, private firms are able to keep their secrets effectively hidden for longer. My research has provided evidence that even senior executives relatively close to these failures miss out on a deep understanding of what happened, partly because they don't want to know.

With the benefit of this book, however, you can acquire the knowledge needed to improve this performance.

NOTES

1. E.g. Samson (1995).
2. Crossley (2011); Trevelyan (2012b).
3. Dunsmore and her colleagues studied engineering students' perceptions of engineering work. The almost complete absence from the research literature of any discussions relating to the financial implications of engineering education suggests that students' views reflect those of faculty. Entrepreneurship is the only aspect of business that has received any attention in the engineering education research literature, and only marginal attention, yet entrepreneurship is itself a small and specialised area of business (Dunsmore, Turns, & Yellin, 2011, p. 339).
4. Note that in Islamic finance, there is a different approach because the idea that a borrower has to pay a fixed amount of interest, whatever happens, is considered contrary to Islam. Islamic finance has a closely related concept called 'mark-up'. The major difference compared with 'Western' systems of finance is that the lender has to accept part of the risk of the borrower being unable to repay the money. In practice, however, the same concepts can be considered to apply for the purpose of understanding this chapter.

5. A detailed study of engineering estimation for construction has been provided by Jacobs (2010). In this study, we also see how engineers set the price for the client, a delicate decision on which the financial viability of an enterprise depends. The price must be attractive enough for the client to accept, and leave enough of a margin to cover unknown unknowns.
6. Hägerby and Johansson (2002).
7. Mason (2000).
8. Work is arranged in shifts when people are required for more than a normal span of working hours. For example, 24-hour site work may be arranged in two shifts of 12 hours or three shifts of 8 hours.
9. PPE – Personal protective equipment such as steel capped boots, helmet, gloves, safety glasses.
10. (Trevelyan, 1992).
11. Cooper (1993) provides a more detailed description of project success factors and decision-making, and Phillips et al. (1999) compare stage gate decision processes in different firms.
12. In the case of robot sheep shearing, there is another possible explanation. Australian shearers exploited the relative inexperience of the Australian Labour government led by Geoff Whitlam in the 1970s and another led by Bob Hawke in the late 1980s to win large pay rises. In about 18 months between 1973 and 1974, the pay for shearing a sheep almost doubled, and increased by almost 50% in 1987. In 1989, we demonstrated our robot shearing sheep reliably removing the fleece in one piece just like a shearer. The robot was relatively slow, due mainly to funding limitations constraining the experimental hardware and software. However, it worked and was widely shown on TV. Since then, the shearers have only demanded a relatively modest cost of living increase in the shearing rate. Other reforms pushed through at the time have provided more flexibility for shearing contractors to hire itinerant New Zealand shearers and adopt other productivity reforms. Certainly, with the benefit of hindsight, one could argue that the decision of the wool industry to exploit shearing robots as a clear and present deterrent to limit unreasonable pay demands from shearers was a prudent commercial decision that has almost certainly yielded huge commercial benefits for the industry. However, this can only continue as long as the Australian community is prepared to allow the wool growing industry to rely on publicly funded health and social welfare programs to deal with occupational health issues that affect many shearers in different ways.
13. This account was based on interviews with engineers who had first-hand experience in this project. The company name, project description, location, and dates have been changed to protect their identities.
14. Known as fluidised bed technology.
15. McNulty (1998).
16. Roberts (2003) discusses the relative merits of each approach in the offshore industry.
17. Love, Mistry, and Davis (2010) discuss the benefits of alliance projects.
18. Lead engineer – has technical expertise in a specific aspect of the enterprise to lead a large team of less experienced engineers.
19. Data from published sources. A similar experience is quoted by Barnes and Jones (2011). Diagram drawn from Figure 1 in McNulty (1998).
20. Information from public seminars by IPA Global, also from the author's own research interviews. Merrow, the CEO of IPA Global, has provided further detailed information in public seminars and in his book (Merrow, 2011).
21. Rizzo (2011).
22. E.g. Boyd (2013). Engineers Australia chose to take no action in response to public revelations on the basis that 'this is a management issue'.
23. E.g. Boyd (2013).

Chapter 12

Negotiating sustainability

This chapter introduces negotiation as a formalised method for solving conflict between two or more individuals or groups. It is the fourth important, combined, collaboration performance for engineers. Earlier chapters introduced teaching (Chapter 8), technical coordination (Chapter 9), and project management (Chapter 10).

This chapter also introduces sustainability, a vexing issue for engineers, and one that nearly always requires negotiated solutions.

Practice concept 80: All engineers negotiate

Negotiation is part of daily life for most engineers. Most of the engineers we interviewed for this research mentioned negotiation as being a significant part of their work as a means to solve conflicts and disagreements on technical issues, contracts, project management, and sustainability.[1]

The aim of negotiation is to arrive at an agreement, preferably in writing, which explains how all of the parties will act in the future. In a technical context, the agreement is a design specification. In some instances, the negotiated agreement may be a contract document, particularly in commercial situations. In other instances, the agreement may a legal document. All parties to the agreement would normally send a confirmation e-mail or sign a copy of the agreement to indicate their acceptance. The negotiated agreement is a kind of script that defines the subsequent collaborative performance that is agreed on as a result of negotiation.

Negotiating with subcontractors and monitoring their work is a recurring theme, as these words of a mechatronic contracting engineer clearly illustrate:

> 'We have to recognise that our subcontractors are usually one-man to three-man companies. They have to pay their bills on time. They have to manage their finances. They will ring you up to ask for early payment. We pay our invoices after 30 to 45 days. But, with a one-man company installing equipment, he may have purchased $10,000 worth of equipment at the beginning of the month and then he has to pay labour and fuel. Then he puts in his invoices and has to wait for 45 days. If we wait 45 days to pay him, he will go broke. You have to learn how to negotiate; otherwise, you will send your contractors broke. Then, suppliers will stop giving them credit and that hurts you in the end. You have to chase them up and encourage them to invoice you so that they don't run out of money or credit.'

Technical, commercial, and sustainability issues all represent difficulties for engineers that involve emotions, and apparently irrational beliefs and opinions. Even 'purely technical' issues demand negotiation skills: they cannot necessarily be solved by logic alone, especially within teams working under time pressure. That is because, particularly among engineers, issues that seem to be purely technical actually carry a hidden agenda of emotions, including lack of self-confidence and a fear of the unknown. The essence of conflict, from a psychological point of view, lies in the perceptions of incompatible interests and discrepant views among the parties involved.[2] Negotiations in all three domains – technical, commercial, and sustainability – are often conducted with varying degrees of mistrust and apprehension between the parties involved.[3]

Negotiation, therefore, is the primary means by which conflict is resolved in engineering practice.

While negotiation is often described in business texts,[4] it is not often appreciated or viewed as a way to solve technical problems. A recent study of technical problem solving in the automotive industry found that successful solutions were always preceded by achieving alignment among the different technical parties – another way of describing negotiation.[5] Engineers almost always face the twin constraints of budget and time which necessitate certain compromises: negotiating ability increases the chances that satisfying solutions can be found. Often, it is even the existence of a problem that is contested, either because one or more parties do not want to or simply cannot foresee future events as other people see them, or because they lack sufficient experience to appreciate the consequences of those events. Another reason for denying the existence of a problem is the belief that it is up to engineers in other technical disciplines to solve it, as this quote from a civil engineer demonstrates:

> *'After the end of the construction boom leading up to 2000, we decided to diversify our cash flow and we bought a partly built water purification plant. Everything proceeded smoothly until the commissioning phase. We hired a mechanical engineer to oversee work on the pumps, pipelines, and valves. An electrical engineer was responsible for instrumentation and controls. A computer software expert was responsible for programming the computers needed for monitoring water quality and guaranteeing that water delivered from the plant would meet the potability requirements. It was a nightmare: the mechanical engineer blamed faults in the instrumentation and controls, the electrical engineer blamed the computer software and the slow response times of the valves, and the computer software engineer blamed the poor quality of information from the instrumentation. Each refused to acknowledge elements of the problem that lay in their own area of responsibility.'*

While negotiation is part of most courses in business schools, technical and sustainability issues introduce problematic aspects that seem to receive less attention in traditional texts, which commonly argue, for example, that one should 'separate people from the problem'.[6] This chapter, therefore, introduces a discussion on ethics, emotions, and value frameworks, all of which impact the types of negotiations needed to resolve technical and sustainability issues.[7] Negotiations on technical and sustainability issues often involve many parties, a situation that is usually only addressed in advanced texts on negotiation.

Technical coordination can involve informal negotiation, so some of the ideas in this chapter can also be helpful for negotiating an agreed upon schedule in that context.

Engineers are involved in various types of negotiations throughout the duration of a project, even before it starts. The first phase of any project, shown at the base of Figure 3.10, involves resolving technical and financial uncertainties with the terms and the level of financial commitment, a subject that was discussed again in Chapter 11 in the context of financial decision-making. This phase always requires a negotiation between the client and the project team to determine the project scope, as described in Chapter 10. However, that chapter left out any descriptions of the negotiation process itself.

Estimating is another key element of a project, as this subsea system engineer explained:

'[How do you estimate the cost when at the same time you're going to negotiate with the supplier to fund part of that themselves?] Well, it depends how you negotiate the contract. With regards to the TSD (a project), we actually negotiate the price book for each individual piece of equipment and also each individual technology qualification activity. Obviously, we're doing that well in advance of when we're going to actually need it, but then you negotiate in rise and fall mechanisms and change mechanisms to take into account changes in material prices, legal prices, etcetera. It's not easy, but—'

'[It sounds like this negotiation takes a lot of your time.] On the commercial side of things, it can do. I mean, I don't get too involved. There is a package manager, say, for this subsea equipment vendor that you're making reference to, there is a package manager who will be responsible for the overall execution of that. The technology qualification, there's only one aspect to that, so he generally heads up the commercial side of things.'

An electronics engineer encountered similar issues in a completely different setting:

'[What happens if you're not expert enough at estimating and you underestimate the time required?] That often happens at the start of a job when you do not have enough background information to quote accurately. There are several possibilities. You can put in a scope change request, particularly if the original scope was written clearly enough. The scope has to be approved by the client at the start of the job and if the client overlooks the omission then they have to pay when the scope is corrected later. You can also follow the normal design change procedure, but there will be some negotiation about who pays for the cost at the other end. Otherwise, you just have to work it in with all the other jobs and gradually make up for the loss.'

Later in the project, even small details require some form of negotiation. Importing supplies into a foreign country can involve frequent negotiations and additional fee payments, at times, as this site engineer revealed:

'We had to grease pockets in the Customs bureau, I forgot to mention that. It was just that we were not getting things through as fast as we needed them. When the customs realised that these parts ... they did not understand what they really were but they knew they were really needed by a foreign company and they just let *them sit there for a few weeks. We had 3 or 4 guys in the capital city who would negotiate with customs, who would go and negotiate with them, or beat their heads against the wall to try and get what we wanted, but things didn't always turn out the way that we hoped. They were company people, Australians. Their job was to get all our containers off the trains and through customs. There were some local people as well.'*

Contract variations (changes) often involve extensive negotiations, as this civil engineering project manager explained:

'Variations are the way we make a profit. We put in a cost price bid, even below cost to win the job, because we know there will be variations on which we can work up the price.'

A mechatronic engineer working on security and access control systems described this as well:

'Most of my estimating work concerns variations to the current contract – the client might want this or that added in or something changed. If the variations mount up to more, then we complain about them and delay payments.'

Towards the end of a project, negotiations focus on unresolved issues where the client thinks that the project team has left things unfinished, whereas the project team members think that they have complied with the client's requirements:

'The problem arises when the client thinks that the work has not been finished. Also, our mechanical client will not pay us because he has not been paid. This might be caused by a problem over which we have no control. When we send an invoice, the client may say "prove that it's finished!" The client cannot see anything because everything is installed in ducts out of sight. Sometimes, you spend a lot of time justifying why a job is half complete and qualifies for a progress payment.'

As explained above, engineers negotiate on technical issues as well. Here is an example from my own experience. A robot presents many conflicting technical requirements. The mechanical structure needs to be as small as possible, have the least possible mass and inertia, and be robust, rigid, inexpensive, and easy to manufacture and assemble. Electrical cables and other services, such as compressed air, need to pass through the mechanical structure in a way that allows them to twist and bend at the mechanical joints without breaking.

Figure 12.1 illustrates the result of a tense negotiation within the team that I was privileged to lead between mechanical and electronic engineers working on the development of robots for shearing sheep in the 1980s. A conventional twisted pair data cable carrying signals for 35 sensors with 7 position and force command settings needed for the Shear Magic robot would have been at least 15 mm in diameter with a minimum bending radius of about 200 mm. Any further bending would have resulted in rapid fatigue failure of the copper wires when the robot moved and the joints flexed. The mechanical designers were extremely reluctant to accommodate a cable of this size. As a result of the negotiation process, everyone realised that a completely different option was needed. A creative solution emerged when the team realised that, at least in electronic terms, the sensor and command signals varied slowly, over time periods of

Figure 12.1   Robot sheep shearing and cabling arrangement. Left: Assembly of the lower end of the "Shear Magic" robot arm in 1985, showing the shielded signal and power supply cables threaded through holes in the structure around one of the mechanical actuators. Right: Robot shearing the neck of a sheep in 1989.

a millisecond or more. Together, they chose an analogue multiplexed solution in which each signal was present every so often on a single shielded signal wire for a predictable few microseconds, with all 45 signals being presented in a fixed sequence sharing the same wire.[8] Other shielded wires carried electric power and the allowable bending radius was reduced to 20 mm by using high-quality aerospace coaxial-shielded wires.[9]

This example illustrates how negotiations of perceived conflicts can also lead to creative solutions, provided the parties know how to constructively negotiate towards a resolution.[10]

Finally, and not to be forgotten, all engineers have experienced the negotiation process with their employer on their terms of employment and salary, even before they have started working as an engineer. In every variation and situation where negotiating is required, understanding the principles of negotiation can help achieve a better result for both parties.

Practice concept 81: Sustainability

'Sustainability' and its sister term 'sustainable development' have become critical concepts for engineers. In a public address, John Grills, CEO of one of the world's largest engineering firms WorleyParsons, suggested that sustainability had become the key focus of engineering projects. '*More than half of our projects face constraints on the supply and disposal of water.*'[11] At the working level, however, many engineers hold a different view: '*I think this sustainability business is not something that engineers need to think about. Engineers should wait for politicians to decide what they want first, and then get on with building it.*' Many of the engineers who participated in our research told us that issues of sustainability present many frustrations.

Before we get any further into discussing sustainability issues, we should answer two important questions: what exactly do sustainability and sustainable development mean? Furthermore, what do they mean for you as an engineer?

Possibly the most widely quoted definition today comes from the report of the 1987 UN-appointed World Commission on Environment and Development, also known as the Brundtland Report:

> *Sustainable development is development that meets the needs of the present without compromising the ability of future generations to meet their own needs. It contains within it two key concepts:*
>
> i) *the concept of 'needs', in particular the essential needs of the world's poor, to which overriding priority should be given; and*
> ii) *the idea of limitations imposed by the state of technology and social organisation on the environment's ability to meet present and future needs.*[12]

Even this widely accepted definition can be difficult to interpret. How do we define the 'needs of the present'? Given that accurately forecasting the future is so difficult, how can we possibly anticipate what future generations will need?

Chapter 7 explains how even simple words can take on quite different meanings in different contexts. Sustainability is no exception: as a concept, it can be interpreted in a wide variety of ways, depending on the people who are speaking and the circumstances in which they find themselves.

Therefore, perhaps the first response by an engineer whenever sustainability is mentioned should be, 'Please explain to me what you mean by that.'

Engineers rely on clients who provide the financial resources for their work. At the same time, engineers have to perform their work in the context of a community. Most of the time, the people who work in an engineering enterprise are also members of the community in which it operates. In the case of government projects, it is the community that provides the financial resources for the engineering work.

Yet another interpretation of sustainability is found in a negotiated agreement itself. An agreement is said to be sustainable if it stands the test of time, meaning that the parties to the agreement continue to support and honour it throughout its life. This is not easily accomplished. Stakeholder interests can change and new stakeholders can emerge. Therefore, a sustainable agreement is likely to be complex; drafting it requires the foresight to anticipate every possible significant future event, including the negotiation of an agreed response to each event if it happens. Sometimes, agreements need to be renegotiated when completely unanticipated events shift the priorities and interests of the parties so much that they are no longer prepared or capable of remaining firm on the original agreement.

As one engineer put it, 'Project management is about doing a project right, but sustainability is about doing the right project'.

This leads to another way to understand sustainability – as 'good engineering practice'.

Even from the brief introduction, we can distinguish some fundamental themes that commonly emerge during discussions on sustainability.

The first is humanity's impact on the planet.

Until the Industrial Revolution began in the 18th century, human development was constrained by a combination of geography and the limitations of muscle power. With favourable geography providing fertile soils and water, ancient civilisations developed modestly sized, durable cities that could be sustained by irrigated agriculture on a few square kilometres of surrounding land. Irrigation schemes were also built along the margins and flood plains of reliable rivers to extend the feasible growing season for crops. Before that, people lived in small groups of hunter-gatherers and spent most of their time searching for food.

With the possible exception of Australian Aborigines, who learnt to manipulate their environment across the vast majority of their continent using fire, as well as the construction of the Great Wall of China (visible from space), humans had a limited visible impact on the physical environment of the planet.

However, the technological advancements that people have developed over the past two centuries using fossil fuel energy have drastically changed our relationship with the planet: humans now manipulate nearly the entire ecosystem. Until the 1950s, engineering enterprises seldom ran into environmental constraints that could not be modified, if necessary. Pumps and pipelines delivered as much water as a project needed, oil and coal provided energy, and waste could be discharged into the local environment with minimal treatment or dispersal measures. Just a few decades later, engineering enterprises now face tight restrictions on water availability, are much more aware of energy-efficient practices, and have implemented elaborate precautions to minimise the impact of operations and waste disposal on the surrounding environment. A new term has emerged – footprint – a composite measure of the human impact of an individual or an engineering enterprise on the global environment.[13] This philosophical evolution over the past few decades reflects a growing awareness that the planet is a closed system and that human activities are now influencing the entire planet in a significant way, including the global climate.

Engineering responses eventually emerged in response to specific environmental degradation issues. In the 19th century, engineers constructed water supply and sanitation systems to remove the collective 'stench'[14] that had characterised cities until that point. In the 1950s, the atmospheric concentrations of gases and soot particles released by burning coal and oil for vehicles and heating buildings in large cities like London rose to sufficiently high levels to cause photochemical smog that restricted visibility to only a few metres. Localised effects like this led to legislation to control air, ground, and water pollution. In 1962, Rachel Carson's book *Silent Spring* appeared, linking the agricultural use of DDT insecticide with the disappearance of songbirds, marking the start of organised concern for the natural environment, leading eventually to the establishment of government departments specifically responsible for protecting the natural environment.[15]

The first global environmental regulatory response came in the 1980s and 1990s after a near catastrophic reduction in ozone concentration was observed over the South Pole region in the winter. This was caused by the emission of hydrocarbon compounds containing chlorine and fluorine, which were used as cleaning solvents and refrigerants.[16] This problem has affected me personally. As spring turns to summer, the ozone-depleted air high above the Antarctic mixes with normal air over equatorial regions, causing wide fluctuations in ultraviolet radiation in middle southern latitudes. On some late spring days, the atmosphere above the southwest corner of Australia is so depleted in ozone that just a few minutes of exposure to sunlight results in sunburn. Hats have become compulsory for children playing outside at school. Since then, the global response has been somewhat effective and the ozone layer is gradually recovering.

Since the 1990s, the accumulation of CO_2 and other 'greenhouse' gases in the atmosphere has become the dominant global environmental issue, although there has not yet been a similarly strong regulatory response because of strong disagreements on what an appropriate response should entail.

The second theme is a reduction in community tolerance of environmental changes and even social changes that have been caused by engineering enterprises.

Until the 1970s, many communities were prepared to accept the dominant visual presence, noise, dust, smoke, and fumes associated with industrial enterprises located in their area. Employment and economic benefits apparently outweighed health risks and other impacts. Until the 1990s, lead pollution from petrol additives[17] was accepted as part of the cost associated with the convenience of having cars in cities. Today in Australia, just 25 years later, even the remotest possibility of lead contamination has been sufficient for the West Australian government to impose tight restrictions on lead carbonate powder transport.[18]

This change is associated with community 'sensitisation'. In this particular instance, the lead carbonate powder mined in Western Australia was originally loaded onto ships using open conveyor belts at a small government-owned port in Esperance, a small and remote town on the southern coast of Western Australia. The toxic powder was often blown over the town by strong winds, but the cumulative effects were only noticed when large numbers of birds died. By that point, lead contamination had spread across the town and many children had accumulated dangerously high concentrations in their bodies. Lead is widely known to be toxic for humans, and people in the town who operated the port actually knew that there was powdered material containing lead blowing in the wind. Despite that awareness, nothing was done to stop contamination until bird deaths were widely and frequently reported. The West Australian government had to cover the cost of the subsequent extensive decontamination work because the port was a state-owned engineering enterprise.

Perhaps the single most significant contributor to community sensitisation was the 1984 Bhopal disaster in India that killed nearly 4,000 people and injured several thousand more as a result of methyl isocyanate gas leaking from a tank at a pesticide manufacturing plant. This and several other similar disasters have eroded community tolerance for engineering enterprises operating among them, with greater awareness of the relative risks to human health, visual appearance, and measurable environmental impact. A more recent example was the BP Horizon explosion and fire in the Gulf of Mexico in 2010 that spread oil pollution across a large part of the Gulf, damaging both the environment and the economies of coastal communities. Another modern example was the hydrogen gas explosions at the Fukushima nuclear power plant caused by the failure of diesel generators needed to power emergency cooling systems as a result of the tsunami caused by the large earthquake off the Japanese coast.[19] Radioactive contamination spread across surrounding farmland and into the ocean, requiring mass evacuations and restrictions on farm produce and seafood consumption. The government, responding to public opinion, took the precaution of closing other nuclear power plants until appropriate safety reviews could be conducted. The resulting electrical power shortages continue to affect the Japanese economy several years later.

As a result of these and many other less well-known disasters, it is now much more difficult for companies to convince governments and local communities to grant approvals for them to operate.

The third theme is a growing awareness that industrialised nations have enriched themselves by exploiting the world's natural resources at a disproportionate rate. It is not possible for the entire world population to use planetary resources in this way: they would be exhausted in a very short time. The Brundtland Report acknowledged the 'essential needs of the world's poor' and thereby drew attention to this fundamental imbalance. Sustainable development, therefore, is also seen by many as a way to redistribute the benefits that result from engineering enterprises more equitably.[20]

All three themes reflect contested views of the future and innovation throughout the world and have sparked ongoing debates about appropriate responses.

There is a wide range of ideas on our ability to innovate. At one extreme, there are people who think that technological advances will enable us to improve our utilisation of the Earth's resources. In their view, industrialised nations are perpetually improving technology and these technology improvements will eventually be made available for the benefit of other nations. Therefore, these other nations and the world's poor will be able to enjoy a much higher standard of living in the future, perhaps even as good as the industrialised world, while still leaving sufficient resources for future generations. At the other end of the spectrum, there are people who focus on our reliance on finite reserves of non-renewable fossil fuels and point out that supplies of suitable fuel for nuclear reactors are also limited, even if people are prepared to accept the risks of using nuclear energy.

Furthermore, there is a diverse spectrum of views on governance. 'Optimists' think that human beings will be able to regulate the use of common resources like the oceans, the atmosphere, and subsurface aquifers. They refer to successful examples, such as emission trading schemes, which have been implemented in the past and to the Montréal Protocol, which resulted in the gradual reduction in the use of chlorofluorocarbon gases and the gradual recovery of the ozone layer. 'Pessimists', on the other hand, point to what they see as the inevitability that common resources will be over-exploited, a situation known as 'Tragedy of the Commons'.[21] This group argues that there will always be people who enrich themselves at the expense of others. They also point to the majority of governments that have limited abilities, if any, to enforce environmental laws in their countries.

There is a diversity of views in terms of predicting the effects of increased concentrations of greenhouse gases on the world's climate. While there is near unanimity among scientists, there are dozens of opinions on how to respond. Many people in industrialised countries see an urgent need for everyone to reduce greenhouse gas emissions. Others think that it would be foolish to harm their economy by adopting more expensive technology to reduce greenhouse gas emissions before everyone else does. Many of these people believe that further industrial development in industrialised countries will lead to innovations able to solve the problem permanently. In their opinion, we should just wait and be patient. Others see economic opportunities in developing low-emission technologies and advocate much higher government spending on research and development. On the other hand, informed people living in low-income countries point out that almost all the additional greenhouse gases currently in the atmosphere have accumulated as a result of industrial growth in high-income countries. They contend that these countries should rapidly eliminate their emissions to allow other countries a chance to develop their economies and accumulate comparable wealth.

Still others view efforts by European countries to achieve international agreement on measures, such as emissions trading, as a strategy to reinforce their own economic advantages, while imposing restrictions on everyone else. France, in particular, has invested heavily in nuclear energy. France, therefore, would pay much less than other countries that rely more on fossil fuels and consequent carbon dioxide emissions into the atmosphere for energy.

There are also conflicting views on the equitable distribution of benefits from engineering and other human developments that reflect the range of views one would find in the political system of any country. There are people who rely on moral judgements to advocate the need to help poor people, particularly those suffering from what many would describe as 'globalised economic rationalism'. According to these individuals, exploitation of natural resources by the industrialised world has enabled overproduction of commodities, which are then 'dumped' in the developing world markets, denying opportunities for local farmers to sell their products at a price that covers their costs. Others think that opening opportunities for free trade and new technologies enables farmers in low-income countries to access much larger markets for their produce and effectively secure higher prices.

As you can see, even though most people agree with notions of sustainability, there is a massive level of disparity in terms of how to respond. This example epitomises the idea of conflicts that can be solved by negotiation.

SUSTAINABILITY ISSUES ARE DIFFICULT FOR ENGINEERS

The research for this book showed that engineers have a number of difficulties dealing with sustainability issues, partly because of several misconceptions discussed in earlier chapters. Many engineers find it difficult to recognise that words have different meanings depending on the context in which they are used. Also, it can be difficult for many engineers to recognise the importance of perceptions: engineers tend to prefer rational argument based on a scientific understanding of issues.[22] In their view, people who overestimate risks that can be scientifically proven negligible are irrational; therefore, their views should not be taken seriously. At the same time, while engineers respect logical arguments based on science, they often fail to state important assumptions that are taken for granted, assumptions that can often be questioned, as explained in Chapter 3 pages 62–63.

As in the previous chapter, a case study can be very instructive.[23] This case study shows how a large engineering project by a major global corporation came close to financial failure because engineers could not understand the importance of perceptions and community trust. This case study illustrates how the firm lost and eventually regained their social licence to operate.

Sustainability issues do not have to be global in context. Local issues can be just as important, as this case study shows.

Kanemex[24] is a multinational chemical processing company. They decided to collaborate with a nationalised titanium mining company to construct a large refinery that would produce titanium dioxide on the outskirts of Oranto, a small European town surrounded by low hills. The town was within trucking distance of several local mines that produced ilmenite, the main mineral source of the titanium.

The first obstacle was to secure enough land for the refinery. Kanemex decided to set up a front company to purchase land; this company announced that farmland was being acquired to grow vegetable oil seeds for an oil refinery. Local people anticipated economic developments and employment opportunities, so they were supportive of the purchases. It did not take long for the front company to acquire enough land on the outskirts of the town. When it was finally revealed that the land would be used for a titanium refinery, some concern was expressed in the local media, but no forthright opposition appeared.

Kanemex secured all the necessary environmental and operating approvals and after three years of construction, the refinery went into operation after commissioning and acceptance testing. As production gradually increased, so did emissions of noise and fumes. Even so, local people were pleased with the influx of educated people into the town, better economic opportunities, improved health clinics, and better quality schools. About half of the refinery workforce relocated to live in the town and the other half commuted from other nearby towns. Gradually, however, previously unseen respiratory illnesses appeared among people living in the town, particularly during the winter. The winter season brought damp, cool, calm nights without any wind to disperse the fumes from the refinery: the fumes seemed to accumulate in the valley between the hills. Even though there were no changes in the operating conditions of the plant, the number of complaints about noise and fumes steadily increased every month. Eventually, publicity in regional newspapers led to a government investigation by the health department. Kanemex responded to accusations that they were poisoning the local population with scientific evidence showing that all chemical emissions from the plant were well below the World Health Organization's maximum permissible limits and that the plant was fully complying with all the government's environmental requirements.

Even though the health department investigation was inconclusive, complaints from the community continued to increase. Local community activists eventually managed to attract the attention of an international environmental NGO, the World Environment Fund (WEF), and the plant became headline news around the world for several weeks. Drawing on talented media professionals, the WEF alleged that even though every chemical emission from the plant was below the maximum permissible limit, together they formed a 'toxic soup' that was 'slowly corroding the lungs of the town's inhabitants'. These words created a much more potent metaphor to describe what townspeople thought was happening, even though the scientific evidence to support this hypothesis was not strong.

Company engineers responded by installing more precise instrumentation in monitoring stations around the town and in several locations around the valley. The results demonstrated that the concentration of chemicals over some parts of the farmland further along the valley was actually higher than the town, although still well below maximum permissible limits. That being said, the incidence of respiratory illness among farmers was well below the reported incidence among townspeople. Therefore, they argued, the cause of the respiratory issues could not be emissions from the plant. At the same time, the company provided scientific lectures for the townspeople on pollution issues to help explain and argue their case. Few townspeople attended these lectures, however.

Local opposition politicians took advantage of the publicity and increased community pressure on the government for stricter environmental monitoring. The regional environmental law enforcement agency (RELEA) responded by ordering even more instrumentation and stricter monitoring of the plant. Complaints continued to rise and the company had to employ specially trained staff to handle the mounting number of complaints.

As spring turned into an unusually hot summer with humid, windless nights, noise once again became a more prominent issue. Plant workers living in the town explained how they all had to wear ear protection while working on the site, even if they were working in buildings that were screened from the worst of the noise. People in the town, particularly those with houses close to the refinery, could hardly sleep with their windows open. However, if they kept their windows closed to reduce the noise, there was little or no ventilation to cool their houses at night. Tempers began to flare.

Eventually, the company realised that relations with the residents would only get worse unless they modified the plant. By then, research and development had found a more economic chemical process with a large reduction in chemical emissions, while a redesign of the plant's cooling system could address the noise issue. The plant was shut down for 18 months at considerable cost in equipment, plant modifications, and lost production: the company could not afford to lay off the operating workforce for fear of worsening community relations with the town.

However, as soon as the company restarted the plant, complaints flooded in and health clinics reported even more respiratory disease incidents. The volume of complaints rose to a level more than ten times higher than before the plant closure. More specialists had to be employed to handle the complaints and the investigations associated with each individual case. Newspaper and TV reporters besieged company offices in the town, at the plant, and in the nation's capital. The regional government, facing an imminent election, decided to respond with even stricter operating conditions limiting the power at which the plant could be operated at night, and increased monitoring for chemical emissions, noise, and other environmental impacts. The company management were at a loss to explain why this had happened. They had invested several hundred million dollars in the plant and had sustained a comparable cost in terms of lost production while they continued to employ the plant's operating workforce on full pay throughout the shutdown. Fearing international repercussions, they had also invested similar amounts to upgrade five other refineries in different countries. They could not understand why complaints had soared when both noise and chemical emissions had been reduced by more than 95%.

Every time they tried to explain why there was no ground for complaint, even more complaints came in. Townspeople had stopped believing statements made by the company, remembering the 'devious scheme' that the company had used to acquire the land for the refinery in the first place. Even when the company brought in independent experts to speak on their behalf, they were ignored by the townspeople. The townspeople had become so sensitised that it took much less irritation than it probably should have for them to initiate a complaint against the company.

Eventually, the company decided to invest another $50 million and offered to acquire the homes of all townspeople living within 2 km of the plant's boundary at 25% above the government valuation of their property. Additionally, the company stated that they would pay for all reasonable relocation expenses. While many people accepted the offer, others resolved to stay, stating that they had made their homes in the town and the company should behave like a responsible neighbour and not force them to move.

Complaints increased again, this time concentrated among people living just beyond the 2 km boundary line. Why, they said, should they not also be entitled to the same offer? Those individuals argued that they were just as badly affected by chemical fumes, using measurements published by the company itself as proof.

Relations between the company and the townspeople on one hand, and the politicians and media on the other, continued to decline. The WEF continued its international campaign, significantly damaging Kanemex's international reputation. The company started to receive letters from their international financiers pointing out that if the problems continued, it might be difficult to renew the loans needed to sustain the company's operations. Even more damaging, relations among the plant workforce started to deteriorate, particularly between workers living in the town and those who commuted from other towns where they did not have to put up with the noise and emissions, and also did not have to speak with angry townspeople every day. Several minor chemical leaks occurred and information reached local media outlets within minutes: employees within the plant itself were alerting local media using cell phones provided by distant relatives to make it harder to trace the calls.

The international head office management of the company decided that a completely different approach was needed. They could see that the financial security of the entire global corporation was being threatened by this one situation, which seemed to be deteriorating more every day. They relocated a senior head-office executive to the town and authorised all necessary steps to eliminate community complaints. For the first time, there was a senior global manager not only living in the town, but also available to meet anyone complaining about the plant face-to-face. The senior executive, who had started with the company as an engineer, insisted on meeting every townsperson that voiced a complaint. All medical cases were arranged to be handled by expert doctors in the regional capital at the company's expense. Community representatives were recruited to join the plant's operating management committee. Their views were taken into account at weekly meetings, which had only minor implications for production and plant operations.

After six months, complaints had almost been completely eliminated, except for those from two individuals who had little standing in the town. The senior executive returned to the company head office, leaving the refinery staff to handle the occasional remaining complaint (although with the same approach of taking every complaint seriously and handling it personally). After 12 months, the newly elected regional government lifted most of the restrictions on plant operations and there were only occasional reports about the plant in the news media.

Three years later, the company sought approval for a major expansion of the plant. Even though they had to face public hearings, no one in the town spoke in opposition to the plant expansion. The approval was granted with only minor environmental restrictions.

Since this episode, there have been further incidents involving the company. A few months after the expanded plant went into operation, dust control measures at the residue evaporation ponds failed and clouds of dust enveloped the town on two successive afternoons. There were many complaints, but the company and the government responded quickly. The government regulator imposed the maximum possible fine for pollution and the plant was shut down for a few days for an investigation. There were reports in the local news media, but the plant went back into operation once again and there were no subsequent complaints.

Losing and regaining a social licence to operate

This case study illustrates the concept known as 'social licence to operate'. Officially, throughout the entire episode except for a few days, Kanemex had all the necessary regulatory approvals in place to operate the plant. However, by the time that the company voluntarily closed the plant to upgrade the chemical process and replace the cooling system to reduce noise emissions, the company had lost its *social licence* to operate. In other words, confidence among the people in the local community had declined to the point where they attributed most of their problems to plant operations. Even though this view could not be supported by scientific evidence, people had no confidence in the assertions that the plant was operating within allowable limits; plant emissions, in their view, were responsible for all the health issues that had emerged in the town.

Eventually, the company regained its social licence to operate, but only at a huge financial cost. The first key step was to meet individual townspeople face-to-face and demonstrate that senior company staff was were taking their problems seriously. The second key step was to share responsibility for plant management decisions with representatives from the town. Rebuilding trust and confidence did not happen overnight. We can only see the real result of that regained trust from the way in which the subsequent approval to expand the plant was granted with minimal complaints.

Four pillars of sustainability

This case study shows that sustainability issues have at least four significant aspects, often referred to as 'pillars' of sustainability: economic, social, environmental, and governance.

Economic sustainability is critical: if the plant cannot make sufficient operating profit to provide reasonable remuneration for the workers, tax and fee payments to the government, and financial returns for its investors, then it will inevitably close. Workers will lose their jobs and governments will forego tax revenue.

Environmental sustainability is also critical. Ultimately, we are all sharing and living in the same environment. Every human activity has some impact on the environment, even though it may be difficult to observe or even measure. Environmental sustainability, therefore, is ultimately a human judgement, that the impact on the environment is acceptably small compared with the benefits obtained by operating the plant, and that any long-term damage can either be remediated, will recover naturally, or is acceptable.

The third pillar is social sustainability, encapsulated in the term 'a social licence to operate'. Perceptions are paramount for this pillar to remain stable: a company that has a social licence to operate is one that has built sufficient confidence among the local community so that people are prepared to entrust plant operations to its management.

Companies reporting 'Triple Bottom Line' (TBL) results focus not only on financial and economic outcomes, but also on environmental and social outcomes.[25]

The fourth pillar is governance, although this is often contested. This case study showed how company governance and community governance played important parts in the story. Community governance encompasses not only politicians, but also regulatory agencies who have to respond to requests from politicians responding to publicity and public complaints, the local community, and organisations within the community, such as news media, community activists, NGOs, and other actors. Some people place governance within the notion of social sustainability.

Stakeholders and influence

The same case study also reveals many of the stakeholders and the influence that they have on each other. Remember that Chapter 10 pages 340–344 introduced the idea of stakeholders, who are individuals or groups of people:

- who are actively contributing to the project,
- whose interests may be affected as a result of project execution or completion, or
- who may also exert influence over the project objectives and outcomes, or the ability of the team to achieve them, should they choose to do so.

In this case study, we can identify most of the stakeholders involved in sustainability issues. Many of the people involved belong to more than one stakeholder group: people can identify themselves with different stakeholders in different contexts. For example, many of the plant employees were also residents of the town being affected by the plant.

The company or proponent

The company, Kanemex, is one of the main stakeholders with possibly more influence than any other. The company can also be referred to as the proponent: the main stakeholder with an interest in the project. Within Kanemex, it is also possible to see different stakeholder groups. As a multinational corporation, the local management of the plant were answerable to an international head office located in a different country. The local workforce had their own interests, which were distinctly different from those of the local management. Within the workforce, again, we can see different stakeholder groups: those who lived in the town formed a separate group from those who commuted from other towns because they identified with other town residents not working for the company. Furthermore, engineers tended to exert much more influence than other workers, particularly because of their technical knowledge. We can learn that within any one stakeholder group, there will be subgroups with different interests and views on the same situation. It is not unusual in a negotiation to find that the major stakeholders have irreconcilable interests – when one gains, the other loses. However, within the major stakeholders, there are almost invariably sufficiently diverse subgroups that make it possible to find common ground that makes a negotiated solution possible.

Another major group in any company that must be considered are the shareholders. Shareholders did not come into this case study, but they almost always exert a strong influence on company operations. In most companies, the shareholders expect to receive regular dividends from profits. The company's share price reflects shareholder expectations about future profits and bad news almost inevitably results in a share price decline. Since large chunks of shares are often held by investment trusts, the people who manage these trusts have a much stronger influence than individual shareholders. In a situation like this case study, senior managers in the company would have needed to spend a lot of extra time reassuring major shareholders and investment trust managers that the situation would be resolved without major financial implications. In the event that there was an extended plant shutdown or the need to replacing critical pieces of the plant, those major shareholders would require even more reassurance. This additional work would have occupied much of the time of senior managers, who would therefore not be able to focus as intensely on other issues.

The local community

People living in Oranto formed the next major stakeholder group. Once again, it is possible to see subgroups of stakeholders within a single community, each of which had different interests. Farmers in the surrounding district had an interest in selling land at a high price to the new business venture in the town. Being located further from the plant than most of the townspeople, they were generally less concerned about emissions from the plant. However, there had been many instances in the past when plant emissions had had more impact on farmers further away from the plant than the local townspeople. Late in the case study, we saw how the company offered a form of compensation to the townspeople closest to the plant by acquiring their property on generous terms. The company sought to create a division within the townspeople, perhaps in an attempt to eliminate most of the complaints by moving people further away from the plant. However, in doing so, the company created an even greater irritation for townspeople who were outside the zone in which compensation was offered. These people had missed out on the compensation offer, but were still affected by the plant operations. This example shows that stakeholder groups are dynamic: they change with time and new stakeholders can emerge to exert different kinds of influence depending on changes in the situation.

As explained in the case study, many of the townspeople worked in the plant. This group would have been composed of people in the permanent workforce who were not affected by the plant shutdown, but would also have included contractors and suppliers in the town who would have been seriously affected by the plant shutdown. On the other hand, it is likely that many local tradespeople and contractors would have benefited from the additional construction work performed during the plant closure. This once again demonstrates that there are always subgroups with different interests and views on the same situation within any single stakeholder group.

Media

While not directly affected by the plant itself, the media had a powerful influence in shaping public awareness outside of the town's normal sphere of influence. Media groups usually have an interest in finding ways to increase their audience, telling stories in a way that people find fascinating to watch and listen to. In deciding what to tell and what not to tell and in order to engage as large an audience as possible, the media helped to portray small groups in Oranto as heroes battling a powerful multinational company. It was a classic David and Goliath story with the chance that the weak and apparently powerless might win in the end. In doing so, they conveyed the townspeople's concerns to the broader community; the media brought the issue to the notice of regional politicians. Ultimately, though, media outlets are still reliant on individuals in the local community to tell their stories. Towards the end of the case study, people in the town gained more influence over the company through their involvement in the plant management committee. They no longer needed the media to exert influence on the company indirectly. Also, the level of irritation caused by plant operations had dropped to the point where the passions aroused among townspeople that could inspire a reporter to write a dramatic story had all but disappeared.

Unions, contractor associations

Though not part of this story, labour organisations almost always exert a strong influence on any project. While unions have been steadily losing influence in most industrialised countries, other labour organisations have gained influence, particularly contractor associations. Companies prefer to hire contractors for certain types of work, such as specialised maintenance that cannot sustain a permanent workforce. Contractors can be laid off on short notice when business declines: laying off permanent staff is always much more difficult and often expensive, since accumulated benefits and entitlements also have to be paid out at the same time. On the other hand, contractors charge more because they need to earn enough to balance the risk of being out of work from time to time. Even though they did not play a part in this particular case study, unions and contractor associations always have an interest in gaining members, which results in more negotiating strength. Times of uncertainty often stimulate an interest among workers to join an association that could give them more negotiating strength.

Regulators

People who worked in regulatory agencies (regulators) exerted a great deal of influence in this case study. Even though Kanemex was operating within the regulations the entire time, the regulators responded to pressure from politicians, who needed to appear to be doing something, by imposing tougher reporting requirements. Under normal circumstances, most regulatory agencies rely on companies to report breaches of the rules. Companies are required to submit a report indicating which rules were breached and providing details of the corrective actions that were taken in response. While regulators can make inspections, the regulatory agencies have limited budgets and they can't monitor every company all the time. What they can do, however, is choose which companies to monitor most closely.

In this instance, the regulators decided to pay more attention to Kanemex, insisting on conducting inspections on short notice, as well as in-depth investigations and additional monitoring. All this imposed significant additional work on the plant engineers, in particular. Engineers would have had to check many aspects of plant operations immediately before an inspection visit: there are almost always day-to-day issues that could attract the attention of inspectors. For example, many plant maintenance operations result in small, accidental, and relatively harmless chemical spills. The vast majority of these are inconsequential, but they would immediately attract the attention of an inspector looking for a reason to justify a visit to the plant. The engineers would have had to check the status of documentation and monitoring instrumentation, as well as ensuring that records of plant operations were complete and up-to-date. Usually, the engineers would have had to organise the regulators' visits, provide safety inductions, show them around the plant, and arrange space for them to work in or even make space for them in their own offices. Then, the engineers would have had to follow up on any issues, no matter how 'inconsequential', that were raised by the regulators during their visits.

Politicians

As explained in the case study, regional politicians were facing an election at the same time that the dispute in the town became more heated. In order to improve their chances of being elected, the politicians needed to gain publicity in local media. Most of the time, local media have far more 'news' and press releases than they can possibly include in the limited space or time available. Therefore, it is not always easy for a politician to gain the attention of local media unless they have done something to upset people, in which case they probably prefer to have less publicity, rather than more. One of the best ways to gain media attention, therefore, is to provide comments on a topic that is already the focus of media attention and consequently in the minds of the audience. That is why, in this case study, we saw how local politicians took advantage of the situation, issuing statements that added to the pressure on the government to do something about the problem. The government politicians, in turn, could only respond by putting pressure on the regulators to do something, which ultimately resulted in direct pressure being put on the engineers.

Financiers

Of all the stakeholders, the ones that had the most powerful influence on the company were its financiers. At the height of the dispute, the financiers were exerting considerable pressure on Kanemex.

Financiers are people who work for the banks that provide the large loans that underwrite the working capital, which in turn enables a company to operate. It is important to remember that engineering enterprises require a great deal of money to be spent long before end users pay for the resulting benefits. Banks provide the money that the enterprise needs to operate until the payments eventually come in. Like shareholders, particularly investment trust managers, financiers have an interest in the long-term profitability of the company. However, financiers also face a different kind of influence. Banks can only operate as financiers because their reputation in the community as trustworthy institutions encourages companies and individuals to deposit money with them. In recent decades, banks have realised that their reputations partly depend on perceptions that they play their part in promoting environmental and social responsibility.[26] If a bank is known to be providing financing for a venture that results in environmental pollution or social disruption, even in another country, senior executives will soon have to be dealing with reporters and a stream of letters from people in the community, many of them bank customers, demanding to know why the bank is supporting these controversial ventures. The banks will therefore be more reluctant to lend money to a company that could cause risk to the bank's reputation in the future. As a result, the company would have to find other lenders who would almost inevitably charge higher interest rates on any money that they lend to the company.

While these issues are often not covered in the media, they often occupy much of the attention of the most senior managers in the company, which is why financiers can be so influential.

Non-government organisations (NGOs)

In the case study, we saw how the World Environment Fund became a stakeholder because, as an international NGO, it had the power to make the relatively small dispute an issue of concern in many other countries. Influential NGOs have dedicated and professional people who work with media outlets to provide them with stories that engage their audiences, particularly in wealthier countries that provide the financing those multinational corporations need to operate. NGOs have learnt how to create publicity that influences the opinions of lenders in the world's leading financial centres. This is another illustration of the way that the situation can quickly evolve and bring in additional powerful stakeholders who were not present at the start of the project.

Government

While governments benefit from tax collections and therefore have an inherent interest in the economic profitability of enterprises, they exert influence in many different ways. In addition, different arms of government can have very different interests. Regulatory agencies are almost always found within government departments, but they may have different interests from the departments in which they are administratively located. Different ministries within governments often have different interests. For example, companies will often deal with a ministry that is responsible for economic development and is therefore able to provide different kinds of incentives for the company to invest in a particular country or region. This ministry often supports a new development, even while the department responsible for environmental protection might be opposing the same venture. Many governments compete with each other, even within the same country, to offer the best incentives to persuade a multinational corporation to make a large investment that creates employment opportunities and export income. That is why different government agencies need to be treated as different stakeholders: while they all exert influence and gain benefits, they exert influence in different ways for different reasons.

Most governments have a department concerned with environmental protection, and it is common to find environmental regulators within the same department. Even this seeming partnership creates conflicts of interest. Promoting responsible behaviour by companies requires regulators and engineers to develop collaborative relationships with each other. Both benefit from this: a greater level of trust enables engineers to gain the confidence of regulators, who are therefore less likely to impose time-consuming processes, such as inspections. At the same time, the same degree of trust enables the regulators to focus their attention where it is really needed, on companies that pose serious environmental concerns. However, other people in the same department will be required to assess applications from the same companies to gain approval for their operations. These people need to be able to demonstrate to politicians and the community that they treat all applications rigorously and that the companies do not have any influence on the process. The collaboration and trust between regulators and engineers can undermine those perceptions in certain situations. Particularly in low-income countries, government salaries are too low to enable civil servants to live in the capital cities in which they work. Therefore, they rely on 'informal means' of raising additional money from companies who need approvals for their operations. These informal payments compromise the integrity of the environmental approval process, but the administrative system would collapse without them.[27]

Other stakeholder groups

Research revealed engineers negotiating with many other stakeholders, including police and law enforcement agencies, indigenous people and their representatives, customs agencies, health ministries and health professionals, and defence forces.

The one remaining stakeholder that consistently needs to be considered is future generations. Unlike other stakeholders, they cannot be represented in negotiations. However, engineers have an ethical responsibility to consider the interests of all children and grandchildren, not only their own. This brings us to the issue of ethics, which is inextricably linked to sustainability.

Practice concept 82: Ethical behaviour supports effective engineering practice

Ethics in engineering is almost always discussed in terms of moral behaviour and values with respect to accepted social norms, religious values, or some other system of ethics, such as professional ethics. In the latter case, engineers who are members of professional organisations can, in theory, be sanctioned for breaches of a professional code of ethics.[28] A typical ethical dilemma quoted in reference texts might concern, for example, the extent to which an engineer should conduct safety checks to protect public safety.[29]

What is not so widely appreciated is that ethical behaviour also has functional value in engineering practice. By this, I mean that ethical behaviour is likely to improve the value of engineering practice in terms of not only economic outcomes, but also social values, safety, environmental values, and social justice.

There are two principal ways in which ethical behaviour supports effective engineering practice.

The first is that ethical behaviour helps with technical coordination, which largely involves exercising influence over other people without having to rely on any formal authority.[30] Even if it were possible to resort to authority, the use of it is likely to compromise the results. Effective technical coordination involves gaining the willing and conscientious collaboration of other people. Resorting to formal organisational authority reduces the extent to which people willingly and conscientiously collaborate.

In terms of Raven and French's explanation of social power, technical coordination relies on referent power, the power that comes from the respect that the coordinator's peers and co-workers have for them.[31] Unethical behaviour inevitably erodes the level of respect a person can expect from other people. While it might still be possible for an unethical individual to coordinate other people, it will require more coordination effort (both time and cost) in terms of negotiating an agreement, monitoring performance, and checking the results. It is much more reliable, easier, and quicker to gain the willing and conscientious collaboration of people who hold you in high regard, as discussed in Chapter 9 pages 289–292 and 315–6.

The second aspect in which ethical behaviour helps effective engineering practice lies in taking the time and trouble to consider the consequences of engineering decisions on other people, both actual and potential. It also means that you, as an engineer, have to put yourself in the mind of the other person and see the situation from their point of view.[32] This takes time and effort because it is first necessary to understand the real interests of the other people who might be affected – the stakeholders. Their interests might not be obvious: it takes time and effort to discover the interests of each stakeholder that can be affected. The second aspect that takes time and effort is to ensure that stakeholders are adequately represented in the decision-making process. If one or more stakeholders cannot represent themselves, either because they cannot understand the decision that needs to be made, they cannot communicate effectively, or they are not able to be present (such as people who are not yet born), then it is necessary to make sure that other people represent their interests. The case study described above shows how this situation can lead to serious problems that can even threaten the economic viability of a large project. The consequences of not considering the implications for one or more stakeholders can be extremely disruptive. Community stakeholders, in particular, can include people who work for your enterprise and who are very knowledgeable about what is happening, including the potential risks. Stakeholders can initiate legal or political action that can cause long and expensive delays or even stop an engineering enterprise completely.

This kind of issue is commonly referred to as social sustainability. Smart enterprises devote considerable resources to build and maintain their reputations as responsible and sensitive towards community concerns, thereby gaining trust, confidence, and implied consent from a community. In other words, they acquire a 'social licence to operate'.

Practice concept 83: Values and value conflicts

The notion of 'values' can provide a useful way to think about some of the conflicts that emerge in the context of sustainability issues. Different people subscribe to different values, can change their values over time, and can even subscribe to different values in different contexts, such as identity.

This section presents a very brief description of a few value frameworks (although there are many), which should be sufficient for you to understand how different people subscribing to different value frameworks can easily come to what can appear to be irreconcilable conflict and disagreement on sustainability issues.

Let's think about some examples.

One framework of values that many people use in the context of business and investment could be described as 'money-centred'. In this value framework, success is represented by the accumulation of money. Money is a proxy measurement for human effort, desires, needs, and even uncertainty. Thus, effort is rewarded by payments of money: effort should only be expended if the monetary reward justifies the form and amount of labour. Desire is represented by price: goods that are more desirable come at higher prices, but people are prepared to pay more money for them due to their increased desire to own or possess them. In this framework, expensive goods are therefore more desirable. Human need is also represented by money: people in greater need are prepared to pay more money relative to what they earn or have saved for the same goods. This value framework is often one in which people keep track of good deeds and bad deeds. Favours typically come with the expectation that they will be repaid. Even uncertainty can be represented in terms of money, as we have seen in previous chapters: the value of future transactions is discounted at a percentage rate proportional to perceived uncertainty. In the absence of other information, many people who might normally subscribe to different value frameworks can take a 'money-centred' view of their investments, measuring their worth in terms of the way they increase in value, capital value appreciation, and the income that they provide. Counting everything in terms of money can be a useful way to simplify one's appreciation for complex activities, activities that have many other aspects that are not necessarily represented by financial transactions or valuations. When people are strongly influenced by a money-centric value framework, they tend to take decisions based on expected gain or loss measured in terms of money. Within this framework, misfortune may be seen as random chance or just bad luck: even good luck will be perceived as random chance.

A different set of values could be called 'moralistic' or 'humanistic', a framework in which honesty, integrity,[33] and altruism[34] tend to govern human decisions. In this framework, people consider that while money has its uses, not all decisions can be made simply in terms of financial gain or loss. Many would consider that monetary values determined by arbitrary human perception, desire, and needs that can change from one day to the next cannot reliably measure the value of aspects of social well-being, such as freedom, security, human relationships, or even natural assets like forests and wilderness areas. People strongly influenced by a moralistic value framework tend to make decisions based on social morals and norms, particularly when it comes to the welfare of people who are less fortunate than they are. They can be strongly influenced by how they imagine a moral and ethical person would perceive their actions. In contrast to a money-centred value framework, favours and good deeds performed for other people do not come with the expectation that they will be repaid. In fact, it is often considered virtuous to perform such deeds without any expectation of reward or even recognition.

Even a simple example can illustrate how actions that make sense in one value framework can seem to be quite illogical in another. Consider, for example, a person thinking in a money-centred framework who benefits from a good deed by another person thinking in a moralistic or humanistic framework. This person may be wondering when and how they have to repay the favour, because in their framework, every good deed comes with the expectation that it will be repaid. To this person, the idea that a good deed is not something to be repaid in the future might seem quite illogical.

Another set of values could be called 'spiritual' or 'religious'. People who are strongly influenced by these values acknowledge the existence of God or spiritual entities that exert a strong influence on human actions and events in the world that others might attribute to random chance. Many of these people subscribe to orthodox versions of a religion in the sense that they subscribe to a framework of behavioural rules that have a divine origin. Others have a more spiritual orientation in the sense that their framework of behavioural rules needs to be applied in the context of meditation, prayer, or some other perceived contact between the individual and God or another spiritual entity. Within this framework, good deeds are seen to be ones that are sanctioned by divinely inspired religious texts (especially for more orthodox individuals) or deeds that appear to be virtuous as a result of prayer or meditation (for more spiritually inclined people). Even suffering can be seen as a positive outcome in certain variations of this value framework. Some people may see suffering or misfortune (in themselves or others) as divinely ordained punishment for past misdeeds or they may see it as a divine message encouraging better behaviour in future. Some people may even see the suffering of others as divine punishment, which means that there is no need to alleviate that form of suffering. For many, performing good deeds counts towards a heavenly reward in the next life.

These value frameworks are not necessarily exclusive. As explained before, it is possible for people to subscribe to different value frameworks at different times and in different contexts. This can lead to internal conflicts (and sometimes external conflicts, as well).

Yet another value framework is one in which resilience, self-reliance, and independence from other people is considered virtuous. People influenced strongly by this framework of values may be reluctant to seek the help of others when they need it, and may also be reluctant to help others when they need assistance. Suffering or misfortune, in this value framework, may be seen simply as a test of strength for an individual. They will see the ability to withstand the suffering and recover from misfortune as an indication of personal strength, which is a virtuous attribute.

It can also be useful to pay attention to other value frameworks that help us understand different ways in which people see themselves in relation to the Earth, our planetary home.

Many people see the Earth's resources as an endowment for human beings that should be used to help build a more advanced human civilisation that will eventually reach a point where that advancement will significantly reduce the need for such resources. Many of these people might even see the Earth's resources as a divine reward for taking the initiative to search for them in the first place, or for the considerable human effort required to locate and extract them. For these people, exploiting the Earth's resources is an obligation, and those that are able to do so before others have the good fortune of being more deserving. We could call this framework of values a 'natural endowment' framework.

Other people see value in the Earth's natural state that existed before human beings had any significant impact. For them, any disturbance or changes made by humans is undesirable, even offensive. Therefore, eliminating or at least minimising human impact on the Earth is virtuous and desirable. We could call this framework of values 'naturalistic', in the sense that people value 'nature' as a force for good in the world that must be protected. For these people, allowing nature to take over and reclaim areas of land previously used by people is desirable. However, people influenced by a natural endowment framework might see this as neglect or abuse.

Still other people might see the Earth's resources as a limited endowment to be shared with other people in a just and equitable way. We can call this an 'equitable endowment framework'. In other words, all people have a proportionate right to use the Earth's resources and share them with future generations. These people face the difficulty of deciding how an appropriate share to set aside for future generations actually might be estimated. A further difficulty is how to take into account the efficiency with which resources should be used. Should people who can only make very inefficient use of a resource be allowed a lesser or greater proportionate share? One could argue that they should be allowed a greater share, because they lack the ability to make efficient use of the resource. On the other hand, one could equally argue that they should be allowed a lesser share, because that lack might force them to learn how to make efficient use of the resource.

Yet another value framework is one adopted by many of Australia's Aboriginal population, which argues that there is no separation between people and the Earth's resources. For these people, every piece of the Earth is a part of the people who inhabit it, and the people are a part of the Earth: the two elements are inseparable from each other. People and animals, even inanimate objects, are merely physical manifestations of spirits that live in the Earth. It is considered virtuous to manage one's 'country',[35] because the resulting improvement will benefit people, animals, and the country itself. It is also virtuous to be completely aware of the state of one's country, including the plants, animals, and water – even the rocks and sand.

Beyond this, there are also social and economic value frameworks that come into play. For much of the 20th century, many people strongly valued 'socialist' ideals, represented by slogans such as 'From each according to his ability, to each according to his need.' Coupled with this was the notion that possessions belonged not to individuals, but to society as a whole, and were to be shared according to need rather than prior, physical possession. With the possible exception of Cuba, most countries that implemented strong versions of socialism have turned away from these ideas, noticing that countries that followed 'free market' and 'capitalist' ideas seem to perform much better, and typically produced much more favourable outcomes for the majority of their citizens.

In the free market value framework, people expect to receive an economic reward for hard work in the form of money, which they can choose to save for the future, spend on assets, or spend on daily consumption and luxuries. Instead of the state deciding on the prices of goods and the rewards for effort, those rates are determined by transactions in markets that are *theoretically* free. In practice, however, countries that adopted free-market frameworks have found that varying degrees of government regulation were desirable; in fact, government intervention has even been necessary from time to time. Most recently, we saw how governments decided to borrow money and spend it to keep people employed after the global financial crisis, following the advice of economists, such as John Maynard Keynes. However, people influenced by strong free market ideas, such as those of the economist Friedrich Hayek, view these interventions as ill-considered and futile, merely postponing inevitable readjustments.

If we now think about contemporary debates on climate change, we can see how these different value frameworks lead to the disagreements that can place major impediments in the path of engineering projects. People influenced by free-market frameworks and money-centred value frameworks make the assumption that markets will efficiently regulate human behaviour and deal with the climate change problem without any intervention being necessary. They argue that as the effects of climate change become more obvious, and the Earth's mineral resources become harder to extract, the financial value placed on minerals and fossil energy resources will rise, making renewable energy options, such as solar photovoltaic panels and wind generators, more economically attractive. They argue that increases in oil prices over the last few years have created an incentive to develop more fuel-efficient vehicles, because people find them more economically desirable. Many people with this value framework even argue that increasing temperatures and the concentration of carbon dioxide in the atmosphere will promote plant growth, making agriculture more productive, on average. They even admit that agriculture may become much less productive in certain regions, but they argue that the overall gain will outweigh these losses. Even those people who think that climate change could be catastrophic in the future argue that governments need to implement simple financial incentives, such as emission trading schemes, in order to promote the responsible uses of common assets, such as the atmosphere and the oceans. These measures introduce real financial externality costs to shift decisions influenced by money-centred values towards more sustainable options.[36]

On the other hand, people with a naturalistic value framework see climate change as extremely damaging to the planet. They attribute this damage to human activity conducted largely in ignorance of the consequences, or at least with a lack of consideration for future generations. In their view, governments need to intervene to protect the natural environment and prevent people from damaging it any further. For these people, economic incentives will never be sufficient: they argue that only governmental control can prevent damage to the natural environment by human beings. Many of these people also urge strong actions to reduce the human population, arguing that this is the only way to protect what is left of the natural environment on the planet.

Many people in low-income countries subscribe to an equitable endowment framework, arguing that wealthy countries in the world have used up their proportionate opportunity to exploit valuable resources, including their quota of carbon dioxide emissions that will remain in the atmosphere for at least 1,000 years. They argue that low-income countries should be allowed to exploit their share of the world's resources until they can develop their societies to a similar level as the currently industrialised, wealthy countries.

There is a gradual accommodation in business to bridge these value frameworks by introducing new ways to conduct financial analysis, mainly by introducing 'externality' costs, in other words, the monetised valuations of environmental, social, and health impacts of engineering enterprises.[37]

Negotiations concerning technical issues are equally influenced by conflicts arising from different value frameworks. Chapter 8 page 272 listed some of the values evident among engineers: technical satisfaction, emergence, elegance, impact, innovation, challenge, and humanitarianism. Just as in the previous example, actions that make sense in an engineering value framework can seem quite illogical when thinking in another framework. This helps to explain why many businesspeople feel very frustrated with the actions of the engineers whom they employ, while the engineers feel equally frustrated with the actions of their employers. Explanations in earlier chapters focused on differences in the use of language, but these differences can also be understood in terms of conflicting value frameworks.

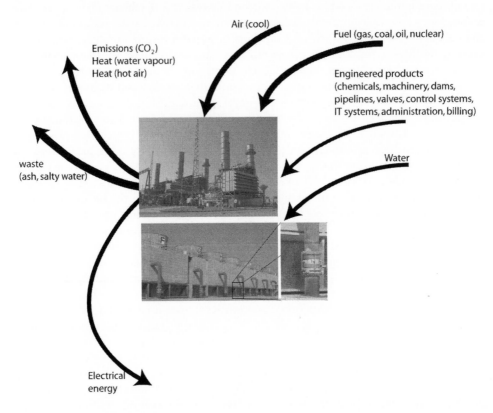

Figure 12.2 Electricity generation: one of the engineered systems that supports human civilisation. Power stations rely on the evaporation of cooling water to disperse waste heat, increasing the concentrations of any salts present in the incoming water.

Many people on the planet now accept that we have to behave differently in the future than we did in the past if human civilisation is to have a reasonable chance of a sustainable future. The difficulty is that, so far, these intentions have not been translated into strong enough action, because people with different value frameworks propose different responses. The world's renewable natural resources, as well as its non-renewable resources, are being depleted at ever-increasing rates, and we are all experiencing declining air quality in the planet's atmosphere.

At the same time, not everyone appreciates how dependent human societies are on the natural environment, just as the future of this planet is increasingly dependent on responsible decision-making by the people who live on it.

What has been interesting for me to observe, as an engineering educator, is the difficulty that engineering students have in understanding this reality.[38] At the start of a fourth-year sustainability course, I asked my students to consider the interdependencies between different aspects of the systems that support human civilisation. I presented the students with a series of diagrams, similar to Figure 12.2. In symbolic form, each diagram portrayed material and other inputs and outputs on which each major aspect of the system depends.

Other diagrams represented systems such as water supply, mining and resource extraction, manufacturing, farming, and forestry. This was not a complete set, of course. However, it was enough for students to start constructing a system diagram portraying all of these engineered systems and their common interdependencies. For example, water supply systems rely on electrical energy for running pumps, while simultaneously, most systems for electricity generation rely on water supply for cooling.

Over several years of teaching this course, none of the 200 or so students in the class remembered to include the atmosphere, the oceans, or the planet itself in their system diagrams. However, as we can see in the example above, the power station cannot operate in the absence of the atmosphere, nor can people breathe. When asked what they might have missed, a few students hesitantly suggested the financial aspect. However, in several years of running this exercise, none had realised that they had omitted these fundamentally vital and obvious parts of the Earth's environment until it was pointed out to them.

This exercise, therefore, helps to expose an inherited value framework that has separated people from the environment that supports us. The vast majority of people on the planet accept that we have to care for the environment and protect it, but do not necessarily see that the environment is part of them, and that they are part of the environment.

Later, I asked the students to participate in a thought experiment. I reminded them that they had all reached the university by travelling along roads leading to the campus. As they did so, they inhaled air contaminated with exhaust emissions from vehicles on the roads. Recognising this, I asked them whether they would be willing to have a small amount of car exhaust gas injected into their lungs through a tube – a precisely equivalent amount corresponding to what they would have inhaled on their journeys to and from the campus. Practically all of them showed considerable revulsion at the thought of this. Yet, in scientific terms, this is the exactly what their lungs experienced every day. It only takes a few seconds for a small amount of poisonous exhaust fumes to become part of us, yet few of us ever think in that way. We dismissively think that the fumes go into the atmosphere and not into our lungs.

One way to understand the actions of my students is to see this as a separation of people from the environment in our value framework. This also reflects the separation of 'mind' and 'body', which follows the beliefs and writings of French philosopher, René Descartes. This is very different from the way that Australian Aboriginal people think about the environment. Their equivalent concept is 'country', the land with which each group of Aboriginal people identify themselves. Aboriginal people regard themselves as being part of 'country', just as 'country' is part of them. They are all spirit children of the spirit mother that is their 'country'. Even though much of Australia appears as a desert to Europeans, to aboriginal people, there is no such thing as unoccupied 'country'. Every stone, every patch of bare sand, the sky and air above, the rocks beneath, every plant, animal, river, and watering hole: all are integral parts of 'country'.[39]

Even though there are roughly 500 recognised language groups among Aboriginal people in Australia, they all seem to share this basic idea of 'country'.

What is particularly remarkable about Aboriginal people is not only the social and linguistic complexity referred to earlier (Chapter 5 page 113), but also the longevity of their civilisation and cultural traditions. Archaeologists began to discover evidence in the 1970s that showed how present-day Aboriginal traditions have varied only slightly from practices that can be traced back to the earliest continental inhabitants who arrived between 40,000 and 50,000 years ago. Through their use of fire to modify the landscape, Aboriginal people were able to improve animal habitats and also make it easier to trap animals, thereby increasing their food supply.[40] We do not know of any other civilisation on the planet that has survived with an intact culture over such a long time period, especially one that has survived the climatic extremes that are present in Australia.

Is it possible that the concept of 'country' and the ideas that Aboriginal people have about the environment in which they live hold lessons for us as we approach the limits of our civilisation's sustainability? Could a different value framework for how we see the environment that supports us be a more effective way to transition to a truly sustainable civilisation?[41] Such a transition is possible, but it might be challenging to achieve. There does appear to be a distinct difference in the results of the value framework of Western civilisation in comparison to those of Australian aboriginal societies. We have abundant evidence that the former is currently on an unsustainable path, whereas the latter has demonstrated at least 40,000 years of sustained existence.

Practice concept 84: Human emotions influence engineering

One recurrent theme in our research evidence is a view among most engineers that emotions have nothing to do with the technical rationality of engineering. Many engineers pride themselves on being 'rational', 'level-headed', and 'unemotional', particularly in contrast to 'other people' who, in their view, are easily swayed by human emotions. The rational basis for engineering is deeply embedded in education. Many engineering degree programmes start with mathematical logic and proofs (though this is often intensely disliked by engineering students). Engineering students, both male and female, form an identity that suppresses attributes such as emotional sensitivity.[42] Unfortunately, research on emotions in the context of organisations is relatively scarce, partly because the idea of suppressing 'undesirable, unprofessional, irrational' emotions has become well-established in Western intellectual circles for two or three centuries.[43]

Research demonstrates that while engineers like to think that they leave their emotions at home, or hanging on the cloakroom hook, their actions demonstrate that emotions play just as powerful a part in the life and work of engineers as they do for other people.

Fear of being seen as ignorant or incompetent lies behind the reluctance of young engineers to seek help and advice from engineers who are more experienced, site supervisors, operators, and technicians.[44] Often, this reluctance will result in behaviour that others judge to be arrogant, simply because a young engineer is trying hard to make it look as if he knows something, when it is obvious to all that he doesn't. This same fear even affects career choices: students about to complete their courses fear that employers will easily detect their weak grasp of technical issues, knowing that they have crammed at the last minute to pass examinations and then forgotten much of what they knew for that short period the time.

Incursions by outsiders into one's home, land, or territory often provoke an emotional response from people. The aftermath of having personal possessions stolen from

one's home brings a sense of loss and insecurity. An incursion into one's physical territory can result in painful emotions for anyone. Territory represented by expertise is an
additional dimension of territory for engineers, so intellectual incursions, or perceived
incursions, can cause similarly hurtful feelings. This can easily happen, particularly
in a complex project with a large number of engineers, possibly working at different
locations. It is not unusual for an engineer to find that decisions have been made by
others that affect their area of responsibility. Here is a rare expression of emotion by
a design engineer:

> 'One day I heard that the downstream[45] designers had completely rearranged the
> beach crossing where the umbilical cables and pipelines come ashore from the
> offshore platform. That was my area of responsibility, but they made the changes
> without telling me anything about it. I felt terrible.'

While engineers may not always show their feelings, their actions and comments
reported in our research demonstrates that they are affected by emotions, often deeply.
There can be few things more hurtful for an engineer than to find that a part of a project
in which they have invested personal energy has suddenly been cancelled. All too often
this can happen without warning or even consultation.

Many engineers have a deep and emotional sense of satisfaction when they see
that a project to which they have contributed has been completed successfully, particularly when there is easily noticeable physical evidence to show others.[46] Several of
the engineers that contributed to our research told us about times that they had shown
their children or grandchildren something that they had been involved with. 'I helped
to build that great big ugly ship in the picture!' one of them said.

Even technical rationality, the essence of engineering, can often be an emotion or
sentiment as explained in Chapter 10 page 362. There is a satisfying *sense* of rationality
that comes with resolving a technical problem but it is actually an emotional response,
because when it comes to detail, we almost always discover endless aspects that have
not, in fact, been resolved. As with value frameworks, scepticism from others can
evoke a defensive emotional response.

'So what?' you might say. 'How does all this affect me?'

First of all, feelings and emotions interfere with human perception abilities. Pleas-
ant, enjoyable, and positive emotions, on one hand, as well as hurtful, negative
emotions, on the other, can significantly erode one's ability to listen accurately and
even to perceive the world around us visually.[47] This increases the likelihood of misunderstanding, and most disagreements in the workplace, even technical disagreements,
can rapidly escalate into destructive conflict.

One of the best ways to become sensitive to the effects of emotions on the people
around you is by observing non-verbal communication, such as body language. If you
can develop the ability to sense the presence of strong emotion that can affect the way
that people behave, then you can more easily predict the results and help to alleviate or
even avoid destructive, open conflict. A mismatch between body and verbal language
can indicate the presence of strong emotion.

Understanding more about emotions can help you become an expert negotiator
much quicker. That means understanding the way that people feel at an emotional
level, as well as understanding the way that they are thinking at an intellectual level.
You should also learn to sense your own emotions. It might start with apprehension,

possibly a slight tightening of the muscles in your abdomen or a tense feeling in your mind, even a headache. Remember that this mental and physical response will impair your own perception abilities. Wait before responding, or simply listen, take notes, and resolve to make a decision or respond at a later time.

PREDICTING THE FUTURE

Sustainability negotiations depend on our ability to predict the future, something that can be very difficult, as the graphs in Figures 11.1, 11.2, and 11.7 demonstrate. Debates on climate change often involve questions on the accuracy of predictions made by the scientists, such as those contributing to the Intergovernmental Panel on Climate Change (IPCC) reports.[48] As engineers, we rely on scientific principles for all our predictions, so we have no choice but to accept the predictions made by climate scientists: it does not make sense for us to be selective in relying on science for making engineering predictions.

Given the amount of effort devoted to climate science predictions today, it is highly instructive to re-examine predictions made in the 1970s on the limits to growth in human activity, predictions that were made with far fewer resources and computing power.[49] The benefit, of course, is that we can compare those predictions with data collected on what has actually happened since then.[50] The results are notable because many of the predictions turned out to be remarkably accurate, as illustrated in Figure 12.3. Three original predictions were made based on different assumptions. The standard model assumed no change from the status quo. A second 'comprehensive technology' model assumed rapid technological advances to improve energy and resource utilisation efficiency and to reduce pollution. A third 'stabilised world' model assumed that population growth would be curtailed. Data collected since and analysed independently matches the predictions of the standardised model. One significant exception is the availability of remaining energy resources: data collected since the 1970s shows that the standard model predicted the most pessimistic estimates published since then, while the comprehensive technology model predicted the most optimistic estimates.

The most concerning aspect of these predictions, backed up by data collected since, is that the standard model predicted a rapid decline in all measures of human progress from about 2040 onwards. The onset of this decline was calculated to be caused by a combination of several factors, including a rapid rise in the investment (both human effort and technology) required to extract any remaining energy resources, accumulating pollution in the atmosphere that reduced the productivity of agriculture, and declining biological resources to support food production for a human population that would still be increasing. According to the original predictions, while the comprehensive technology model predicts that the collapse can be deferred until the 2090s, collapse can only be avoided by eventually stabilising the world's population.

When the original predictions were published in 1972, they were widely criticised. When I read the original publications, I was equally critical due to weaknesses in the computational methods used to solve the many interconnected equations in the prediction models. The limited accuracy of the calculations could have caused large errors in the results. Now, with the benefit of independent analysis and data confirming the predictions, we have to take those earlier predictions much more seriously. In essence, they provide strong evidence supporting other scientific findings claiming that

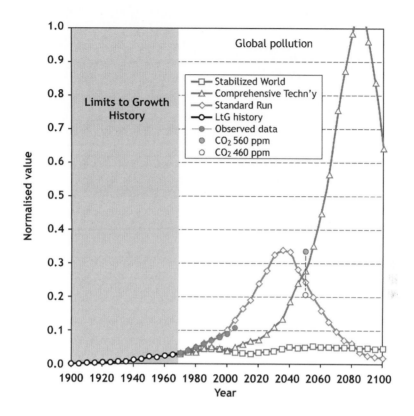

Figure 12.3 Graph comparing the estimates of global pollution with observed data collected since the predictions were made.[51] Refer to the text for explanation of the different prediction models.

we have gone beyond the limits of our planet's ability to support our civilisation, unless we make large changes in the way that we use the planet's resources.

It is often difficult for engineers to bring future predictions into sustainability, or even commercial, negotiations. As Chapter 11 has shown, any prediction of future economic conditions is almost certainly going to be wrong. Many people with a money-centred value framework will be reluctant to accept forecasts of the future, preferring to rely on their own intuitive 'feel'. The following example shows how engineers can make a business case without having to rely on explicit predictions. This quote comes from a food-processing engineer:

'It took ages, but we finally convinced our management to install a reverse osmosis plant so that we could recycle process water, enormously reducing our consumption and also significantly reducing biological nutrient discharge into the waste stream. Both of these were significant gains for environmental sustainability. It took several years, because the management would not compromise on seeking a 25% return on investment. We tried several ways to make the necessary business case before it was finally accepted. Sometime after we installed the reverse osmosis plant, the city water supply was drastically curtailed because of drought. Our

competitors had to close down their production lines because they were consuming too much water: they had to import supplies and still sell them at a loss to maintain their market presence. We were able to continue at full production and made huge profits, repaying the reverse osmosis investment by several hundred percent. In retrospect, we should have based our case on the risk of water restrictions: it would have been accepted much more readily.'

This shows how engineers can make a case for sustainable solutions by arguing in terms of risk reduction. Renewable energy supplies, for example, are unlikely to be economically competitive with fossil fuel energy due to the cost of energy storage. However, by considering the risk that energy costs could suddenly and unpredictably rise as they have done in the past, sometimes by a factor of four, renewable energy can be justified as a risk mitigation measure. A further factor is that environmental regulations are almost invariably tightened every few years, becoming more restrictive. Often, this requires expensive plant modifications. As explained in earlier chapters, it is usually much cheaper to make provisions for these types of plant modifications in the original design rather than trying to improve the environmental performance of an existing plant when the original design did not allow for future modification.

Engineers are on much safer ground by predicting larger future uncertainty, rather than trying to accurately predict the future.

PREPARING TO NEGOTIATE

This outline is based on Fisher and Ury's methods.[52]

Success in negotiation depends more on preparation. Success in negotiation relies on forward planning, research and preparation, awareness of the interests of the other parties, and having alternative plans if negotiation is either not possible or not advisable.

Who are the parties in the negotiation?

The case study earlier in this chapter demonstrates how there are many stakeholder groups involved in a typical conflict over sustainability issues. Technical conflicts within a team of engineers, or those involving two or more teams, can also affect several stakeholders. Most commercial negotiations, on the other hand, typically have two stakeholders: the client or customer and the supplier.

When there are many stakeholders, it would be unusual for all of them to be represented. Often, this is impractical: future generations and other stakeholders may not be able to be present or may not be able to represent themselves at the negotiating table. There may be language issues or, alternatively, there may be significant inequalities in terms of negotiating power that make this impractical. That is why engineers often have to ensure that the views of stakeholders who are not at the table are still represented. Otherwise, the aim of the negotiation, a negotiated agreement that everyone is comfortable with, will remain out of reach.

Leaving out important stakeholders can have disastrous consequences. For example, in our research, we heard about a small town in a remote desert region with about 50,000 inhabitants. The governing council of the town employed several engineers for aspects of their operations, including sanitation, roads, waste disposal, electric power, and the fishing harbour. Water supply was not included specifically among any of the engineers' responsibilities. The governing council had not designated an engineer to be responsible for water supply because it was unclear whether this was the responsibility of the regional government or the governing council of the town. About two years later, the town ran out of water. Although a water supply reservoir had been constructed, no one had monitored the rate of evaporation. Early one summer, after a particularly dry winter, the water remaining in the reservoir was too salty to drink. For many months afterwards, water had to be transported over 400 km in trucks to keep the town habitable. Furthermore, the inhabitants had to pay a very high price for the water to cover the transportation costs.

A similar situation was reported in the research literature during the collaborative design of a high-energy particle accelerator that was part of a multinational physics research project.[53] While there were physicists and engineers with designated responsibility for all the major equipment modules, including the ultra-high vacuum system, focusing magnets, electricity supply, radioactive particle sources, and detectors, only one had been given responsibility for the shielding wall. It had seemed an uninteresting, mundane task and no one had expressed interest in taking on the responsibility. This shield separated detectors and magnets that could withstand intense radiation from other electronic modules that could not function in the presence of such radiation. With the design phase well under way, this omission was finally noticed and a young mechanical engineer was allocated what seemed to be the simple task of designing the shielding wall. After several weeks of careful inspection of drawings of the components that had already been designed, this engineer discovered that the space originally allocated for the shielding wall had been punctured in many places. It turned out that the space needed for several equipment modules on both sides of the wall had been underestimated. The equipment designers had requested and were granted additional space, but in doing so, many modules now passed through the space that had been allocated for the shielding wall. What started as an apparently simple task turned into a highly complex negotiation between stakeholders located in several different cities. This case study demonstrates how important technical issues have to be adequately 'represented' in technical negotiations. The consequence of not doing so is that they will be ignored and then, later, when they have to be dealt with, the solutions will be much more complex because many conflicting constraints have already been accepted.

The important lesson to be learnt here is that in any collaborative technical undertaking, all the critical technical elements need to be adequately represented when decisions are made. In another study, we heard how none of the most senior engineers responsible for the design of a very large multibillion dollar oil and gas project had any experience with reliability engineering. Reliability engineering was considered to be a specialist discipline for which consultants were hired. These consultants reported to a mid-level systems engineer who then reported to a lead engineer with responsibility for reliability and many other system-wide functions, such as data communications, logistics, and controls and instrumentation. As a result, reliability considerations had been downgraded in importance, because the lead engineer could not adequately represent the issues being raised by the consultants or negotiate on behalf of 'reliability'.

The same lesson applies to sustainability issues. In the case study, we saw how the relationship between the townspeople and the company improved once the townspeople's complaints were taken seriously, even though they might have had no provable scientific basis. This can mean direct representation: including town representatives on the plant management committee was one of the key steps taken to resolve the conflict.

As in project management, one of the first steps is to identify the stakeholders who need to participate in the negotiation, either being directly represented at the negotiation table or at least having their interests taken into account. When it comes to sustainability, future generations almost always need to be represented and their interests must be taken into account. However, there may be other stakeholders who cannot necessarily participate directly, or whose interests need to be taken into account if the outcome is to be successful and sustainable. In this context, we also need to consider outcomes that are sustainable in the sense that once an agreement has been reached, all stakeholders will continue to respect the agreement, even if they had not actually been present in the negotiations.

Once the stakeholders have been identified, the next step is to understand every stakeholder's interests, values, requirements, and expectations. This process is outlined in Chapter 10 pages 341–2. It is crucially important, whenever possible, to record everything in the language and terminology used by each stakeholder. Often, different stakeholders have common interests, but cannot necessarily see that due to their use of different concepts, terminology, and language. Conversation and dialogue is the fastest and most effective way to build your own understanding about each of the stakeholders; in other words, a series of discovery learning performances is necessary, as outlined in Chapter 7 pages 209–10 and elaborated in Chapter 8 pages 258–261. However, this is not always possible. Sometimes, the only way to reach this understanding is through researching written documents or past actions. Understanding the interests of future generations can only come through a carefully built understanding of sustainability principles, along with listening to engineers, scientists, and others who can provide insight into the future.

Fisher and Ury focus on interests as one of the best ways to work towards negotiated agreements. However, values, requirements, and expectations also play a critical part.

Major differences in values between different stakeholders may flag potential conflicts that can be very difficult to resolve. For example, contemporary debates on climate change highlight value conflicts within many industrialised countries, as explained earlier. While some stakeholders adopt the endowment values framework and regard the Earth's resources as their right, or even as their duty to exploit responsibly, others see them as an endowment to pass on to future generations with only minimal exploitation – or even none at all. Still others argue for the right of all people to have an equitable proportion of the Earth's resources for themselves and their future generations.

Expectations highlight issues that can cause difficulties during negotiations, but can often be resolved through a gradual process of adjustment. For example, a contractor might expect to be able to make a large profit on supplying equipment that is needed urgently for a very large project. The contractor might expect that the urgency will provide the opportunity to charge a higher price, knowing that any delay will result in higher costs for the client. However, by working with the contractor and focusing on his interests, such as reputation and the likelihood of future business, the contract expectations can be gradually adjusted, saving costs in the short-term and promoting a better business relationship in the long-term.

Focus on interests, not positions

In the preparatory steps leading up to face-to-face negotiations, it is important to focus on understanding the interests, values, requirements, and expectations of each stakeholder. By doing so, you will have established the foundation of a negotiated solution. Most negotiated solutions gain acceptance from stakeholders because the interests of most, if not all, stakeholders are advanced by the agreement: they all gain something.

Stakeholder requirements, as explained before, indicate the likely constraints within which the negotiations need to be conducted. Expectations indicate the kind of results that each stakeholder would like to achieve, but they probably know that they are not going to get everything that they want. Values are also important to monitor: large differences in apparent values will influence the language and the ideas that stakeholders talk about in the negotiation. It can be much more difficult to reach an agreement between stakeholders that are heavily influenced by different value systems. Also, it is likely that the interests expressed by each stakeholder would need to be translated or rephrased into language that makes sense within the dominant value framework used by other stakeholders.

Some stakeholders will mention particular requirements and these help to clarify constraints on negotiation. Requirements can be negotiable or non-negotiable. For example, in a negotiation for space within the design of a new object, some of the space requirements will be non-negotiable. Physical components that have already been confirmed as part of the design dictate some of these space requirements, both in terms of accommodating the component and also providing access for assembly and maintenance. An example of a negotiable requirement might be the choice of cables and data communication options, which must fit within spaces remaining between mechanical, structural, and other components.

Fisher and Ury quote a useful example of using an interest-based approach to help negotiations succeed. After fighting a war in 1967, Israel occupied the entire Sinai Desert, which had originally been a part of Egypt. Eventually, they withdrew, but they reoccupied the Sinai Desert in 1973 during another conflict. During subsequent peace negotiations in 1977, Israel refused to withdraw their tanks, because they did not want Egyptian tanks to be stationed in Sinai, easily able to roll into the heart of Israel without warning. Egypt insisted that every inch of the Sinai be returned. The solution lay in identifying their mutual interests:

Egypt wanted sovereignty: Sinai must be in Egypt.
Israel wanted security.

The solution, then, was to arrange for a large part of Sinai that was adjacent to Israel to be declared a de-militarised zone: Israel would have a sizable buffer area to prevent a surprise attack, but the Egyptian flag would be flying over the desert once again.

In buying a car, the interests might be:

Buyer: Needs a reliable car for transport and possibly prestige.
Seller: Needs money from the commission on the sale, as well as after-sales service business.

Look for common interests such as:

Security, certainty, and resolution of a problem or disagreement
Good business relationship
Known financial situation
Satisfaction (and possibly prestige) obtained by resolving the problem

Once interests have been identified, particularly common interests, it is far easier to devise options from which both sides can benefit.

Be wary of the zero-sum game, however.

Win-win vs zero-sum

It is easy to look at the negotiation process in terms of 'gaining or losing ground'. Every time you concede something on an issue being negotiated, one way to look at it is as 'giving something away'. This is the 'Zero-Sum' game – whatever you lose, others win. Thinking in these terms is not helpful: unless you can think in terms of interests, you are unlikely to successfully negotiate an agreement.

In fact, you have not gained or lost anything before a negotiated agreement has been reached. Therefore, any result that is better than what you would have by not negotiating at all represents a win for you. Since the other parties will only remain in the negotiation if they are better off by doing so, they are also winning something. This means that a properly conducted negotiation always leads to a win-win position, compared with a failure to negotiate at all.

Best alternative to a negotiated agreement

Negotiation is not always successful. You may not be able to agree on an outcome that all parties can accept. However, you are much more likely to be happy with the outcome if, before negotiating, you decide on your 'bottom line' or your 'best alternative to a negotiated agreement'.

You cannot decide this without some informed thinking about the consequences of any failure to negotiate. Therefore, you need to do some research first.

Example:
Peter has found a new car that he likes, and he thinks he can persuade his parents to pay most of the cost. The salesman tells him the price after accounting for the value of Peter's old car, which he will trade in. Peter knows that he can negotiate because there is some flexibility in deciding the 'trade-in value'. However, Peter also knows his friend Simon has already offered $3,400 for the old car, so Peter wants to be sure of getting a better deal from the salesman before he agrees to buy the new car.

Peter's *best alternative to a negotiated agreement* is to sell the car to Simon for $3,400.

Peter might also visit another car showroom to see some different new and second-hand cars, and to ask about their opinion on the trade-in value of his old car.

Peter can then enter a negotiation to purchase the car he really likes, knowing that he has alternatives to negotiating with the salesman.

The crucial step is to decide on the best alternative to the current negotiation. It is best to adopt one or more indicators or criteria that can tell you how much better the proposed agreement is compared with your best alternative to negotiation. Make sure that you use clearly defined criteria; otherwise, you can be swayed by emotion or pressure to reach a solution.

The third step is to decide on your opening position, also known as your opening bid.

The opening position

It helps if all parties in a negotiation can state their opening positions, or preferred outcomes, even though they initially may seem to be a long way apart. However, since nearly every successful negotiation involves all parties giving up something of apparent value, the opening bid may allow for this.

Once again, the key is careful planning and preparatory research before the negotiation starts, and careful listening once negotiations commence.

To write the opening position, you must have some idea of a likely negotiated agreement. In simple price terms, let's take the example a bit further:

Peter expects to pay about $15,000 for his car after a trade-in value of about $5,000. His parents will contribute $12,000, which he can pay back over three years, interest-free. He knows the salesman will give him a larger trade-in than the value of the car because the stated price of the new car includes an allowance that enables the dealer to quote a generous trade-in valuation.

The dealer knows that Peter's old car, as a trade-in, has a resale value of about $3,000 after allowing for his own expenses. He suggests an initial trade-in value of about $2,000, saying that the car needs a major mechanical overhaul before he (the dealer) can legally offer it for sale.

If the dealer were to suggest an opening position of $20,000 (i.e. zero trade-in value), Peter would walk away disgusted. If he offers $16,000 (trade-in $4,000), Peter may ask for a few extras to clinch the deal (such as air-conditioning), which would mean that the dealer would not make any profit from the deal.

In other situations, the relationship between the opening bids can be quite different. For instance, when buying carpets in Pakistan (from my own experience) the dealer normally asks for about two or three times the eventual sale price. The buyer will normally respond with a suggested maximum of about one quarter what the dealer has suggested. Thus:

Buyer: What's your best price for this nice carpet?
Dealer: About 240,000 rupees sir . . . it's the very best quality, sir, you have made an excellent choice.
Buyer: Well, that's a very high price. I know about carpets, and I know that's way too high. I wouldn't normally pay more than 60,000 for it.

They will eventually settle on a price roughly in the range of 80,000–110,000 rupees.

In complex negotiations, the opening position needs to address all the factors that relate to the interests of each stakeholder. For example, in an employment negotiation, an opening position might address the following issues:

- Salary and salary packaging
- Leave entitlements, freedom to choose leave timing and duration

- Office accommodation (quality, location)
- Car parking
- Freedom (or otherwise) to choose subordinate staff
- Duration of contract, timing of performance reviews
- Performance criteria
- Relocation allowance
- Commencement date
- Working hours – flexibility
- Timing

Good timing is essential. It takes time to negotiate a good deal for both parties. Any time limit will restrict the scope for making numerous small adjustments to the agreement and could result in an overall failure to reach a satisfactory agreement.

If one party is under time pressure to finalise an agreement, then they will often be worse off than a party that can afford to wait.

Practice exercise 20: Role playing

One way to effectively prepare for a multiparty negotiation is to role-play the negotiation beforehand. Ensure that every party represented at the negotiation is played by someone in your organisation who is prepared to research and learn about their interests, values, requirements, and expectations. Then, the negotiation discussions are played out, simulating the actual negotiations still to come, with every participant trying to do their best for the party and other stakeholders that they represent.

WORKING FACE TO FACE TOWARDS AN AGREEMENT

Once negotiations commence, each of the parties will be discussing aspects of the negotiation either face-to-face or through e-mail correspondence, or some combination of both.

Listening – learning the interests of each stakeholder

This is the most important phase of any negotiation. Careful listening (i.e. active listening where there is a dialogue between the parties to understand each other) is essential in this phase. You should have received (preferably in writing) the opening positions of the other parties before this phase commences.

Remember to write the interests of each stakeholder using the same words as they themselves have used. Then, if necessary, translate these words into equivalent words and ideas used by the other stakeholders.

The reason for establishing the interests of the other parties is to determine:

- Common interests – directions that benefit all or most of the parties
- Group interests – issues on which two or more parties share a degree of common interest
- Individual interests – issues that are important for only one stakeholder

These interests need to be explored through dialogue. It will be helpful for all parties to understand everyone's respective interests, so there is nothing to be gained by secrecy at this stage. Any interests that are not disclosed at this stage are likely to interfere with later negotiations, unless a stakeholder does not want the negotiations to succeed. It is worth establishing that one of the common interests is a successful outcome for the negotiations.

It is quite possible that one or more of the negotiators will not have thought about the process in this way, and it may take time for them (and you) to establish what their interests are.

Agenda

Once all relevant interests are better understood by all parties, you will be able to decide an agenda for the negotiations. The agenda needs to cover all the issues on which agreement is needed, but it is more useful to establish agreements on common interests first. Set the more difficult interests to the side for the time being – another way to describe this is to 'park' these issues. The agenda will need to be agreed on by all the parties. There is no need to reach agreement on issues in the order dictated by the agenda.

Group issues can be explored by side-discussions between the parties, where common interests can be explored more efficiently than in a full negotiating session.

Creative solutions

There needs to be time to think about creative solutions. Creative solutions often require lateral thinking and will usually not be evident from an understanding of the stakeholder interests alone.

It is useful to leave at least a week for thinking about creative solutions. Once a solution has been proposed, it will also take some time to clarify the details for each of the parties, and then they will need time to assess the benefits.

Negotiating tactics

Sometimes, negotiations can become 'bogged down' on minute details or simply through losing sight of the common interest of reaching an agreement. There are some tactics that can help to strengthen your own position, or overcome a 'stalemate', but these must be used with caution. Some of these tactics may be seen as aggressive by some parties.

Referring to a higher authority

Before moving towards a new negotiating position, you should indicate what the negotiating position would be, but inform the other parties that you are not authorised to make this kind of offer. Suggest that only the 'owner' or 'the director' could authorise such a 'reduction' in 'our' position. This has the effect of suggesting to the other side that you are trying to go beyond your bottom line and whatever you subsequently suggest will be your final bottom line.

Imposing a false time limit

Suggest that in, say, two hours time you have to leave to catch a plane. This puts pressure on the other parties to move more quickly and possibly make a rash move towards a more attractive offer for you. It only works if the other party needs to reach an agreement before you leave.

However, the time limit has to be credible, and you run the risk that you may have to leave the negotiation if the time limit is reached, just to maintain your credibility.

Playing for time

This is the opposite of the former tactic. If you are not under pressure to reach a negotiated agreement, or if you think that the other parties *are* under a time constraint, then you can attempt to delay each step of the process. This often conveys the impression that you are a very powerful negotiator, provided that you maintain consistency.

Sometimes, playing for time happens by default when the people in your own organisation cannot make up their mind about the next move. This can suggest that you are playing for time to the other parties.

Threatening to walk out ... or actually walking out

If you think that other parties have not moved significantly and you want to make them move, you could consider threatening to walk out. This sends the message that, 'We really don't think you are making an effort to negotiate in good faith, and we are thinking of calling off the negotiations in favour of some alternative'.

Your bluff may be called, however, and you may have to walk out ... which can be a useful last resort when the other party has not moved for some time, or has not moved fast enough to make further negotiation worthwhile.

If, on your way out, the other side makes any approach that suggests you should stay, you know that they don't want you to walk out. At this stage, it is best to keep walking, but be prepared to stop just before you drive off (assuming that you're using a car) and relent. However, when relenting, it is essential to obtain a positive indication of a major move from the other party (or parties), or you will have achieved nothing.

Bringing in your expert

It is useful to take an expert with you. This is entirely sensible in the first place, because you need someone to do the background 'calculations' to let you have a better idea of where you are in relation to your best alternative at all times.

However, it can be a useful tactic to take a person with you who is not an expert, but who pretends to be. It shows that you are serious, and if you make some remark that suggests your knowledge is not what it ought to be, the other parties may not be prepared to pursue their advantage if they think you have a better-informed person with you to advise you.

FRAMING THE AGREEMENT

With common interests identified and conflicting interests resolved, the one remaining task is to write the agreement. As explained before, a sustainable negotiated agreement may be a complex document, as it has to take into account many possible events, specifying how each party will respond if any of these events occur. Lawyers can be very helpful in drafting such agreements and many commercial negotiations will involve lawyers at the negotiating table.

One of the most important aspects in gaining consent from all of the parties to accept the agreement is the choice of language. One way to improve the chance of consent is to accept the language of parties that have the most sensitivity on any particular issue, even though the language may not be technically correct. It may sound more logical to adopt a scientifically rational basis for engineering enterprise negotiations. However, the important aspect of a negotiation is to secure the agreement from all parties. It is not uncommon for small issues of wording in the final agreement to take up much of the negotiation time, long after all the major issues of conflict and disagreement have been resolved. The appropriate choice of language can help reduce the amount of time spent on these issues.

SUMMARY

This chapter has only been an introduction on sustainability issues and how conflicts that engineers encounter in the workplace can be solved through negotiation. Negotiation skills are critically important for an expert engineer and should be practised and developed throughout your career.

Sustainability issues are likely to become more, not less, challenging for expert engineers in the future. While we can all debate the accuracy of future predictions, there is already enough evidence that we need to make significant changes to the way we consume the Earth's resources, and the ways we distribute the subsequent benefits. The argument I have presented in this book is that these challenges present great opportunities for expert engineers who can not only devise solutions, but also deliver them on time, within budget, and with the promised performance, safety, and environmental impact.

NOTES

1. Bucciarelli (1994, pp. 51, 146–149).
2. Borrego, Karlin, McNair, and Beddoes (2013).
3. Suchman (2000).
4. E.g., Lewicki, Saunders, and Barry (2006).
5. Itabashi-Campbell & Gluesing (2014); Suchman, 2000).
6. Fisher & Ury (2011).
7. Lewicki et al. (2006, Ch. 16).
8. A serial digital communication bus, such as CAN, was also considered, but the technology was not commercially available at the time in a sufficiently miniaturised form (Trevelyan, 1992).
9. Another example is the accommodation of a shielding wall for components in a high-energy particle accelerator and collider, reported by Dominique Vinck (2003, Ch. 1).
10. An extended discussion on this with references to psychology and organisation literature can be found in Borrego et al. (2013).
11. Address to WA Division Engineers Australia Centre for Engineering Leadership and Management, March 2010.
12. Brundtland (1987, Ch. 2).

13. E.g. http://www.footprintnetwork.org/en/index.php/GFN/.
14. Stench – unpleasant smells of human waste and excreta which were previously thrown into the street outside houses.
15. Carson (1962); Hardisty (2010, Ch. 2).
16. Known as 'The Montreal Protocol'.
17. There were other sources, such as tin-lead solder, used for joining metals and some paints, that greatly extended the durability and improved the appearance of wood in buildings.
18. http://www.abc.net.au/news/2013-03-27/epa-gives-lead-exports-go-ahead-through-fremantle/4598080.
19. Paradoxically, the reactors were being closed down to prepare for decommissioning. Even though a study had pointed out the possibility that the reactor cooling systems might be damaged by a tsunami, it was considered that the risk was acceptable, as the plant was about to be shut down.
20. Lucena (2008, 2010); Lucena *et al.* (2010); United Nations Development Programme (UNDP, 2013).
21. Hargroves & Smith (2005).
22. On this issue, it is interesting to observe how many engineers, particularly in Australia and the USA, are reluctant to accept the predictions of climate scientists or even appreciate the difficulties faced by climate scientists in making predictions. These same engineers, however, rely on scientific predictions for almost every aspect of their work.
23. Chilvers and Bell have shown how sustainability issues impinge on the core values of an engineering firm and how they have to be compromised when faced with the realities of client expectations (2014).
24. Fictitious names and places have been used to protect the participants' identities.
25. Hardisty (2010, Ch. 2).
26. Bob Dudley, CEO of BP at the time of the 2010 BP Horizon rig explosion and fire caused by the Macondo well blowout in the Gulf of Mexico, said that BP came perilously close to complete financial collapse because suppliers were demanding cash up front and bankers were reluctant to renew loans to keep the company operating (BBC Money Programme, March 3, 2011).
27. See Chapter 13 for issues that affect engineering practice in low-income countries and pages 471–3 for a discussion on corruption. While the effects are different in low-income countries, corruption occurs in all countries.
28. Engineers Australia code of ethics, http://www.engineersaustralia.org.au/ethics. ASCE (American Society of Civil Engineers), ASME (American Society of Mechanical Engineers), IEEE (Institute of Electrical and Electronic Engineers), NSPE (National Society for Production Engineers), and most other engineering professional organisations have similar codes.
29. Davis provides several detailed case studies (1998).
30. Trevelyan (2007).
31. Chapter 9; Raven's retrospective review (1992) of the theory that first appeared in the 1960s.
32. One of the prerequisites advocated by Nussbaum (2009) for an education that promotes human liberty and is also essential for an expert engineer.
33. Acting in accordance with moral values and justice: note that the word 'integrity' has quite a different meaning in the context of engineering systems, as explained in the online glossary.
34. Performing deeds that help other people more than oneself.
35. Gammage (2011) and many other works on Australian Aboriginal culture and traditions.
36. Hardisty (2010).
37. For fully detailed methods and applications, Hardisty (2010, Ch. 3–7) has several case studies.
38. Carew and Mitchell have systematically analysed student attitudes (2002).

39. Gammage, 2011 (Ch. 5 pp. 139–154).
40. Gammage, 2011 (Ch. 6 pp. 155–186).
41. Hardisty, with extensive experience of engineering project decision-making, reports only slow progress towards real change in decision-making by industrial firms towards more sustainable alternatives (2010).
42. Godfrey (2003).
43. Gooty, Gavin, & Ashkanasy (2009); Muchinsky (2000). Several writers have argued for the treatment of emotional issues in engineering education, for example: Ravesteijn, de Graaff, and Kroesen (2006), Riemer (2003), Scott and Yates (2002), and Toner (2005).
44. D. M. S. Lee (1986).
45. An oil and gas engineering term that usually refers to onshore processing. 'Upstream' refers to the infrastructure required to get the oil and gas out of the ground, the offshore platforms, and pipelines.
46. Engineers' values are discussed in Chapter 8 page 272.
47. Psychological testing demonstrates this effect in the laboratory (J. M. G. Williams, Mathews, & MacLeod, 1996). See also Chapter 5.
48. Reports available from http://ipcc.ch/.
49. Meadows, Randers, Meadows, & Behrens (1974).
50. Turner (2008).
51. Adapted from Turner (2008, Figure 10).
52. Fisher & Ury (2011).
53. Vinck (2003, Ch. 1).

Chapter 13

Great expectations

'Take nothing on its looks; take everything on evidence. There's no better rule.'
—Charles Dickens, Great Expectations

This chapter is written especially for engineers living and working in the Third World. However, it is almost certain that every engineer will collaborate with engineers or clients in the Third World at some time in their career. Certain types of engineering work, such as detailed design and analysis, are routinely outsourced to specialist firms that are often located in countries such as Mexico, India, Pakistan, and the Philippines. Therefore, any book on engineering practice would be incomplete without including an attempt to consider the wider range of settings in which engineers might reasonably find themselves working. There are also lessons in this chapter that are applicable to all engineers. It is often easier for an outsider to appreciate your own culture: in the same way, you learn more about your own culture by understanding another culture. Furthermore, the Third World holds exciting and potentially profitable opportunities for any engineer with sufficient expertise.

This chapter is divided into four parts. The first introduces some of the important economic factors that characterise the Third World, including a short section showing how corruption can be encountered anywhere, even in the most unexpected places. The second section explains similarities and differences that distinguish engineering practice in South Asia (representing the Third World) from Australian engineering practice. The third part includes various success stories, how we discovered some truly expert engineers, and discusses factors that seem to explain the great success of mobile telecommunications in Third World countries relative to most other engineering enterprises. Finally, the chapter ends by pointing out great opportunities for future generations of engineers, and makes suggestions on how to start working on them in the present.

The research that led to this book started with observations and first-hand experience from employing engineers in Pakistan. One of our research conclusions is that the greatest future opportunities for engineers lie in the Third World, as this chapter will explain.

DEVELOPMENT AND THE THIRD WORLD

Historically, the First World encompassed the well-established industrialised countries, also known as 'The West', while the Second World consisted of Russia, China, and their allied states in Eastern Europe and Asia. The Third World was the category into which the remaining countries on the planet were placed. Recently, the definition of the Third World has been slightly clarified as the group of countries with the lowest Human Development Index (HDI).[1] These countries are also called 'low-income countries' (LICs), as there is a strong correlation between income and HDI.

The chapter title reflects the great expectations among approximately 60% of the world's population who endure subsistence living in the Third World, people who, thanks to the propagation of the Internet, can now understand and aspire to a First World-quality of life, at least for their children, if not for themselves. Most likely, you also had great expectations that an engineering career would bring great rewards, at least when you were studying engineering. However, many students graduate and are rapidly disappointed by the fruitless search for a satisfying engineering job. Others have jobs, but their career prospects are constrained by the organisational circumstances that they find themselves in, as described towards the end of this chapter.

I also have great expectations for you due to the knowledge that you can gain from this book. As an expert engineer, your career will bring great rewards. This chapter also explains how; with enough expert engineers working towards important goals, no one needs to live in poverty any longer.

It is important to understand how large-scale engineering has been a critical factor in human development. In the 17th and 18th centuries, engineers developed machines that harnessed fossil fuel energy, greatly extending the human capacity to modify our living environment and produce food, leading to the Industrial Revolution in Western Europe and America. These machines were rated in terms of horsepower: horses, along with other slower animals, had provided the only available extension to human muscle power for the previous few thousand years. Scientific advances made possible by the Industrial Revolution led to electric power, communications, transportation vehicles, chemical and biochemical technology, and eventually information technology, enabling huge improvements in human productivity. These improvements, in turn, liberated most people from the drudgery that had formerly been required for producing food and acquiring basic necessities. In many developing countries, 60–70% of the population still work in subsistence agriculture to feed their families. Today, less than 2% of the population of an industrialised country are needed to produce water and food for everyone. The productivity improvements from these advances provided the additional human capacity needed to improve education, health care, social welfare, and governance: services that continue to distinguish a First World lifestyle.

Due to the fact that the earliest machines in the Industrial Revolution were relatively inefficient, European nations started to run short of resources, so several drew on their colonies to obtain what they needed for industrialisation. That option is no longer available for other countries seeking to industrialise their economies, and improvements in technology have enabled similar transformations with greatly reduced material resource requirements. Countries such as Chile, Korea, and Malaysia have transformed themselves into industrial economies with First World living standards. However, there are two major barriers for transformation in most LICs today.

First, our entire global society still needs large improvements in resource utilisation efficiency to enable the entire population of this planet to enjoy comparable living standards. Wealthy countries still use a disproportionate share of resources and in the previous chapter, we reviewed evidence demonstrating that we cannot maintain even our current rate of total resource consumption indefinitely.

Second, many LICs are making slow progress, if any, towards improving the quality of life for most of their people. Contemporary development theories attribute the wealth disparity to differences in governance, economic policy, and health, so it is not surprising that most interventions have focused on capacity development in these aspects.[2] Our research has demonstrated that the relatively high cost of engineering in LICs is another, more significant factor impeding development that has not yet been recognised. This chapter presents explanations for this finding based on our research.

There is also evidence that engineering work that is outsourced to firms in LICs is often more expensive and challenging than first anticipated.[3]

A further difficulty is that engineering is only recognised by the human development community as a significant activity within the industrialised world.[4]

Development economists, such as Jeffrey Sachs, have recognised that the most critical measures needed to transform societies with extreme poverty are:

1.   Agricultural inputs (e.g. fertiliser, water harvesting, irrigation) and produce storage, including roads and transport for people and materials
2.   Investment in basic health: clinics, medicine
3.   Investment in education
4.   Power, transportation, and communication services
5.   Safe drinking water and sanitation (without which, item 2 is ineffective)[5]

Although not explicitly acknowledged, all of these rely on engineering investments, even education and healthcare. Without sufficient productivity levels, a society will not have sufficient human resources for health care, education, welfare, and governance. More than any other factor, these basic resource constraints can explain why so many LICs experience governance issues.[6] Engineering, more than any other intervention, is critical for improving human productivity.

From 1997 onwards, I was forced to confront evidence that confounded my initial expectations on the cost of engineering in the developing world. I had expected that labour-intensive engineering research and development work would be less expensive, since we could hire engineers for 25–35% of Australian salaries.[7] I was employing Pakistani engineers to design and construct prototype equipment to demonstrate improved methods for landmine clearance. These engineers were unable to match expectations based on my experience with similarly experienced engineers in Australia. Neither the level of qualification (bachelor's or master's degree), nor the country in which it was gained (Pakistan, the UK, the USA, or Australia) appeared to make any difference. Visiting local industrial firms led me to the realisation that one has to adopt a different level of expectation in order to make realistic performance predictions for engineering work in South Asia.[8]

However, it was very difficult to explain the differences when compared with Australia. There seemed to be little difference in the level of understanding in terms of fundamental technical issues. There was, however, a significant difference in the ability of local firms to supply many specialised engineering components and materials. Few South Asian technicians worked from drawings in the way that their Australian counterparts would, yet they were just as skilled in fitting and machining. The most significant qualitative difference between Pakistani and Australian engineers was in aspects that one might describe as 'practical skills', but it was very difficult to articulate exactly what this term meant.

I soon realised that the same issue affected locally produced engineering products and services. I was confronted with a number of unanswerable questions. Why, for example, were the extraordinary engineering achievements that characterised the Mughal period in South Asia (16th–18th centuries) apparently not feasible today? While huge constructions like the Red Fort in New Delhi are still perfectly straight and aligned 400 years after they were built, why is it that new buildings today cannot even be constructed with similar precision? Why is it that engineers from South Asia perform as well as any others when they work (as many do) for firms in Europe and America, yet they seem unable to perform at the same level in their home environment?

Around 2001, I focused my attention on water supplies and sanitation for government schools in villages on the fringes of Rawalpindi. I began to appreciate the real significance and social value of effective engineering, because it was conspicuously absent in that area. A village family insisted that I have tea with them and proudly showed me their new hand pump, inside their little walled compound, for which they had paid $1,500 (in this village, the water lay 30–50 metres below a layer of hard rock). I was astonished: these were poor people – how could they possibly afford that? Before they had their pump, it took the women an hour or so to carry buckets of water back from the village well.

In today's money, the value of their time worked out to roughly $0.30 per hour.[9] Before they installed their hand pump, each time they went to their well, the women from that household managed to carry home about 17 litres of water and it took them another hour to boil it. In terms of the value of their time and fuel, the cost of that water was about $35 per tonne.[10]

In a Pakistani city, you can have water delivered in 20-litre bottles. The cost works out to $50–70 per tonne but it's more reliable, cleaner, and safer to drink.

Across South Asia, city water supplies, which are run by engineers, are intermittent. In most areas, water flows from the main water supply pipe for an hour or so every other day: the cost of a water service works out to about $20 a year. However, that water is almost certainly contaminated by sewage that seeps in through thousands of broken and half-repaired connections that run through open sewers. The water pressure in the supply pipe is negligible, except for a few hours when the local engineers turn on the supply in your area of the city every other day. The pipe pressure can be negative relative to atmospheric pressure, because many consumers install their own pumps to extract the last of the water from the supply pipes. You can drink the water: it is chlorinated, but this is often insufficient. When I first started employing people in Islamabad, I soon found that those who drank the local water were frequently absent from work due to illness. Providing clean water did not eliminate illness completely, but it did reduce the incidence of illness. Illness is also an economic penalty: even if you don't require treatment from a doctor, it still reduces your earning ability.[11]

Sanitation mostly relies on open drains that are often blocked by garbage. Heavy rains bring sewage water onto the streets and into domestic compounds.

Most costs are invisible, because they are represented by a loss of human capacity in terms of time and opportunities. There is no charge for water from public purification plants and women's labour for carrying water is regarded as a free household

Figure 13.1 Women carrying water in a Pakistani refugee camp.

commodity. The notion that water could be far cheaper in a wealthy country seems beyond comprehension as it is not easy to perceive economic costs that are not direct cash payments. According to WHO reports, in 2004, India and Pakistan had already met their Millennium Development Goals for providing drinking water, providing for about 90% of their population: the remainder of the population was far from the large cities. However, the definition of 'improved water source' was specifically crafted to make the goal achievable. There is no mention of water availability or cost.[12]

As economists would predict, the total economic costs associated with different water sources in a community turn out to be fairly similar. Whether you pay for bottled water to be delivered, arrange for your servants to carry water from a local public purification plant, stand for an hour or so in a queue at a city water depot at 4.30 am and pay a bribe to ensure that someone delivers water the same day, install your own pump, and then filter and boil city water, the cost is roughly the same when all expenses, including time and opportunity costs, are taken into account.

When visiting India and Pakistan, I have watched women up and down the country spending a large portion of their days carrying water and performing other essential maintenance tasks for their families and animals. The effort/cost required to obtain the bare minimum 10 litres of water a day that people need for survival can be calculated as equivalent to more than 10% of the GDP.[13] The cost of meeting the same requirement in an industrialised country is almost negligible, thanks to effective engineering in water supply enterprises. Compared with industrialised countries, the performance of large

Figure 13.2 Water delivery in Hyderabad, India.

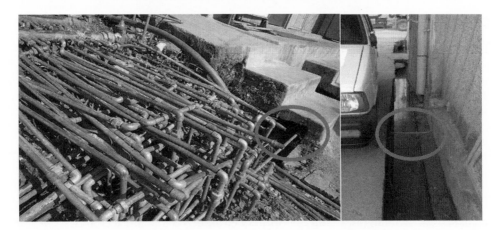

Figure 13.3 Left: Water supply connections in Hyderabad, India. Right: House connection in Rawalpindi, Pakistan. Water connections run through open sewers.

engineering investments in urban water supply is poor, both for the owning public and the end users.

Medical professionals have told me that most of the illness cases they deal with in Rawalpindi every day could be prevented by washing hands before eating and washing hands and the anus after using a toilet. However, that is already a strict requirement for Muslims, who dominate the local population. Why, then, do most people avoid what is both a health and religious requirement? The cost of water could certainly explain why: it is difficult for these people to afford enough water even to wash their hands.

The cost of other services that rely on engineering (such as electricity supply and construction), taking into account total costs and comparable end user service quality, also tend to be higher than in Australia, though the differences are not as great as for potable water, which is heavy to transport.

Electric distribution networks are still built using the same technology as industrialised countries employed in the early years of the 20th century, and are largely funded by concessionary loans from agencies, such as the World Bank. However, in a social environment with relatively weak governance, revenue collection that relies on manual meter reading is easily subverted, particularly by wealthy customers.[14] Governments have been unable to raise electricity tariffs to an economically sustainable level that covers the full cost of generation and distribution infrastructure, so most low-income countries continue to provide large subsidies for electric power. Wealthy people who install conventional air conditioners enjoy a disproportionate benefit from these subsidies and also have a strong influence on governments. Therefore, electric power is intermittent in a multitude of countries. In many parts of South Asia, one can expect at least 12 hours of power interruptions every day, especially during hot weather.

Electricity utilities typically have insufficient generation or transmission capacity to meet the demand, and also may not collect sufficient revenue to pay for fuel, particularly in hot, summer weather when wealthy customers use air conditioners, or in winter when low water levels constrain hydroelectric generation capacity. Overloaded distribution networks with aged and decaying equipment impose further restrictions on supply capacity. A generator is essential for anyone who needs continuous power, which significantly increases the average cost of electric power.[15] Even people who cannot afford a generator are affected by indirect costs. Without being able to rely on refrigeration, uneaten food has to be discarded: it cannot be safely kept for eating later. Fresh food has to be brought in daily, as it cannot easily be stored. India and Pakistan grow three times as much food as is actually consumed: the rest is lost in storage, processing, and distribution. Commodities purchased through shops and markets are more expensive because the shops and their suppliers have to run generators. Then, there is the opportunity cost when productive work has to be suspended due to the lack of lighting or electric power. As in the case of water supply, the direct costs seem low, but the indirect costs are unexpectedly high.[16]

Many would argue that these difficulties reflect political and economic challenges: they would assert that there are no 'technical engineering' issues without readily available solutions. However, electricity and water utilities are predominantly engineering enterprises: the core staff members are professional engineers with university training. Transport, communications, and construction also rely on engineering to a similar extent, although seldom with monolithic enterprises that characterise water and electricity distribution.

Taken together, these performance differences in engineering enterprises provide a powerful explanation for on-going poverty in LICs. These factors help to explain lower productivity and high costs, particularly for water, which is essential for survival.

The next part of the chapter examines various patterns in engineering practice that can explain enterprise performance differences.

Practice concept 85: Engineering is not the same everywhere

Students, many engineers, and most other people think that …	Research demonstrates that …
Engineering is based on applied scientific principles that are universal. Engineering, therefore, is the same everywhere.	Engineering relies on extensive technical collaboration that depends on social interactions and economics. Since social interactions and economics are strongly influenced by the prevailing social culture, engineering outcomes are also strongly influenced by the prevailing social culture. Compared with industrialised countries, a given engineering enterprise providing the same service quality can be much more expensive in equivalent currency terms.

What could explain the difficulties with water and electricity distribution compared with industrialised countries? Many see engineering as simply 'applied science', comprised of hard and fast principles that apply equally in Perth, Pune, and Peshawar. The cost and service quality difference is ascribed by some to a 'lack of resources', or 'subversion by many of the self-interested social elite'.[17] For several years, I struggled to understand why basic engineering was clearly not (and is still not) working in South Asia. Electricity and water utilities rely on substantial contributions from engineers and there is little need for advanced technology that is beyond the reach of LICs.

Is it possible, therefore, that the practical skill shortcomings that I encountered first-hand were being replicated on a grand scale in these large enterprises, influencing the relative level of benefits from a given engineering investment? Is it possible to identify specific skill shortcomings?

The research that led to this book actually originated with these challenging questions. When the research started, there were remarkably few published accounts of systematic research examining the work of ordinary 'everyday' engineering in industrialised countries and none from the developing world.[18] Most of this research has been based on interviews and fieldwork in Australia as the main example of the industrialised world to provide a reasonable 'baseline' for comparison with South Asia. Gradually, the objective of our research became clearer: a conceptual and detailed framework for engineering practice that could explain observations on performance differences in different settings and disciplines. This book attempts to provide this framework in a form that can help engineers learn from the research.

One view of engineering that emerged is its role as a lever that amplifies the effect of a given economic investment, yielding human productivity improvements. Engineering reduces the human effort, energy, and material resources needed to provide a given level of benefits. When seen only as the application of scientific and mathematical principles, there is a tacit assumption that a given engineering investment will essentially yield the same benefits wherever it is applied. Data on potable water and electric power costs, coupled with observations of engineering practice in South Asia and Australia, reveal that this assumption is questionable. The most promising explanation from our research is that the political economy in which it is situated and the prevailing social culture shape patterns of collaboration that are needed for engineering practice, which in turn influence the performance of engineering enterprises, causing large differences in the benefits and results of a given investment.

We compared engineering practice in three settings: water supply utilities, metal manufacturing, and telecommunications.[19] These observations revealed the mediating influences of social hierarchy, the application of engineering management systems, system knowledge representation, community education and skill development, and the entanglement of engineering networks with political power structures. These influences can explain many of the large performance differences.

In terms of enterprise performance, the South Asian manufacturing companies employing the engineers participating in this research were domestic leaders, but were not yet internationally competitive at the time (2008–2010). With labour costs around 7–10% of their total expenses, their performance was largely determined by plant availability, throughput, and product quality, all of which were compromised by factors discussed below.[20] While labour represented a higher proportion of expenses for Australian firms (10–20%), engineers were more effective in maintaining high quality, availability, and throughput.

The success of mobile telecommunications in practically every low-income country, however, is a notable exception to the average performance of engineering enterprises. Observations explained later in the chapter help to demonstrate that certain socio-technical choices influence collaboration patterns and almost entirely eliminate performance differences. In Pakistan, for example, thriving commercial enterprises provide mobile calls with high service quality that cost as little as US $0.05 per hour of talk time and international calls at US $0.02 per minute, along with network and mobile Internet coverage extending to the remotest settlements in the Himalayan mountains. In Australia, the cost is between $0.05 and $0.10 per minute, although the country is much more sparsely populated, with far fewer handsets within the range of any one base station, and it is therefore more expensive to provide mobile telecommunications.

Space engineering in China and India might also be cited as success stories, but more information on relative cost is needed to draw conclusions with any confidence.

Can corruption explain cost differences in engineering performance?

Many would cite corruption as a significant difference affecting the cost of engineering in LICs. Many low-income countries, such as Nigeria, Pakistan, Indonesia, and India, are notorious for corruption. While corruption is significant, it can only explain 10–15% of the cost difference,[21] yet the water supply and electricity cost differences uncovered by our research are much greater. Corruption, therefore, can be ruled out as the major explanation for these cost differences. Some have argued that corruption can be seen as an effective means for expediting business in otherwise slow, costly, and inefficient government bureaucracy and justice systems.[22] Of course, a detailed comparison should also include the relatively high cost of monitoring, compliance, and audit processes required to contain the influence of human dishonesty in industrialised country enterprises.

As engineers, you will almost certainly encounter situations where corruption is possible, because you will be responsible for spending large amounts of money.

At this point, you might be asking yourself: what exactly do we mean by corruption?[23]

Here is an example from my personal experience. I was working on a project for which we needed a 130-square metre workshop constructed with various services, including compressed air, three-phase electrical power, and water. It was to be constructed alongside a small single-story office building on the main campus of a large biological and agricultural research organisation.

We approached the campus building department, who prepared detailed plans and arranged for quotes[24] from three different construction firms. When the quotes arrived, we were astonished at the cost estimates: they were all around three times higher than what we had expected. At first, we thought that the building department had special requirements that would apply to all new campus buildings, such as energy efficiency and the provision for future air conditioning. We asked why the cost was so high, but we did not receive a clear explanation.

I was concerned that the three construction companies who had supplied quotes might have some association with the campus building department staff, as they were all located near the campus. I arranged for quotes from three other companies that were much further away from the campus. Their quotes were far closer to our original cost estimates.

Subsequent investigation revealed that there had been collusion between the building department manager and the three construction companies. For several years prior, there had been a number of large construction projects at the campus, but these had all been completed. The manager, however, had become accustomed to a high-spending lifestyle by arranging with the construction firm owners to add a small amount to the quoted cost of each building and then divide the money between them. All three firms cooperated and the manager arranged for each to win contracts in return for supplementing his income. That year, there were no large construction projects at the campus. Therefore, the manager and the construction firm owners had to try and meet the cost of their normal lifestyle expenses from our shed construction budget, along with one or two other small jobs. The construction firms went bankrupt when the campus demanded repayment of their accumulated losses and the building department manager spent a few years in prison.

This all happened in Australia, not Pakistan or Indonesia. The take-home lesson is that similar practices can be encountered anywhere. With a student, I also became involved with another, much more serious, case involving academic fraud at a North American university in which several people were shot and killed by an enraged victim of the fraud. My student was also a victim: he found that his thesis work had been published by academics at this university who claimed to have done the work themselves. Through an amazing coincidence, I met a student from the same university where the shooting had occurred who had hacked into university computers and secured evidence of very large financial irregularities. We provided this evidence to assist an international police investigation, which subsequently uncovered evidence of tens of millions of dollars in financial fraud by the same group of academics who had published my student's work.

Certain types of large engineering construction projects in my own part of Australia are reportedly affected by large-scale corruption. When I have asked why the relevant governmental anti-corruption agencies are not actively investigating these activities, I have been informed that they are fully occupied with investigations of dishonest behaviour within government agencies: they do not have sufficient resources or jurisdiction to investigate private sector corruption.

There are important lessons here. First, there are always people who will act dishonestly, if given the incentive, opportunity, and a large enough probability of escaping any negative consequences. Since collaboration is an essential aspect of engineering, dishonesty is always a threat that can undermine collaboration and confidence. Second, even with a tempting incentive, dishonest behaviour can be moderated by reducing opportunities, as well as by visible supervision and monitoring, which lowers the chances of escaping the possible consequences. Both measures require resources: higher human productivity in industrialised societies enables more resources to be allocated.[25]

In a low-income country that has fewer available resources for monitoring behaviour, corruption can be reduced by reducing opportunities for dishonest interference through the effective use of information technology, as demonstrated by the success of mobile phone systems.

Similar aspects in engineering work

Before detailing the differences, it is worth describing aspects of engineering practice that are similar in South Asia and Australia.

At least on a superficial level, there are many similarities in the ways that engineers perform their work. Most check their e-mail upon arriving at their cubicle in an open-plan office space each morning, whether they work in Chennai or Sydney. Most likely, an engineer has less space in Chennai, simply because the land is much more valuable in relation to the notional value produced by local engineers. Engineers working in industrial zones on the urban fringes or small regional centres enjoy more space, perhaps even a shared office with their own desk. Senior engineers are more likely to have individual offices. Most engineers have either a laptop or a desktop computer, often both, and many take their work home with them via the laptop. Depending on the scale and nature of the enterprise, they all use similar software applications. Wherever they work, engineers make frequent use of their mobile phones, and many can access e-mails on their phones while they are on the move.

Our research evidence revealed that most engineers enjoyed their work and often worked for approximately 45 hours per week, not including the 8–10 hours of work-related socialising, which is often an important aspect of their jobs. For most engineers, it is hard to focus on any one task: each day is punctuated with frequent interruptions in the form of phone calls, meetings organised on short notice, and casual conversations with co-workers, visitors, suppliers, and contractors.[26] Depending on their role in the organisation or project, many engineers spend an hour or so every day inspecting aspects of work being performed on-site or on the shop floor in factories.

Using almost exactly the same language, regardless of where they are, engineers complain about the amount of routine administrative 'paperwork' they have to do and report that 'real engineering', such as technical calculations and design, is a relatively small component of their job. Reflecting on their education, nearly all engineers complained that it was mostly theoretical and that they had experienced limited opportunities to see the industrial relevance of their studies. Engineers in South Asia commonly believe that their counterparts in the industrialised world have had much better opportunities to appreciate the industrial relevance of engineering science during their education. However, in reality, the differences are relatively minor.

DIMENSIONS OF DIFFERENCE IN ENGINEERING PRACTICE

The differences primarily reflect the pervasive influence of social culture and the local economy. Six different dimensions of difference characterise the effects on engineering practice. These dimensions emerged from systematic research on engineers in Australia, Pakistan, India, and Brunei.[27] Therefore, there may be other dimensions of difference that would only be apparent from observing engineers in other countries.

Hierarchy influences coordination and access to distributed knowledge

Particularly in South Asia, there is a much larger socio-economic difference between the different participants in an engineering enterprise, such as between management and day labourers, than one would find in Australia. Even between more closely related social strata, for example, between engineers and senior technicians, there is still a stark difference in social status that would be much less evident in Australia. An engineer in India or Pakistan is regarded as a member of the social elite, someone who stands above 'the teeming masses' and who commands respect, simply because of their position in life. Even senior technicians will display deference to a young engineer. Caste (in India) and language can further accentuate the difference: engineers think in English and they speak in a mixture of English and the national language (Hindi in India, Urdu in Pakistan), randomly swapping from one language to another, and possibly also including their mother tongue (e.g. Punjabi, Marathi) from another part of the country. Technicians and other technical staff predominantly converse in a dialect of the local mother tongue and may only have an elementary grasp of the national language and very little mastery of English. In Australia, on the other hand, even though English is a second language for many engineers and technicians, everyone will be fully proficient in 'workplace' English. Engineers and technicians are much more likely to treat each other as equals, or near equals. On some occasions, many technicians will openly display contempt for young and inexperienced engineers.

Chapters 7 and 9 explained the significance of collaboration, particularly technical coordination: gaining the willing and conscientious collaboration of other people outside the lines of formal authority. This is the dominant aspect of practice for most engineers. However, the necessity to defer to the social (and workplace) hierarchy makes this much more complicated in South Asia. It takes a great deal of skill and experience to exert influence that can persist for a sufficient amount of time in the presence of hierarchical control over human actions. An engineer seeking the help of a drafter in Australia would not necessarily have to involve anyone else in the management hierarchy. The drafter would probably let their supervisor know of the arrangement, but only if a significant amount of work was involved. In most instances, the engineer would be expected to take the initiative. In South Asia, on the other hand, taking that sort of initiative would be more risky and would require extensive preparation. The engineer would have to have a strong pre-existing relationship with their supervisor, as well as the drafting supervisor, to be able to directly approach the drafter for help. A much more extensive and complex set of relationships is needed to enable direct coordination with the people who perform technical work, and these demand a substantial investment of time.

A production engineer at a manufacturing plant described how the influence of hierarchy can interfere with more organic collaboration:

> 'We started a new idea in our assembly plant called a cellular manufacturing system. We organised small cells and the participants were from all levels of the company, including executives from different departments like maintenance. [And did it work out?] No, it didn't last long. [Why not?] Because I would say that here, the culture is such that people do not like change, so they were not receptive to it and secondly, not all departments were involved in it that you needed to execute a job. For example, if there was a problem and the maintenance department had to fix it, they would say, "Okay we have indented the parts and when we get them you would have the thing fixed". Of course, the indent would have to be processed through the administration department and they would take their own time. Of course, if the cell had been given both responsibility and authority, it would have worked, but the authority was not given to them, only the responsibility. That's why it did not work.'

The stronger influence of social hierarchy in South Asia, therefore, makes it more difficult for an enterprise to work effectively with distributed knowledge, because it is more difficult to coordinate skilled contributions when they are needed. Social barriers also inhibit the flow of knowledge – even the awareness of knowledge. For example, when an engineer is explaining the requirements for technical work that needs to be performed, technicians will stand and listen in silence. It is rare for clarifying questions to be asked, because asking a question admits the possibility that one might not have listened properly in the first place. If the technicians know something important that the engineer does not, this knowledge will almost certainly not be volunteered: to do that would risk exposing the engineer's lack of knowledge, possibly causing a 'loss of face' for the engineer. In many cultures, public loss of face can precipitate intense emotional disturbance with unpredictable consequences.

Deep emotions were seldom far from the surface and often emerged at unexpected moments: I witnessed an engineer confronted with what seemed like a minor difficulty who left his job (and secure income) on the spot, for fear of losing face. It is possible that these underlying tensions will be associated with the huge and immutable disparities in wealth and opportunity for people at different levels of South Asian society.

Engineers in South Asia occupy a privileged niche, but they are still set at a distance below the top of the social hierarchy. Nearly everyone maintains subservience to the social hierarchy, which is reflected in the firm's organisational structure, as well as to a large personal network of extended family and relatives, friends, and acquaintances. Together, the hierarchy and one's personal network provide the only social safety nets that can protect an individual from a personal catastrophe and the loss of earning capacity. Getting a job invariably requires access to someone with influence, often combined with a significant financial payment as well. Most people are only a short step away from destitution. Therefore, loyalty to the social hierarchy and one's personal network takes precedence over everything else, including productive work.

The differences are not absolute: it is a question of degree. The social safety net is much stronger in Australia, so one can afford to take greater social risks. The subservience to social hierarchy that we see in South Asia is also present to a degree in Australia, but the hierarchy is less visible, less obvious, and less influential. It is not uncommon, for example, for senior engineers in an Australian firm to wear clothing and drive cars that are almost indistinguishable from technicians and tradespeople. Most Australians address each other by their first names and may interrupt while the other person is speaking, even if they are more senior. On occasions, one can see the same in South Asia. The difference is one of degree and the degree of difference changes between different settings.

Another effect of South Asian hierarchy, apart from limiting the access to distributed knowledge, is the centralisation of decision-making at the most senior levels of the enterprise. Even a factory manager may have negligible financial authority in South Asia. At one manufacturing plant that turned over tens of millions of dollars every year, the managing director had signing authority for roughly US $3,000. At a cement plant with a similar turnover rate, the managing director had authority to spend only US $200. In Australia, even relatively junior engineers have much greater financial responsibility and flexibility. Many of the larger engineering firms use a risk-based decision hierarchy that allows for practically all decisions to be delegated, freeing up the time of the most senior staff to focus on the most critical decisions with the greatest financial consequences for the firm.

In an Australian metal manufacturing company, an electrical engineer explained how he made most decisions for himself and only informed his boss later:

> '*My boss wants me to bring solutions, rather than problems.*'

Centralising decisions introduces delays and the risk of misunderstanding, as another Australian senior engineer explained:

> '*If you work through the traditional lines of authority, it's going to take too much time and may even get forgotten.*'

Social power, autonomy, and supervision

Raven and French's description of social power and influence[28] discussed in Chapter 9 pages 293–294 provides a useful way to understand the next key dimension of difference that distinguishes South Asian and Australian engineering practice. Data from our research supported earlier findings that reported a greater use of coercion and reward in South Asia, whereas in Australia, referent power tends to be favoured in the form of technical coordination.[29] For example, a senior engineer in India talked about motivating junior engineers:

> '*When it comes to training, I prefer to set a goal and let them work freely towards it. When they do well, I say nothing, but when they don't perform, I use the rod freely.*'

The greater use of coercion and reward therefore requires a correspondingly larger presence of supervision: the visible means of monitoring behaviour in order to punish or reward. Engineers in South Asia reported that supervision has to be almost continuous in order to achieve useful results. In the absence of supervision, work often stops completely. For example, an engineer working as a production superintendent explained how he spent 90% of his time supervising a squirrel cage motor production line operated by six workmen and two supervisors.

> 'I fear the workmen will hide rejects because they are afraid they will lose money, they won't get their quality bonus. We rely on self-inspection using simple Go/No-Go gauges. That's why I have to be there nearly all the time.'

A production line development engineer in the same plant spent most of his time supervising 11 qualified technicians and 15 day labourers hired to move material and perform other 'unskilled' tasks.[30] The technicians were constructing a new production line for rotary compressors.

Continuous presence does not have to be physical: the mobile telephone enables a degree of supervision from a distance. An engineer, the owner of a small company supplying specialised textile machinery, explained how he spent much of his 17-hour working day on the telephone, a lot which occurred while being driven through congested city traffic as he moved between his factory, office, client sites, marketing consultants, and financiers.

> 'I would say I make at least 200 calls a day, all of them follow-ups ... I have 14 educated people and 80 unskilled. If I put these 14 just on brainwork, that would create havoc. I don't allow them to think they'd have to plan themselves. I allot people to their tasks and keep monitoring progress. I keep shouting at them morning and afternoon and review their progress twice a day.'

Once again, personal presence can be just as important in an Australian situation. This engineer was explaining how his presence as a subcontractor influenced productivity:

> 'When they see you're around, they make sure they have the right people working on your jobs, and they give just a little bit more priority to your work than the other work that's piling up and also running late. If you're not there, then they'll give your work less attention and won't see it as being as urgent.'

As in the case of social hierarchy, the difference between South Asia and Australia is one of degree: in South Asia, supervision is, relatively speaking, more important and time-consuming. It is also more demanding, as we shall see later.

In Australia, engineers and technicians are entrusted to perform their work with minimal supervision. Supervision is seen as an expensive form of overhead to be deployed only when absolutely necessary. Engineers and technicians receive extensive training to enable them to work largely unsupervised. Production workers, on the other hand, will have a greater need for supervision. However, a production supervisor, often educated as an engineer, spends most of his time performing complementary support work, such as planning logistics and maintenance to minimise disruptions to production. Direct face-to-face interaction with production workers, while still important, is only part of a supervisor's responsibility.[31]

In South Asia, work usually stops in the absence of supervision. Without the opportunity to clarify and resolve uncertainty that is inherent in verbal instructions, action is usually deferred in the absence of a supervisor who can assume a large portion of responsibility and accountability for mistakes. The one-way verbal explanation that is routinely used to instruct workers lower in the hierarchy can leave the listener with a significant amount of uncertainty, regardless of the language or situation. One only has to remember trying to sit down after a lecture and then applying what one has heard (or misheard) to appreciate this. However, in South Asia, language differences further compound the uncertainty, and the consequences of taking inappropriate action are far more serious, especially if valuable material would be ruined as a result. As explained above, discourse that involves an engineer, a supervisor, and workers may have to encompass three or four languages, whereby technical details are easily lost in successive translations without anyone realising until it is too late.

In comparison to a commercial manufacturing plant, hierarchy and power relations in the context of a South Asian water utility are infinitely more complex. While engineers work within an elaborate bureaucratic hierarchy within the utility, they are also directly responsible to local agencies of political power in the community. Describing a young depot engineer investigating an apparently illegal water connection, Coelho wrote:[32]

> 'What was interesting to me in this incident was the intense dilemma the engineer was thrown into by the seemingly straightforward problem of an illegal connection (requested for a powerful person in the community). She had to negotiate a labyrinth of plots constituted by rumours, illicit acts, and transgressive collaborations in order to enact or exert her own agendas of personal survival, responsibility to her workers and colleagues, and a wider official accountability. She was also caught in a classic bureaucratic conundrum where, as head of the unit, she was also the newest kid on the block with at best a shallow grasp of local geographies and histories of power and collaboration. All these needed to be unravelled in order to act effectively, or at least safely.'

This shows how engineers in public utilities, therefore, also have to contend with the continually changing local community power structure within which their organisation and operations are couched. This makes their collaboration work even more time-consuming and complex than it is in a manufacturing plant. Every individual within the organisation will have different connections with this local political power structure.

Organisational processes and management systems

Much of the distributed knowledge that supports an engineering enterprise is accessed with the help of processes and procedures that formalise interactions between engineers and other actors to ensure that essential interactions take place that might otherwise be overlooked. It is not uncommon for engineers to regard much of the work associated with procedures as administration or 'not real engineering', yet most still appreciate their importance.

The relative strength of organisational processes is another of the dimensions that differentiate engineering practice in South Asia, Brunei, and Australia.

In Chapter 7 page 230, we saw how many Australian enterprises use formal procedures and check lists requiring engineers to work through documents, sitting with client engineers and technicians, line-by-line and page-by-page, following checklists to make sure that every aspect of every detail is discussed and checked.

This is an instance of distributed cognition, aided by formal procedures that embody knowledge accumulated within the enterprise, to ensure that knowledge is exchanged through social, face-to-face interaction.

The engineers that we interviewed in South Asia reported few organisational procedures being in place, and often they reported making them up as needed, such as this refinery operations engineer:

> 'I don't have fixed procedures ... each person in this position will institute different procedures to suit himself ... [he] will design the procedures to suit the requirements as he sees fit.'

Gradually, I realised that engineers in South Asia make much less use of 'embodied knowledge' in the form of procedures and documents than their counterparts in Australia. Even large firms seemed to impose fewer formal processes on engineers in the form of procedures. The notion of building on embodied knowledge and experience from other people represented by procedures seldom surfaced. The invisibility of such embodied knowledge is one possible explanation for difficulties expressed by many engineers interviewed in South Asia in terms of understanding the apparent superiority of engineering firms from industrialised countries. This issue surfaced in an interview with a senior consulting engineer:

> 'Private sector clients often are reluctant to pay extra for a consultant ... Interestingly, foreign institutions such as the World Bank are prepared to pay 6% to a foreign consultant but only 2% or 3% to a local consultant (like us) for infrastructure projects.'

After the interview, I spent some time touring the design office with junior engineers in the firm. It seemed dimly lit, but relatively clean, with piles of rolled drawings and box files heaped in corners. Most engineers had their own cubicles with computer workstations and access to shared telephones. Individual directors had their own offices with secretaries sitting just outside. Compared to many similar offices, it was relatively spacious and uncrowded. There was no documentation storage and indexing system, so much of the useful documentation was relatively inaccessible: access relied on someone's memory of where they might last have left a document. Nor was there a systematic change management process to keep track of the reasons for design changes. While documentation and configuration management systems are perceived by Australian engineers as an administrative overhead, a non-technical issue, the access that they provide to embodied knowledge underpins a significant part of their competitive advantage.

A lack of formal documentation is common on the shop floor, even in a well-organised South Asian firm. A production superintendent explained that in his factory, the manufacturing engineering department provides the tooling needed for his production line:

> 'Their only input is a drawing of the original component that we want to make: they have skilled toolmakers. They don't make any drawings of the tooling itself.

It is a largely undocumented process. Of course, it makes it difficult to maintain quality and consistency in the product.'

Much of the manufacturing is outsourced to street-side vendors who perform basic cutting and machining work on materials that they either purchase themselves or are supplied by the enterprise. Vendors are often selected by a purchasing department on the lowest-priced tender with an agreed upon rate per piece produced. Their working methods are even less formal than the larger factory enterprises that hire them. Drawings and documentation are often ignored: most of the vendors cannot reliably read drawings and specification documents. As a result, quality control for engineers in the more organised enterprise tends to be a constant fire-fighting occupation.

A quality assurance engineer described his experience working with vendor-supplied components in this way:

'It is a very stressful job. All day long I am faced with different problems, from vendors or the assembly section. I never have the time to find myself in a relaxed condition.'

A senior development engineer echoed those comments:

'There is a constant stream of "crises" concerning part quality through the company. This seems to be an accepted state of being.'

The respective purchasing departments do not have any method or data to take into account the real economic consequences in terms of engineering time cost, production disruptions, and low-quality products. They make their decisions to award vendor contracts on price and firmly resist informal 'meddling' by engineers who, in their opinion, would lead the company into bankruptcy if they were given a free hand in purchasing. In a large number of firms in diverse specialties, this is 'an accepted state of being'.

The end result is a loss of accumulated knowledge that can only be represented with written documents or computer records. I have observed many other instances, far too numerous to describe in this chapter. While South Asian engineers are aware that well-organised firms tend to use more documentation, none of them connected contemporary informality with knowledge or information loss.

In the public utility sector, documentation takes on quite a different significance. A senior utility engineer explained his experience with household connections:

'[You were doing household connections?] House connections . . . we used to submit as a plan, how they are going to connect it, which mean we have to connect it. We used to have plans from state department, but nowadays we don't have plans for house connections.'

Gradually, over time, the water distribution utility had moved from formal plans and documentation to an informal system based on outdated plans, ad hoc adaptations, bypasses, and connections arranged verbally with depot engineers with varying degrees of documentation.[33]

In a later section, we will see how, in the case of mobile phone enterprises, knowledge of the system is encoded in software, where it is relatively resistant to the influence of political power.

Labour cost perceptions

The societal perception that labour is relatively cheap exerts a subtle influence on decision-making. Engineers in India and Pakistan reported that labour costs represented about 7–9% of the total cost in engineering work compared with about 15% in Australia. However, what became apparent was that when it came to decisions, engineers in both regions seemed only to perceive the direct cost of labour, rather than the total cost, which included supervision and other indirect costs. Some people were able to perceive these costs, at least qualitatively, such as this senior design engineer at a Pakistani manufacturing plant:

> 'With non-engineering personnel, we would employ three instead of one for non-commercial considerations, political considerations. This does not apply at all in the case of engineering staff. However, these additional staff require considerable additional supervision and this impacts on the work of engineers here.'

However, here is a more typical response to the idea of investing capital to reduce labour cost:

> 'I would need to invest millions of rupees to reduce my labour cost. If it is less than 10% of my overall cost, why would I think of doing that unless I had the money sitting there?'

This engineer is talking about machines as an alternative to human labour: machines can replace labour, but only by way of a considerable investment. His comment reflects the view of labour as a relatively cheap commodity in South Asia, perceived as cheaper than investing in automated machines.

Here is a comment from a Canadian engineer working as an expatriate landmine clearance supervisor in Cambodia, responding to a suggestion that improved tools would improve the productivity of the local deminers:

> 'Those guys only cost us $120 a month. What's the point of improving their tools if it would cost as much as their salary to replace their tools every month?'

Table 11.4 on page 392 helps to demonstrate that indirect costs are much higher than the direct salary cost of labour in a low-income country, as much as 13–15 times higher, in fact. In this instance, the improved tools could have doubled the productivity of the deminers, achieving major savings when indirect costs are factored in. However, the engineer was unable to appreciate this opportunity because he was focusing only on the direct salary payment component.

This quote illustrates how misperceptions of labour costs drive inappropriate decision-making on tools, supervision, and training. In an industrialised country environment, however, even though the engineers still tend to focus only on the direct salary cost, ignoring significantly larger indirect costs, the perception that labour is expensive still persists. Appropriate tools, high-quality materials, supervision, and training are seen as ways to make the best use of labour, and achieving high productivity is seen as being important. It is often tempting to focus on 'labour productivity', the amount of output produced for a given number of hours of labour input. It is also more meaningful to take 'multifactor productivity' into account, which is the amount of output produced for a given amount of labour and capital inputs.[34] In a typical engineering enterprise, whether it is in South Asia or Australia, labour is usually only a small percentage of the overall operating cost. Typically, capital costs form the largest component of expenses, including depreciation and opportunity costs. Therefore, multifactor productivity is a much more appropriate measure of performance.

Even though engineers in both South Asia and Australia focus mainly on direct labour salary costs, the effect on decision-making in South Asia is much greater because the direct salary costs are a much smaller proportion of the total labour expenses. None of the engineers we interviewed in our research showed any understanding of this reality. The results show up in performance: whereas machine utilisation in the factories that we observed in South Asia varied between 30% and 50%, utilisation in similar Australian settings varied between 60% and 90%. This factor helps to explain why, for example, industrialised countries with high hourly labour costs still dominate manufacturing. Even though China is now seen as the centre of world manufacturing, profit margins tend to be very low, well below levels that would be acceptable in an industrialised country.[35]

Engineering, supervisory, and technical skills

The supervision work performed by the superintendent working on the squirrel cage motor production line described above had an additional dimension beyond the sheer amount of time spent monitoring the production workers. The production workers were day labourers paid about US $100 per month, and under restrictive labour laws, they could only remain with the firm for a few months. They were practically unskilled when engaged for the first time. Much of the superintendent's work was training these workers to operate simple machine tools, and also devising simple manufacturing steps that were within their capabilities. As soon as they had reached a modest level of skill, new unskilled labourers replaced them. Some had previous machine operating experience, but the majority required continuous training and skill development from the superintendent who explained:

> 'They need direct on-the-line training minimum for the first month, sometimes they damage the equipment as well, and they waste materials.
>
> 'Take an example, the rotor diecasting step. The operator has to pour molten metal from the furnace. If he doesn't shut the door immediately, the temperature of the molten metal drops and the next pour will be substandard – one of the squirrel cage bars may be broken because the liquid metal has not penetrated far enough. These are simple things, but when you understand that neither the workers nor the supervisors have a technical education to enable them to understand, even if they are good we cannot keep them or they have to become union and permanent.'

In an industrialised country setting, supervisors acquire extensive technical training before employment (for example, in Germany) or have university engineering qualifications. They play a key role in manufacturing and perform complex scheduling and process optimisation work, as well as managing the production workforce.[36] In South

Asia, most of the production supervisors in the factories that we observed only had on-the-job training. The superintendent continued:

'The supervisors make mistakes because they don't have a technical background and they don't know the process. . . . They come from a worker background . . . Each of them may know one machine but they don't understand the whole production line. The supervisors are permanent, under the union, and require very careful handling. If you want a worker to do something, you tell the supervisor but he may refuse to do anything . . . Supervisors who come in from two-year technical training courses are still mostly unable to read drawings.'

He explained how the firm was reluctant to have too many permanent workers who had to be employed under a union agreement. It was almost impossible to fire them for misconduct. They were also paid about US $400 per month, including allowances for welfare, shoes, clothing, medical care, and the education of their children – several times more than the day labourers. Therefore, the company tried to employ as many day labourers as possible. The production superintendent and other mid-level engineers were paid about US $1,200 per month.

The young engineers told us that they spend most of their time on work that is far below their technical capacity – *'under level responsibilities'* – and spend much of their time trying to develop the skills of supervisors and line workers. Another engineer in the same plant developing a new production line explained his training role:

'We picked up these technicians three years ago. Initially they had no experience and we had to train them. They were there from the beginning, we had to train all of them, how to set up the machines, how to fix teething problems, how to get materials.'

Domal's field study in a similar plant revealed similar conditions with inexperienced junior engineers in charge of manufacturing cells, often with unionised machine operators, who took little or no notice of their requests, and casual day labourers who could not read, and often could not understand what the engineers were telling them.[37]

It is important to remember that these comments came from engineers working in export-competitive manufacturing plants. At another manufacturing plant operating with the protection of significant import tariffs, an even greater proportion of the workforce was composed of casual production workers. With 2,500 employees in the factory, I was curious to ask how all their training could be accomplished in one small room. An engineer explained:

'We only have 120 skilled workers, the rest are unskilled and don't require any training. We only need to train 10 or 12 people every year.'

Walking around the plant, I noticed how the 'unskilled' production workers actually needed to develop extraordinary skills simply to survive without serious injuries from the glowing metal billets rattling along chutes onto the forging machines they were operating, all of which was being done without any protective clothing or even safety guards on the machines.[38]

The more experienced engineers commented on the degree of training and attitudes required by junior engineers who arrived with few appropriate skills, even though they had graduated from institutions with similar engineering education curricula as in Australia. A specialist machine tool designer explained it this way:

> 'These engineers, they have rough roots, you need to mould them. I give them general training for six months on the shop floor, machining and making things. It should be two years, but I cannot hold them that long. Then they go to the assembly shop for six months, just watching.'

Family and social background was also considered to be an important attribute. The machine tool designer continued:

> 'I observe them – can he come up with some suggestions? He should be able to question what he is seeing. He needs to come from an appropriate environment, a social family – the work mind is important, his attitude to work. . . . He has to be a practical engineer, he has to be out on the shop floor. In his mind and hands alone he cannot be a good boss. With his family, if no one is looking for his money, I can take a gamble. I won't take sons of rich people, or the sons of IAS[39] officers. If the father's scene is different, the son cannot be totally different. Easy money produces spoiled children.'

He was expressing a preference for an engineer who realised that curiosity, hard work, and the confidence to be able to question the *status quo* is needed for success. A highly respected senior product development engineer explained the same ideas in a different way:

> 'Before I take a guy I have a very, very fine filter. I only take in a certain kind of person. Really, the interview is not concerned with engineering. It is his social background and his thinking processes and his hobbies ... or lack of. I don't ask them what a crankshaft is or what an engine is, anything remotely connected with engineering. He can get that from a book. But he must have the right social background, and an inquisitive mind. It is a very fine filtration before he gets to me ...
>
> 'What kind of family he comes from, that's very important.
>
> 'Somebody in government service, at a very low level, a clerk or something, if his son is applying for work with me, he will probably be useless. It probably sounds horrible, what I am saying. On the other hand, I have this immensely powerful, economically successful person, his son, he will probably be useless as well. Probably somewhere in between you have a person with a family background with a work ethic, that you have to work for a living; generally those are the people, including grade 17 and grade 18 government officers, doctors, or alternatively, if it's in the private sector, medium-income groups where they know you have to work for a living, where that fundamental idea has set in and they have had sufficient awareness of the world and the way the world thinks.'

In Australia, there is a universal and predominantly state-funded education system that enables almost anyone with the aptitude and determination to graduate as an engineer. In South Asia, a wealthy family background is almost essential for a young person to gain sufficient education through private tuition to graduate as an engineer. While there is intense competition for a few thousand places at the few Indian Institutes of Technology and other respected public institutions, a wealthy family can afford tuition at a private engineering college. Both of these engineers had realised that an engineering qualification alone is not a sufficient predictor of success: the inherited wealth that opens the opportunity to study engineering can also contribute to the notion that one is born to rule, rather than to work for a living.

The product development engineer went on to explain how he developed both technical and people skills in his engineers. They came to his department from within the firm, meaning that they had already spent a few years in the firm. On the technical side, without prompting, he mentioned how young engineers needed to understand the purpose behind drawings, comments that echo almost exactly the same issues as engineers described in Australia:

> *'You see, to me, the drawing is not just a few lines. It is the technology of manufacturing that part. I think that once you explain that to your team, that the geometry of the part dictates how you're going to hold that part, the way you dimension it determines the way you machine it, then they start thinking along appropriate lines.'*

He understood that young engineers work with distributed expertise, but without using that term. He explained how much effort he devoted to teach young engineers how to collaborate effectively with others in the firm. Here, he was explaining his method of evaluating their performance:

> *'One of (the attributes) is the way he holds meetings. You know, here we have a way of holding meetings that produces no answers, they go in unprepared and they come out with no decisions. I review the progress that he has achieved during the year in that respect. This aspect is very important. If you are with a group of people, you must give them ownership of what your thought processes are and you must carry them with you and you must get at least 80% of what you would like them to decide at the end of that one hour or half an hour.*
>
> *'How his technical progress has been. How well he conforms to systems. Innovativeness. Now that is very difficult to evaluate. How he interacts with people and gets work out of them, things like that. I score each one of these attributes. I do that with the guy sitting in front of me and he is at perfect liberty to argue it out with me. It is very important that that guy must know, otherwise he would never get better.*
>
> *'In the rest of the firm, it's not done that way. The so-called thing is filled in in the absence of that person.'*

The last comment underlined the contrast between his approach and the practice with other engineers in the same firm.

> *'It is hard to maintain an island of excellence in a sea of mediocrity.'*

The gaps between engineering education and practice in industrialised countries have received frequent attention in engineering education literature.[40] While some reports have pointed to similar issues in South Asia,[41] this research has revealed subtleties and difficulties for novice engineers far beyond those experienced by novices in an industrialised country setting.

In Australia, for example, novices can mostly rely on being able to learn from highly experienced senior engineers, production supervisors, foremen, and even tradespeople in workplaces. With few language or social barriers, a network of distributed expertise can help them negotiate the leap into proficient practice.

In the absence of well-run, state-funded industrial education and training, novice engineers in South Asia, on the other hand, find themselves having to train their own workforce, not just once, but many times over. At the same time, they are confronted with learning the intricacies of social interactions with their peers, seniors, and the other actors that confront them in a much more complex workplace. Graduates from engineering schools in South Asia emerge almost completely unprepared for these challenges. These observations explain the critical importance of effective teaching skills for engineers.

Specialist suppliers

South Asian engineers, compared with their Australian counterparts, receive much less support from specialist engineering supply companies.

Specialist engineering suppliers fulfil a critical role in engineering practice: they provide education for practicing engineers on the vast array of specialised components, software, and materials that make engineering possible. Most engineers encounter only a few basic materials and components in their education. Engineering schools that can afford them will provide a basic introduction to AutoCAD, MATLAB, and possibly LabVIEW software. Particularly in LICs, many engineering schools have to rely on free public domain software versions with limited capability, and will have few, if any, representative hardware components for students to work with beyond the most basic.

High-quality (and often more expensive) components also come with extensive documentation and design guides to help engineers reliably select the most appropriately-sized component in a given product range. The same documentation may also provide engineers with shortcut design techniques, or even free software, to help them with common design calculations. For example, the Maxon Company provides comprehensive design guides and selection software for their high-performance electric motors.[42] There are countless electric motor manufacturers, but most provide very sparse documentation. Maxon is an example of a more expensive specialised product, backed up with extensive information resources. This helps engineers working on new product development: because the documentation exists, and is freely available, engineers searching the Internet are more likely to choose a Maxon motor for early product development. Then, they may try competing products at a later date, seeking the cheapest possible solution for mass manufacture. Maxon's strategy is to provide a product that few others can match in terms of technical performance. In many cases, a cheaper motor will provide adequate performance for the resulting commercial product, but in enough instances, the resulting product will incorporate a Maxon motor. Other electric motor manufacturers adopt different strategies: they may rely on sales representatives, trade shows, or simply word of mouth.[43]

This is all part of effective product marketing. The component or material manufacturers know that novice engineers will know little or nothing of their products. Therefore, they have to provide workplace education for engineers to learn about their products in order to create awareness that they exist, as well as information on how to apply them effectively. A product will only gain a good reputation if it is applied appropriately. For that reason, suppliers have to make sure that engineers can apply their products correctly. In most cases, this requires extensive personal contact by technically educated sales representatives who are often engineers themselves. Some companies prefer their sales representatives to work one-on-one with customers, while others run workplace product seminars or provide in-service training courses, often charging fees because these are effective and highly valued by customers. Suppliers know that an Internet presence is not sufficient in itself: personal contact is as important for securing sales as it is for effective education.

Compared with Australia, the market for specialist engineering product suppliers is much more challenging. As explained already, most engineers do not have the authority, financial skills, or the strong influencing skills to guide purchase decisions on technical products. A combination of the strongest relationship between the supplier and the purchasing department staff and the lowest price is most likely to win the contract. In South Asia, engineers are found in small numbers relative to the population size. Unless the owners have studied engineering, many small firms would not even have an engineer on staff; instead, they rely on self-trained technicians. Larger firms, as explained, keep their engineers well away from purchasing decisions. Therefore, it can be difficult for sales representatives to find and work with engineers.

As a result, relative to Australian engineers, South Asian engineers are less likely to have the support of technical sales engineers from suppliers and are less likely to be able to travel to trade shows. This means that they have fewer opportunities to learn about so many specialised products that could help them provide better value for their employers and clients.

Six degrees of separation

On each of the six dimensions of difference, we found a clear separation in engineering practice between South Asia and Australia among the engineers we interviewed and observed. They were mostly working in civil engineering construction, continuous process chemical plants, metal manufacturing, electricity supply, water supply, and telecommunications. The degree of separation will vary with individual capabilities, settings, disciplines, industries, and firms. The least degree of separation occurred in telecommunications: more on that sector will be covered later.

Earlier chapters have shown that engineering practice depends on distributed knowledge and effective, coordinated collaboration. Engineering also depends on technical ideas surviving more or less intact through multiple reinterpretations by the people who implement them. Comparisons between South Asia and Australia demonstrate that there are patterns of social behaviour in Australia that make it easier to support an elaborate engineering enterprise that relies on distributed knowledge. In South Asia, on the other hand, long-established patterns of behaviour inhibit informal flows of information and collaboration such that distributed knowledge is less likely to be accessible. Language differences, weaknesses in education for workers, difficulties in reading technical documents, and social inhibitions all increase the chances that technical ideas will be interpreted quite differently from the engineers' intentions in the absence of close supervision. The need for continuous and close supervision, in turn, decreases productivity and adds large indirect costs at the same time: supervisors are relatively expensive to employ, and there is little skill development and training in place to establish effective supervision methods. With inappropriate beliefs on the relative cost of labour, these problems are further magnified. The lack of training and education for what is perceived as 'cheap labour' results in the underutilisation of expensive capital in the form of land, buildings, plants, and machinery. The resulting indirect costs of these decisions are not well understood in the work place.

Together, these factors represent a strong explanation for the performance differences we observed between South Asia and Australia. However, there are fascinating exceptions to this general pattern.

DISCOVERING EXPERT ENGINEERS

At this stage, a reader could be forgiven for suspecting that I am arguing that certain social cultures can perform engineering better than others. Employment patterns provide strong evidence that individual engineers from South Asia perform as well as engineers from industrialised countries, because they readily find employment in engineering enterprises in Europe, America, and other industrialised societies. It would be easy to conclude, therefore, that the evidence presented so far suggests that there are social or cultural factors in their home environments that prevent South Asian engineers from being effective when they work for firms at home. In the course of my research, however, I found (by chance) a small number of 'expert' South Asian engineers with an outstanding record of domestic success, thereby contradicting this hypothesis.

Until then, I had been faced with an apparent contradiction. As explained before, most engineers in South Asia earn 25–35% of the salaries enjoyed by their counterparts in industrialised countries. Many engineers appeared to be employed for work that did not rely on their technical education to a large extent, and a significant number were unemployed, around 15%, according to anecdotal evidence. Engineers were often found to be seeking offshore employment opportunities, particularly in the Middle East and Europe, confident that they would be able to earn more than they could at home. With a simplistic understanding of labour markets, this situation suggested that there was a surplus of skilled engineering labour in South Asia. However, the relatively high cost of drinking water and energy caused mainly by supply intermittency, as well as relatively high manufacturing costs, suggested that there was a shortage of engineering skills in the South Asian market.

When I had the opportunity to discuss this with a leading labour market economist,[44] he explained the notion of marginal product that I described in Chapter 1. He suggested two possibilities. Either I had disproved the basic theories of labour market economics, or I was looking in the wrong places. The latter alternative suggested that engineers in South Asia were contributing relatively low marginal product, and since there is a global market for engineering skill, engineers in industrialised countries were able to contribute a significantly higher marginal product. Luckily, for economic theoreticians, this second possibility turned out to be correct.

A few months later, I met the CEO of a small engineering supply company selling high-value, specialised engineering components and materials in Pakistan. Knowing that I was interested in interviewing engineers working in commercial firms, he offered to introduce me to several who were long-established clients.

Some of these engineers did not fit the normal pattern that I had previously encountered in Pakistan. First of all, they were able to effectively gain the collaboration of people across their organisations, particularly with accounts staff in purchasing departments who would normally seek the lowest price for engineering supplies. These engineers had managed to persuade staff in their firms to purchase the highest priced materials and components on the market. The engineers had recognised that the productivity gains from using these products far outweighed the initial purchase cost, particularly when used by skilled technicians who had received in-service training from this specialised supply company. One technique for doing this was to draw up detailed technical specifications to guide purchase decisions that ruled out all the alternative suppliers with lower priced products. However, it was not as simple as merely producing technical documents. They also had to gain the support and collaboration of many other people throughout their company, including senior management.

The second aspect that distinguished these engineers was their high remuneration. I did not include a question on salaries or remuneration benefits in my research questionnaire, because I knew that it was normally a confidential issue. However, I found that I was able to ask about this informally, after the interview in more casual conversation once the recorder had been stopped and I was no longer taking notes. Later, when I calculated the value of their remuneration packages as annual salaries in Australian dollar terms, I was amazed. These engineers were earning salaries that were higher than equivalent engineers in Australia, typically by 15–30%.

The third aspect that distinguished these engineers was that they voluntarily explained the commercial value of their technical contributions in several different ways during the interviews. In contrast with most of the engineers I had interviewed before, they explained the commercial benefits of their work in terms of reducing risks and increasing the utilisation of capital assets. They articulated a much clearer understanding of worker productivity and indirect costs. Some were also engaged in a constant struggle to change the social culture, at least within their own workplace in the firm. They encouraged their engineers to spend time gaining the confidence of production workers so they would talk openly about the challenges they faced on the shop floor (I had similar findings with construction workers on-site, in the case of civil engineers), thereby breaking down normal social barriers that prevent engineers from learning via the insights of manual workers.

Other significant findings included:

- Some had completed their education in an industrialised country, such as Britain or the USA, but not all of them. Overseas education did not seem to be a requirement, though there were some advantages perceived by engineers. The main advantage they mentioned was that they thought they had developed the courage and ability for independent thinking and were more willing to challenge ideas advanced by their lecturers and other experts.
- Some, but not all, had work experience in an industrialised country, or with companies based in industrialised countries.

- All the expert engineers had developed high-level collaboration skills: they had learnt technical coordination skills (though they did not describe their skills that way), project management, and negotiation skills.
- Most saw the importance of being able to develop the skills and abilities of the people working around them; in other words, they appreciated and utilised informal teaching skills.
- All the engineers had acquired detailed and extensive knowledge of engineering component and service supply companies and their abilities, such as local subcontractors (or vendors), equipment suppliers, and even international supply firms who did not necessarily have offices or representatives in the same country.
- Some, but not all, had completed MBA degrees or studies in finance.
- Most expressed their frustration in working with other engineers and managers who could not appreciate the elements of estimating costs, risks, and likely benefits of engineering proposals. They complained of 'short-sighted' decisions that often impaired their ability to produce effective commercial results.

Reflecting on this later, I realised that I had stumbled on a small number of expert engineers whose value was readily appreciated by their employers. The salary differential had to be sufficient to retain these engineers, because they were critically important people in their organisations. They would have easily been able to find rewarding engineering work in an industrialised country if they had the motivation to look there. The salary differential confirmed the local shortage of engineering skill. What it also showed was that the vast majority of engineers did not have comparable skills and were therefore contributing much less commercial value for their employers: their marginal product was less, which meant that their salaries were comparatively lower.

While reflecting on my interview and field notes, I began to realise that I needed to carefully study the differences between this small number of 'expert' engineers and the much larger number of other engineers whom I had interviewed. Around this time, Vinay Domal had made a courageous decision to work on a PhD research project with me, comparing engineering practice in metal manufacturing in Australia and India.[45] He identified consistent differences between engineering practices in the two different settings, providing the framework for the earlier discussion in this chapter. His work, together with results provided by several other students working on engineering practice research, helped to provide much of the data that eventually led to this chapter.

This leads us to two valuable conclusions.

1. This small group of expert engineers has demonstrated that it is possible for engineers in a low-income country to earn salaries higher than engineers performing comparable work in industrialised countries.
2. They have demonstrated that it is possible to produce outstanding commercial results in the South Asian environment by understanding how to work within the culture, and modifying the local culture when necessary. At the same time, they all had to work extremely hard to achieve this.

Part of the aim of this book is to make their knowledge and understanding much more widely available, and also to place this within a theoretical understanding that may make it easier for others to learn.

MOBILE TELECOMMUNICATIONS – A NEW START?

In contrast to most other engineering enterprises in South Asia, mobile phone enterprises have demonstrated extraordinary commercial success. This success has been replicated in most low-income countries around the world. This seems to have happened by accident: mobile phone technology and the enterprises to support it were developed for the industrialised world in the 1980s and 1990s.[46]

Since their introduction in the mid-1990s, mobile telephone networks have become well established in South Asia with access prices well below the available rates in many industrialised countries; for example, it currently costs around US $0.05 per hour of talk time in Pakistan. Most adults and many younger people carry Internet-enabled phones with them at all times, even in the most remote areas of the Himalayan Mountains.[47] Mobile phone networks have eclipsed public landline telephone monopolies and have such a reliable reputation that mobile phone credit often serves as an informal payment settlement system in many low-income countries.

Compared to water and electricity utilities, mobile telephone networks provide high-quality service at a low cost and also provide profits for investors. What insights can we gain into engineering practice in South Asia that might help to explain such large and apparently sustainable beneficial outcomes?

Mobile phone enterprises in South Asia adopted the same technology and enterprise processes that were originally designed for industrialised countries. Some of these choices seem to have provided some critical differentiating factors that enabled these privately owned utilities to become commercial successes in developing countries, in contrast to the publicly owned landline monopolies that they have displaced.

Briefly summarising the earlier discussion in this chapter, we have identified six factors differentiating engineering practice in South Asia compared with Australia:

i)   Influence of organisational and social hierarchy makes coordination and access to distributed knowledge more challenging.

ii) Using coercive and reward forms of social power rather than referent power limits autonomy and increases the need for continuous supervision, thereby imposing higher costs.

iii) Relatively weak organisational processes and management systems reflect less recognition of, and reliance on, the experience from earlier generations, and allow political and other external influences to undermine organisational knowledge and information.

iv) Inaccurate labour cost perceptions lead to lower productivity and higher indirect costs from the low utilisation of capital assets and the need for supervision.

v)   Inappropriate engineering, supervisory, and technical skills: engineers spend much of their time supervising and attempting to train low-skilled workers. University education provides these engineers with little or no useful background, with the result that they are almost completely unskilled for the work they are expected to perform.

vi) The small number of specialist technical suppliers limits the opportunities for engineers to build their technical expertise and knowledge of high-value tools, materials, and components that they could apply in their work.

It is instructive to examine some of the critical features of mobile phone enterprises to understand how they (perhaps inadvertently) have worked around these issues. These features provide starting points for any new engineering enterprises aspiring to commercial success in a low-income country.

1. Mobile telephones have provided lasting economic benefits for end users, mainly in the form of time saving. For example, a tenant farmer many kilometres from the nearest economic centre can send a text message to his seed and fertiliser suppliers to find out when his consignments will be ready for collection. He can also negotiate advance sales of produce with processors, which are several hours away by bicycle or donkey cart, without having to leave his farm. A few text messages and phone calls replace entire days of exhausting travel. Even though the hourly value of time for a tenant farmer is small, the savings are significant and more than justify the cost of mobile phone access.

    We learnt from this that it is important to understand the economic value of time, explained in Chapter 1 page 12. One way to measure the social benefits of engineering improvements that result in time saving is to calculate the average value of time saved per person.

    Mobile phone sales have been promoted by emphasising the links between the latest mobile phone technology and keeping up with fashion trends. As a result, up-market mobile phones released in low-income countries have as much or even more of the latest technology as phones released in Europe and the USA.

2. The technology permits incremental extensions of network capacity once the initial installation of antenna towers, exchanges, microwave, and optical fibre links has been completed. Although the upfront capital investment is still large, it is within the capacity of financial institutions and investors in developing countries. (As demonstrated in the Indian example, the size of capital investment is usually beyond the reach of the government. Initially, the Indian government wanted to retain a controlling share of ownership in critical infrastructure but it has since allowed a much greater proportion to be held by private and foreign owners.)

    The cost of capital for investment in many low-income countries is higher, particularly those with less stable governments, on-going insurgencies, or a reputation for arbitrary changes in government policy that can make life difficult for investors. This means that a higher discount rate applies in calculating NPV (explained in Chapter 11 pages 379–384). Deferring capital investment as long as possible, even if the costs are higher, makes economic sense. Therefore, it is important to aim for a technical design that allows for capacity upgrading in gradual stages, long after the initial investment. Capital spending on capacity upgrades has to be deferred as long as possible until the extra capacity is really needed.

3. Prepaid scratch cards with encrypted access code numbers enable the efficient collection of large numbers of small payments by phone users.

    The mass consumer market in low-income countries can be very large because of high populations. In South Asia, in particular, the high fertility of the soil enables high population density: a large number of people can live in a relatively small area.

However, this does not apply to all low-income countries. In Africa, for instance, population density is much lower, thereby increasing the relative cost of distribution and logistics.

Technical solutions that enable large numbers of small payments for services to be collected efficiently in these environments, as well as a secure IT solution that minimises opportunities for fraudulent access to credit, are both important components for commercial success. The important parameter is the proportion of the sale price that has to be devoted to security and safeguards against human dishonesty.

There are important technical design issues here. Weighing scales and petrol pumps are ubiquitous in human societies today. What are the technical design and regulatory features needed to reassure customers buying petrol (by volume) or food (by weight) that they are receiving the correct quantity of product for which they have paid? If customers lose confidence in the accuracy with which goods are being measured, then they will perceive a need to have their own measuring equipment with them, introducing economic inefficiencies caused by lost time and investment in measuring equipment.

4. Centralised payment processing enables efficient sales tax revenue collection for the government. The collection of tax revenue is a significant difficulty in many low-income countries. For example, in Pakistan, it was recently reported that in 2011, less than a third of national parliamentarians had filed tax returns to declare income beyond their low official salaries.[48]

Markets have proved to be more efficient regulators of human behaviour than government. However, although we mostly dislike it, history has shown that governments need to regulate markets and provide information to help make markets more efficient and accessible for people. Otherwise, people with special access to information can develop an unfair advantage and power over everyone else. Also, most people seem to have more trust in governments to provide essential services, such as water supply, sanitation, gas, and electricity.

Governments have to sustain themselves through tax collection. Once again, economic efficiency is important. Collecting taxes can impose a significant administrative burden if it is distributed across large numbers of tax collection points, as is the case with sales taxes, such as in the USA, or GST, and VAT in other countries.[49] Any opportunities to centralise tax collection or reduce the number of tax collection points can result in significant productivity gains by reducing the amount of work required for the administration, as well as opportunities for the loss of revenue through corruption.

The IT systems associated with the mobile phone networks provide an opportunity for governments to collect taxes with minimal administrative overhead.

5. The service is not accessible by users unless they have paid in advance or have provided a reliable line of credit.

In South Asia, the vast majority of users access networks with prepaid cards. Unlike water and electricity, users cannot access the service without paying in advance.

This feature is vital for providing confidence for investors who build the infrastructure and operate the mobile phone enterprises. Like most engineering enterprises, the investors receive their return on their original investment long after the money they contributed has been irrevocably spent. For example, mobile phone towers and optical fibre cables require substantial capital investment. If they were not used, it would not be possible to retrieve more than a small proportion of the original capital invested. Unlike land, optical fibre cables in the ground do not have much inherent value if they are not being used.

From this, we learn that it is critical to provide the investors with confidence that they will not lose money over the long-term. Uncertainty over the rate of return – the level of profit that they make on their original investment – is less important than eliminating the possibility that they might lose most of the money that they have invested. One way to provide this level of assurance is to ensure that payments are collected for services provided. Using automated systems reduces the cost of measures required to protect payments from being diverted as a result of dishonest behaviour.

6.   Information technology has thus far enabled the implementation of effective barriers that raise the cost of fraudulent intervention by individual users. Even though mobile phones are an extension of the Internet, they are protected with special access controls that have restricted unauthorised access to an acceptably low level.

In comparing mobile phone enterprises with water and electricity utilities, we can see that the cost of influencing meter readers, even engineers, is well within the resources available to medium- and large-scale businesses, as well as wealthy individuals. People with high levels of social power can persuade engineers to install illegal connections for which they do not have to pay. The same power prevents these connections from being properly documented, because the information system depends entirely on human inputs.

In summary, technological choices appropriate for industrialised countries have also served to protect investors, government revenue collection, and end users in developing country environments.

Interviews and observations of telecommunication engineers have revealed additional significant differentiating factors that help to explain why mobile phone enterprises have succeeded in comparison to water and energy utilities.

In water and electricity utilities, even in manufacturing, hands-on aspects of production and service delivery work are performed by people with limited education and who have little or no influence in the social hierarchy. Having to get one's hands dirty at work or engaging in physical labour is considered an indicator of low social standing.

In the telecommunications sector, on the other hand, apart from the initial installation of towers, antennas, standby generators, and connecting cables, hands-on service delivery, production, and maintenance work is performed by highly skilled technicians, most of whom have obtained degree qualifications in an English-speaking environment. In Pakistan, for example, there was a surge of demand for private tertiary education in electrical and computer engineering and computer science in the second half of the 1990s and early 2000s. Initially, this demand was driven by the prospect of migration to Europe and or America, where IT skills seemed to be in strong demand at the time. In many cases, what passed as a degree in computer science was little more than a practical hands-on introduction to the use of spreadsheets, word processors, e-mail, and databases. As a result, people with good computer operating skills and reasonable English were readily available in the employment market and were paid only a moderately higher salary compared to other clerical staff.

This reservoir of people with 'applied' IT skills provided an important skills base for mobile communications.

Much of the work in the mobile telecommunications sector requires technicians to configure digital exchanges, apply software patches and upgrades, and configure network routers. Online certifications from various tech companies, such as Microsoft and Cisco, have provided additional skills.

The mobile phone systems that they were supporting simultaneously provided technicians with on-the-job access to a wide circle of friends. When confronted with a fault that they could not solve on their own, they could always resort to their network for assistance: someone would know the answer.

Level 1 field service engineers explained some aspects of their technical responsibilities:

'It is hard to diagnose a fault. I have to think of the justified logical evidence and locate the fault with precision.'

'The procedures for fault tracing are set down in international standards for the company. I have to trace fault events in hexadecimal code (a computer code for representing numbers). Where the exchange sends an invalid message, then I have to trace through the software to find out the problem for myself.'

At Level 2, the engineers also have supervision responsibilities, as this excerpt demonstrates.

'First I sit with my computer and carry out patch administration. These are software corrections that have to be applied to the systems operating in the various digital exchanges. I receive the patches from Germany and I issue them to the field technicians. I then check on the final status of the patches once they have been installed in the exchanges.'

More extensive studies in Brunei and Australia[50] have confirmed that technical aspects of telecommunications engineering practice largely involves incremental network capacity expansion through equipment procurement, installation of cables and equipment in buildings, and the use of software to configure network switching. High-level mathematical analysis and modelling expertise, which comprises a large percentage of telecommunications engineering degree courses in industrialised countries, play only an indirect role.

Another engineer explained how he was required to comply with company procedures and standards that had been laid down in Europe. In addition, he had travelled to Europe for training courses.

'[Tell us about fixed procedures you are required to follow.] Yes, as laid down in the company standards. The company software centres decide very detailed procedures to be followed for every aspect of the work. This is done in Germany and the software development centre in Ljubjana in Slovenia.'

In contrast with the other utilities and manufacturing, there was little, if any, disparity between the engineers and technicians performing hands-on work in terms of social hierarchy. Their informal social networks blurred the distinctions between them and they freely exchanged technical expertise when needed. The social barriers that we observed interfering with the successful enactment of distributed expertise in water utilities and manufacturing did not appear to the same extent in telecommunications work. Fluency in technical English coupled with comprehensive implementation of standards and procedures, as well as e-mail access to specialised international communities of practice, enabled engineers to draw on similar embodied expertise as their European counterparts:

> 'I use internet, intra-net, specifications, procedures. The Internet is helpful for "personal grooming", improving my background knowledge. The MultiTel[51] intranet (we call it sharenet) is really useful. We post technical questions on the "MultiTel Everywhere" net and someone who knows the answer will respond. We study the draft international standards for interfaces between telephone systems, you know, between us and companies like Siemens and Alcatel, etc. Also we have sales presentations using Netmeeting, and we listen on the phone using a conference call. We use that for some sales training courses.'

In contrast to the water utilities, mobile telecommunication engineers had little or no contact with end users of the service other than through their own social network and family. Operational knowledge of the system is encoded in software. Engineers seemed to have no idea of the credit of individual customers, nor any ability to influence the level of service that an individual customer received. Unyielding financial control is maintained through software, allowing little, if any, space for political and other social influences, protecting both end users who have contributed advance payment for service and investors who have paid for the infrastructure. In terms of the water utilities, on the other hand, operational knowledge of the system is diffused among outdated documents and the memories and histories of power and influence within and around the enterprise, far beyond the reach of reasonable technical and financial restraint. As a result, individual engineers and technicians can readily influence the service level provided for powerful individuals or groups of customers. Furthermore, there is little or no protection from socially irresponsible manipulation beyond the reach of governance. An unauthorised water connection can be arranged by working within the human networks of power and influence that permeate a water utility organisation and, once installed beneath the ground, the connection remains out of sight and fades from memory. On the other hand, software systems that form part of the infrastructure of mobile phone systems make it extremely difficult if not impossible to arrange such a connection or service unless payment records or credit status exists within the system.

ENGINEERING OPPORTUNITIES

Living in the Third World with an elementary understanding of the local economy, both formal and informal, as well as the invisible economy of people's time, opens up a multitude of exciting opportunities for an engineer. It is essential to understand how local people see the world around them to translate an idea into a commercial opportunity.[52] However, it is almost impossible to see these opportunities unless you are faced with them every day. Even as a frequent traveller to LICs, you can remain completely oblivious to possibilities that can turn into large opportunities. The following sections present two opportunities where engineers can make a large difference to the lives of hundreds of millions of people – perhaps even billions. It was only by living in LICs that I came to understand the very real possibility of transforming the lives of so many people.

Water and electricity supplies

Learning from the success of mobile telecommunications in LICs, it should be possible to transfer the same factors that enabled that success to water and electricity supply utilities.

In an environment where the most essential commodity required for human existence is so expensive, a reliable engineered supply for potable water that can meet the demand at a significantly reduced cost could have extreme commercial success and result in huge social benefits at the same time. If the cost can be reduced to around $7 per tonne, about one tenth of the cost of water in 20-litre bottles, although still well above industrialised country prices, the overall benefit would be profoundly significant. One option would be to emulate the transformation of Phnom Penh's water supply system, which supplies bulk water at a low cost. By gaining the confidence of the international community, the Cambodian government has been able to attract the necessary grants and significant loan funding to construct a new infrastructure and rebuild the water supply enterprise.[53] However, there are other alternatives. Mobile phone enterprises did not replace the existing government-owned landline monopolies; instead, by offering a much better alternative service, people chose to adopt mobile phones because they provided better value for their money.

Transposing the key mobile phone success factors for water and power utilities requires the following:

1. Value for money: providing a reliable alternative service at a cost well below the current alternatives.
2. Capacity has to be added incrementally to defer capital expenses; smaller localised schemes are more likely to be successful in the early stages.
3. Existing prepaid scratch cards and centralised revenue processing systems from the telecommunications industry can be used with only minor re-engineering.
4. Defence measures that deter fraudulent intervention can be transferred from the telecommunications sector.

The key remaining factor is to ensure that the service is only provided when the user has sufficient credit. However, since water is a physical commodity that is essential for sustaining life, a different approach is needed. One can survive without a mobile phone, but not without water.

There have been several attempts, particularly in South Africa, to use prepaid meters in order to reduce 'non-revenue water', which prevents people from accessing the service unless they have established credit.[54] However, so far, these efforts have not been successful. Firstly, there have been many technical problems with the service. Electronic metering devices have been damaged by lightning and power surges, while batteries have also proven to be insufficiently reliable. A requirement for both electronics and mechanical skills could not be met by normal maintenance technicians who had no training in electronics. Instead of one person attending to the issues, two people were required to deal with faults, and electronics technicians cost much more to employ than traditional water supply mechanical technicians. Local authorities that were responsible for the water supply saw prepaid meters as a means of increasing their revenue, rather than improving the overall supply and they met strong resistance from local communities. Lower than expected production meant that the meter cost was higher than what was originally forecasted. Finally, there was no provision for a minimum 'social supply' of water for customers who could not afford to pay due to changed circumstances. For example, a pregnant woman who was living on her own and unable to fix a broken toilet would quickly exhaust her credit and would therefore be unable to access any further water.

In many societies, water is seen as a 'public good' and people hold their governments responsible for maintaining adequate supplies. In some cultures, it is not socially acceptable to charge a fee for water. However, the reality is that most people in the world pay to have water transported to their homes. The engineering challenge is to be able to regulate water flow to maintain adequate service for those who cannot afford to pay, while simultaneously providing a cost-effective service that provides sufficient revenue to cover the cost of meeting social needs and also supplies an appropriate return for investors.

These difficulties can be overcome by redefining a water and electricity service in two parts. The first part, an essential service, is a guaranteed minimum service level that is provided for every customer connection as a community service obligation. The second is an optional additional service for which payment is necessary, with rates depending on the consumption pattern.

The key technical component, therefore, is the metering device. This is a technology challenge for an inventive expert engineer: a reasonably accurate metering device for water that guarantees a minimum service level and delivers additional service provided that credit is sufficient. The metering device must be highly reliable, tamper-proof, inexpensive, and easy to maintain or replace. The reliability requirement for providing the guaranteed minimum service is critical: lives will quite literally depend on this.

Tamper-proof systems for electricity metering are newly emerging and this technology is opening up opportunities for commercial investors to bypass existing electricity networks, investors who can be confident that they will recover their investment with a reasonable profit due to technological interventions.[55]

I look forward to hearing news that this global challenge has been overcome.

Air conditioning

A large proportion of the world's population live in climates with hot and humid weather for several months of the year. Even with acclimatisation, human productivity is significantly reduced during the hottest and most humid conditions. There is also strong evidence of impaired decision-making and heightened emotions.[56]

Conventional air conditioning, as developed for industrialised countries, particularly the USA, provides conditioned air to maintain comfortable living conditions within an enclosed space. However, it is not feasible to provide air conditioning for most of the world's population using this technology, because we have no means of providing sufficient electrical energy. For example, even the most efficient split system air conditioner requires a continuous supply of approximately 1800 watts to keep a bedroom cool in a typical South Asian house. Furthermore, the doors and windows have to be kept closed and the air in the room is not refreshed, except by natural ventilation through gaps in doors and windows, which reduce thermal efficiency. Most dwellings in low-income countries are not insulated, so most of the energy is used to cool the walls, ceiling, and floor – not the people.

Air conditioners like this have crippled power grids in many LICs.

Paradoxically, traditional building architecture often provides a much more comfortable airflow without needing air conditioning, but most people do not have the means to reconstruct their homes.

Space air conditioning based on solar and geothermal-powered absorption cycle technology is a long-term solution, but these systems require large upfront capital investment and efforts to refine the technology for commercially feasible operating results are progressing rather slowly.

Indirect evaporative cooling also offers large energy reductions. This technology can be used to cool hot outside air without increasing the moisture content, yet it still exploits the latent heat absorbed by evaporating water to achieve the desired temperature reduction. This leads to significant energy savings, especially for cooling make-up air, which is hot air from the outside that needs to be cooled before entering a building's air conditioning system.

Another attractive pathway towards long-term sustainability could be technologies that allow the air conditioning to be focused only where it is needed, just on the people, rather than seeking to cool the entire space within a building or even a whole room within a building. Displacement air conditioning has been used for several decades for large buildings, which relies on cool dehumidified air being denser than the ambient air. The huge mosques in Saudi Arabia, for example, provide a cool environment for worshippers by maintaining a layer of cool air that is only two or three metres deep. The air conditioning even extends into sheltered open-air spaces.

Even greater energy efficiency is possible, however.

Only certain parts of the human body need to feel cool to create a sensation of comfort. At night, our bodies prefer to be in a warm, humid environment under bedding, which are physical conditions that we would consider very uncomfortable for working in. However, we only feel comfortable at night if our face and neck are exposed to cool air. The temperature at which we feel comfortable depends on the surrounding climate and varies with the season and location. In Lahore, for example, people feel comfortable at a temperature of 28°C in summer, whereas people in Chicago feel comfortable at 22°C.

By localising the air conditioning effect just where it is needed, a small air cooler can provide personal comfort for two people using only 250 watts of power, using up to 90% less energy than conventional air conditioning. This is the same principle as a low-power bedside reading lamp that focuses the light just where it is needed. It would take much more power to light up the whole room, but you only require the light on a small area. This sort of small cooler can be run from a modest solar photovoltaic (PV) panel or a battery backup power supply when grid power is not available. People who

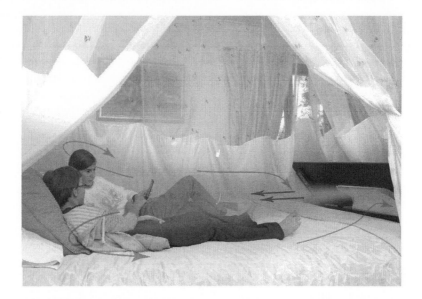

Figure 13.4 Early prototype localised air conditioner in use: cool air has been shaded blue to show the localised flow. Normally, the curtains around the bed would be closed for night-time use.[57]

have to endure hot, sticky, summer nights can sleep comfortably without worrying about an expensive electricity bill at the end of the month.

Devising a compact, portable, quiet, and low-power device to deliver the air precisely where it is needed is not easy. When used in combination with a small air cooler, a specially designed fabric enclosure can retain a layer of cool air above a bed providing comfortable sleeping conditions even when the room temperature and humidity reach sweltering levels. New techniques for measuring very slow airflows and tiny sub-Pascal pressure differences help to optimise the design of the air cooler, the enclosure, and air flow control ducts.

Localised air conditioners work with the doors and windows open, except in windy conditions. They can easily be used in offices, campsites, shops, and street-side stalls.

Conceptually similar products were invented in the 1930s, and more recently in China but have not yet been commercially successful, because precision in controlled airflow is essential to achieve reliable cooling where it is needed, which leaves users feeling relaxed and comfortable.

A similar concept emerged from the design school at Delft University in the Netherlands, based on a 400 Watt split system air conditioner. Cool air from the evaporator cascades down onto a double bed from an overhead canopy designed to ensure that no condensation drips on the bed's occupants. Initially, the product was designed for luxury resorts on the Caribbean islands, but with a warming climate and demands for healthier air conditioning solutions, the European market has become the main focus. This device is so quiet that some users find the silence disturbing. People are often accustomed to an air conditioner that masks the sounds of a typical urban residential environment in summer.

Similar technologies are appearing for office environments, particularly for older buildings that cannot be adapted for centralised air conditioning with high standards of thermal insulation.

Localised air conditioning can provide health benefits as well. Conventional technology that cools an entire room requires doors and windows be kept closed. Popular split system air conditioners provide no make-up air, so dust and biological contaminants can remain circulating in the room. Localised energy-saving air conditioning technologies escape from the need to keep doors and windows closed, and offer a personal air supply that is cleaned and filtered to remove contaminants. With insect protection being added to various solutions, new air conditioning technologies could bring large sustainability and health benefits to major populations in hot and humid climates. The high indirect cost of intermittent electricity mentioned before provides a stronger economic incentive for energy efficiency in LICs than in industrialised countries. There are very large and potentially profitable opportunities for engineers to devise similar improvements in the future.

LESSONS FOR ENGINEERS

Now that you have nearly reached the end of this book, let's summarise the opportunities you might have that will allow you to change the way you work as an engineer. I will base this on two situations in which I have observed engineers working in South Asia. However, the same lessons apply for engineers working anywhere.

Scenario 1: City-based design office

Tara is a design engineer working in an engineering design company based in a large South Asian city. She prepares detailed designs for reinforced concrete foundations based on European construction codes, such as DIN1045-1 (2001). She receives sketches with indicative dimensions from engineers and architects in France, Germany, and some other countries, and by working with drafters, she produces detailed construction drawings that her clients pass on to construction firms to obtain their quotations. Usually, she will also have drawings of the construction site to provide details of the ground conditions, but occasionally, she will only have surveyors' sketches and reports of soil conditions from a geotechnical consultant. Sometimes, she receives detailed drawings from construction companies and she has to check them for compliance with design codes and then prepare new drawings with changes, if necessary. She has never visited Europe, but she has seen plenty of movies – mostly Indian movies shot in European locations. She is hoping to save up enough money to travel there, but she has younger siblings and parents to support as well, so it is taking a long time to put aside money for herself. She takes crowded buses to and from work, taking up to 90 minutes each way through congested traffic, and she usually does not return home until well after dark. She has been doing this work for five years, and is gradually learning to take more responsibility and relies less on senior engineers to explain what she has to do. She has a reasonably up-to-date PC computer running AutoCAD and some other software programs that she uses for calculations and for reviewing drawings from the drafters. She earns the equivalent of US $1,600 monthly, including a bonus that depends on feedback from her clients and how well the firm has performed over the past 12 months.

Scenario 2: Plant maintenance

Dubey works as an electrical plant maintenance engineer in a cement plant in a rural location. He lives at the plant colony, in a house in a residential area owned by the company on the edge of a small town that is not far from the plant. The company also provides transport to and from the plant every day. He works under the supervision of the plant maintenance manager and performs electrical maintenance work: planning the work and preparing schedules and budgets for annual maintenance during the biannual plant shutdowns, as well as supervising work in progress, mainly work being performed by electrical technicians around the plant. Some of the work is performed by engineers and technicians who work for contractors, such as the company that maintains the specialised process control and instrumentation system designed and installed by a European firm four years earlier. Dubey is also investigating computerised maintenance management software systems, as he thinks that the current system is clumsy to use and causes a lot of problems, such as issuing too many duplicated inspection work orders, and creating three or four separate maintenance jobs on different parts of the same equipment in the same location, even though it would be much better to combine them into a single maintenance job done by the same maintenance team. Originally from a large city, he finds life in this small country town to be too quiet. He receives a lower salary than his friends working in the large cities, only making approximately US \$1,100 every month, but his living costs are almost negligible, so he can save some for himself, and is also helping to support two brothers studying engineering at a prestigious private university and a sister studying journalism in France.

He has no financial authority: every payment, no matter how small, has to be approved by his supervisor – the plant maintenance manager. Even the plant maintenance manager has to refer anything more than \$80 to the plant general manager. He uses a 6-year-old PC computer for e-mail, although there is a fast Internet connection. At home, using a computer is impractical because of frequent power surges, brownouts, and interruptions, so he uses a smartphone with a large screen for web browsing and·e-mail. He is engaged to be married, and his wife will join him after their wedding, which is happening in a few months.

Both Tara and Dubey see promotions as something that might happen in the distant future. Tara's firm is family-owned and all the senior positions are held by family members. This makes it difficult, because only the owner's son-in-law is reasonably competent in a technical sense. The owner's brother lives in Germany: he has strong technical skills, and works part-time as a civil engineer in Heidelberg, Germany, spending the rest of his time visiting clients to develop the business and seeking new clients in Germany and eastern France. Tara's career progression will probably depend on finding a similar job in a larger firm, but that will also depend on her familial responsibilities.

Dubey is one of five engineers in the plant. He works alongside the mechanical plant maintenance engineer. The plant general manager is the most senior engineer on-site. The other engineer is the plant operations manager: he works with two production superintendents, but neither of them has engineering qualifications. Although he could move up to become plant maintenance manager one day, his present boss is at least 25 years away from retirement and the plant general manager is only a few years older. He works at one of four cement plants owned by the company and the others are in much less desirable locations.

Although both individuals are earning reasonable salaries, many of their classmates migrated to Europe or Canada, some to the USA, and they are earning far more than Tara and Dubey could ever dream of making in their careers, even if they did get those distant-future promotions.

How do you think that Tara and Dubey could take advantage of the ideas in this book? What changes would you suggest for each of them? What are the options that they could explore to develop their careers beyond the confines of their present situations? How could they start to contribute substantial commercial value in a way that would enable them to gain significant rewards for themselves, as well as helping their respective societies?

Perhaps the words of Paul Polak might be helpful; his company develops affordable technologies for low-income societies:

- Go see the problem for yourself.
- Talk to the people and listen when they explain their needs.
- Observe the context.
- Think and act big.
- Think like a child.
- Do the obvious.
- Ask, has it been invented already?
- Design to a critical price target.
- Aim to improve the lives of at least one million people.
- Create and follow a three-year plan.
- Keep learning from customers.
- Keep thinking positively, don't get discouraged.
- Customer benefits exceed costs in no more than 12 months.
- Sell more than a million.
- Get the people that you're working for to help you do the design. They have lots of ingenuity: they just lack knowledge, certain expertise, and know-how.
- Design a new box to think outside of.
- Small is beautiful – it just needs to be marketed.
- Reduce the weight: more material means more cost.
- Eliminate redundancy: if it's cheap it doesn't matter if it fails.
- Move forward by designing backward from the requirement.
- Make sure the system is infinitely expandable.
- Dealers need 12% commission.
- Develop products that encourage the development of technicians in villages served by the product.
- Make a Bollywood movie to sell your product.[58]

Gaining influence and respect

To achieve results as an engineer, you will need to influence other people, particularly people who enjoy much higher social status than you currently do. You will seldom have organisational authority and, as explained in Chapter 9, resorting to authority is often counterproductive. Therefore, your influence will depend on referent power associated with the respect and trust that other people have in your abilities and the personal relationships you develop with other people.

This means maintaining high ethical standards of behaviour. Engineers have explained to me that nearly all corruption and dishonesty arises out of the need to hide a lack of competence. In my personal experience, and in the interviews with the engineers in my research, I have learnt that it is almost impossible to distinguish the boundary between corruption and incompetence. Often, the only way to distinguish one from the other is intent, which can be almost impossible to discern.

Therefore, one of the best ways to earn the trust and confidence of the people that you work with is by building the skills described in this book: earlier chapters explain how to do this. Another senior manager explained, 'Always plan on delivering something a little better than you promised, a little earlier than you promised. It is better to be careful with promises so that results exceed expectations rather than the reverse.' Try and anticipate future requests so you can be well prepared: that way, you will usually only have to make modest adjustments to previously prepared resources and will still be able to exceed expectations.

Financial information can be hard to find

Almost anything you suggest to enlarge the scope of your work will almost certainly require you to influence financial decision-making in your firm.

Both engineers in the scenarios above face a common issue: it is often hard for engineers to obtain data on the true financial situations of their companies or projects. Few engineers, particularly in South Asia, are trusted with any financial responsibility or even information about finance unless they own the company. Many company owners, both large and small, are reluctant to reveal the true financial state of affairs to anyone except a few trusted staff. One reason is that most local company owners negotiate an arrangement with their local tax collector, a senior officer in the taxation department, to present a set of 'official' accounts showing considerably less profit in order to reduce their tax payments. Naturally, there is a fee for this service that is proportional to the degree of tax reduction achieved.

One company owner I met during my research explained how he had tried to regularise his tax affairs because of his personal distaste for the corruption in which he had become a participant. He told his tax collector that he wanted to pay his full share of tax so that he could run his company with a clear conscience. The tax collector was 'helpful' and explained how he would need to present 'corrected' accounts for the previous year because there would need to be an explanation for the sudden jump in taxable company profits from one year to the next. Since there had been no significant investment in the company, the only explanation that would satisfy the auditors would be accounting errors in the previous year's figures. Of course, similarly corrected accounts would need to be submitted for several years before that. Then, additional tax would be payable for these previous years, with penalties for failing to declare income received and additional penalties for late payment of tax with compound interest. The tax collector estimated a total payment for this moral 'correction', which was more than the entire company was worth. 'Of course,' he said, 'it might be easier just to continue with the present arrangement. Please let me know … Oh, by the way, if you do decide to continue with the present arrangement, please could you have my fee available in cash next Thursday?'

During my research, I discovered other reasons for company owners to hide the true state of their finances. Many owners do not want their shareholders (in the case of companies listed on the stock exchanges) or even family members (in the case of smaller private companies) to learn the true financial situation of the company, as they might then demand a larger share of profits. Many company owners do not account for liabilities,[59] such as depreciation of their assets, accrued benefits payable to their employees, or the future cost of remediating pollution of land or ground water caused by company operations. In my research, even in companies where financial information was widely disseminated, I realised that profits were being significantly overestimated because there was no provision for these liabilities. Cash seems to be pouring in, but the apparent profit can often be illusory. Bankers explained to me that they are trying their best to persuade firms to adopt international accounting standards and fully account for these liabilities. However, if they were to make this a requirement for lending, they would lose out on business opportunities, because other banks would still be lending to firms without insisting on these requirements.

Therefore, to make a legitimate business case for engineering improvements, an engineer will need to estimate the real financial situation of the company. Background information in Chapters 1 and 11 should be helpful. The Internet has made it much easier to access helpful information, not just locally, but also throughout the world. Of course, fluency in written English is still vital to access this information. Engineers' blogs, discussion forums, LinkedIn, and similar professional networking sites provide access to an international community of engineers from whom valuable data can be obtained. Equipment suppliers can provide information on prices of machinery. However, in making a business case, it is also essential to understand how a business owner will see the same set of circumstances.

One senior engineer explained to me how he had tried to argue that new excavation machinery made much more commercial sense than the traditional local strategy of buying old second-hand machinery from European agents at low prices.

> 'However, it has not been possible to convert this into a first-class contracting organisation company in Pakistan: it has been much more difficult than I expected.
>
> '[What you think of the main reasons for this?] The main reasons are financial: the owner of the company is not prepared to purchase the best equipment or hire the best engineers. Instead, he hires mediocre engineers and buys second-hand equipment, or worse, repairs the equipment that we already have.'

He had shown how new machinery would be much more reliable, would be consistently available for site work, and require significantly less maintenance. The company would be able to make greater profits by avoiding production losses caused by frequent machine breakdowns. It would no longer be necessary to keep a separate collection of spare parts for every machine in a fleet in which no two machines were alike. Maintenance fitters would only have to learn how to repair one type of machine. Another company owner had exclaimed to me, 'Listen, the moment one of those fitters touches a new machine, it is ruined. Look what happens to my new cars: the first time they go for any repairs, I have to replace them because they're ruined. Those fitters simply don't know how to do their jobs properly.' The engineer was making a reasonable case; however, the company owner may have implicitly recognised the simultaneous need for much more highly skilled maintenance support staff that were appropriately equipped with expensive tools and workshop facilities. Recruiting, training, and retaining such highly skilled workers would be a significant challenge, not to mention expensive.

Take opportunities when they present themselves

The last piece of advice is to seize opportunities that will inevitably present themselves. However, to do this, you will need to be prepared.

You will need to develop the essential skills that were presented in earlier chapters. Almost any engineering job presents a wealth of opportunities to do this.

However, you will also need to ensure that you can set aside time to follow up on an opportunity. If you are stretched for time to meet your job expectations, then you will not be able to take advantage of opportunities when they appear. If you make a regular habit of using part of your time (say 20%) to develop your skills every week, then your employer will be happy. Then, when an opportunity appears, you can also use this time to follow up and take advantage of the opportunity to move into the next exciting part of your career.

LOW-INCOME COUNTRY ISSUES – JOB SEEKING

If you have graduated in a low-income country, you may face special challenges in finding engineering employment. In many similar societies, families see engineering as one of the three professions that enable upward mobility, along with law and medicine. If your family is already well connected, then you will have plenty of contacts to help you find a rewarding engineering job. However, many young engineers lack contacts and connections, and find it hard to even locate engineering jobs, let alone land a job. In many countries, in the absence of influential friends and family, you can be expected to pay several years of your salary to obtain a 'comfortable' job.

You may be envious of your counterparts in a country like Australia, where engineers are in such high demand. However, even in a country like Australia, engineering jobs elude many graduates. In fact, in the 2006 census, when there was an acute shortage of engineers, only around 60% of engineering graduates were working in engineering-related employment.[60] About a quarter of them were in management positions that *could* have had some connection with engineering. While unemployment among engineering graduates was low, many were frustrated and had resorted to driving taxis or working in shops or clerical jobs, as they were unable to find engineering jobs. Meanwhile, companies complained that they could not find enough engineers. The reason was that companies want engineers with experience,[61] partly because books like this one had not been made available to help graduates know what they need to learn outside the somewhat limited range of their university courses.

Chapter 14 provides suggestions for seeking work that apply anywhere. However, in most LICs, there are some additional important factors.

The first is language. English is basically the universal language of engineering. It is the default language of the Internet, and most information resources are in English.

In addition, there is another equally important factor. There are important engineering concepts described in English words that do not exist in some other languages. For example, concepts of risk and uncertainty are difficult to describe in Arabic. In Japanese, it can also be difficult to express a definite opinion.

Therefore, fluency in English is highly valuable for you in order to help you become an expert engineer. It will also greatly improve your employment prospects. Achieving an IELTS score of 7.5 or more should be one of your primary objectives.[62] One way that you can improve your English fluency is by listening to Internet radio stations. There are now many Internet radio stations and blog sites where you can not only listen to programmes, but also download transcripts of the programmes.[63]

Beyond English, you will also need fluency in your national language, which may not be your mother tongue – the language you speak with your parents.

To become an expert engineer, you also need to be able to access distributed knowledge. Much of the valuable knowledge you will need is in the heads of experienced manual workers, people who 'get their hands dirty' in the course of their work. In many LICs, these workers often speak their own dialect or language, and are often seen to have low social status. To access their knowledge, you need to be able to speak and understand their languages, too. You also need to be able to build social relationships with manual and technical workers so that they are comfortable and confident enough to talk openly with you. One of the best ways to do that is to work as a trade or technical assistant, perhaps sweeping the floor in a factory or roadside workshop, or even volunteering.

Your family may complain that they did not pay for your university or college education just so you could work as a factory labourer. However, this can be a valuable learning experience while you are looking for an engineering job. You will not be able to become an expert engineer without access to information, help, and insights from manual and technical workers.

The next issue is honesty. As an engineer, you can only enjoy a rewarding career if other people trust you with their money. You can only gain this respect and trust by being consistently honest, even if you miss out on attractive work in the process. In countries where many people need to be dishonest just to earn enough money to feed a family, this can be very difficult. Even if you are honest, if the people you work with are known for their dishonesty and lack of integrity, their reputation will also affect your reputation.

Above all, remember that when you do become an expert engineer, one day, your true value and potential will be appreciated by others. Until then, you need to keep reminding yourself that anyone can become an expert engineer, but achieving that goal takes persistence, hard work, and a fierce determination to succeed. Never forget that. Your persistence will be rewarded in the end.

NOTES

1. The Human Development Index is calculated from the geometric mean of normalised life expectancy, expected and actual years of education, and Gross National Income per capita (United Nations Development Programme, 2013, pp. 144–147). While there have been remarkable recent improvements in many countries, much of the improvement

reflects upper-middle-class economic empowerment while others may have seen much less improvement.

2. E.g. Srivastava (2004).
3. Author's interviews, see also a detailed study of outsourced engineering analysis in the automotive industry (Bailey, Leonardi, & Barley, 2012).
4. Even though there are many reports of major engineering projects that address poverty reduction (e.g. Singleton & Hahn, 2003), engineering is hardly mentioned in recent mainstream human development reports, except in the context of providing IT products and services for industrialised countries (e.g. United Nations Development Programme (UNDP), 2013). A special report on engineering and development was commissioned to try and help change this perception (Marjoram et al., 2010). The recent E4C website may also help to change this perception (https://www.engineeringforchange.org/home).
5. Sachs remarked with surprise concerning the high cost of engineered products, such as fertiliser, in Third World subsistence agriculture, although he offered no explanation (2005, Ch. 12 pp. 234–5).
6. Analysis by the author showed that government spending on policing per capita in Pakistan's 2006 budget was little more than 1% of Australian spending on policing.
7. See http://www.mech.uwa.edu.au/jpt/demining/ for further details.
8. South Asia includes India, Pakistan, Bangladesh, Sri Lanka, Nepal, Sikkim, Bhutan, and the Maldives.
9. See explanation on the value of time provided in Chapter 1. Value of time estimation is often used to help evaluate costs and benefits of social investment (IT Transport Ltd, 2002). The high cost of water in low-income societies has been noted before (e.g. Kayaga & Franceys, 2007; Neuwirth, 2005, p. 79; Reddy, 1999).
10. Trevelyan (2005a, 2005b).
11. Anand (2011).
12. World Health Organization (2006).
13. Based on 10 litres per person per day at a cost of US $30–70 per tonne.
14. For a comprehensive account of power theft, see Ong (2013).
15. Typical diesel generation cost in 2013 can be estimated at US$0.30–0.45 per kW-hr for a 25% utilisation factor using data from (Bannerji, 2006) and allowing for changes in currencies and oil prices. Metered power costs in 2013 were around US $0.10 per kW-hr.
16. Lucena and Schneider have argued that traditional engineering problem-solving approaches are flawed when working with these issues, because it is difficult for such diverse perspectives to be taken into account (Lucena & Schneider, 2008).
17. Coelho (2004, 2006).
18. Barley (2005); Trevelyan & Tilli (2007). See, as an example, Alice Lam's informative and detailed comparisons of engineering in Japan and Europe (Lam, 1996, 1997). Domal provided a detailed review of comparative literature (Domal, 2010).
19. Domal (2010); Trevelyan (2014a).
20. The manufacturing issues we observed appear to be representative of South Asia as a whole and these issues may help to explain recently reported economic difficulties in that sector (The Economist, 2012).
21. Evidence from research interviews suggests that corruption (dishonest behaviour) accounts for 15% of the cost for well-managed Pakistani projects (though it can be higher). The contractor typically pays a fee of 3–5% to have the contract awarded, and another 3–5% to receive timely contract payments. The remainder covers inducements needed to run the project without major delays. A survey of nearly 2,000 Indonesian firms in 2002 yielded an average cost of 10% (Patunru & Wardhani, 2008).
22. Hill (2011).

23. For comprehensive information on combating corruption, see reports by the ICC Standing Committee on Extortion and Bribery (1999) and Transparency International (2000).

24. Quote, short form of quotation, in this context a written cost estimate for the construction. As long as there are no variations in the scope, the building will be constructed for the given cost estimate.

25. Corruption was eliminated in the Hong Kong Airport construction project, but this required an independent commission with 1,350 highly skilled legal and technical staff (Rooke & Wiehen, 1999).

26. Perlow (1999).

27. Some of this material has been adapted from Trevelyan (2014a), while other contributions came from Domal (2010) and S. S. Tang (2012).

28. Raven (1992).

29. Ansari and his colleagues suggested that in Indian society, there is a 'ready-made rule-book' for power relationships, with an established superior-subordinate relationship. They then observed that sanctions, including presumably legitimate position power and coercive/reward power, work better in Indian society, while friendship (personal reward and referent power) would be less effective in India, as compared to Western societies (Ansari *et al.*, 1984, cited in Raven, 1992).

30. Day labourers at the time were hired at the front gate of the factory each morning. Government regulations stipulated that a day labourer could not be hired by the same factory for more than 180 days consecutively and could not be hired for more than 180 days in a year, without being made a permanent employee and therefore eligible for extensive additional remuneration and benefits through compulsory union membership.

31. Mason has described industrialised country production supervisor roles in detail (2000).

32. Coelho (2004).

33. Coelho (2006).

34. D'Arcy & Gustafsson (2012).

35. For example, Linden reported that industrialised country component suppliers were achieving margins of 15–25%; Chinese assembly plants manufacturing iPods in 2005 were achieving a margin of 3%. Out of the selling price of US $300, only a few dollars of value were added in China.

36. Bennell (1986); Mason (2000); Prais & Wagner (1988).

37. Domal (2010). See also a report from Sudan on a railway workshop (Ketchum, 1984).

38. Safety is another important issue that is not addressed in detail in this chapter that requires approaches that recognise the local culture and governance constraints (e.g. Koehn, Kothari, & Pan, 1995).

39. Indian Administrative Service – the most senior and privileged ranks of the Indian government service.

40. Martin et al (2005); Perry et al. (2012); Shuman et al. (2005); Spinks et al. (2007); Dahlgren et al. (2006); Darling & Dannels (2003).

41. Blom & Saeki (2011).

42. www.maxon.com.

43. Technical sales engineering work has been described by Darr. These accounts demonstrate some of the contributions that sales engineers make in educating the engineers who buy their products (Darr, 2000, 2002).

44. Richard B. Freeman (2003) Personal communication. Professor Freeman from the National Bureau for Economic Research at Harvard University kindly took the time to discuss this issue in depth with me, and his comments eventually led to some of the ideas presented in this chapter.

45. At the time, we could not find other researchers studying similar issues. This made it much more challenging for Domal to demonstrate that he had met the basic requirement of a PhD, an original contribution to knowledge in a particular field (Domal, 2010).

46. The first public demonstration of mobile phone technology was in 1973 by Motorola. It is worth noting that it took around 30 years for this technology to become widely available around the world following this first public demonstration. It is widely believed that, in the 20th and 21st centuries, technology is advancing faster than at any other stage of history. However, close examination of the diffusion of other technologies, such as electric power generation in the 1880s and 1890s helps to disprove this. It still takes decades for new technology to become widely available throughout the world.

47. In Pakistan, mobile phone companies contribute approximately 3% tax on their income to provide funding to service remote areas that are not commercially viable. The service area contracts are awarded to particular commercial networks by a government agency.

48. The Nation, 'Tax cheat politicians named' (2013).

49. GST: Goods and Services Tax, VAT: Value Added Tax.

50. S. S. Tang (2012).

51. Multinational telecommunication equipment supplier pseudonym.

52. http://extreme.stanford.edu/, ideas also from Dr. Paul Polak (http://www.d-rev.org/). While there have been many proposals for commercial products that improve life for poor people, often referred to as 'base of the pyramid' (BoP), it can be challenging to realise full commercial success.

53. Leong Ching (2010).

54. Bajaj (2011).

55. Trevelyan (2013).

56. Hsiang, Burke, & Migual (2013).

57. Companies working on similar technology include Close Comfort (a start-up founded by the author); HIVAP in Western Australia; Task Air, a Queensland start-up; and Evening Breeze based near Delft University in the Netherlands.

58. Polak (2008).

59. Liability: an amount of money that has to be paid out at some future time. For example, depreciation represents an amount of money that will be paid in the future to replace an asset when it reaches the end of its useful life.

60. Trevelyan & Tilli (2010).

61. Trevelyan (2012a).

62. IELTS – International English Language Testing System, similar to TOEFL (Test of English as a Foreign Language).

63. E.g. ABC Radio National, http://www.abc.net.au/rn/. The Engineering Commons, http://theengineeringcommons.com/, provides regular commentaries and interviews on engineering. See also Chapter 6 on listening skill development.

Chapter 14

Seeking work

By the time you reach this chapter, if you have not already obtained a job in the engineering field, you have probably not been seeking engineering work in an effective manner. Throughout this chapter, I will explain why this might be the case as well as what you can do to change it. Obviously, I cannot guarantee that you will get a job as a result of reading this. I can, however, guarantee that your chances of obtaining a job will be much greater if you follow the advice given in this chapter. If you manage to follow this advice and you are still unable to find a job that will satisfy you, please take the time to write about your experiences; I can use your feedback to improve the next edition.

As in previous chapters, it is useful to start with common misconceptions about seeking engineering work.

Misconception 16: Engineering jobs are always advertised

Misconception 17: You cannot get an engineering job without relevant experience

Students and many engineers often think that . . .	Expert engineers know that . . .
Companies advertise for engineers when they need them. Most jobs require experience.	Many if not most jobs are obtained through informal networking. Employers hope that they will meet experienced applicants but know that they will have to train them otherwise.

Few students, in my experience, think about applying for engineering jobs until towards the end of their penultimate year of studies. For most, their first experience is applying for an internship, sometimes known as a professional practicum. This is often a compulsory part of an engineering degree course.

There are many factors that focus the minds of engineering students on job advertisements as the 'normal' way to get a job. In reality, that's not how most engineers get hired.

Career advisers emphasise the need to prepare a CV (curriculum vitae) or résumé[1] taking into account the particular requirements listed in a job advertisement. Larger companies visit university campuses and often provide exciting presentations in order to attract the best students to apply for jobs with them. The students need to be motivated; many of these companies expect students to spend the equivalent of two or more days of full-time work for a single job application. Aside from the preparation of a CV, students are expected to complete extensive online psychometric tests. Students selected into a shortlist of applicants can then expect a full day consisting of tests and other activities in which their performance will be observed by company staff. Responding to each advertisement, for a student, can be the equivalent of a major project assignment – a significant additional commitment on top of full-time study.

Last but not least, practically all of the available texts that discuss the transition from study to work devote extensive space to the preparation of a CV, writing a job application in response to an advertisement, and preparing for an interview.

Few if any of these texts mention the informal job market: jobs that are not advertised for a variety of reasons. This can account for 80% or even 90% of available jobs. The only way to find these jobs is through networking: going out to meet other engineers and people who work with engineers. This takes time and effort, of course. However, as we shall see later, this time can be spent on useful learning that will contribute to your career as an expert engineer. It is never wasted effort.

Later in the book, I will also explain how you can start your own business – that's another opportunity that is easy to miss.

Another misconception worth mentioning is the notion that graduates can't apply for most jobs that are advertised because the jobs require relevant experience. In reality, when a company has a job opening as a result of the resignation of an engineer, they will advertise the job hoping to receive applications from engineers who have, at the very least, an equal amount of experience to the person who has just left. The engineer might have started as a graduate with no experience at all. However, after three or four years with the company, they would have developed to a reasonable level of professional ability. Naturally the company would prefer to hire somebody with the same kind of experience or even better. However, if no one with appropriate experience applies, the company may have to employ the best available. That's why it is always worth applying for an engineering job, even if the advertisement requires two or three years of experience. It would probably be inappropriate for a graduate to apply for a job requiring extensive experience, perhaps 10 years or more. However, whenever a job is advertised for a senior engineer it can be worth asking whether there are also openings for less experienced engineers.

LOOKING FOR WORK

Looking for an engineering job can be one of the most emotionally draining times in your entire career (it certainly was for me). I graduated with a master's degree with an invention in my hands, a new form of electronic flight display that could be used to help train pilots. I had had a wonderful time developing a digital flight simulator, an ancient precursor to the ubiquitous computer games today. Unfortunately the major aircraft simulator company that had expressed a strong interest in my ideas went bankrupt the day after they promised to find a job for me.

It took months for me to find my first engineering job. I became depressed and very irritable, difficult to live with.

This chapter will help you spend your time productively while you are looking for an engineering job. If you work hard, it will probably help you get a job sooner because the information and skills that you will acquire will enhance your employability.

First, let's review four possible reasons why you have not already found a job.

The first could be an economic recession. At the time of writing this book, the USA and much of Europe is enduring a severe economic downturn with unemployment rates ranging between 8% and 25%. Many companies who would usually have taken on graduate engineers have stopped recruiting completely. Many engineering companies are facing bankruptcy, particularly small firms. Even engineers who have employment may find themselves working part-time or at least being paid only a fraction of their normal salary.

Unfortunately, while it is tempting to blame your difficulty in securing an engineering job on an economic recession due entirely to factors well beyond your control, unemployment is not an option for you. With an engineering degree, you are among the most highly skilled and educated in your society and there is no reason for you to not be employed by somebody. You may not have all the skills and knowledge that they require, the work might not be what you imagined or even what you have been looking for, but even with a 20% unemployment rate, you are among the most employable of all young people. There are always jobs for engineers.

Even in an economic recession, when the economy is stagnant, people still need engineers. People still need food, water, sanitation, transport, communications, consumables, heating, cooling, maintenance, repairs, and replacements. Metal keeps corroding, roads keep cracking, pipes keep becoming blocked, buildings keep decaying, engines, transmissions, and generators keep breaking down, software keeps needing updating and expanding, insulators on power lines keep cracking. Every bit of this happens regardless of the economic situation. All of this creates work for engineers.

Thus we have eliminated the first possible reason for your difficulty in finding work.

There are three other possible reasons. The first is that you have been using an ineffective method to find yourself an engineering job. The second is that you have not been able to recognise opportunities for paid employment. The third reason is that you have not yet seen the opportunity to start your own independent engineering business.

Apart from working through the other chapters of this book, you need to carefully read and follow the steps outlined in this chapter. The last part of the chapter describes homework that you can do while looking for employment opportunities. This work will help to improve the chances of gaining an engineering job when you find an opportunity.

If you faithfully follow all the advice given in this chapter, and you have kept detailed records of your experiences and still have not found work, I invite you to contact me directly. I would like to hear from you.

HIDDEN JOB MARKET

Up to 80% of all jobs are not advertised, especially in times of economic hardship. It can be daunting for an employer to work through 500 applications that arrive the day after a job is advertised on the internet. Each application has between 20 and 50 pages of text to read. The employer might think of interviewing five applicants, but it will not be possible to be certain that they chose the best five from the many applicants. For this reason, many employers wait for someone to visit and ask if there is work available. Companies can save a lot of money by not bothering to advertise a job, particularly in tough economic times.

This is called the 'hidden job market'.[2] The hidden job market also includes jobs like these:

a) Work that needs to be done but no one in the company human resources department knows that yet because the engineers have been too busy to tell them.
b) Work that they already know about but have been too busy to bother with recruiting engineers for yet.
c) Work that they know about but they are not sure what kind of position to advertise in order to attract suitable applicants.

You need to know about ways to find these jobs.

Read the newspapers and check internet job advertisements like, 'Highly experienced engineering or mining project manager to set up a team of ...'. Think about approaching the company as a prospective junior member of their new team. Before you do this, find out as much as possible about the company. Visit their website and read their annual reports, brochures, and publications. If you cannot obtain information on them from an Internet search, visit the company office and ask for a copy of their printed annual report and brochures. Wait a few weeks and try visiting the company, following the advice on how to prepare that is provided later in this chapter. By then, the new senior manager has probably started work and you should ask to meet him (or her).

Obtain a notebook (size A4 or A5) and write your name and contact details on the front page. This will be highly valuable for you: you need to make sure it can be returned if you misplace it. Use this notebook to record everything you do in terms of seeking work: all the names of the people you meet, their contact details, and everything you see or hear that is related to your search.

If you are a young engineer looking for experience and find an advertisement like, 'Project engineer required, 3 years' experience in mining or automation ...' think about the people looking for a new engineer. Most likely, the engineer who just left had 3 years of experience and they are hoping to find someone just like that. Don't hesitate to apply: they will take someone experienced if they can find one, but if they can't it will be your opportunity to gain experience.

Look through the business pages of the newspapers and note the names of companies that have announced plans for expansions or special projects. Then think about the types of smaller companies that may be involved in those projects and think about approaching them. Ask them what kinds of work they are hoping to win in the near future. Always find out as much as possible about each company as you can before making an approach about working there.

Most industries have free magazines (or e-mail/web newsletters) that advertise products used by engineers in that field. An example is, 'What's New in Process Engineering'. Make sure you subscribe to these magazines and read about the products in your area.

As a graduate or novice engineer, or even an engineer who has recently arrived from another place, you will need to gather as much information as you can on the capabilities of local companies, both engineering companies and the firms that supply services and products that they use. Use directories such as the 'Yellow Pages' to compile a list of firms.

One of the most valuable areas of knowledge for any practising engineer is awareness of which firms supply which products. This also means awareness about the support they provide by keeping commonly used parts in stock and having technical experts to help you make the best use of their products. The only way to acquire this knowledge is by meeting people in these firms. Trade exhibitions are a wonderful place to start. However, there is little substitute for taking the time to visit the firms and learn about the products and services that they provide. Taking the time to build up this knowledge while you are looking for work is a great way to invest your time in productive learning.

Here are some suggestions for engineering product suppliers; look up firms that supply these items:

- Communications equipment
- Concrete additives, plasticisers
- Drafting and design services
- Drainage pipes, silt and grease traps, strainers
- Electric cables and wiring
- Electrical connectors
- Electrical switchgear
- Electronic components
- Fasteners
- Fluid flow meters
- Geo-textiles
- Hydraulic fittings and connectors
- Lighting
- Lightning protection
- Pneumatic cylinders and valves
- Powdered metals
- Power line insulators
- Product and industrial design
- Security equipment
- Steel, plastic, copper tubing
- Structural steel supplies
- Tarpaulins
- Underground cabling systems (directional drilling)
- Welding supplies

There are several kinds of engineering suppliers:

Manufacturers: They will supply either directly or through local representatives. Local representatives charge higher prices, particularly for small orders, but provide technical advice, locally available stock available for immediate delivery, and sometimes even training for engineers on how to use their products effectively.

International supply houses: These include businesses such as RS components or Element14 (electrical, electronic, mechatronic components). These firms provide a combination of a huge range of components with local stock and a rapid supply of small numbers of parts. Prices are much higher than buying in bulk through a web-based agent, but they have local representatives and offer good technical support.

Local in-country engineering product sales organisations: These companies often sell a wide range of engineering products with a high level of technical expertise and support. Prices are higher, but come with good service, and stocks of spare parts are held for many years to cover ongoing maintenance and repair. Sometimes these organisations secure exclusive agreements with suppliers such that the product can only be obtained within a specific country through the local agent. However, internet communications make enforcement of these agreements steadily more difficult.

Consultancies: These firms provide design or specialist consulting services and often (but not always) provide specialised products as well to support the kinds of techni-cal solutions they offer. For example, a consultancy may provide machine condition monitoring advice, and also may sell specialised instrumentation that they use themselves.

Take the time needed to carefully prepare yourself before you visit any of these firms.

BUILDING YOUR NETWORK OF CONTACTS

Keep building your network of contacts. Most engineers find jobs through contacts, though many apply for advertised jobs as well.

Ask everyone in your family if they know anyone with connections to engineering projects. An uncle or aunt may not be an engineer, but they may be able to introduce you to one who can be your starting point. Even if that person is, for example, a civil engineer, they will probably know people who can connect you with mechanical or electrical engineers working in areas where mechatronic, instrumentation, control, automation, or other similar classes of engineers work as well. Ask them to introduce you to other people connected with your areas of engineering.

Keep in contact with your class colleagues – many of them will be able to suggest contacts for you.

Join the organisation for engineers in your discipline. For example, in Australia there is a single organisation that represents all engineering disciplines, 'Engineers Australia'. Go to meetings. At each meeting, make a point of getting to know one or two other members and asking them to suggest people you can call. You will find most members to be very helpful as they have all been through the same experience as you.

Go to as many conferences and trade shows as you can. When visiting companies, ask them which shows they go to, especially when they show off their own products or services.

You can try asking for work experience. However, you need to realise that the company still has to pay money to support you even though you don't receive a salary. Someone has to supervise and train you, and your work will require other forms of support, such as office and work space for which they pay rent. They have to cover you under their insurance policy as well, just in case you are injured at work. If you take on unpaid work, make sure there is a clear agreement on how long unpaid work will last. If there is no pay at the end of that period, leave immediately: don't hang around just because you have been assured of paid work 'coming soon'. I recommend that you don't work unpaid for more than three weeks. If they have not employed you by then, they do not appreciate the value you can provide and you will do better by spending time on other work described in this chapter.

PREPARING YOUR CV AND RÉSUMÉ

Most companies reject 75–95% of the applications they receive because of one of the following reasons:

- The covering letter was either missing or did not mention the needs of the particular company.
- The CV (résumé) was poorly formatted or contained obvious spelling errors.
- The descriptions of the work that the applicant had previously performed did not give a clear idea of their capabilities. For example, the applicant might have stated, 'I was involved in the installation of solar photovoltaic systems.' Instead of this, describe the actual work you were required to do, such as, 'I had to liaise with clients and installation contractors to arrange the best time to install solar photovoltaic systems. I had to confirm these arrangements just prior to installation to make sure that both clients and contractors were ready and prepared for the installation work.' Describe the actual work that you performed even if it does not seem to involve what you think is engineering work.

Arrange for a local expert to review your CV before you send it to a company to seek work.

Learn that written communication skills are one of the two most important attributes for an engineer. The other one is face-to-face communication skills. Companies will employ an engineer with less than perfect spoken English if they think they are technically competent and able to communicate effectively. However, a poor CV will generally result in your application not even being considered.

There are many good books on preparinga résumé or CV. Here are some suggestions that should be available through your local library or online bookstore:

- Millar, D. C. (Ed.). (2011). *Ready for Take-off*. Upper Saddle River, New Jersey, USA: Prentice Hall (Pearson Higher Edn).
- Wells, D. J. (1995). *Managing Your First Years in Industry: The Essential Guide to Career Transition*. Piscataway, New Jersey: IEEE.

While books can provide you with a lot of helpful guidance, you should seek feedback on your résumé, preferably from at least one senior engineer with experience in hiring engineers and from a human resources professional or careers adviser as well. The books also provide tips on interview skills.

If you are interviewed by a company, remember to prepare well in advance.

Practice your listening and reading skills as explained in Chapter 6. Companies tend to display little tolerance for job applicants who have not read the company requirements carefully, and when being interviewed, even on the telephone, do not listen carefully to questions from the interviewer.

Remember to interview your employer at the same time as they interview you. Ask them about the types of work that they will expect you to do, training opportunities, supervision, career development, and your professional competency development. An interview is as much an opportunity for the company to learn about you as it is for you to learn about the company. If you come away from an interview realising that this is not the company you want to work for, you can still maintain a good relationship and save them trouble by letting them know that you do not plan to accept any appointment they might offer.

Here is a list of some important attributes that engineering companies look for in prospective job applicants. Read this list carefully and address each of these items in your résumé and during your interview with the company.

Leadership

An engineer always has to be a leader, even in a purely technical role without management responsibilities. A leader is someone who is able to understand what needs to happen, and then inspire other people to follow. You can demonstrate your capacity for leadership in many different extracurricular roles such as helping to organise activities in clubs and societies, working for an NGO, and working for a voluntary organisation, particularly in fundraising. You can also demonstrate your capacity for leadership through your part-time work record. For example, you may have acted as a shift supervisor in a restaurant. You don't have to be working in an engineering capacity to exercise leadership skills. You don't even have to have a job. Remember to describe these experiences briefly in your résumé.

Teamwork

As you have learnt in this book, engineering relies on being able to coordinate your own work and collaborate effectively with other people. It is important that you demonstrate that you can work as part of a team, and not necessarily in a leadership role. Once again, there are many ways to do this. You may have played in a sporting team or worked on a committee to organise an event. Even if you have worked as part of a team in a restaurant, you can use that to demonstrate your ability for teamwork.

Initiative

Exploring the informal and hidden job market presents many opportunities to display your initiative. Demonstrate that you have taken the time and trouble to find out about engineering organisations and that you have managed to get yourself past the company reception and speak with technical people. That demonstrates initiative. You display initiative when you initiate effective actions without waiting for instructions or requests from other people.

In many cultures, demonstrating initiative can seem unnatural, stepping out of line with accepted social practice. In these cultures, it is common for younger people to remain silent and respectful in the presence of older, more experienced people. Acting with initiative in these cultures requires different methods from the more open, less hierarchical European, Australian, or American cultures. However, our research demonstrates that one of the big differences among engineers who have become experts in developing countries is that they are able to take the initiative.

Persistence

Demonstrating persistence is important in two ways. First, as explained in Chapter 4, persistence is a vital quality in deliberate practice, the main pathway to becoming an expert in any field. Second, persistence is highly valued by engineering employers.

Whatever your circumstances, if you can demonstrate that you have overcome disadvantages or major setbacks such as an illness, this will help to demonstrate to an employer that you have persistence. Demonstrating that you have read this book from cover to cover will also be a significant achievement in terms of persistence. Don't forget to mention that.

Reliability and responsibility

Describe circumstances in which other people have placed their trust in you. For example, if you have acted as treasurer for a club or society, and the members have trusted you with their money, that is a sign that you are both respected and trusted, and demonstrates that you can be depended upon. It may be a time when your employer asked you to act as a shift leader or manager, even for a single night. It can be that someone has entrusted you with a responsibility to teach people who are less experienced. You may have taken on a tutoring position at a university. Naturally, it helps to have a reference letter from people who have placed their trust in you. Remember to describe these circumstances in your résumé.

Notice that none of the special attributes listed above have any direct link with technical engineering work. They apply to any and all forms of work, either paid or while acting as a volunteer. You might imagine that an engineering employer would ask you more about your technical abilities than these other attributes. You would be quite incorrect in this assumption. Engineering employers are generally confident about your technical abilities simply because you have a degree qualification in engineering or equivalent experience, which is sufficient evidence of your ability to learn about technical issues when required. That is why employers prefer to focus on these other important attributes when they are selecting someone for an engineering job.

Local work experience

While you are searching for a job, taking part-time work with any organisation can provide you with useful experience. This can be valuable even if it is not specifically engineering experience, especially if it allows you to gain some background in sales, marketing, administration, occupational health, or the safety and supervision of staff. Employers will be more likely to offer an interview if you have local employment experience like this.

Try to avoid full-time work if you can afford to. Looking for, and preparing for an engineering job can take a lot of time during normal working hours, particularly when you call your contacts or need to visit companies.

It is vital that you demonstrate how you have been able to learn from experience and your surroundings. Even if you are working in a fast food restaurant, there are lots of opportunities to learn how to handle incoming supplies, orders in progress, finance records, staff performance records, and many other aspects of running a business. Not only is it important for you to learn about these aspects, it is equally meaningful for you to be able to demonstrate to a prospective employer that you are willing and able to learn in these situations. Learn to write brief paragraphs for your résumé explaining what you have been able to learn in different situations. Remember to have examples ready for your interview; do not count on the memory of the interviewers. While they will have looked at your résumé, they will not necessarily be able to remember all the details.

PREPARE BEFORE VISITING A COMPANY

The way you approach a company makes a big difference to the chances of obtaining work with them.

Before you approach the company, think about what your objectives are:

a)   Career? Long term or short term?
b)   Vacation or professional internship employment – is money important?
c)   Engineering experience?In a particular field?
d)   General work experience?Would you be prepared to work unpaid for a limited time?
e)   Part-time employment to help gain experience and financial support?

You must learn about a company before approaching them. Look at their website, and call in at the head office and ask if you could have (or at least read) a copy of their annual report, brochures, etc. When you do this, ask for the names (and job titles) of

  i)   The chief executive
 ii)   The most senior engineer in the organisation
iii)   The person responsible for recruitment or human resources

Call or write a letter asking to visit the company to learn about the products and services they provide. It is important to mention that first. You can also mention the possibility of a career, work experience (preferably paid), or part-time employment. Be specific about what you are looking for and in the order of your preference. Make sure you briefly describe the skills that you can offer that are relevant to the company's operations, and show that you have done your research on the company.

Ask for them to select a suitable time and date. Make sure you clearly state what times you would be available to meet them.

Address the letter to the senior engineer or chief executive – they will pass it on to the most appropriate person to handle your letter.

If you do not hear from the firm within two weeks (longer over holiday breaks), telephone the company to ask when you could expect a reply and ask if they could suggest a time for you to call in.

Follow up with telephone calls each week until you receive a response. Make sure you note the name of the person to whom you are speaking in each case, the date and time, and whether there is a direct telephone number you can call to reach that person without having to go through the company switchboard.

Unless the company contacts you by e-mail or asks you to use e-mail, never use e-mail to approach the company or follow up on application letters. Always call by telephone, or visit the company yourself if at all possible.

When visiting a company

Make sure you are well dressed. A suit is not always essential, but whatever you wear must be clean, pressed, and smart, with closed shoes. Make sure you wear strong, closed-toe shoes along with a shirt with long sleeves and trousers if there is a possibility of seeing workshop or site operations – you will need this to comply with safety requirements.

Use the following self-introduction:

'I will soon be working as an engineer and would like to learn about your products and services. Before I start work, I have more time to learn about engineering products and the companies that supply them in this region.'

Do not start by asking for, or about, employment. The reason for this is simple. Asking about employment gives the staff a chance to say something like, 'Sorry, we only advertise jobs on the internet – we don't answer queries about jobs here.' Asking for information about their products and services makes it much harder to send you away. Only companies that you would not want to work for will send away someone who may soon be buying their products and services.

Ask to speak with one of their engineering staff or a technical sales representative because you need to discuss technical issues.

Ask about the range of products they hold in stock locally or in the same country, any stock that can be made available on site within 24 hours anywhere in the country. Ask about delivery times for items not held in stock, including the time needed for customs clearance. Ask about the availability of technical data sheets; usually these are provided online by the original manufacturers, but you often need help to locate them.

Explain that you need to learn, and ask for material that you can study at home and, if possible, a quick look at some actual product samples.

The engineer or technical sales representative will probably ask what kind of work you expect to be doing. Describe the kind of position you would really like to get, but let them know that you are still undecided.

If the technical sales representative doesn't ask you about the work you will be doing, then ask 'Who do you think I should be talking to about getting engineering work in areas like ...?'

You could even ask them for alternative suggestions for interesting places to work.

Keep a detailed record of the names and contact details for all the people you meet: you never know when this information will be useful.

Learn about the companies you visit and keep a list of their products and services.

Listen carefully and learn what your prospective employer wants to learn about you. Remember to take notes and ask for clarification if you are not sure that you have understood something correctly.

There may even be openings to work as a technical sales representative. This might not sound like the kind of job you're looking for, but if you are having trouble finding jobs in your area it is a great way to start building an extensive network of contacts. It can be great fun, and technically challenging as well.

If the company staff do not treat you well, make a point of recording the details. In a year or two, when you are working, and you come across the company again, you can turn this to your advantage as you will be able to say, 'I remember your people very well. They probably don't remember me, but I can certainly remember how badly your people treated me. You're going to have to make a really attractive offer to make up for the way I was treated last time!'

Remember that senior staff in most progressive companies are always looking for good people, even if they don't have work immediately.

Make a point of telephoning the people you meet after 3–4 weeks to ask whether the company has won the contracts they were hoping to get and whether they are thinking of taking on new people to do the work.

Even if you are not offered employment or work experience, ask if someone could show you around the company so you can learn more about how the company works and their products.

If the response is, 'Sorry, we don't have anything at the moment', remember to follow up later, after another 3–4 weeks.

If the response is an outright rejection, ask for some comments or feedback on your approach to the company or your application letter or CV. Remember, if you are treated rudely or in an offhand manner, this is not necessarily because of something you have done wrong. It may just be a company or organisation that treats most people badly.

Here is some advice from a student on obtaining work:

a)   Get engineering-related work experience as early as possible, even if it is unpaid and only consists of sweeping the floor or making coffee. Long before I was given highly paid and interesting employment, I had worked for almost nothing in several other short-term jobs in engineering organisations.

b)   Use your contacts. All of them. Anyone may be able to help: your church, local pub, your dad's or mum's work colleagues, local sporting clubs, etc. If you can say, 'Brian Chew suggested I speak with you', you have a head start. That's how I got my first job working as a factory hand.

c)   Marks are not everything, in fact they count for much less than you might think. If you have adequate marks, the emphasis is on your personal, communication, and leadership abilities and experience.

d)   If your marks are not fantastic (i.e. you have failed many units), do not bother about advertised positions. You will have a much better chance with contacting smaller companies. Be persistent and do not include a copy of your results with your application letter and CV. However, be prepared to work for a small amount of money and perhaps even for free.

HOMEWORK

Even while you are looking for work, there is an enormous amount to be learnt. You don't have to be employed to start learning from experience. This section will require you to obtain information from many different people. Practice doing this. Learn how to develop a friendly relationship with people to build a degree of trust and confidence. Learn how to seek their willing collaboration to provide you with the information you need, and even volunteer information beyond what you asked for. The skills you develop to do this will serve you well on your way to becoming an expert engineer.

Obtain another notebook (size A4 or A5) and write your name and contact details on the front page. Record all your homework, observations, reflections, predictions, costs, and data. This notebook should be separate from your job-seeking notebook.

By spending two to three hours a day building your knowledge in your own local community, you will be developing valuable skills that will help you begin working at an engineering job more quickly.

Start by observing everything that has been built around you systematically. Although you can focus on the engineering discipline in which you have the most interest, it is important that you also observe products and infrastructure created by other engineering disciplines. You also need to observe the engineered services provided in your area including water supply, drainage and sanitation, electric power, roads, buildings, communications, and transport.

Take photographs of representative installations, clearly showing their state of repair, particularly if there are obvious faults. Also make a point of preparing sketches, from real life if possible, otherwise from photographs. Sketches can illustrate technical ideas much more clearly because you can eliminate visual clutter that cannot be removed from photographs.

Evaluate how well-existing products and infrastructure are meeting the needs of people. To do this, you need to ask the people who actually make use of them; it is not sufficient to rely on your own opinion. Keep your discussions informal and sociable, ask about interests such as which sporting teams they follow. Ask them about what they think about the value of the services and products they use. Keep notes about what people say in response to your questions. Prepare detailed records of any infrastructure that clearly needs to be repaired, particularly roads, drains, water supplies, and electricity connections. Even if it is working satisfactorily, note the condition of the infrastructure. Try and evaluate how much longer it will continue to work without the need for further maintenance. Whenever you find maintenance work being performed, ask permission from the maintenance workers to watch and learn about the work they are undertaking. Engage them in casual conversation; you will be surprised at how much you can learn.

Think about the products and services that you would need to make use of in an engineering job that you are likely to find yourself in. Search for information about these products and services on the Internet. Accumulate your own library of technical data sheets and service specifications, costs, delivery times, and applicable standards.

One of the attributes that impresses engineers who might be reviewing your application or interviewing you will be your ability to display your working knowledge of relevant standards. By networking with other engineers, you will be able to find out which standards are considered important in the industry that you would like to work in.

Workplace safety standards and local employment laws and regulations are some of the most important for engineers to know about. These are usually available from government websites.

If you are still a student, and your university provides access to online standards documents, make a point of downloading your own copies before you finish your studies. Some organisations that issue standards in electronic form provide PDF documents that 'self-destruct' after a time limit. There is a good reason for this; it is important that engineers always access the current version of the standard. It is all too easy to file away an outdated standard document and then use it by mistake. However, this can be annoying if you are a student and you want to retain your own copy to develop your working knowledge of relevant standards. Therefore, it is best to print your own copy and keep it as a printed document or convert it to a scanned PDF.

If you no longer have access to online standards, use your network. Most engineers will be happy to share at least the important parts of relevant standards to help you improve your knowledge.

Which standards apply in your industry will vary enormously from one country and setting to another. These are some of the major organisations that provide standards for engineers.

- ASME – American Society of Mechanical Engineers
- ASCE – American Society of Civil Engineers
- IEEE – Institute for Electrical and Electronic Engineers (USA)
- ASTM – American Society for Testing Materials
- ASHRAE – American Society for Heating Refrigeration and Air-conditioning Engineering
- DIN – German national standards authority
- DNV – Norwegian national standards authority – standards used widely in the offshore oil and gas industry
- API – American Petroleum Institute – standards used widely in the oil and gas industry
- IEC – International Electrotechnical Commission
- SAI Global – source for Australian Standards
- ISO – International Standards Organisation
- BSI – source for British standards

Teach yourself how to use macro programming for software such as Microsoft Office, Visual Basic, and Acrobat. Word, Excel, and Outlook all support macro programming that enables you to automate many tasks and prepare intelligent document templates. Many engineering organisations make extensive use of macros in routine engineering documents to ensure that engineers deal with all of the issues required for a particular task. If you can demonstrate that you have been able to learn macro programming, even at a basic level, you will significantly improve your job prospects in an engineering organisation. Remember that engineers tend to write software in languages such as Visual Basic for Excel more often than the languages that you may have learnt at university such as C, C++, Java, and Matlab. Another useful programming environment to master is 'Python', which enables you to make many computerised tasks fully automatic. All the software you need is readily available at a modest cost. You can find extensive self-study materials online.

Through local contacts, news media, and directories such as 'Yellow Pages', learn about small contracting firms in which you would need to perform various kinds of support work needed in engineering. This includes earthmoving and excavation, cable laying, security systems, fencing, lighting, and transport. Learn about material suppliers and tool suppliers. Visit all of these firms to establish personal contacts. Either obtain business cards and note the details, or write down names and telephone numbers and, if possible, e-mail addresses.

Find out how to perform due diligence checks on companies through firms such as Dunn and Bradstreet.

Before you engage the services of a contractor, you will need to find out about their financial status, whether they have the financial capacity to perform the work. Remember that engineering often requires money to be spent long before the benefits are obtained. The same applies at all levels of an engineering enterprise. A contractor will need to spend time and money before receiving payment for the work. If they do not have sufficient time or financial resources (such as the ability to borrow the necessary funds from a bank) they will not have sufficient credit to obtain the materials and components they need to be able to perform the work, even to pay their own subcontractors. On top of that, they may go bankrupt and leave you with half completed work that, legally, still belongs to the people to whom they owe money, not your enterprise. A bankrupt contractor leaves you with additional costs and delays. Therefore, monitoring the financial capacity of contractors is a necessary part of engineering. Sometimes it is necessary to persuade contractors to submit invoices for payment early so they will receive money in time. Many contractors, particularly small firms or individual workers, can provide excellent technical results but cannot manage their financial affairs well.

By securing a good understanding of the costs of labour and materials, you can pick up the background knowledge needed for this. For example, visit a hardware store and learn about the costs of all the items needed to complete a house: pipes, taps, toilets, doors, hinges, door handles, and electrical fittings like lights, power outlets, ceiling and ventilation fans, circuit breakers, and wiring. Develop a comprehensive list of all the items and calculate the cost of the components and materials. Find some building contractors and learn about the cost of the labour needed to install these components and materials, and don't forget the cost of hiring someone to supervise the work. This takes time and effort, but you will find the exercise valuable practice for your engineering career.

Watch construction or other engineering-related activities that you can see from your neighbourhood. There may be trains on a nearby railway or a port with ships being loaded or unloaded. Every fast food restaurant or bakery represents a manufacturing operation from which you can learn much about engineering practice. Learn to make accurate forecasts. For example, estimate

i)   time and distance needed for a train to come to a stop at a station (or a bus at a bus stop),

ii) time needed for the doors to open,

iii)   time needed for a given number of passengers to leave the train,
iv)   time needed for a given number of waiting passengers to enter the train,
v)   additional waiting time,
vi)   time needed for the doors to close,
vii) time and distance needed to accelerate to the normal running speed, and
viii) total space needed.

Learning to make relatively accurate predictions quickly, without having to rely on detailed calculations, helps you further develop abilities you will need as an engineer.

Learn about logistics by working in a bakery or fast food restaurant. Find out how much of each material the enterprise needs to purchase in advance, how it is stored, how long it can be stored, how much is used, how much has to be discarded because it is no longer fit to be used. Learn how the manager figures out what needs to be done and when so that the bread or cooked food is ready just when customers arrive.

Learn about the economics of the business. Learn how much the materials cost and how much the labour costs. Remember all the components of labour costs, not just what each person is paid, but also taxes, the cost of the building and space they need to work in, the cost of having someone to supervise, manage, and train people, the cost of someone to clean up waste material, insurance, the cost of administration and accounting, and all the other items that form part of the cost of employing people. Learn approximately what the customers pay for the baked bread or ready-to-eat food.

All this is about learning to be observant.

Watch for and anticipate breakdowns. Learn to anticipate when something is going to fail. It might be a toilet cistern in the home, or a loose piece of roofing that will blow away the next time strong winds come. It might be overhead electrical wires or poles that have corroded or a noisy wheel bearing on a car or another vehicle.

While you are waiting for the failure to occur, calculate the likely cost of the failure. This requires more forecasting ability. Anticipate the disruption likely to be caused by the failure, for example, the cost of repairs and the cost of not having the service performed by the failed part. Anticipate the time needed to get repairs organised and performed. Anticipate what other activities will have to stop or be relocated while the repairs are being performed. Calculate the cost of providing the same service by other means, including any extra travelling time. If there is unpaid labour or extra travelling or waiting time, calculate the cost of this time. A rule of thumb for the value of unpaid time is between half and two-thirds of the money that person would earn if they were working at the time. Child labour can be assessed at between half and two-thirds of the local earning rate for women. Consider several possible times at which the failure might occur, for example at the worst possible or least inconvenient times, and see what difference this makes to the cost.

Next, calculate the total cost of making the repair at the most convenient possible time. If this cost is less than the cost of failure, even at the least inconvenient time, talk to the owner about making the repair or replacement before failure happens. Talk to someone with experience with this kind of repair work to confirm your estimates.

By now, if it is a relatively small cost, you may be able to at least organise the repairs or even perform the repairs yourself. This might involve asking someone to lend you the money needed to buy the materials, if not paying for the work to be done.

Another way to do this is to arrange with the client to pay for materials and labour as the materials actually arrive and the work is done.

Once the repairs are complete, arrange to have them inspected by someone with experience as a final check.

If you manage to do this, congratulations! You are now almost in business as a self-employed engineer.

Look back at what you have done.

You identified the need, predicted what needed to be done, how much it would cost, what the benefits would be, and you negotiated with the client to get the work done. You arranged the finance, materials, and possibly the labour. You arranged for the work to be performed and for the results to be checked.

All that's left is to get the client to pay you enough to cover the cost of doing the work, along with a modest payment for the time and trouble you have taken to get everything organised.

As long as you manage to get paid, you are really in business. At the same time, you have learnt all the basic elements of conducting engineering work. With practice, you can learn different ways to arrange the payment. One way is to arrange for the client to pay for each item, with an additional percentage to cover the cost of your time and effort to make the arrangements. Another is to forecast the likely cost and negotiate for the client to pay a fixed amount when everything is completed, making sure that you build in a reasonable amount for your time and effort, and also for accepting the risk that you can forecast the actual cost accurately and get the client to pay on time!

You will find that even these small undertakings take lots of time to organise, especially at the start. It will be difficult to earn a reasonable income at first, but not impossible, provided that you always remember to build in the cost of your time.

Never take on work unless the client is willing (and able) to pay!

Remember to keep detailed records so you can compare the predictions you made with the actual results. Learn to improve your predictions with practice.

Seek feedback from your clients as well. Once you have completed two or three small repair jobs, go back and check to make sure that the clients are happy with the results. If you make a mistake, apologise, arrange for someone more capable to perform the work if needed, and do your best to restore the client's trust and confidence in your abilities.

You might be reading this and see yourself working in advanced technology, aerospace, or software engineering; you may well be telling yourself, 'This kind of basic repair work is not the kind of thing I need to learn about for that kind of engineering. I would be wasting my time. I will be better off reading books about high strength materials, electron beam machining, or operating systems'.

You might be better off doing that. It is possible.

However, having been in just that situation, waiting months for a job in the aerospace industry, I can tell you that learning about the business of engineering will be highly rewarding. When I eventually started work in aerospace engineering I learnt very quickly that the most advanced technology engineering (which I was fortunate enough to have a small part in) depends on some of the simplest and most basic engineering at the same time. Complex systems usually fail for the simplest possible reasons, precisely because it is so easy to overlook the cost (or space and time needed) to perform a simple repair or modification.

Remember to continue your search for engineering jobs at the same time. Prioritise your time so you learn basic engineering skills at the same time.

Learning about basic engineering, and the skills you need to perform it well, will always help you start your career as an expert engineer in any field or discipline. Practice these skills by improving your ability to observe accurately, listening to and learning from clients and people with experience, making accurate predictions about technical performance, costs, and benefits, negotiating for payment, and writing about your experiences. By all means, read interesting books, preferably out loud (to the wall in your room) to improve your ability to write well.

With time, practice, and effort, your homework will either lead to a business that you can build for yourself or provide you with some of the most valuable learning experiences possible with which to build your career as an expert engineer.

NOTES

1. Pronounced 'rezioomay'. There are many books that can help you with writing one (e.g. Millar 2011).
2. The material on the hidden employment market has been adapted (by permission) from contributions by Ginetta Papaluca, CareerConnect – Fremantle Career Development Centre, http://www.challengertafe.wa.edu.au/careers.

Chapter 15

Conclusion

Figure 15.1 below is one of the ways I have summarised the ideas in this book. Expert performance in engineering practice, in its essence, requires a combination of technical and financial foresight and planning as well as the technical collaboration performances required to convert plans into reality. These layers in turn are supported by three foundations: engineering and business science, tacit ingenuity, and accurate perception skills. Unfortunately, our current education systems focus on only the first of these three foundations, as represented in the diagram by shading to indicate the extent to which each underpinning aspect is addressed. While tacit ingenuity can accumulate through education, it is not encouraged by appropriate incentives. Mass education has attenuated the possibilities for students to develop effective interpersonal skills. As Chapter 6 explained, expert performances are not possible without the three fundamental perception skills of listening, reading, and visual perception, which are hardly addressed at all in formal education today.[1]

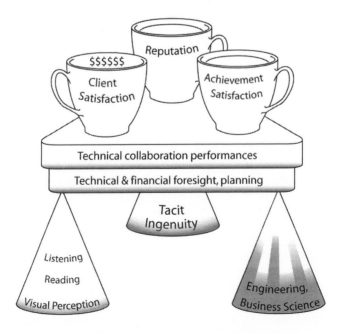

Figure 15.1 Engineering practice.

This platform in turn supports the three main outcomes of expert practice: reputation, client satisfaction, and intrinsic achievement satisfaction. Client satisfaction, of course, brings financial rewards. Reputation brings clients. And achievement satisfaction keeps you going when clients or external circumstances let you down.

What becomes clear from the research is the extent to which engineering practice relies on human performances, particularly collaboration performances. While engineering science and even many aspects of business science are universal, human performance is inevitably influenced by culture and location. Therefore, outcomes from engineering practice are also dependent on culture, setting, and location in ways that also depend on the choice of technologies and organisations.

LEARNING FROM THIS BOOK

In the introduction to this book, I set out an ambitious agenda. I remain confident that with a sufficient number of expert engineers, the sustainability of human civilisation can be assured and the degrading poverty in which billions of people still live can be mostly eliminated. However, writing a book is only a very small step towards achieving these goals.

I have written this book with the expectation that it will help many engineers learn how we can improve engineering performances, knowing that it is possible for anyone to become an expert engineer whose performances are recognised as outstanding by their peers. One of the aims of the book has been to provide terminology to describe the majority of what we actually do, which is the part that many engineers still dismiss as 'not real engineering, not design, calculations, and technical problem-solving.' Chapter 7 explains how engineering relies on collaboration, without which engineers can contribute little of value. The terminology and maps provided can help navigate these spaces that today remain confusing and challenging for many engineers.

The need for this book became apparent when I tried explaining our research to small groups of engineers. While some of the concepts were readily grasped, many remained out of reach, partly because at that time I only had access to half-formed explanations. However I came to realise that these engineers had developed deeply seated beliefs that seemed to be blocking their ability to learn. Imagining that more experienced engineers would have come to appreciate how misplaced these notions are, we collected survey responses from 160 engineers with 1–25 years of experience. Analysis showed that these beliefs were remarkably robust and did not significantly change with experience.[2] On this basis, therefore, it is sufficient to refer to these assumptions as misconceptions and accept that the realities of engineering practice as revealed by research studies may not align well with the beliefs of many practising engineers.

Examples of these beliefs examined in light of research evidence in Chapter 3 include the following:

- Engineering is a hands-on occupation.
- Engineers are naturally logical (relative to other people).
- Facts are more objective when stated in terms of numbers.
- Engineers are problem solvers (relative to other people).

Why do these beliefs seem to be so persistent?

First, it is not easy to learn about engineering practice. Research studies on the realities of engineering practice are still rare compared with all other fields of enquiry relevant to engineering. There are currently only about 20 researchers around the world working on such studies.[3] The ones that are available have appeared in diverse publication media rarely encountered by practising engineers or engineering educators. Even for those actively seeking knowledge, these publications are hard to find.

Second, these beliefs appear to be perpetuated from one generation of engineers to another without being questioned. Several of these premises, such as the primacy of design and technical problem solving, may also have been inadvertently reinforced by philosophical discussions on engineering,[4] studies on the beliefs of engineering educators,[5] or books that have attempted to describe professional practice for aspiring students, based largely on the views of engineering educators.[6]

I expect that some readers will have difficulty accepting the research evidence presented in this book that exposes these misconceptions. However, it is my hope that enough readers will be persuaded that these beliefs need to be questioned, and that it will therefore be possible to overcome these learning barriers.

As explained in Chapter 8, any teaching performance has to take prior learning into account. These misconceptions present rocky ground upon which to start cultivating an expert engineering performance. Presenting as much of the research evidence as is currently available can help provide nutrition for the learning that will be needed. Practice concepts in each chapter of this book provide a framework for learners to start appropriating this evidence and building new ways for them to understand engineering practice.

While there are many engineers who are comfortable learning for themselves by reading books like this, I expect that the vast majority of engineers will need more help. Assuming that the material in this book is of potential value to these engineers, it is worth raising the question as to how this help could be provided. Beyond self-study there are four possibilities:

- undergraduate engineering education,
- full-time postgraduate engineering education,
- part-time postgraduate engineering practice education, and
- workplace learning supervised by mentors.

At this point it is worth remembering that until the 1980s many, if not most, engineers started their careers in public sector enterprises like public works departments, defence force engineering organisations, or power and water utilities. Most of my peers did just that, and complained bitterly about endless bureaucracy and paperwork as they learnt about specifications, contracts, maintenance, and running small projects that seemed insignificant. In retrospect, with the help of the research leading to this book, we can see that they were learning engineering practice without necessarily being aware of what they were learning. Urgent research is needed before we lose this generation of engineers to understand more about their experiences and subsequent practice. By the end of the 1990s, most governments had decided to outsource their engineering work to private sector organisations and it is now apparent that valuable learning opportunities may have been lost as a result.

I have reservations about using this book for undergraduate or even full-time postgraduate engineering courses. There may be too many obstacles.

First of all, there are so few engineering educators who have sufficient understanding of practice to understand why the issues in this book are important. Like other aspects of engineering, engineering practice teaching needs to be led by researchers who study engineering practice. As explained earlier, there is currently only a handful of researchers studying this topic.

Not only is there a lack of research-based understanding, there is also a critical lack of experience of practice in engineering schools. It might be helpful to contrast engineering with medicine. Practically all the teaching in medical schools beyond the halfway point in the curriculum is delivered by people who practice medicine every day in the wards of teaching hospitals. In contrast, few if any engineering educators have extensive experience of engineering practice beyond the confines of research laboratories.[7]

Next, university performance incentives reward individual intellectual efforts: grades for students and publications for educators in leading research-based universities. In contrast, as we have seen through the research results, engineers work in an environment where success is determined not by our individual intellectual performances, but by the way our technical insights enable us to influence others to achieve consistent results that match expectations. As long as performance incentives in formal education remain so misaligned with those that apply in engineering practice, learners will be faced with implicit contradictions between what they are trying to learn and how they are rewarded for that learning.

Likewise, it is difficult to break away from the supremacy of the written word in formal education. Mass education that characterises universities today requires a focus on written forms of learning assessment that contrast sharply with the largely verbal culture and unwritten knowledge that characterises engineering practice. There are alternatives. Performances can be assessed, although such assessment can be more time-consuming. Instead, tacit knowledge accumulation from laboratory experiences can be assessed using online testing instruments, the results of which predict abilities such as troubleshooting and fault diagnosis much better than conventional written assessments.[8]

Finally, and particularly in undergraduate and full-time postgraduate engineering education, engineering practice would most likely comprise a small and isolated component of a curriculum dominated by traditional engineering science studies. Unless these latter studies can incorporate the cooperative learning environments that characterise engineering practice, where social interactions are taught and practised, staff and students will develop similar resistance to engineering practice studies as they do so often today with communication studies.[9] Engineering practice is likely to be best learnt in an environment that encourages distributed cognition that mirrors workplace practice, as explained in Chapter 5 pages 140–146 and Chapter 8 pages 276–277. The changes needed to achieve this are unlikely, given that current performance indicators for engineering educators emphasise scientific research.[10]

It is possible that part-time postgraduate studies for engineers who are practising at the same time could have a greater chance of success. Students in this environment might also have the opportunity to perform small research studies to expand our knowledge of engineering practice by learning to observe their own practice and that of their workplace peers. These research studies would then reinforce coursework learning, even though the coursework learning would have to be assessed largely through written responses. The online appendix describing the research methods provides details on how these studies could be performed.

I have, however, come to the conclusion that our best educational opportunities lie in the workplace. Senior engineers who have been able to understand the material presented in this book might have the best chance of educating less experienced engineers.[11] It is my hope that by providing terminology and a theoretical framework in each chapter, it will make it easier for them to explain the ideas in this book. Chapter 8 includes material on effective teaching that should be helpful. Senior engineers, acting as mentors, can then observe whether and how the younger engineers are responding by listening carefully to the ways in which they reflect on what they have learnt and how it has worked for them in their practice.

FURTHER RESEARCH STUDIES

There are many ideas in this book for which the evidence from our research is still relatively tenuous. I hope that researchers will find intriguing challenges in testing many of these ideas.

The research studies that led to this book attempted to answer the question 'What Do Engineers Really Do?' in many different settings and contexts.[12] The description of the research methods in the online appendix includes references to other published studies that have helped inform this book, and new studies continue to appear.[13] There are many other settings in which engineering is practised that are waiting to be examined by researchers. These include chemical, environmental, petroleum, geotechnical, and military engineering. Engineering consulting firms also need to be examined more closely, although there are several informative studies to provide starting points.[14] Engineering in cultural contexts such as China, Russia, Germany, Africa, the Middle East, Iran, Indonesia, and South America needs research to build a more broadly based understanding of how regional cultures influence technical collaboration, and how engineers might achieve more consistent results in those settings.

The next logical question is 'Why do engineers do what they do?' That will embrace issues such as value frameworks, identity, motivation, sense making, incentives, and workplace culture. If it turns out that we would like engineers to behave differently, perhaps with different value frameworks, the next question would then be how we might begin to achieve such a transformation.

One of the most promising developments has been the emergence of engineering education as a field of research because a significant number of engineering educators are learning to use and appreciate social science research methods.[15] These people have therefore acquired many of the necessary tools to pursue further research on engineering practice. At the same time, however, it is unfortunate that so many studies on engineering education take our knowledge of engineering practice for granted.[16] There have been several bold calls for reform in engineering education in recent decades, all based on similarly tenuous understandings of engineering practice.[17] I have argued that the slow progress of reform might reflect an instinctive apprehension among other engineering educators that there are significant gaps in the arguments. In particular, there is our lack of understanding of engineering practice and how it is influenced by educational approaches. In this book, I have drawn attention to some issues that could be further investigated by researchers to improve that understanding.

For example, several of the misconceptions presented in this book may stem from statements repeated by engineering educators or gaps in textbook accounts of engineering practice provided to students. Since these misconceptions have the potential to interfere with learning the realities of the practice, it would be valuable for rigorous research to test whether this is the case or not. It would also be useful to understand how we might better educate undergraduate students in terms of their technical collaboration performances. Problem-based learning (PBL) approaches founded on limited scope engineering projects have the potential to engage undergraduate students in authentic learning environments.[18] It would be beneficial to understand how far student conceptions of practice shift as a result of these pedagogies.

Finally, the links between engineering education and subsequent practice need to be explored in more detail across many different cultural settings.

ON GENDER IN ENGINEERING AND WHY WE DO ENGINEERING

Some readers may notice a prominent gap in the treatment of engineering practice in this book: the gender issue. It is clearly understood that female engineers fare less well than their male counterparts, particularly in Europe, Australia, Canada and America.[19] Participation rates seem to be much higher in some Middle Eastern countries such as Turkey and Iran. Despite decades of research, the reasons for this difference remain as unclear as the efficacy of interventions to help more women succeed as engineers.

I have not addressed the gender issue in this book for two reasons beyond the lack of clear conclusions from research on this issue.

First, we did not observe any significant differences in the way that women practice engineering as opposed to men. We included women as participants in our research on engineering practice, in roughly the same proportion as they participate in the relevant practice settings. This finding might seem to conflict with Fletcher's study that revealed women performing more 'relationship' and 'community-building' work.[20] However, the emphasis of our research was more on the kinds of work that engineers perform, and less on how much of each particular kind of work women perform. Fletcher's study provided only limited data on the full spectrum of the work being performed by her participants. Several quantitative studies with large samples have demonstrated strong gender differences in employment outcomes for female engineers, of course. Women typically have less responsibility, supervise fewer engineers, often work in positions requiring more relationship skills, and are paid less for similar work.[21]

The second reason is that the need for an improved understanding of engineering practice seems equally significant for both male and female engineers.

That being said, the research for this book has led to some questions that may interest gender researchers.

Chapters 1 and 11 discussed what seems to be a critical gap in the understanding of practice by many engineers, and in nearly all of those who participated in our studies. Few engineers seem to have developed more than a very superficial understanding of the reasons why engineering is valuable in human society, particularly its economic value. In contrast, the social and economic values of other professions such as teaching, medicine, and law, are widely appreciated.

In most Western countries over the last 5–6 decades, women have all but eliminated the historic gender gaps in professions like architecture, law, teaching, and medicine. Why not engineering? It is possible that these two issues are related – the lack of a clearly understood social purpose of a particular profession and the extent to which women pursue that profession.

It may be worth asking whether women, perhaps unconsciously, tend to assert their presence in professions where the social benefits are more clearly understood. The relative lack of understanding about the social benefits of engineering among engineers may well be driving women away from the profession, and undermining the motivation for women to study physical sciences and mathematics in school to open up admission to engineering courses. Perhaps, in countries like Iran and Turkey, women have internalised a clearer understanding of the social benefits of engineering, such as nation-building or improving access to basic services such as water, sanitation, and transport.[22]

In the meantime, professional engineering associations in Western countries could use the material in this book to improve the ways they describe the contributions of engineers in their societies. An improved understanding of the social and economic value and specific benefits of engineering investment would be helpful for all engineers, judging by the results of our research. Above all, a clearer understanding of engineering practice might help overcome the identity crisis that lies at the heart of contested engineering education priorities.[23]

THE CHALLENGE AHEAD – REGAINING RESPECT FOR ENGINEERING

In many countries, engineers have lost the respect with which they were once held. Evidence for this appears regularly in correspondence reproduced in professional engineering media. Engineers seek to regain what they see as lost respect by advocating the use of special titles, extending education qualifications, or tightening entry requirements for professional associations. Engineers complain that they are underpaid, that they are passed over for promotion by others without engineering qualifications, that they find themselves working for managers without engineering qualifications or knowledge, and that their opinions are not sought on matters with engineering implications.

In the course of the research for this book, I have also heard from frustrated business owners, senior managers and government ministers who have only limited respect for the ability of engineers to deliver valuable outcomes.

Reflecting on the knowledge of practice, some of which is described in this book, knowledge that has been invisible, even lost, and also the less than satisfactory recent record in delivering project results that match expectations as described in Chapter 11, engineers need to rebuild respect among investors, clients and governments. As explained in Chapter 1, engineers' remuneration, also a mark of respect, is strongly influenced by "marginal product" or value creation. The evidence collected in our research suggests that most engineers could benefit from a much clearer understanding on how their work contributes value for their clients and society at large. While more research is needed, as outlined above, it is likely that value perceptions strongly influence the "intuitive choices that engineers make through their working days on what they think they need to do and when to do it. A clearer understanding that takes into account the value perceptions of investors and clients, as well as the wider community, might help engineers achieve outcomes that help to regain the respect that contemporary generations of engineers have been yearning for. Mastery of the performances described in Chapters 7–12: technical collaboration performances such as discovery learning, teaching, technical coordination, project management and negotiation will also help. Finally, we could recognise how improving our listening, reading, seeing and critical thinking skills, and giving more respect to the equally valid ways in which others think and see the world, might also go some way towards regaining the respect we seek from others.

CONTINUING THE CONVERSATION

I hope that this book has met at least some of the expectations that I set out in the introduction and that its contents help you enjoy a more rewarding engineering career as a result.

Chapter 6 explains that much like listening and visual perception, accurate reading demands a conversation. Although I cannot be present for most of you while you are reading this book, I look forward to hearing more from you, particularly on aspects of the book that need improved explanations. If you achieve some success that you think this book might have contributed towards, I would be grateful to hear about it. By the time this book is released, there will be an interactive website available to facilitate and continue these conversations.

NOTES

1. Lucena & Leydens (2009).
2. Robinson (2013).
3. Much of their work has already been cited in earlier notes, with more in the appendix (online) describing our research methods. Of course, there are large communities of researchers working on related issues such as design studies, project management, and studies of scientists and their interactions with wider societies. Perhaps because most engineers are much harder to observe, being located far from university campuses and often behind walls to safeguard commercial or defence secrets, studies examining all aspects of what engineers do are comparatively rare, and tend to be conducted by researchers who have little contact with undergraduate engineering schools. Additionally, many studies on design, for example, have focused on the cognitive processes involved in design, rather than asking what designers spend their time doing.
4. E.g. Koen (2009).
5. E.g. Pawley (2009); Sheppard, Colby, Macatangay & Sullivan (2006).
6. E.g. Beakley, Evans & Keats (1986); Burghardt (1995); Hansen & Zenobia (2011); Jensen (2005); Wells, (1995); Wright (2002); Yuzuriha (1998).

7. Cameron and his colleagues performed a detailed study in Australia (Cameron, Reidsema & Hadgraft, 2011). Anecdotal evidence reveals a similar situation in both research-based and teaching-specialised engineering schools around the world.

8. Tacit knowledge measuring instruments emerged from a dispute among psychologists searching for psychometric tests that predict job performance (Cianciolo et al., 2006; Sternberg et al., 1995). Razali adapted these techniques and demonstrated how they could be used to assess tacit knowledge accumulation in engineering education (Razali & Trevelyan, 2012; Trevelyan & Razali, 2011).

9. Emilsson & Lilje (2008); Paretti (2008).

10. Quinlan provides an extended discussion on this (2002).

11. Workplaces can provide rich opportunities for learning with appropriate support (Marsick & Watkins, 1990; Rainbird, Fuller & Munro, 2004; Rooney et al., 2014).

12. Research studies were conducted in Pakistan, India, Brunei, and Australia (Domal, 2010; Tan & Trevelyan, 2013; S. Tang & Trevelyan, 2009; S. S. Tang, 2012).

13. E.g. Henriksen (2013); Johri (2012); Stevens, Johri & O'Connor (2014); B. Williams et al. (2014).

14. Tan's study includes a comprehensive reference list (Tan, 2013).

15. Heywood's book and more recently Johri and Olds' *Handbook of Engineering Education Research* provide comprehensive accounts of the field (Heywood, 2005; Johri & Olds, 2014). Jørgensen (2007) provides a history of engineering education.

16. For example Borrego et al. open their paper with 'Teamwork is the predominant mode of engineering professional practice', citing references to accreditation criteria rather than studies of engineering practice. Accreditation criteria tend to reflect understandings of practice by engineering educators which, as we have seen, are tenuous at best. An examination of engineering practice studies might have prompted the authors to question their opening statement. Similar claims with respect to certain team design projects are made by exponents of the CDIO (Conceive-Design-Implement-Operate) curriculum (Crawley, Malmqvist, Östlund & Brodeur, 2007).

17. Trevelyan (2012c).

18. Kolmos & De Graaff (2007); Savin-Baden (2007).

19. Faulkner (2009); Hermwati & Luhulima (2000); Male, Bush & Murray (2009); Male, Chapman & Bush (2007); McIlwee & Robinson (1992); Mills, Mehrtens, Smith & Adams (2008); Powell, Bagilhole & Dainty (2009).

20. Fletcher (1999).

21. Mills et al. (2008).

22. Lucena (2005, 2010).

23. Rosalind Williams pointed to an identity crisis within engineering in her provocative essay (2003).

Guide to online appendices

All the appendices for this book are available at the online website for this book provided by the publisher.

Research methods: Why we can trust the results

Interview questionnaire

List of participants

Classification of engineering activities and specialist knowledge

Classification of practice concepts and misconceptions

Glossary of confusing engineering terms

A guide to the language of accounting

Seeing is believing: colour illustrations

Listening skills worksheet

Learning to see by sketching

Life cycle costing guide

Practice Quiz 2 self-evaluation guide

Finding the beam loadings self-evaluation guide

Practice Exercise 9 self-evaluation guide

Practice Exercise 16 self-evaluation guide

Practice Exercise 19 self-evaluation guide

References

Ahern, A., O'Connor, T., McRuairc, G., McNamara, M., & O'Donnell, D. (2012). Critical thinking in the university curriculum – the impact on engineering education. European Journal of Engineering Education, 37(2), 125–132.

Akerlof, G. A., & Kranton, R. E. (2005). Identity and the Economics of Organizations. The Journal of Economic Perspectives, 19(1), 9–32.

Ambrose, S. A., Bridges, M. W., Lovett, M. C., DiPietro, M., & Norman, M. K. (2010). How Learning Works. San Francisco, California, USA: Jossey-Bass (Wiley).

American Society for Engineering Education (ASEE). (2013). Transforming Undergraduate Education in Engineering: Phase 1: Synthesizing and Integrating Industry Perspectives. Retrieved November 5, 2013, from http://www.asee.org/TUEE_PhaseI_Workshop Report.pdf

American Society of Civil Engineers. (2008). The Vision for Civil Engineering in 2025 (Vol. 2008, pp. 114). Reston, Virginia, USA: ASCE Steering Committee to Plan a Summit on the Future of the Civil Engineering Profession in 2025.

Anand, N. (2011). Pressure: The PoliTechnics of Water Supply in Mumbai. Cultural Anthropology, 26(4), 542–564.

Anderson, K. J. B., Courter, S. S., McGlamery, T., Nathans-Kelly, T. M., & Nicometo, C. G. (2010). Understanding engineering work and identity: a cross-case analysis of engineers within six firms. Engineering Studies, 2(3), 153–174.

Ansari, M. A., Kapoor, A., & Rehana, A. (1984). Social power in Indian organizations. Indian Journal of Industrial Relations, 20, 237–244.

Ashforth, B. E., & Mael, F. (1989). Social Identity Theory and the Organization. Academy of Management Review, 14(1), 20–39.

Ashforth, B. E., Sluss, D. M., & Saks, A. M. (2007). Socialization tactics, proactive behavior and newcomer learning: Integrating socialization models. Journal of Vocational Behavior, 70(2), 447–462.

Aster, R. (2008, March 1–8). Lessons learned from developing new engineering managers at JPL. Paper presented at the IEEE Aerospace Conference, Big Sky, Montana, USA.

Badawy, M. K. (1983). One More Time: How to Motivate Your Engineers. In R. Katz (Ed.), Managing Professionals in Innovative Organizations (pp. 27–36). Cambridge, Massachusetts: Ballinger.

Badawy, M. K. (1995). Developing Managerial Skills in Engineers and Scientists: Succeeding as a Technical Manager (2nd ed.): Van Nostrand Reinhold.

Bailey, D. E., & Barley, S. (2010). Teaching-Learning Ecologies: Mapping the Environment to Structure Through Action. Organization Science Articles in Advance, 1–25.

Bailey, D. E., Leonardi, P. M., & Barley, S. (2012). The Lure of the Virtual. Organization Science, 23(5), 1485–1504.

Bailyn, L., & Lynch, J. T. (1983). Engineering as a life-long career: Its meaning, its satisfactions, its difficulties. Journal of Occupational Behaviour (pre 1986), 4(4), 263–283.

Bajaj, M. (2011). Application of Advanced Utility Metering in the Developing World. (B.Eng thesis), The University of Western Australia, Perth.

Bannerji, R. (2006). Comparison of options for distributed generation in India. Energy Policy, 34, 101–111.

Barley, S., & Bechky, B. A. (1994). In the Backrooms of Science: the Work of Technicians in Science Labs. Work and Occupations, 21(1), 85–126.

Barley, S., & Orr, J. (Eds.). (1997). Between Craft and Science: Technical Work in US Settings. Ithaca, New York: Cornell University Press.

Barley, S. R. (2005). What we know (and mostly don't know) about technical work. In S. Ackroyd, R. Batt, P. Thompson & P. S. Tolbert (Eds.), The Oxford Handbook of Work and Organization (pp. 376–403). Oxford: Oxford University Press.

Barnes, A. R., & Jones, R. T. (2011, October 2–5). Cobalt from Slag – Lessons in Transition from Laboratory to Industry. Paper presented at the 50th Conference of Metallurgists, Montreal.

BBC Money Programme. (2011). BP: $30 Billion Blowout. Retrieved February 14, 2014, from http://www.bbc.co.uk/programmes/b00vys0q; http://www.abc.net.au/4corners/content/2011/s3160312.htm

Bea, R. (2000). Human and Organizational Factors in the Design and Reliability of Offshore Structures. (PhD thesis), The University of Western Australia, Perth.

Beakley, G. C., Evans, D. L., & Keats, J. B. (1986). Engineering: An Introduction to a Creative Profession (5 ed.). New York: Macmillan.

Bechky, B. A. (2003). Sharing Meaning Across Occupational Communities: The Transformation of Understanding on a Production Floor. Organization Science, 14(3), 312–330.

Becker, H. S., & Carper, J. (1956). The Elements of Identification with an Occupation. American Sociological Review, 21(3), 341–348.

Begel, A., & Simon, B. (2008, March 12–15). Struggles of New College Graduates in their First Software Development Job. Paper presented at the SIGCSE'08, Portland, Oregon, USA.

Bella, D. A. (2006). Organizational Distortions and Failures: A Method to Expose Them. Journal of Professional Issues in Engineering Education and Practice, 132(1), 18–23.

Bennell, P. (1986). Engineering Skills and Development: The Manufacturing Sector in Kenya. Development and Change, 17(2), 303–324.

Bennington, G., & Derrida, J. (1993). Jacques Derrida. Chicago: University of Chicago Press.

Bigley, G. A., & Roberts, K. H. (2001). The incident command system: high-reliability organizing for complex and volatile task environments. Academy of Management Journal, 44(6), 1281–1299.

Blandin, B. (2012). The Competence of an Engineer and how it is Built through an Apprenticeship Program: a Tentative Model. International Journal of Engineering Education, 28(1), 57–71.

Bloch. (2009). Cooperation is key: the operator's role in achieving equipment reliability. Cinde Journal: Uptime Magazine, 30(2), 18–20.

Blom, A., & Saeki, H. (2011). Employability and Skill Set of Newly Graduated Engineers in India. World Bank Policy Research Working Paper 5640. Retrieved June 17, 2011, from http://papers.ssrn.com/sol3/papers.cfm?abstract_id=1822959#

Bolton, R. (1986). People Skills. New York: Touchstone Books.

Borrego, M., Karlin, J., McNair, L. D., & Beddoes, K. (2013). Team Effectiveness Theory from Industrial and Organizational Psychology Applied to Engineering Student Teams: A Research Review. Journal of Engineering Education, 102(4), 472–512.

Boyd, T. (2013, December 20). Still lucky, but useless at settling projects. Australian Financial Review, p. 1.

Bransford, J. D., Brown, A. L., Cocking, R. R., Donovan, M. S., & Pellegrino, J. W. (Eds.). (2000). How People Learn: Brain, Mind, Experience and School. Washington DC, USA: National Academy Press.

Brown, A. L., Ash, D., Rutherford, M., Nakagawa, K., Gordon, A., & Campione, J. C. (1993). Distributed expertise in the classroom. In G. Salomon (Ed.), Distributed cognitions: psychological and educational considerations (pp. 188–228). Cambridge: Cambridge University Press.

Brundtland, G. H. (1987). Report of the World Commission on Environment and Development: Our Common Future (pp. 300). New York: United Nations.

Bucciarelli, L. L. (1994). Designing Engineers. Cambridge, Massachusetts: MIT Press.

Burghardt, M. D. (1995). Introduction to the Engineering Profession. New York: Harper Collins College Publications.

Busby, J. S. (2006). Failure to Mobilize in Reliability-Seeking Organizations: Two Cases from the UK Railway. Journal of Management Studies, 43(6), 1375–1393.

Busby, J. S., Chung, P. W. H., & Wen, Q. (2004). A situational analysis of how barriers to systemic failure are undermined during accident sequences. Journal of Risk Research, 7(7–8), 811–826.

Busby, J. S., & Strutt, J. E. (2001). What limits the ability of design organisations to predict failure? Proceedings of the institution of Mechanical Engineers, 215 Part B, 1471–1474.

Button, G., & Sharrock, W. (1994). Occasioned practices in the work of software engineers. In M. Jirotka & J. A. Goguen (Eds.), Requirements Engineering: Social and Technical Issues (pp. 217–240). London: Academic Press Ltd.

Buzan, T., & Buzan, B. (1993). The Mind Map Book. London: BBC Books.

Cameron, I., Reidsema, C., & Hadgraft, R. G. (2011, December 5–7). Australian engineering academe: a snapshot of demographics and attitudes. Paper presented at the Australasian Conference on Engineering Education, Fremantle.

Carew, A., & Mitchell, C. A. (2002). Characterizing undergraduate engineering students' understanding of sustainability. European Journal of Engineering Education, 27(4), 349–361.

Carson, R. (1962). Silent Spring. Boston, USA: Houghton Mifflin.

Chilvers, A., & Bell, S. (2014). Professional lock-in: Structural engineers, architects and the disconnect between discourse and practice. In B. Williams, J. D. Figueiredo & J. P. Trevelyan (Eds.), Engineering Practice in a Global Context: Understanding the Technical and the Social (pp. 205–222). Leiden, Netherlands: CRC/Balkema.

Christiansen, F. V., & Rump, C. (2007). Getting it right: conceptual development from student to experienced engineer. European Journal of Engineering Education, 32(4), 467–479.

Cianciolo, A. T., Matthew, C., Sternberg, R. J., & Wagner, R. K. (2006). Tacit Knowledge, Practical Intelligence and Expertise. In K. A. Ericsson, N. Charness, P. J. Feltovich & R. R. Hoffman (Eds.), The Cambridge Handbook of Expertise and Expert Performance (pp. 613–632). New York: Cambridge University Press.

Coelho, K. (2004). Of Engineers, Rationalities and Rule: and Ethnography of Neoliberal Reform in and Urban Water Utility in South India. (PhD thesis), University of Arizona, Tucson.

Coelho, K. (2006). Leaky sovereignties and Engineered (Dis)Order in and Urban Water System. In M. Narula, S. Sengupta, R. Sundaram, A. Sharan, J. Bagchi & G. Lovink (Eds.), Sarai Reader 06: Turbulence (pp. 497–509). New Delhi: Centre for the Study of Developing Societies.

Cohen, A. R., & Bradford, D. L. (1989). Influence without authority: The use of alliances, reciprocity, and exchange to accomplish work. Organizational Dynamics, 17(3), 5–17.

Cohen, A. R., & Bradford, D. L. (2005). Influence Without Authority (2nd ed.). New York: John Wiley & Sons.

Collins, H. M. (2010). Tacit & Explicit Knowledge. Chicago: University of Chicago Press.

Collins, H. M., & Evans, R. (2002). The Third Wave of Science Studies: Studies of Expertise and Experience. Social Studies of Science, 32(2), 235–296.

Collinson, D., & Ackroyd, S. (2005). Resistance, misbehaviour, and dissent. In S. Ackroyd, R. Batt, P. Thompson & P. S. Tolbert (Eds.), The Oxford Handbook of Work and Organization (pp. 305–326). Oxford: Oxford University Press.

Cooper, R. G. (1993). Winning at New Products: Accelerating the Process from Idea to Launch (2nd ed.). Reading, Massachussetts: Addison-Wesley.

Cousin, G. (2006, December). An introduction to threshold concepts. Planet, 17, 4–5.

Crawley, E. F., Malmqvist, J., Östlund, S., & Brodeur, D. R. (2007). Rethinking Engineering Education: The CDIO Approach. New York: Springer.

Cropanzano, R., & Mitchell, M. S. (2005). Social Exchange Theory: An Interdisciplinary Review. Journal of Management, 31(6), 874–900.

Cross, J., Bunker, E., Grantham, D., Connell, D., & Winder, C. (2000). Identifying, monitoring and assessing occupational hazards. In P. Bohle & M. Quinlan (Eds.), Managing Occupational Health and Safety: a multi-disciplinary approach. Sydney: Macmillan.

Crossley, M. (2011). Business Skills in Engineering Practice. (B.Eng thesis), The University of Western Australia, Perth.

D'Arcy, P., & Gustafsson, L. (2012). Australia's Productivity Performance and Real Incomes. Reserve Bank of Australia Bulletin(June Quarter), 23–35.

Dahlgren, M. A., Hult, H., Dahlgren, L. O., Segerstad, H., & Johansson, K. (2006). From Senior Student to Novice Worker: Learning Trajectories in Political Science, Psychology and Mechanical Engineering. Studies in Higher Education, 31(5), 569–586.

Darling, A. L., & Dannels, D. P. (2003). Practicing Engineers Talk about the Importance of Talk: A Report on the Role of Oral Communication in the Workplace. Communication Education, 52(1), 1–16.

Darr, A. (2000). Technical Labour in an Engineering Boutique: Interpretive Frameworks of Sales and R&D Engineers. Work, Employment and Society, 14(2), 205–222.

Darr, A. (2002). The technization of sales work: an ethnographic study in the US electronics industry. Work, Employment and Society, 16(1), 47–65.

Davenport, T. H. (1997). If only HP knew what HP knows.

Davis, M. (1998). Thinking Like an Engineer. New York, USA: Oxford University Press.

Dekker, S. (2006). The Field Guide to Understanding Human Error. Aldershot, UK: Ashgate.

Domal, V. (2010). Comparing Engineering Practice in South Asia and Australia. (PhD thesis), The University of Western Australia, Perth.

Doron, R., & Marco, S. (1999). Syllabus Evaluation by the Job-analysis Technique. European Journal of Engineering Education, 24(2), 163–172.

Dowling, D. G., Carew, A., & Hadgraft, R. G. (2009). Engineering Your Future: Wiley.

Downey, G. L. (2009). What is engineering studies for? Dominant practices and scalable scholarship. Engineering Studies, 1(1), 55–76.

Draper, S. W. (2009a). Catalytic assessment: understanding how MCQs and EVS can foster deep learning. British Journal of Educational Technology, 40(2), 285–293.

Draper, S. W. (2009b). What are learners actually regulating when given feedback? British Journal of Educational Technology, 40(2), 306–315.

Duderstadt, J. J. (2008). Engineering for a Changing World: a roadmap to the future of engineering practice, research and education. Ann Arbor, Michigan.: The University of Michigan.

Dunsmore, K., Turns, J., & Yellin, J. M. (2011). Looking Toward the Real World: Student Conceptions of Engineering. Journal of Engineering Education, 100(2), 329–348.

Eccles, J. S. (2005). Subjective task value and the Eccles et al Model of Achievement Related Choices. In A. J. Elliot & C. S. Dweck (Eds.), Handbook of competence and motivation. New York: The Guildford Press.

Edwards, B. (2012a). Drawing on the Artist Within: Touchstone.

Edwards, B. (2012b). Drawing on the Right Side of the Brain (4th ed.): Tarcher.

Emilsson, U. M., & Lilje, B. (2008). Training social competence in engineering education: necessary, possible or not even desirable? An explorative study from a surveying education programme. European Journal of Engineering Education, 33(3), 259–269.

Engeström, Y. (2004). The new generation of expertise: seven theses. In H. Rainbird, A. Fuller & A. Munro (Eds.), Workplace Learning in Context (pp. 145–165). London: Routledge.

Eraut, M. (2000). Non-formal learning and tacit knowledge in professional work. British Journal of Educational Psychology, 70, 113–136.

Eraut, M. (2004). Transfer of knowledge between education and workplace settings. In H. Rainbird, A. Fuller & A. Munro (Eds.), Workplace Learning in Context (pp. 210–221). London: Routledge.

Eraut, M. (2007). Learning from other people in the workplace. Oxford Review of Education, 33(4), 403–422.

Eraut, M., Alderton, J., Cole, G., & Senker, P. (2000). Development of knowledge and skills at work. In F. Coffield (Ed.), Differing Visions of a Learning Society (Vol. 1, pp. 231–262). Bristol, UK: The Policy Press.

Ericsson, K. A. (2003). The Acquisition of Expert Performance as Problem Solving: Construction and Modification of Mediating Mechanisms through Deliberative Practice. In J. E. Davidson & R. J. Sternberg (Eds.), The Psychology of Problem Solving (pp. 31–83). New York: Cambridge University Press.

Ericsson, K. A. (2006). An Introduction to Cambridge Handbook of Expertise and Expert Performance: Its Development, Organization and Content. In K. A. Ericsson, N. Charness, P. J. Feltovich & R. R. Hoffman (Eds.), The Cambridge Handbook of Expertise and Expert Performance (pp. 3–19). New York: Cambridge University Press.

Ericsson, K. A., Krampe, R. T., & Tesch-Römer, C. (1993). The Role of Deliberate Practice in the Acquisition of Expert Performance. Psychological Review, 100(3), 363–406.

Evans, A. G. T. (2001). C. Y. O'Connor: His Life and Legacy: University of Western Australia Press.

Evans, R., & Gabriel, J. (2007, October 10–13). Performing Engineering: How the Performance Metaphor for Engineering Can Transform Communications Learning and Teaching. Paper presented at the 37th ASEE/IEEE Frontiers in Engineering Education Conference, Milwaukee, Wisconsin.

Ewenstein, B., & Whyte, J. (2007). Beyond Words: Aesthetic Knowledge and Knowing in Organizations. Organization Studies, 28(5), 689–708.

Faulkner, W. (2007). Nuts and Bolts and People. Social Studies of Science, 37(3), 331–356.

Faulkner, W. (2009). Doing gender in engineering workplace cultures 1. Observations from the field. Engineering Studies, 1(1), 3–18.

Felder, R. M., & Brent, R. (2007). Cooperative Learning. In P. A. Mabrouk (Ed.), Active Learning: Models from the Analytical Sciences (Chapter 4). Washington, D. C.: American Chemical Society.

Ferguson, E. S. (1992). Engineering and the Mind's Eye. Cambridge, MA, USA: MIT Press.

Fisher, R., & Ury, W. (2011). Getting to Yes: Negotiating an Agreement Without Giving. In: Penguin Books.

Fletcher, J. K. (1999). Disappearing Acts: Gender, Power, and Relational Practice at Work. Cambridge, Massachusetts: MIT Press.

Florman, S. (1976). The Existential Pleasures of Engineering. New York: St Martin's Press.

Florman, S. (1987). The Civilized Engineer. New York: St Martin Press.

Florman, S. (1997). The Introspective Engineer. New York: St Martin's Press.

Foucault, M. (1984). The Order of Discourse: Inaugural lecture at College de France December 1970. In M. Shapiro (Ed.), Language and Politics (pp. 108–138). Oxford: Blackwell.

Fuller, A., & Unwin, L. (2004). Expansive learning environments: integrating organizational and personal development. In H. Rainbird, A. Fuller & A. Munro (Eds.), Workplace Learning in Context (pp. 126–144). London: Routledge.

Gainsburg, J. (2006). The Mathematical Modeling of Structural Engineers Mathematical Thinking and Learning, 8(1), 3–36.

Gainsburg, J., Rodriguez-Lluesma, C., & Bailey, D. E. (2010). A "knowledge profile" of an engineering occupation: temporal patterns in the use of engineering knowledge. Engineering Studies, 2(3), 197–219.

Galloway, P. (2008). The 21st Century Engineer: A Proposal for Engineering Education Reform. Washington DC, USA: ASCE.

Gammage, B. (2011). The Biggest Estate on Earth: How Aborigines Made Australia. Sydney: Allen & Unwin.

Gherardi, S. (2009). The Critical Power of the 'Practice Lens'. Management Learning, 40(2), 115–128.

Gibbs, G., & Simpson, C. (2004). Conditions Under Which Assessment Supports Students' Learning. Learning and Teaching in Higher Education, 1, 3–31.

Godfrey, E. (2003). The Culture of Engineering Education and its Interaction with Gender: A Case Study of a New Zealand University. (PhD thesis), Curtin University, Perth.

Goldman, S. L. (2004). Why we need a philosophy of engineering: a work in progress. Interdisciplinary Science Reviews, 29(2), 163–178.

Goold, E. (2014). Mathematics in engineering practice: tacit trumps tangible. In B. Williams, J. D. Figueiredo & J. P. Trevelyan (Eds.), Engineering Practice in a Global Context: Understanding the Technical and the Social (pp. 245–279). Leiden, Netherlands: CRC/Balkema.

Gooty, J., Gavin, M., & Ashkanasy, N. M. (2009). Emotions research in OB: The challenges that lie ahead. Journal of Organizational Behavior, 30(6), 833–838.

Gorman, M. E. (2002). Types of Knowledge and Their Roles in Technology Transfer. Journal of Technology Transfer, 27(3), 219–231.

Gouws, L. (2013). The Extent and Nature of Problems Encountered in Maintenance: The Story Behind the Text. (PhD thesis), The University of Western Australia, Perth.

Gouws, L., & Gouws, J. (2006, 10–14 July). Common pitfalls with SAP™-based Plant Maintenance Systems. Paper presented at the 2006 World Congress on Engineering Asset Management, Surfer's Paradise, Queensland.

Gouws, L., & Trevelyan, J. P. (2006, 10–14 July). Research on influences on maintenance management effectiveness. Paper presented at the 2006 World Conference on Engineering Asset Management, Surfer's Paradise, Queensland.

Grant, K. P., Baumgardner, C. R., & Shan, G. S. (1997). The Perceived Importance of Technical Competence to Project Managers in the Defense Acquisition Community. IEEE Transactions on Engineering Management, 44(1), 12–19.

Grinter, L. E. (1955). Report of the Committee On Evaluation of Engineering Education (Grinter Report). Journal of Engineering Education, 44(3), 25–60.

Gupta, A., Sharda, R., & Greve, R. A. (2010). You've got email! Does it really matter to process emails now or later? Information Systems Frontiers (Online).

Guzzomi, A. L., Maraldi, M., & Molari, P. G. (2012). A historical review of the modulus concept and its relevance to mechanical engineering design today. Mechanism and Machine Theory, 50(1), 1–14.

Hager, P. (2004). The conceptualization and measurement of learning at work. In H. Rainbird, A. Fuller & A. Munro (Eds.), Workplace Learning in Context (pp. 242–257). London: Routledge.

Hägerby, M., & Johansson, M. (2002). Maintenance Performance Assessment: Strategies and Indicators. (MSc), Linköping Institute of Technology, Linköping.

Hamilton, W. (1973). Socrates, in Plato: Phaedrus & Letters Vii And Viii (W. Hamilton, Trans.). London: Harmondsworth: Penguin.

Han, A. S. (2008). Career development and management in the engineering industry. (B.Eng thesis), The University of Western Australia, Perth.

Hansen, K. L., & Zenobia, K. E. (2011). Civil Engineer's Handbook of Professional Practice. Hoboken, New Jersey, USA: ASCE Press, John Wiley & Sons, Inc.

Hardisty, P. E. (2010). Environmental and Economic Sustainability. Boca Raton, Florida, USA: CRC – Taylor & Francis.

Hargroves, K. C., & Smith, M. H. (Eds.). (2005). The Natural Advantage of Nations (Paperback ed.). London: Earthscan.

Hartley, S. (2009). Project Management: Principles, Processes and Practice (2nd ed.). Sydney, Australia: Pearson.

Henriksen, L. B. (2001). Knowledge management and engineering practices: the case of knowledge management, problem solving and engineering practices. Technovation, 21, 595–603.

Henriksen, L. B. (Ed.). (2013). What did you learn in the real world today? Aalborg, Denmark: Aalborg University Press.

Hermwati, W., & Luhulima, A. S. (2000). Women in Science, Engineering and Technology (SET): A Report on the Indonesian Experience. Gender, Technology and Development, 4(1), 87–100.

Heywood, J. (2005). Engineering Education: Research and Development in Curriculum and Instruction. Hoboken, New Jersey: John Wiley & Sons, Inc.

Heywood, J. (2011). A Historical Overview of Recent Developments in the Search for a Philosophy of Engineering Education. Paper presented at the Philosophy of Engineering and Engineering Education, Rapid City.

Heywood, J., Carberry, A., & Grimson, W. (2011). A Selected and Annotated Bibliography of Philosophy In Engineering Education. Paper presented at the Philosophy of Engineering and Engineering Education, Rapid City.

Hill, H. (2011). Corruption Harms Development? It's Not That Simple. Global Asia, 6(4).

Hobbs, A. (2000). Human Errors in Context: a Study of Unsafe Acts in Aircraft Maintenance. (PhD thesis), University of New South Wales, Sydney.

Horning, S. S. (2004). Engineering the Performance: Recording Engineers, Tacit Knowledge and the Art of Controlling Sound. Social Studies of Science, 34(5), 703–731.

Hoy, D. (1985). Jacques Derrida. In Q. Skinner (Ed.), The Return of Grand Theory in the Human Sciences (pp. 45–64). Cambridge: Cambridge University Press.

Hsiang, S. M., Burke, M., & Migual, E. (2013). Quantifying the Influence of Climate on Human Conflict. Science, 341(1235367), 1–14.

Hubert, M., & Vinck, D. (2014). Going back to heterogeneous engineering: the case of micro and nanotechnologies. In B. Williams, J. D. Figueiredo & J. P. Trevelyan (Eds.), Engineering Practice in a Global Context: Understanding the Technical and the Social (pp. 185–204). Leiden, Netherlands: CRC/ Balkema.

Huet, G., Culley, S. J., McMahon, C. A., & Fortin, C. (2007). Making sense of engineering design review activities. Artificial Intelligence for Engineering Design, Analysis and Manufacturing, 21(3), 243–266.

Hutchins, E. (1995). Cognition in the Wild. Cambridge, Massachusetts, USA: MIT Press.

Hysong, S. J. (2008). Technical skill in perceptions of managerial performance. Journal of Management Development, 27(3), 275–290.

ICC Standing Committee on Extortion and Bribery. (1999). Fighting Bribery: a corporate practices manual. Paris, France: International Chamber of Commerce.

Iqbal, S. T., & Bailey, B. P. (2006). Leveraging characteristics of task structure to predict the cost of interruption. Paper presented at the Proceedings of the SIGCHI conference on Human Factors in computing systems Montréal, Quebec, Canada. http://portal.acm.org/citation.cfm?id=1124882

Iqbal, S. T., & Bailey, B. P. (2008). Understanding changes in mental workload during execution of goal-directed tasks and its application for interruption management. ACM Transactions on Computer-Human Interaction (TOCHI), 14(4), Article 21.

IT Transport Ltd. (2002). The Value of Time in Least Developed Countries: Final Report (pp. 113).

Itabashi-Campbell, R., & Gluesing, J. (2014). Engineering problem-solving in social contexts: 'collective wisdom' and 'ba'. In B. Williams, J. D. Figueiredo & J. P. Trevelyan (Eds.), Engineering Practice in a Global Context: Understanding the Technical and the Social (pp. 129–158). Leiden, Netherlands: CRC/ Balkema.

Itabashi-Campbell, R., Perelli, S., & Gluesing, J. (2011, June 27–30). Engineering Problem Solving and Knowledge Creation: An Epistemological Perspective. Paper presented at the IEEE International Technology Management Conference, San Jose, CA, USA.

Jackson, T., Burgess, A., & Edwards, J. (2006). A simple approach to improving email communication. Communications of the ACM, 49(6), 107–109.

Jackson, T., Dawson, R., & Wilson, D. (2003). The Cost of Email Interruption. Journal of Systems & Information Technology, 5(1), 81–92.

Jacobs, R. J. (2010). An Investigation into the Practice of Engineering Estimating and Tendering. (B.Eng thesis), The University of Western Australia, Perth.

James, W. (1905). The Sentiment of Rationality. London: Longmans Green & Co.

Jensen, J. N. (2005). A user's guide to engineering. Upper Saddle River, New Jersey, USA: Prentice Hall College Publishing.

Johnson, C. (1997). Derrida. London: Pheonix.

Johnson, D. W., & Johnson, F. P. (2009). Joining Together: Group Theory and Group Skills (10th ed.). Englewood Cliffs, New Jersey, USA: Prentice Hall: Simon & Schuster.

Johnson, D. W., & Johnson, R. T. (1999). Making cooperative learning work. Theory into Practice, 38(2), 67–73.

Johnston, S., Lee, A., & McGregor, H. (1996). Engineering as a Captive Discourse. Techné: Journal of the Society for Philosophy and Technology, 1(3–4), (http://scholar.lib.vt.edu/ejournals/SPT/v1_n3n4/Johnston.html, accessed 8 Mar 2005).

Johri, A. (2012). Learning to demo: the sociomateriality of newcomer participation in engineering research practices. Engineering Studies, 4(3), 249–269.

Johri, A., & Olds, B. M. (2014). Cambridge Handbook of Engineering Education Research. New York: Cambridge University Press.

Jonassen, D., Strobel, J., & Lee, C. B. (2006). Everyday Problem Solving in Engineering: Lessons for Engineering Educators. Journal of Engineering Education, 95(2), 139–151.

Jørgensen, U. (2007). History of Engineering Education. In E. F. Crawley, J. Malmqvist, S. Östlund & D. R. Brodeur (Eds.), Rethinking Engineering Education: The CDIO Approach (pp. 216–240). New York: Springer.

Kahneman, D. (2011). Thinking, Fast and Slow. London: Penguin, Allen Lane.

Kass, M., Witkin, A., & Terzopoulos, D. (1988). Snakes: Active Conour Models. International Journal of Computer Vision, 1(4), 321–331.

Katz, S. M. (1993). The Entry-Level Engineer: Problems in Transition from Student to Professional. Journal of Engineering Education, 82(3), 171–174.

Kayaga, S., & Franceys, R. (2007). Costs of urban utility water connections: Excessive burden to the poor. Urban Policy, 15, 270–277.

Kendrick, T. (2006). Results Without Authority: Controlling a Project when the Team doesn't Report To You. New York: AMACOM Books: American Management Association.

Ketchum, L. (1984). Sociotechnical Design in a Third World Country: The Railway Maintenance Depot at Sennar in the Sudan. Human Relations, 37(2), 135–154.

Kidder, T. (1981). Soul of a New Machine. New York: Avon Books.

Kilduff, M., Funk, J. L., & Mehra, A. (1997). Engineering Identity in a Japanese Factory. Organization Science, 8(6), 579–592.

Klein, R. L., Bigley, G. A., & Roberts, K. H. (1995). Organisational Culture in Higher Reliability Organisations: An Extension. Human Relations, 48(7), 771–793.

Kletz, T. (1991). An Engineer's View of Human Error (2nd ed.). London: Institution of Chemical Engineers, VCH Publishers.

Knoke, D., & Yang, S. (2008). Social network analysis (2nd ed.). Thousand Oaks, California, USA: Sage.

Koehn, E., Kothari, R. K., & Pan, C.-S. (1995). Safety in Developing Countries: Professional And Bureaucratic Problems. Journal of Construction Engineering and Management, 121(3), 261–265.

Koen, B. V. (2009). The Engineering Method and its Implications for Scientific, Philosophical and Universal Methods. The Monist, 92(3), 357–386.

Kolmos, A., & De Graaff, E. (2007). Process of Changing to PBL. In E. De Graaff & A. Kolmos (Eds.), Management of Change: Implementation of Problem-Based and Project-Based Learning in Engineering (pp. 31–43). Rotterdam/Taipei: Sense Publishers.

Korte, R. (2007). A review of social identity theory with implications for training and development. Journal of European Industrial Training, 31(3), 166–180.

Korte, R. F., Sheppard, S. D., & Jordan, W. (2008, June 22-26). A Qualitative Study of the Early Work Experiences of Recent Graduates in Engineering. Paper presented at the American Society for Engineering Education, Pittsburgh.

Kotta, L. T. (2011). Structural conditioning and mediation by student agency: a case study of success in chemical engineering design. (PhD thesis), University of Cape Town, Cape Town.

Kraut, R. E., & Streeter, L. A. (1995). Coordination in Software Development. Communications of the ACM, 38(3), 69–81.

Lam, A. (1996). Engineers, Management and Work Organization: A Comparative Analysis of Engineers' Work Roles in British and Japanese Electronics Firms. Journal of Management Studies, 33(2), 183–212.

Lam, A. (1997). Embedded Firms, Embedded Knowledge: Problems of Collaboration and Knowledge Transfer in Global Cooperative Ventures. Organization Studies, 18(6), 973–996.

Lam, A. (2000). Tacit Knowledge, Organizational Learning and Societal Institutions: An Integrated Framework. Organization Studies, 21(3), 487–513.

Larsson, A. (2007). Banking on social capital: toward social connectedness in distributed engineering design teams. Design Studies, 28(6), 605–622.

Latour, B. (2005). Reassembling the Social: an Introduction to Actor Network Theory. Oxford: Oxford University Press.

LaToza, T. D., Venolia, G., & DeLine, R. (2006, May 20–28). Maintaining Mental Models: A Study of Developer Work Habits. Paper presented at the ICSE'06 – International Conference on Software Engineering, Shanghai.

Lawson, D. (2005). Engineering Disasters: Lessons to be Learned. London, UK: Professional Engineering Publishing Ltd.

Layton, E. T. (1991). A Historical Definition of Engineering. In P. T. Durbin (Ed.), Critical Perspectives on Nonacademic Science and Engineering (pp. 60–79). Cranbury, New Jersey, USA: Associated University Presses.

Lee, D. M. S. (1986). Intellectual, motivational, and interpersonal qualities and NOT academic performance predicted performance on first job. IEEE Transactions on Engineering Management, EM33(3), 127–133.

Lee, D. M. S. (1994). Social ties, task-related communication and first job performance of young engineers. Journal of Engineering and Technology Management, 11, 203–228.

Lee, G. L., & Smith, C. (1992). Engineers and Management: International Comparisons. London: Routledge.

Leonardi, P. M., Jackson, M. H., & Diwan, A. (2009). The enactment-externalization dialectic: rationalization and the persistence of counterproductive technology design practices in student engineering. Academy of Management Journal, 52(2), 400–420.

Leong Ching. (2010). Transformation of the Phnom Penh Water Supply Authority. Retrieved January 21, 2014, from http://www.iwawaterwiki.org/xwiki/bin/view/Articles/TransformationofthePhnomPenhWaterSupplyAuthority

Leveson, N. G., & Turner, C. S. (1993). An Investigation of the Therac-25 Accidents. Computer, 26(7), 18–41.

Lewicki, R. J., Saunders, D. M., & Barry, B. (2006). Negotiation. New York: McGraw-Hill.

Lionni, L. (1970). Fish Is Fish. New York: Scholastic Press.

Litzinger, T. A., Van Meter, P., Firetto, C. M., Passmore, L. J., Masters, C. B., Turns, S. R., . . ., Zappe, S. E. (2010). A Cognitive Study of Problem Solving in Statics. Journal of Engineering Education, 99(4), 337–353.

Love, P. E. D., Mistry, D., & Davis, P. R. (2010). Price Competitive Alliance Projects: Identification of Success Factors for Public Clients. Journal of Construction Engineering and Management, 136(9), 947–955.

Lucas, D., Mackay, C., Cowell, N., & Livingston, A. (1997). Fatigue risk assessment for safety critical staff. In D. Harris (Ed.), Engineering Psychology and Cognitive Ergonomics (Vol. 2, pp. 315–320). Aldershot, UK: Ashgate.

Lucena, J. C. (2005). Defending the Nation: US Policy-making to Create Scientists and Engineers from Sputnik to the War against Terrorism. Lanham, Maryland, USA: University Press of America.

Lucena, J. C. (2008). Engineers, development, and engineering education: From national to sustainable community development. European Journal of Engineering Education, 33(3), 247–257.

Lucena, J. C. (2010). What is Engineering for? A Search for Engineering beyond Militarism and Free-markets. In G. L. Downey & K. Beddoes (Eds.), What is Global Engineering Education For? The Making of International Educators, Part 1 (Vol. 1, pp. 361–384). London: Morgan and Claypool.

Lucena, J. C., & Leydens, J. A. (2009). Listening as a Missing Dimension in Engineering Education: Implications for Sustainable Community Development. IEEE Transactions on Professional Communication, 52(4), 359–376.

Lucena, J. C., & Schneider, J. (2008). Engineers, development, and engineering education: From national to sustainable community development. European Journal of Engineering Education, 33(3), 247–257.

Lucena, J. C., Schneider, J., & Leydens, J. A. (2010). Engineering and Sustainable Community Development. London: Morgan and Claypool.

Luzon, M. J. (2005). Genre Analysis in Technical Communication. IEEE Transactions on Professional Communication, 48(3), 285–295.

Lynn, L. H. (2002). Engineers and Engineering in the US and Japan: A Critical Review of the Literature and Suggestions for a New Research Agenda. IEEE Transactions on Engineering Management, 49(1), 95–106.

Male, S., Bush, M. B., & Chapman, E. (2009, December 7–9). Identification of competencies required by engineers graduating in Australia. Paper presented at the Australasian Association for Engineering Education Annual Conference, Adelaide.

Male, S., Bush, M. B., & Chapman, E. (2011). Understanding generic engineering competencies. Australasian Journal of Engineering Education, 17(3), 147–156.

Male, S., Bush, M. B., & Murray, K. (2009, December 6–9). Gender typing and engineering competencies. Paper presented at the Australasian Association for Engineering Education Annual Conference, Adelaide.

Male, S., Chapman, E., & Bush, M. B. (2007, December 11–13). Do female and male engineers rate different competencies as important? Paper presented at the Australian Association for Engineering Education Conference, Melbourne.

Malone, T., B., Anderson, D. E., Bost, R. J., Jennings, M., Malone, J. T., McCafferty Denise, B., … Terry, E. (1996). Human Error Reduction through Human and Organisational Factors in Design and Engineering of Offshore Systems. In R. Bea, R. Holdsworth, D. & C. Smith (Eds.), 1996 International Workshop on Human Factors in Offshore Operations: Summary of Proceedings and Submitted Papers (pp. 138–169). New York: American Bureau of Shipping.

Marjoram, T., Lamb, A., Lee, F., Hauke, C., & Garcia, C. R. (Eds.). (2010). Engineering: Issues, Challenges, Opportunities for Development. Paris, France: United Nations Educational, Scientifi c and Cultural Organization.

Marsick, V. E., & Watkins, K. (Eds.). (1990). Informal and Incidental Learning in the Workplace. London: Routledge.

Martin, R., Maytham, B., Case, J., & Fraser, D. (2005). Engineering graduates' perceptions of how well they were prepared for work in industry. European Journal of Engineering Education, 30(2), 167–180.

Marton, F., & Pang, M. F. (2006). On Some Necessary Conditions of Learning. The Journal of Learning Sciences, 15(2), 193–220.

Mason, G. (2000). Production Supervisors in Britain, Germany and the United States: Back from the Dead Again? Work, Employment and Society, 14(4), 625–645.

McGregor, H., & McGregor, C. (1998). Documentation in the Australian Engineering Workplace. Australasian Journal of Engineering Education, 8(1), 13–24.

McGregor, H. T., Marks, G., & Johnston, S. (2000). Recognising Workplace Learning: The UTS Combined Degree – Bachelor of Engineering, Diploma in Engineering Practice. Paper presented at the 5th Asia Pacific Conference on Cooperative Education.

McIlwee, J. S., & Robinson, J. G. (1992). Women in Engineering: Gender, Power, and Workplace Culture. New York: State University of New York Press.

McNulty, T. (1998). Developing Innovative Technology. Mining Engineering, 50(10), 50–55.

Meadows, D. H., Randers, J., Meadows, D. L., & Behrens, W. W. (1974). The Limits to Growth: A report for the Club of Rome's Project on the Predicament of Mankind (2nd ed.). New York: Universe Books.

Mehravari, D. (2007). Systematic Checking in Engineering Design. (B.Eng thesis), The University of Western Australia, Perth.

Meiksins, P., & Smith, C. (1996). Engineering Labour: Technical Workers in Comparative Perspective. London: Verso.

Merrow, E. W. (2011). Industrial Megaprojects: Concepts, Strategies, and Practices for Success. New Jersey: John Wiley & Sons.

Meyer, J. H. F., Land, R., & Baillie, C. (2010). Threshold Concepts and Transformational Learning. Rotterdam: Sense.

Millar, D. C. (Ed.). (2011). Ready for Take-off. Upper Saddle River, New Jersey, USA: Prentice Hall (Pearson Higher Edn).

Mills, J., Mehrtens, V., Smith, E., & Adams, V. (2008). An Update on Women's Progress in the Australian Engineering Workforce (45 pp.). Canberra: Engineers Australia.

Minneman, S. L. (1991). The social construction of a technical reality: Empirical studies of group engineering design practice. (PhD thesis) Stanford, USA.

Mintzberg, H. (1973). The Nature of Managerial Work. New York: Harper and Rowe.

Mintzberg, H. (1994). Rounding out the Manager's Job. Sloan Management Review, 36(1), 11–26.

Mintzberg, H. (2004). Managers not MBAs: A Hard Look at the Soft Practice of Managing and Management Development. London: Prentice Hall, Financial Times, Pearson Education.

Mitcham, C. (1991). Engineering as Productive Activity: Philosophical Remarks. In P. T. Durbin (Ed.), Critical Perspectives on Nonacademic Science and Engineering (pp. 80–117). Cranbury, New Jersey, USA: Associated University Presses.

Moore, G. E. (1959). 1200 Case Studies of Engineering Motivation. IRE Transactions on Education, 82–84.

Muchinsky, P. M. (2000). Emotions in the workplace: the neglect of organizational behavior. Journal of Organizational Behavior, 21(7), 801–805.

Mukerji, C. (2009). Impossible Engineering: Technology and Territoriality on the Canal du Midi. Princeton: Princeton University Press.

Nair, S., & Trevelyan, J. P. (2008, May 27–29). Current maintenance management methods cannot solve engineering asset maintenance history data quality problems. Paper presented at the ICOMS Asset Management Conference, Fremantle, Western Australia.

NASA. (2003). Chapter 8: History as Cause: Columbia and Challenger Report of the Columbia Accident Investigation Board (pp. 195–204): NASA.

Neuwirth, R. (2005). Shadow Cities: A Billion Squatters in a New Urban World. New York: Routledge.

Newport, C. L., & Elms, D. G. (1997). Effective Engineers. International Journal of Engineering Education, 13(5), 325–332.

Nicol, J. (2001). Have Australia's major hazard facilities learnt from the Longford disaster? Canberra: Engineers Australia.

Nonaka, I. (1994). A Dynamic Theory of Organizational Knowledge Creation. Organization Science, 5(1), 14–37.

Nowotny, H., Scott, P., & Gibbons, M. (2003). INTRODUCTION: 'Mode 2' Revisited: The New Production of Knowledge. Minerva, 41(3), 179–194.

Nussbaum, M. (2009). Education for profit, education for freedom. Liberal Education, 95(3), 6–13.

Ong, S. (2013). Power Theft in India: Investigations, analysis and suggestions for a socio-technical approach in rural Maharashtra. (B.Eng thesis), The University of Western Australia, Perth. Retrieved from http://www.mech.uwa.edu.au/jpt/eng-work/publications.html

Oosterlaken, I., & Hoven, J. v. d. (Eds.). (2012). The Capability Approach, Technology and Design (Vol. 5): Springer.

Orr, J. (1996). Talking About Machines: An Ethnography of a Modern Job. Ithaca, New York: Cornell University Press.

Paretti, M. (2008). Teaching Communication in Capstone Design: The Role of the Instructor in Situated Learning. Journal of Engineering Education, 97(5), 491–503.

Parkinson, C. N. (1957). Parkinson's Law. London: Buccaneer Books.

Patunru, A. A., & Wardhani, S. B. (2008). Political Economy of Local Investment Climates: A Review of the Indonesian Literature. Retrieved January 8, 2014, from www2.ids.ac.uk/gdr/cfs/pdfs/PatunruLitRev.pdf

Pawley, A. L. (2009). Universalized Narratives: Patterns in How Faculty Members Define "Engineering". Journal of Engineering Education, 98(4), 309–319.

Perlow, L. A. (1999). The Time Famine: Towards a Sociology of Work Time. Administrative Science Quarterly, 44(1), 57–81.

Perlow, L. A., & Bailyn, L. (1997). The Senseless Submergence of Difference: Engineers, Their Work, and Their Careers. In S. Barley & J. Orr (Eds.), Between Craft and Science: Technical Work in US Settings (pp. 230–243). Ithaca, New York: Cornell University Press.

Perlow, L. A., & Weeks, J. (2002). Who's helping whom? Layers of culture and workplace behavior. Journal of Organizational Behavior, 23, 345–361.

Perry, T. J., Danielson, S., & Kirkpatrick, A. (2012, April 10–13). Creating the Future of Mechanical Engineering Education: The ASME Vision 2030 Study. Paper presented at the Pan American Convention of Engineering XXXIII, UPADI 2012, Havana, Cuba.

Petermann, M., Trevelyan, J. P., Felgen, L., & Lindemann, U. (2007, August 28–31). Cross-Cultural Collaboration in Design Engineering – Influence Factors and their Impact on Work Performance. Paper presented at the International Conference on Engineering Design ICED'07, Paris.

Phillips, R., Neailey, K., & Broughton, T. (1999). A comparative study of six stage-gate approaches to product development. Integrated Manufacturing Systems, 10(5), 289–297.

Pielstick, C. D. (2000). Formal vs. Informal Leading: A Comparative Analysis. Journal of Leadership & Organizational Studies, 7(3), 99–114.

Polak, P. (2008). Design for Affordability. Paper presented at the American Association for Engineering Education (ASEE), Pittsburgh, Pennsylvania. http://www.d-rev.org/

Polanyi, M. (Ed.). (1966). The Tacit Dimension. Garden City: Doubleday.

Powell, A., Bagilhole, B., & Dainty, A. (2009). How Women Engineers Do and Undo Gender: Consequences for Gender Equality. Gender, Work and Organization, 16(4), 401–428.

Prais, S., & Wagner, K. (1988). Productivity and management: the training of foremen in Britain and Germany. National Institute Economic Review, 120, 34–47.

Project Management Institute. (2004). A Guide to the Project Management Body of Knowledge (3rd ed.). Newtown Square, Pennsylvania, USA: Project Management Institute, Inc.

Purves, D., Shimpi, A., & Lotto, R. B. (1999). An Empirical Explanation of the Cornsweet Effect. The Journal of Neuroscience, 19(19), 8542–8551.

Quinlan, K. M. (2002). Scholarly Dimensions of Academics' Beliefs about Engineering Education. Teachers & Teaching: Theory and Practice, 8(1), 41–64.

Rainbird, H., Fuller, A., & Munro, A. (2004). Workplace Learning in Context. London: Routledge.

Ravaille, N., & Vinck, D. (2003). Contrast in Design Cultures: Designing Dies for Drawing Aluminium. In D. Vinck (Ed.), Everyday Engineering: An Ethnography of Design and Innovation (pp. 93–118). Boston: MIT Press.

Raven, B. H. (1992). A Power/Interaction Model of Social Influence: French and Raven Thirty Years Later. Journal of Social Behavior and Personality, 7(2), 217–244.

Ravesteijn, W., de Graaff, E., & Kroesen, O. (2006). Engineering the future: the social necessity of communicative engineers. European Journal of Engineering Education, 31(1), 63–71.

Razali, Z. B., & Trevelyan, J. P. (2012). An Evaluation of Students' Practical Intelligence and Ability to Diagnose Equipment Faults. In M. Y. Khairiyah, A. A. Naziha, M. K. Azlina, K. S. Y. Sharifah & M. Y. Yudariah (Eds.), Outcome-Based Science, Technology, Engineering, and Mathematics Education: Innovative Practices (pp. 328–349). Hershey, Pennsylvania: IGI Global.

Reason, J., & Hobbs, A. (2003). Managing Maintenance Error: Ashgate.

Reddy, V. R. (1999). Quenching the Thirst: The Cost of Water in Fragile Environments. Development and Change, 30(79–113).

Rescher, N. (2001). Philosophical Reasoning: A Study in the Methodology of Philosophizing: Wiley-Blackwell.

Riemer, M. J. (2003). Integrating emotional intelligence into engineering education.

Rizzo, P. J. (2011). Plan to Reform Support Ship Repair and Management Practices. Canberra, Australia: Department of Defence, Ministerial and Executive Coordination and Communication Division.

Roberts, D. (2003, July). Contracting/commercial strategy for the 21st Century. Offshore, July Issue, 112.

Roberts, K. H. (1990). Some characteristics of one type of high reliability organisation. Organization Science, 1(2), 160–176.

Roberts, K. H., & Rousseau, D. M. (1989). Research in Nearly Failure-Free, High-Reliability Organisations: Having the Bubble. IEEE Transactions on Engineering Management, 36(2), 132–139.

Roberts, K. H., Stout, S., & Halpern, J., J. (1994). Decision Dynamics in To High Reliability Military Organisations. Management Science, 40(5), 614–624.

Robinson, S.-J. (2013). A study of the experience of engineers in relation to the skills learnt in their tertiary education how they relate to their workforce responsibilities and perceptions about their role in the workforce. (B.Eng thesis), The University of Western Australia, Perth.

Rogers, G. F. C. (1983). The Nature of Engineering. London: Macmillan.

Rolls Royce. (1973). The Jet Engine (3rd ed.). Derby, UK.

Rooke, P., & Wiehen, M. (1999). Hong Kong: The Airport Core Programme and the Absence of Corruption – Transparency International Working Paper: Transparency International.

Rooney, D., Willey, K., Gardner, A., Boud, D., Reich, A., & Fitzgerald, T. (2014). Engineers' professional learning: through the lens of practice. In B. Williams, J. D. Figueiredo & J. P. Trevelyan (Eds.), Engineering Practice in a Global Context: Understanding the Technical and the Social (pp. 265–280). Leiden, Netherlands: CRC/Balkema.

Roy, S. K. (1974). Management in India. New Delhi: Meeakshi Prakashan.

Sachs, J. (2005). The End of Poverty. London: Penguin.

Salas, E., Rosen, M. A., Burke, C. S., Goodwin, G. F., & Fiore, S. M. (2006). The Making of a Dream Team: When Expert Teams Do Best. In K. A. Ericsson, N. Charness, P. J. Feltovich & R. R. Hoffman (Eds.), The Cambridge Handbook of Expertise and Expert Performance (pp. 439–453). New York: Cambridge University Press.

Samson, D. (Ed.). (1995). Management for Engineers (2 ed.). Melbourne: Longman Australia.

Savin-Baden, M. (2007). Challenging Models and Perspectives of Problem-Based Learning. In E. De Graaff & A. Kolmos (Eds.), Management of Change: Implementation of Problem-Based and Project-Based Learning in Engineering (pp. 9–30). Rotterdam/Taipei: Sense Publishers.

Schön, D. A. (1983). The Reflective Practitioner: How Professionals Think in Action: Basic Books Inc., Harper Collins.

Schwartz, N., & Skurnik, I. (2003). Feeling and Thinking: Implications for Problem Solving. In J. E. Davidson & R. J. Sternberg (Eds.), The Psychology of Problem Solving (pp. 263–290). New York: Cambridge University Press.

Scott, G., & Yates, W. (2002). Using successful graduates to improve the quality of undergraduate engineering programmes. European Journal of Engineering Education, 27(4), 363–378.

Shainis, M. J., & McDermott, K. J. (1988). Managing without authority: The dilemma of the engineering manager and the project engineer. Engineering Management International, 5(2), 143–147.

Shannon, C. E. (1948). A Mathematical Theory of Communication. Bell System Technical Journal, 27, 379–423, 623–656.

Shapiro, S. (1997). Degrees of Freedom: The Interaction of Standards of Practice and Engineering Judgement. Science, Technology and Human Values, 22(3), 286–316.

Sheppard, S. D., Colby, A., Macatangay, K., & Sullivan, W. (2006). What is Engineering Practice? International Journal of Engineering Education, 22(3), 429–438.

Sheppard, S. D., Macatangay, K., Colby, A., & Sullivan, W. (2009). Educating Engineers. Stanford, California: Jossey-Bass (Wiley).

Shippmann, J., Ash, R., Carr, L., Hesketh, B., Pearlman, K., Battista, M., ..., Sanchez, J., I. (2000). The practice of competency modelling. Personnel Psychology, 53(3), 703–740.

Shuman, L. J., Besterfield-Sacre, M. E., & McGourty, J. (2005). The ABET "Professional Skills" – Can They Be Taught? Can They Be Assessed? Journal of Engineering Education, 94(1), 41–55.

Singleton, D., & Hahn, N. (2003). Engineers can help alleviate poverty. Engineers Australia, May 2003, 40–43.

Smith, F. (2010, November 9). Time to disconnect, unplug and do some real thinking. Australian Financial Review.

Smith, K. A., Sheppard, S. D., Johnson, D. W., & Johnson, R. T. (2005). Pedagogies of Engagement: Classroom-Based Practices. Journal of Engineering Education, 94(1), 87–101.

Sonnentag, S., Niessen, C., & Volmer, J. (2006). Expertise in Software Design. In K. A. Ericsson, N. Charness, P. J. Feltovich & R. R. Hoffman (Eds.), The Cambridge Handbook of Expertise and Expert Performance (pp. 373–387). Cambridge, UK: Cambridge University Press.

Spinks, N., Silburn, N. L. J., & Birchall, D. W. (2007). Making it all work: the engineering graduate of the future, a UK perspective. European Journal of Engineering Education, 32(3), 325–335.

Srinivasan, M. V., Zhang, S. W., & Bidwell, N. J. (1997). Visually mediated odometry in honeybees. The Journal of Experimental Biology, 200, 2513–2522.

Srivastava, P. (2004). Poverty Targeting in Asia: Country Experience of India ADB Institute Discussion Paper No. 5 (pp. 53). Tokyo: Asian Development Bank Institute (ADBI).

Standards Australia. (1999). Australian and New Zealand Standard 4360:1999 Risk Management.

Standards Australia. (2009). AS/NZS ISO 31000:2009 Risk Management – Principles and Guidelines: Standards Australia and Standards New Zealand.

Standards Australia. (2013). SA/SNZ HB 89:2013 Risk management – Guidelines on risk assessment techniques Standards Australia and Standards New Zealand.

Stanovich, K. E. (2003). The Fundamental Computational Biases of Human Cognition: Heuristics that (Sometimes) Impair Decision Making and Problem Solving. In J. E. Davidson & R. J. Sternberg (Eds.), The Psychology of Problem Solving (pp. 291–342). New York: Cambridge University Press.

Steiner, G. (1975). After Babel: Aspects of Language and Translation. Oxford: Oxford University Press.

Sternberg, R. J., Wagner, R. K., Williams, W. M., & Horvath, J. A. (1995). Testing Common Sense. American Psychologist, 50(11), 912–927.

Stevens, R., Johri, A., & O'Connor, K. (2014). Professional Engineering Work. In A. Johri & B. M. Olds (Eds.), Cambridge Handbook of Engineering Education Research. Cambridge: Cambridge University Press.

Stewart, D. M., & Chase, R. B. (1999). The impact of human error on delivering service quality. Production and Operations Management, 8(3), 240–263.

Suchman, L. (2000). Organising Alignment: a Case of Bridge-Building. Organization, 7(2), 311–327.

Tan, E. (2013). How do architects perceive their interactions with engineers and the service quality of these engineers? (Ph.D.), The University of Western Australia, Perth

Tan, E., & Trevelyan, J. P. (2011, December 5–7). Problem Comprehension is the Key to Client Problem Solving. Paper presented at the Australasian Association for Engineering Education Annual Conference Fremantle.

Tan, E., & Trevelyan, J. P. (2013). Understanding of clients' perspective of engineers' service quality can inform engineering education and practice. Paper presented at the Australasian Association for Engineering Education Conference, Gold Coast.

Tang, S., & Trevelyan, J. P. (2009, December 7–9). Engineering Learning and Practice - a Brunei Perspective. Paper presented at the Australasian Association for Engineering Education Annual Conference, Adelaide.

Tang, S., & Trevelyan, J. P. (2010, December 6–8). The Impact of Socio-Cultural Differences on the Management of Technical Error: Are Engineering Graduates Aware of This? Paper presented at the Australasian Association for Engineering Education Annual Conference, University of Technology, Sydney.

Tang, S. (2012). An Empirical Investigation of Telecommunication Engineering in Brunei Darussalam. (PhD thesis), The University of Western Australia, Perth.

Tax cheat politicians named. (2013, December 13). The Nation. Retrieved from http://www.nation.com.pk/national/13-Dec-2012/tax-cheat-politicians-named

Tenopir, C., & King, D. W. (2004). Communication Patterns of Engineers. Hoboken: IEEE Press – Wiley & Sons Inc.

Terkel, S. (1972). Working: People Talk About What They Do All Day and How They Feel About What They Do. London: Wildwood House Ltd.

The Economist. (2012, August 11). Manufacturing in India: the marsala mittelstand. The Economist.

Toner, M. (2005, June 2005). Emotional intelligence is worth pursuing. Engineers Australia, 68–69.

Transparency International. (2000). TI Source Book 2000: Confronting Corruption: the Elements of a National Integrity System: Transparency International.

Trevelyan, J. P. (1992). Robots for Shearing Sheep: Shear Magic: Oxford University Press.

Trevelyan, J. P. (2000). Reducing Accidents in Demining. Paper presented at the Standing Committee of Experts in Mine Action Technologies, 2nd Meeting, Geneva. http://www.mech.uwa.edu.au/jpt/demining/reports.html

Trevelyan, J. P. (2002). Technology and the landmine problem: practical aspects of landmine clearance operations. In H. Schubert & A. Kuznetsov (Eds.), Detection of Explosives and Landmines: Methods and Field Experience (Vol. 66, pp. 155–164).

Trevelyan, J. P. (2005a). Drinking Water Costs in Pakistan. from http://www.mech.uwa.edu.au/jpt/pes.html

Trevelyan, J. P. (2005b). Pakistan: Expensive Labour and High Costs. from http://www.mech.uwa.edu.au/jpt/pes.html

Trevelyan, J. P. (2007). Technical Coordination in Engineering Practice. Journal of Engineering Education, 96(3), 191–204.

Trevelyan, J. P. (2010a, October 27–30). Engineering Students Need to Learn to Teach. Paper presented at the 40th ASEE/IEEE Frontiers in Education Conference, Washington, DC.

Trevelyan, J. P. (2010b). Reconstructing Engineering from Practice. Engineering Studies, 2(3), 175–195.

Trevelyan, J. P. (2011). Are we accidentally misleading students about engineering practice? Paper presented at the 2011 Research in Engineering Education Symposium (REES 2011), Madrid.

Trevelyan, J. P. (2012a). Submission to Senate Enquiry on the Shortage of Engineering and Related Employment Skills. Canberra: Australian Parliament.

Trevelyan, J. P. (2012b). Understandings of Value in Engineering Practice. Paper presented at the Frontiers in Education 2012, Seattle.

Trevelyan, J. P. (2012c, December 3–5). Why Do Attempts at Engineering Education Reform Consistently Fall Short? Paper presented at the Australasian Association for Engineering Education Annual Conference, Melbourne.

Trevelyan, J. P. (2013). Pakistan has a bright energy future. The News. Retrieved from http://www.thenews.com.pk/Todays-News-9-181003-Pakistan-has-a-bright-energy-future

Trevelyan, J. P. (2014a). Observations of South Asian Engineering Practice. In B. Williams, J. D. Figueiredo & J. P. Trevelyan (Eds.), Engineering Practice in a Global Context: Understanding the Technical and the Social (pp. 223–244). Leiden, Netherlands: CRC/ Balkema.

Trevelyan, J. P. (2014b). Towards a Theoretical Framework for Engineering Practice. In B. Williams, J. D. Figueiredo & J. P. Trevelyan (Eds.), Engineering Practice in a Global Context: Understanding the Technical and the Social (pp. 33–60). Leiden, Netherlands: CRC/Balkema.

Trevelyan, J. P., & Murphy, P. (1996). Fast Vision Measurements using Shaped Snakes. Paper presented at the International Conference on Advanced Robotics and Computer Vision (ICARCV96), Singapore.

Trevelyan, J. P., & Razali, Z. B. (2011). What do students gain from laboratory experiences? In A. K. M. Azad, M. E. Auer & V. J. Harward (Eds.), Internet Accessible Remote Laboratories: Scalable E-Learning Tools for Engineering and Science Disciplines (pp. 416–431): IGI Global

Trevelyan, J. P., & Tilli, S. (2007). Published Research on Engineering Work. Journal of Professional Issues in Engineering Education and Practice, 133(4), 300–307.

Trevelyan, J. P., & Tilli, S. (2008, June 20–22). Longitudinal Study of Australian Engineering Graduates: Preliminary Results. Paper presented at the American Society for Engineering Education Annual Conference, Pittsburgh.

Trevelyan, J. P., & Tilli, S. (2010). Labour Force Outcomes for Engineering Graduates in Australia. Australasian Journal of Engineering Education, 16(2), 101–122.

Turner, G. M. (2008). A comparison of The Limits to Growth with 30 years of reality. Global Environmental Change, 18, 397–411.

Tversky, A., & Kahneman, D. (1974). Judgement under uncertainty: heuristics and biases. Science, 185(4157), 1124–1131.

United Nations Development Programme (UNDP). (2013). Human Development Report 2013 – The Rise of the South: Human Progress in a Diverse World. New York: United Nations Development Programme,.

Vincenti, W. G. (1990). What Engineers Know and How They Know It: Analytical Studies from Aeronautical History. Baltimore: The Johns Hopkins University Press.

Vinck, D. (Ed.). (2003). Everyday Engineering: An Ethnography of Design and Innovation. Boston: MIT Press.

Walther, J., Kellam, N., Sochacka, N., & Radcliffe, D. F. (2011). Engineering Competence? An Interpretive Investigation of Engineering Students' Professional Formation. Journal of Engineering Education, 100(4), 703-740.

Waring, A., & Glendon, I. A. (2000). Managing Risk: Critical Issues for Survival and Success in the 21st Century. London: International Thompson Business Press.

Webb, M. (1995). Death of a Hero: the strange suicide of Charles Yelverton O'Connor. Early Days: Journal and Proceedings of the Royal Western Australian Historical Society, 11(1), 81–111.

Weick, C. (1987). Organizational Culture as a Source of High Reliability. California Management Review, 29(2), 112–127.

Weick, C., & Roberts, K. H. (1993). Collective Mind in Organizations: Heedful Interrelating on Flight Decks. Administrative Science Quarterly, 38, 357–381.

Wellington, A. M. (1887). The Economic Theory of the Location of Railways. New York: John Wiley.

Wells, D. J. (1995). Managing Your First Years in Industry: The Essential Guide to Career Transition. Piscataway, New Jersey: IEEE.

Wenger, E. (2005). Communities of Practice: A Brief Introduction. Retrieved May 9, 2008

Wenke, D., & Frensch, P. A. (2003). Is Success or Failure at Solving Complex Problems Related to Intellectual Ability? In J. E. Davidson & R. J. Sternberg (Eds.), The Psychology of Problem Solving (pp. 87–126). New York: Cambridge University Press.

Whyte, J. (2013). Beyond the computer: Changing medium from digital to physical. Information and Organization, 23(1), 41–57.

Whyte, J., Ewenstein, B., Hales, M., & Tidd, J. (2008). Visualizing Knowledge in Project-Based Work. Long Range Planning, 41(1), 74–92.

Whyte, J., & Lobo, S. (2010). Coordination and control in project-based work: digital objects and infrastructures for delivery. Construction Management and Economics, 28(6), 557–567.

Wilde, G. L. (1983). The skills and practices of engineering designers now and in the future. Design Studies, 4(1), 21–34.

Willers, A. (2005). Report on Fatigue Management and Shiftwork Workshop (pp. 2). Sydney: ANSTO.

Williams, B., & Figueiredo, J. (2014). Finding workable solutions: Portuguese engineering experience. In B. Williams, J. D. Figueiredo & J. P. Trevelyan (Eds.), Engineering Practice in a Global Context: Understanding the Technical and the Social (pp. 159–184). Leiden, Netherlands: CRC/Balkema.

Williams, B., Figueiredo, J. D., & Trevelyan, J. P. (Eds.). (2014). Engineering Practice in a Global Context: Understanding the Technical and the Social. Leiden, Netherlands: CRC/Balkema.

Williams, J. M. G., Mathews, A., & MacLeod, C. (1996). The Emotional Stroop Task and Psychopathology. Psychological Bulletin, 120(1), 3–24.

Williams, R. (2003, January 24). Education for the Profession Formerly Known as Engineering. The Chronicle of Higher Education, p. B12.

Winch, G. M., & Kelsey, J. (2005). What do construction project planners do? International Journal of Project Management, 23, 141–149.

Wong, W. L. P., & Radcliffe, D. F. (2000). The Tacit Nature of Design Knowledge. Technology Analysis and Strategic Management, 12(4), 493–512.

World Health Organization. (2006). Meeting the MDG drinking water and sanitation target: the urban and rural challenge of the decade.

Wright, P. H. (2002). Introduction to Engineering (3rd ed.). Hoboken, New Jersey, USA: Wiley & Sons.

Youngman, M., Oxtoby, R., Monk, J. D., & Heywood, J. (1978). Analysing Jobs. Farnborough, Hampshire, UK: Gower Press.

Yuzuriha, T. (1998). How to Succeed as an Engineer. Vancouver, Washington, USA: J&K Publishing.

Keyword index

Index of people and organisations

Printed and bound by CPI Group (UK) Ltd, Croydon, CR0 4YY

25/10/2024

01779124-0001